ALGEBRAIC AND GEOMETRIC
METHODS IN STATISTICS

This up-to-date account of algebraic statistics and information geometry explores the emerging connections between the two disciplines, demonstrating how they can be used in design of experiments and how they benefit our understanding of statistical models and in particular, exponential models. This book presents a new way of approaching classical statistical problems and raises scientific questions that would never have been considered without the interaction of these two disciplines.

Beginning with a brief introduction to each area, using simple illustrative examples, the book then proceeds with a collection of reviews and some new results by leading researchers in their respective fields. Parts I and II are mainly on contingency table analysis and design of experiments. Part III dwells on both classical and quantum information geometry. Finally, Part IV provides examples of the interplay between algebraic statistics and information geometry. Computer code and some proofs are also available on-line, where key examples are also developed in further detail.

ALGEBRAIC AND GEOMETRIC METHODS IN STATISTICS

Edited by

PAOLO GIBILISCO

EVA RICCOMAGNO

MARIA PIERA ROGANTIN

HENRY P. WYNN

CAMBRIDGE UNIVERSITY PRESS
Cambridge, New York, Melbourne, Madrid, Cape Town, Singapore, São Paulo, Delhi

Cambridge University Press
The Edinburgh Building, Cambridge CB2 8RU, UK

Published in the United States of America by Cambridge University Press, New York

www.cambridge.org
Information on this title: www.cambridge.org/9780521896191

First published 2010

Printed in the United Kingdom at the University Press, Cambridge

A catalogue record for this publication is available from the British Library

ISBN 978-0-521-89619-1 Hardback

This volume is dedicated to
Professor Giovanni Pistone
on the occasion of
his sixty-fifth birthday

Contents

Contents

Contributors

Satoshi Aoki
Department of Mathematics and Computer Science, Kagoshima University, 1-21-35, Korimoto, Kagoshima 890-0065, Japan

Enrico Carlini
Department of Mathematics, Polytechnic of Turin, Corso Duca degli Abruzzi 24, 10129 Torino, Italy

Yuguo Chen
Department of Statistics, University of Illinois at Urbana-Champaign, 725, S. Wright Street Champaign, IL 61820 USA

Ian H. Dinwoodie
214 Old Chemistry Building, Box 90251, Duke University, Durham, NC 27708-0251 USA

Adrian Dobra
Department of Statistics, University of Washington, Seattle WA 98195-4322 USA

Stephen E. Fienberg
Department of Statistics, Machine Learning Dept. and Cylab, Carnegie Mellon University, Pittsburgh, PA 15213-3890 USA

Roberto Fontana
Department of Mathematics, Polytechnic of Turin, Corso Duca degli Abruzzi 24, 10129 Torino, Italy

Kenji Fukumizu
Institute of Statistical Mathematics, 4-6-7 Minamiazabu, Minatoku, Tokyo 106-8569, Japan

Paolo Gibilisco
Department S.E.F.E.M.E.Q., University of Roma Tor Vergata, Via Columbia 2, 00133 Rome, Italy

Frank Hansen
Department of Economics, University of Copenhagen, Studiestrde 6, 1455 Kbenhavn K, Denmark

Patricia Hersh
Department of Mathematics, Indiana University, Bloomington, IN 47405-7000 USA

Serkan Hoşten

> Department of Mathematics, San Francisco State University, 1600 Holloway Avenue, San Francisco, CA, 94132 USA

Daniele Imparato

> Department of Mathematics, Polytechnic of Turin, Corso Duca degli Abruzzi 24, 10129 Torino, Italy

Anne Krampe

> Fakultät Statistik, Technische Universität Dortmund, 44221 Dortmund, Germany

Sonja Kuhnt

> Technische Universiteit Eindhoven, P.O. Box 513, 5600 MB Eindhoven, The Netherlands

Reinhard Laubenbacher

> Virginia Bioinformatics Institute, Virginia Polytechnic Institute and State University, Washington Street, MC 0477, USA

Guy Lebanon

> Colleges of Science and Engineering, Purdue University, 250 N. University Street, West Lafayette, IN, 47907-2066, USA

Hugo Maruri-Aguilar

> Department of Statistics, London School of Economics, London WC2A 2AE, United Kingdom

Roberto Notari

> Department of Mathematics, Polytechnic of Milan, Via Bonardi 9, 20133 Milano, Italia

Giovanni Pistone

> Department of Mathematics, Polytechnic of Turin, Corso Duca degli Abruzzi 24, 10129 Torino, Italy

Fabio Rapallo

> Department DISTA, University of Eastern Piedmont, Via Bellini, 25/G, 15100 Alessandria, Italy

Eva Riccomagno

> Department of Mathematics, Genoa University, Via Dodecaneso, 35, 16146 Genova, Italia

Alessandro Rinaldo

> Department of Statistics, Carnegie Mellon University, Pittsburgh, PA 15213-3890 USA

Maria Piera Rogantin

> Department of Mathematics, Genoa University, Via Dodecaneso, 35, 16146 Genova, Italia

Aleksandra B. Slavković

> Department of Statistics, Pennsylvania State University, State College, PA USA

Brandilyn Stigler

> Mathematical Biosciences Institute, The Ohio State University, 231 West 18th Avenue, Columbus, OH 43210, USA

Raymond F. Streater
> Department of Mathematics, Kings College London, The Strand, London WC2R 2LS, United Kingdom

Seth Sullivant
> Department of Mathematics, Harvard University, One Oxford Street, Cambridge, MA 02138 USA

Akimichi Takemura
> Department of Mathematical Informatics, University of Tokyo, Bunkyo, Tokyo 113-0033, Japan

Barbara Trivellato
> Department of Mathematics, Polytechnic of Turin, Corso Duca degli Abruzzi 24, 10129 Torino, Italy

Henry P. Wynn
> Department of Statistics, London School of Economics, London WC2A 2AE, United Kingdom

Anna Jenčová
> Mathematical Institute, Slovak Academy of Sciences, Stefanikova 49, SK-84173 Bratislava, Slovakia

Ruriko Yoshida
> Department of Statistics, University of Kentucky, 805A Patterson Office Tower, Lexington, KY 40506-0027, USA

Yi Zhou
> Machine Learning Department, Carnegie Mellon University, Pittsburgh, PA 15213-3890 USA

Preface

Information Geometry and Algebraic Statistics are brought together in this volume to suggest that the interaction between them is possible and auspicious.

To meet this aim, we couple expository material with more advanced research topics sometimes within the same chapter, cross-reference the various chapters, and include many examples both in the printed volume and in the on-line supplement, held at the Cambridge University Press web site at www.cambridge.org/9780521896191. The on-line part includes proofs that are instructive but long or repetitive, computer codes and detailed development of special cases.

Chapter 1 gives a brief introduction to both Algebraic Statistics and Information Geometry based on the simplest possible examples and on selected topics that, to the editors, seem most promising for the interlacing between them. Then, the volume splits naturally in two lines. Part I, on contingency tables, and Part II, on designed experiments, are authored by researchers active mainly within Algebraic Statistics, while Part III includes chapters on both classical and quantum Information Geometry. This material comes together in Part IV which consists of only one chapter by Giovanni Pistone, to whom the volume is dedicated, and provides examples of the interplay between Information Geometry and Algebraic Statistics.

The editors imagine various entry points into the volume according to the reader's own interests. These are indicated with squared boxes in Figure 0.1. Maximum likelihood estimation in models with hidden variables is revisited in an algebraic framework in Chapter 2 (S. E. Fienberg *et al.*) which is supported by a substantial on-line section, including Chapter 22 (Y. Zhou) where the role of secant varieties for graphical models is detailed. Chapter 3 (A. Slavkovich and S. E. Fienberg) gives old and new geometric characterizations of the joint distribution on $I \times J$ contingency tables and can be used to gain familiarity with algebraic geometric jargon and ideas common in Algebraic Statistics. The next two chapters present fast algorithms for the computation of Markov bases in model selection (Chapter 4 by A. Krampe and S. Kuhnt) and under strictly positive margins (Chapter 5 by Y. Chen *et al.*), while Chapter 6 (E. Carlini and F. Rapallo) defines a class of algebraic statistical models for category distinguishability in rater agreement problems. The algebraic notion of index of complexity of maximum likelihood equations is used in Chapter 7 (S. Hoşten and S. Sullivant) for bivariate data missing at random. This part of the volume ends with Chapter 8 by S. E. Fienberg and A. Dobra.

Part II considers the two technologies of Algebraic Statistics most employed in design and analysis of experiments. Chapter 12 (R. Fontana and M. P. Rogantin) uses the game of sudoku to review polynomial indicator functions and links to Part I via the notion of Markov bases. This link is developed for a special case in Chapter 13 (S. Aoki and A. Takemura). This chapter should appeal to a reader acquainted with the classical theory of experimental design. Chapters 9, 10 and 11 develop in different settings the ideas and techniques outlined in the first part of Chapter 1: Chapter 9 (H. Maruri-Aguilar and H. P. Wynn) argues that algebraic sets can be used as repositories of experimental designs; Chapter 10 (R. Laubenbacher and B. Stigler) presents an application to the identification of biochemical networks from experimental data; and Chapter 11 (E. Riccomagno and R. Notari) considers designs with replicated points.

The Information Geometry part of the volume starts with Chapter 14 (R. F. Streater) which provides a gentle and short, though comprehensive, introduction to Information Geometry and its link to the theory of estimation according to Fisher. It keeps as far as possible the analogy between the classical and the quantum case. It extends to the purely quantum case in Chapter 15 (R. F. Streater) which, together with Chapter 16 (A. Jenčová), provides an extension to the quantum case of the statistical manifolds modelled on an Orlicz space. Also, Chapter 20 (F. Hansen) deals with quantum Information Geometry. A construction of a statistical manifold modelled on a Reproducing Kernel Hilbert Space is presented in Chapter 18 (K. Fukumizu), where the application to the theory of estimation is based on a suitable class of likelihood functions defined point-wise. Chapter 19 (D. Imparato and B. Trivellato) extends the standard non-parametric exponential model by considering its limit, developing ideas in Chapter 21. An application of classical information geometry for text analysis is developed by G. Lebanon in Chapter 17.

Chapter 1 includes a glossary of terms from Algebraic Geometry that are recurrent in the volume.

The editors thank the authors for providing interesting papers, the many referees who helped with the peer-reviewing, our publisher CUP and the ever patient and capable Diana Gillooly. Some chapters in this volume were first presented to the conference 'Mathematical explorations in contemporary statistics' held in Sestri Levante on 19–20 May 2008. Some chapters were also presented at the opening workshop of the 2008–09 SAMSI Program on Algebraic Methods in Systems Biology and Statistics, 14–17 September 2008.

This volume is dedicated to Giovanni Pistone on the occasion of his sixty-fifth birthday. We are grateful for his discreet and constant support.

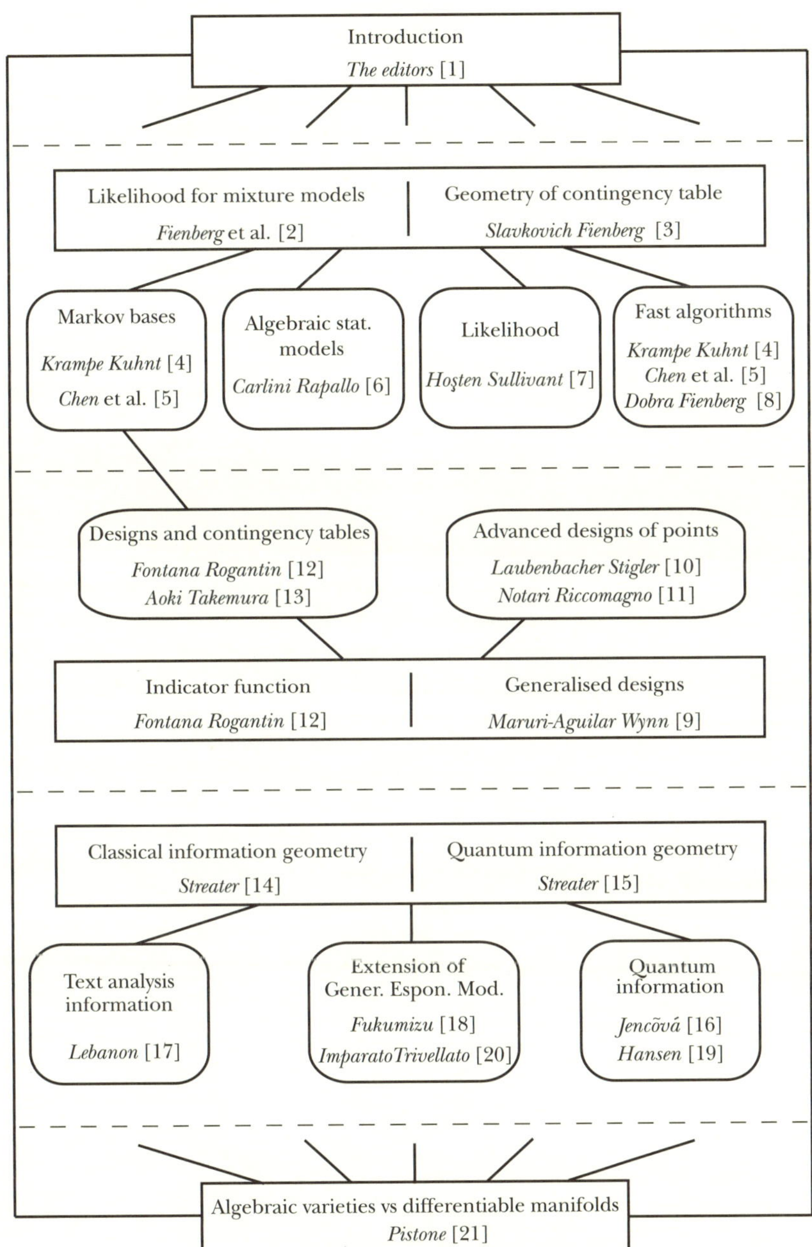

Fig. 1 Layout of the volume.

Frequently used notations and symbols

$\mathbb{N}$	natural numbers
$\mathbb{Z}$	integer numbers
$\mathbb{Q}$	rational numbers
$\mathbb{R}$	real numbers
$\mathbb{C}$	complex numbers
$\mathbb{R}_{>0}$	strictly positive real numbers
$\mathbb{R}_{\geq 0}$	non-negative real numbers

$\mathrm{E}_p[X]$	expectation of a random variable X w.r.t. the probability measure p
$\mathrm{Cov}_p(X, Y)$	covariance of X, Y w.r.t. p
$\mathrm{Var}_p(X)$	variance of X w.r.t. p
M_n	space of the $n \times n$ matrices with complex entries
$\mathrm{K}(p, q)$ or $\mathrm{KL}(q\|p)$	Kullback–Leibler relative entropy
I_X or $\mathcal{I}_X$ or G (resp. I_f or $\mathcal{I}_f$ or G)	Fisher information of X (resp. the density f)
$(\Omega, \mathcal{F}, \mu)$	measure space
$\mathcal{M}_>$, $\mathcal{M}_>(\mu)$ (resp. $\mathcal{M}_\geq$, $\mathcal{M}_\geq(\mu)$)	space of strictly positive (resp. non-negative) densities w.r.t the measure μ

$\lvert \cdot \rvert$	cardinality of a set
k	number of factors
n	number of observations
$\mathcal{D}$	design
$\mathcal{K}[x_1, \ldots, x_k]$	set of polynomials in $x_1, \ldots, x_k$ with coefficients in $\mathcal{K}$
$\mathrm{I}(f_1, \ldots, f_l)$ or $\langle f_1, \ldots, f_l \rangle$	ideal generated by the polynomials $f_1, \ldots, f_l$
$\mathrm{I}(\mathcal{D})$	ideal of the points in the design
$\mathbb{R}[x_1, \ldots, x_k]/\mathrm{I}(f_1, \ldots, f_l)$	quotient space modulo $\mathrm{I}(f_1, \ldots, f_l)$
$\mathrm{NF}(f, I)$	normal form of f w.r.t. I
A or A_T	constraint matrix

1

Algebraic and geometric methods in statistics

Paolo Gibilisco, Eva Riccomagno, Maria Piera Rogantin, Henry P. Wynn

1.1 Introduction

It might seem natural that where a statistical model can be defined in algebraic terms it would be useful to use the full power of modern algebra to help with the description of the model and the associated statistical analysis. Until the mid 1990s this had been carried out, but only in some specialised areas. Examples are the use of group theory in experimental design and group invariant testing, and the use of vector space theory and the algebra of quadratic forms in fixed and random effect linear models. The newer area which has been given the name 'algebraic statistics' is concerned with statistical models that can be described, in some way, via polynomials. Of course, polynomials were there from the beginning of the field of statistics in polynomial regression models and in multiplicative models derived from independence models for contingency tables, or to use a more modern terminology, models for categorical data. Indeed these two examples form the bedrock of the new field. (Diaconis and Sturmfels 1998) and (Pistone and Wynn 1996) are basic references.

Innovations have entered from the use of the apparatus of polynomial rings: algebraic varieties, ideals, elimination, quotient operations and so on. See Appendix 1.7 of this chapter for useful definitions. The growth of algebraic statistics has coincided with the rapid developments of fast symbolic algebra packages such as CoCoA, Singular, 4ti2 and Macaulay 2.

If the first theme of this volume, algebraic statistics, relies upon computational commutative algebra, the other one is pinned upon differential geometry. In the 1940s Rao and Jeffreys observed that Fisher information can be seen as a Riemannian metric on a statistical model. In the 1970s Čencov, Csiszár and Efron published papers that established deep results on the involved geometry. Čencov proved that Fisher information is the only distance on the simplex that contracts in the presence of noise (Čencov 1982).

The fundamental result by Čencov and Csiszár shows that with respect to the scalar product induced by Fisher information the relative entropy satisfies a Pythagorean equality (Csiszár 1975). This result was motivated by the need to minimise

Algebraic and Geometric Methods in Statistics, ed. Paolo Gibilisco, Eva Riccomagno, Maria Piera Rogantin and Henry P. Wynn. Published by Cambridge University Press. © Cambridge University Press 2010.

1

relative entropy in fields such as large deviations. The differential geometric counterparts are the notions of divergence and dual connections and these can be used to give a differential geometric interpretation to Csiszár's results.

Differential geometry enters in statistical modelling theory also via the idea of exponential curvature of statistical models due to (Efron 1975). In this 'exponential' geometry, one-dimensional exponential models are straight lines, namely geodesics. Sub-models with good properties for estimation, testing and inference, are characterised by small exponential curvature.

The difficult task the editors have set themselves is to bring together the two strands of algebraic and differential geometry methods into a single volume. At the core of this connection will be the exponential family. We will see that polynomial algebra enters in a natural way in log-linear models for categorical data but also in setting up generalised versions of the exponential family in information geometry. Algebraic statistics and information geometry are likely to meet in the study of invariants of statistical models. For example, on one side polynomial invariants of statistical models for contingency tables have long been known (Fienberg 1980) and in phylogenetic algebraic invariants were used from the very beginning in the Hardy–Weinberg computations (Evans and Speed 1993, for example) and are becoming more and more relevant (Casanellas and Fernández-Sánchez 2007). While on the other side we recall with Shun-Ichi Amari[1] that 'Information geometry emerged from studies on invariant properties of a manifold of probability distributions'. The editors have asked the dedicatee, Giovanni Pistone, to reinforce the connection in a final chapter. The rest of this introduction is devoted to an elementary overview of the two areas, avoiding too much technicality.

1.2 Explicit versus implicit algebraic models

Let us see with simple examples how polynomial algebra may come into statistical models. We will try to take a transparent notation. The technical, short review of algebraic statistics in (Riccomagno 2009) can complement our presentation.

Consider quadratic regression in one variable:

$$Y(x) = \theta_0 + \theta_1 x + \theta_2 x^2 + \epsilon(x). \tag{1.1}$$

If we observe (without replication) at four distinct *design points*, $\{x_1, x_2, x_3, x_4\}$ we have the usual matrix form of the regression

$$\eta = \mathrm{E}[Y] = X\theta, \tag{1.2}$$

where the X-matrix takes the form:

$$X = \begin{pmatrix} 1 & x_1 & x_1^2 \\ 1 & x_2 & x_2^2 \\ 1 & x_3 & x_3^2 \\ 1 & x_4 & x_4^2 \end{pmatrix},$$

and Y, θ are the observation, parameter vectors, respectively, and the errors have

---

[1] Cited from the abstract of the presentation by Prof Amari at the LIX Colloquium 2008, Emerging Trends in Visual Computing, 18th-20th November 2008, Ecole Polytechnique.

zero mean. We can give algebra a large role by saying that the design points are the solution of $g(x) = 0$, where

$$g(x) = (x - x_1)(x - x_2)(x - x_3)(x - x_4). \tag{1.3}$$

In algebraic terms the design is a zero-dimensional variety. We shall return to this representation later.

Now, by eliminating the parameters θ_i from the equations for the mean response: $\{\eta_i = \theta_0 + \theta_1 x_i + \theta_2 x_i^2,\ i = 1, \ldots 4\}$ we obtain an equation just involving the η_i and the x_i:

$$-(x_2 - x_3)(x_2 - x_4)(x_3 - x_4)\eta_1 + (x_1 - x_3)(x_1 - x_4)(x_3 - x_4)\eta_2$$
$$-(x_1 - x_2)(x_1 - x_4)(x_2 - x_4)\eta_3 + (x_1 - x_2)(x_1 - x_3)(x_2 - x_3)\eta_4 = 0, \tag{1.4}$$

with the conditions that none of the x_i are equal. We can either use formal algebraic elimination (Cox *et al.* 2008, Chapter 3) to obtain this or simply note that the linear model (1.2) states that the vector η belongs to the column space of X, equivalently it is orthogonal to the orthogonal (kernel, residual) space. In statistical jargon we might say, in this case, that the quadratic model is equivalent to setting the orthogonal cubic contrast equal to zero. We call model (1.2) an *explicit* (statistical) algebraic model and (1.4) an *implicit* (statistical) algebraic model.

Suppose that instead of a linear regression model we have a Generalized Linear Model (GLM) in which the Y_i are assumed to be independent Poisson random variables with means $\{\mu_i\}$, with log link

$$\log \mu_i = \theta_0 + \theta_1 x_i + \theta_2 x_i^2, \quad i = 1, \ldots, 4.$$

Then, we have

$$-(x_2 - x_3)(x_2 - x_4)(x_3 - x_4) \log \mu_1 + (x_1 - x_3)(x_1 - x_4)(x_3 - x_4) \log \mu_2$$
$$-(x_1 - x_2)(x_1 - x_4)(x_2 - x_4) \log \mu_3 + (x_1 - x_2)(x_1 - x_3)(x_2 - x_3) \log \mu_4 = 0. \tag{1.5}$$

Example 1.1 Assume that the x_i are integer. In fact, for simplicity let us take our design to be $\{0, 1, 2, 3\}$. Substituting these values in the Poisson case (1.5) and exponentiating we have

$$\mu_1 \mu_3^3 - \mu_2^3 \mu_4 = 0.$$

This is a special variety for the μ_i, a toric variety which defines an implicit model. If we condition on the sum of the 'counts': that is $n = \sum_i Y_i$, then the counts become multinomially distributed with probabilities $p_i = \mu_i / n$ which satisfy $p_1 p_3^3 - p_2^3 p_4 = 0$.

The general form of the Poisson log-linear model is $\eta_i = \log \mu_i = X_i^\top \theta$, where $^\top$ stands for transpose and $X_i^\top$ is the i-th row of the X-matrix. It is an exponential family model with likelihood:

$$L(\theta) = \prod_i p(y_i, \mu_i) = \prod_i \exp(y_i \log \mu_i - \mu_i - \log y_i!)$$

$$= \exp\left(\sum_i y_i \sum_j X_{ij} \theta_j - \sum_i \mu_i - \sum_i \log y_i!\right),$$

where y_i is a realization of Y_i. The sufficient statistics can be read off in the usual way as the coefficients of the parameters θ_j:

$$T_j = \sum_i X_{ij} y_i = X_j^\top Y,$$

and they remain sufficient in the multinomial formulation. The log-likelihood is

$$\sum_j T_j \theta_j - \sum_{i=1}^{n} \mu_i - \sum_{i=1}^{n} \log y_i!$$

The interplay between the implicit and explicit model forms of algebraic statistical models has been the subject of considerable development; a seemingly innocuous explicit model may have a complicated implicit form. To some extent this development is easier in the so-called *power product*, or *toric* representation. This is, in fact, very familiar in statistics. The Binomial(n, p) mass distribution function is

$$\binom{n}{y} p^y (1 - p)^{n-y}, \; y = 0, \ldots, n.$$

Considered as a function of p this is about the simplest example of a power product representation.

Example 1.2 (Example 1.1 cont.) For our regression in multinomial form the power product model is

$$p_i = \xi_0 \xi_1^{x_i} \xi_2^{x_i^2}, \quad i = 1, \ldots, 4,$$

where $\xi_j = e^{\theta_j}$, $j = 0, \ldots, 2$. This is algebraic if the design points $\{x_i\}$ *are integer.* In general, we can write the power product model in the compact form $p = \xi^X$. Elimination of the p_i, then gives the implicit version of the toric variety.

1.2.1 Design

Let us return to the expression for the design in (1.2). We use a quotient operation to show that the cubic model is naturally associated to the design $\{x_i : i = 1, \ldots, 4\}$. We assume that there is no error so that we have exact interpolation with a cubic model. The quadratic model we chose is also a natural model, being a sub-model of the saturated cubic model. Taking any polynomial interpolator $\tilde{y}(x)$ for data $\{(x_i, y_i), \; i = 1, \ldots, 4\}$, with distinct x_i, we can quotient out with the polynomial

$$g(x) = (x - x_1)(x - x_2)(x - x_3)(x - x_4)$$

and write

$$\tilde{y}(x) = s(x)g(x) + r(x),$$

where the remainder, $r(x)$, is a univariate, at most cubic, polynomial. Since $g(x_i) = 0$, $i = 1, \ldots, 4$, on the design $r(x)$ is also an interpolator, and is the unique cubic interpolator for the data. A major part of algebraic geometry, exploited in algebraic statistics, extends this quotient operation to higher dimensions. The design $\{x_1, \ldots, x_n\}$ is now multidimensional with each $x_i \in \mathbb{R}^k$, and is expressed as the unique solution of a set of polynomial equations, say

$$g_1(x) = \ldots = g_m(x) = 0 \tag{1.6}$$

and the quotient operation gives

$$\tilde{y}(x) = \sum_{i=1}^{m} s_i(x)g_i(x) + r(x). \tag{1.7}$$

The first term on the right-hand side of (1.7) is a member of the *design ideal*. This is defined as the set of all polynomials which are zero on the design and is indicated as $\langle g_1(x), \ldots, g_m(x) \rangle$. The remainder $r(x)$, which is called the *normal form* of $\tilde{y}(x)$, is unique if the $\{g_j(x)\}$ form a Gröbner basis which, in turn, depends on a given *monomial ordering* (see Section 1.7). The polynomial $r(x)$ is a representative of a class of the quotient ring modulo the design ideal and a basis, as a vector space, of the quotient ring is a set of monomials $\{x^\alpha, \ \alpha \in L\}$ of small degree with respect to the chosen term-ordering as specified in Section 1.7. This basis provides the terms of e.g. regression models. It has the *order ideal* property, familiar from statistics, e.g. the hierarchical property of a linear regression model, that $\alpha \in L$ implies $\beta \in L$ for any $\beta \leq \alpha$ (component-wise). The set of such bases as we vary over all term-orderings is sometimes called the *algebraic fan* of the design. In general it does not give the set of all models which can be fitted to the data, even if we restrict to models which satisfy the order ideal property. However, it is, in a way that can be well defined, the set of models of minimal average degree. See (Pistone and Wynn 1996) for the introduction of Gröbner bases into design, (Pistone *et al.* 2001) for a summary of early work and (Berstein *et al.* 2007) for the work on average degree.

Putting all the elements together we have half a dozen classes of algebraic statistical models which form the basis for the field: (i) linear and log-linear explicit algebraic models, including power product models (ii) implicit algebraic models derived from linear, log-linear or power product models (iii) linear and log-linear models and power product models suggested by special experimental designs.

An explicit algebraic model such as (1.1) can be written down, before one considers the experimental design. Indeed in areas such as the optimal design of experiments one may choose the experimental design using some optimality criterion. But the implicit models described above are design dependent as we see from Equation (1.4). A question arises then: is there a generic way of describing an implicit model which is not design dependent? The answer is to define a polynomial of total degree p as an analytic function all of whose derivatives of higher order than p vanish. But this is an infinite number of conditions.

We shall see that the explicit–implicit duality is also a feature of the information geometry in the sense that one can consider a statistical manifold as an implicit object or defined by some parametric path or surface.

1.3 The uses of algebra

So far we have only shown the presence of algebraic structures in statistical models. We must try to answer briefly the question: what real use is the algebra? We can divide the answer into three parts: (i) to better understand the structure of well-known models, (ii) to help with, or innovate in, statistical methodology and inference and (iii) to define new model classes exploiting particular algebraic structures.

1.3.1 Model structure

Some of the most successful contributions of the algebra are due to the introduction of ideas which the statistical community has avoided or not had the knowledge to pursue. This is especially true for toric models for categorical data. It is important to distinguish two cases. First, for probability models all the representations: log-linear, toric, power product are essentially equivalent in the case that all probabilities are restricted to be *positive*. This condition can be built into the toric analysis via the so-called *saturation*. Consider our running Example 1.2. If ξ is a dummy variable then the condition $p_1 p_2 p_3 p_4 v + 1 = 0$ is violated if any of the p_j is zero. Adding this condition to the conditions obtained via the kernel method and eliminating v turns out to be equivalent to directly eliminating the ξ in the power product (toric) representation.

A considerable contribution of the algebraic methods is to handle boundary cases where probabilities are allowed to be zero. Zero counts are very common in sparse tables of data, such as when in a sample survey respondents are asked a large number of questions, but this is not the same as zero probabilities. But we may in fact have special models with zero probabilities in some cells. We may call these models *boundary models* and a contribution of the algebra is to analyse their complex structure. This naturally involves considerable use of algebraic ideas such as *irreducibility, primary decompositions, Krull dimension* and *Hilbert dimension.*

Second, another problem which has bedevilled statistical modelling is that of identifiability. We can take this to mean that different parameter values lead to different distributions. Or we can have a data-driven version: for a given data set (the one we have) the likelihood is locally invertible. The algebra is a real help in understanding and resolving such problems. In the theory of experimental design we can guarantee that the remainder (quotient) models (or sub-models of remainder models), $r(x)$, are identifiable given the design from which they were derived. The algebra also helps to explain the concept of *aliasing*: two polynomial models $p(x)$ and $q(x)$ are aliased over a design $\mathcal{D}$ if $p(x) = q(x)$ for all x in $\mathcal{D}$. This is equivalent to saying that $p(x) - q(x)$ lies in the design ideal.

There is a generic way to study identifiability, that is via elimination. Suppose that $h(\theta)$, for some parameter $\theta \in \mathbb{R}^u$ and $u \in \mathbb{Z}_{>0}$, is some quantity of interest such as a likelihood, distribution function, or some function of those quantities. Suppose also that we are concerned that $h(\theta)$ is over-parametrised in that there is a function of θ, say $\phi(\theta) \in \mathbb{R}^v$ with dimension $v < u$, with which we can parametrise the model but which has a smaller dimension than θ. If all the functions are polynomial we can write down (in possibly vector form): $r - h(\theta) = 0$, $s - \phi(\theta) = 0$, and try to eliminate θ algebraically to obtain the (smallest) variety on which (r, s) lies. If we are lucky this will give r explicitly in terms as function of s, which is then the required reparametrisation.

As a simple example think of a 2×2 table as giving probabilities p_{ij} for a bivariate binary random vector (X_1, X_2). Consider an over-parametrised power product model for independence with

$$p_{00} = \xi_1 \xi_3, \quad p_{10} = \xi_2 \xi_3, \quad p_{01} = \xi_1 \xi_4, \quad p_{11} = \xi_2 \xi_4.$$

We know that independence gives zero covariance so let us seek a parametrisation in terms of the non-central moments $m_{10} = p_{10} + p_{11}$, $m_{01} = p_{01} + p_{11}$. Eliminating the ξ_i (after adding $\sum_{ij} p_{ij} - 1 = 0$), we obtain the parametrisation: $p_{00} = (1 - m_{10})(1 - m_{01})$, $p_{10} = m_{10}(1 - m_{01})$, $p_{01} = (1 - m_{10})m_{01}$, $p_{11} = m_{10}m_{01}$. Alternatively, if we include $m_{11} = p_{11}$, the unrestricted probability model in terms of the moments is given by $p_{00} = 1 - m_{10} - m_{01} + m_{11}$, $p_{10} = m_{10} - m_{11}$, $p_{01} = m_{01} - m_{11}$, and $p_{11} = m_{11}$, but then we need to impose the extra *implicit* condition for zero covariance: $m_{11} - m_{10}m_{01} = 0$. This is another example of implicit–explicit duality.

Here is a Gaussian example. Let $\delta = (\delta_1, \delta_2, \delta_3)^{\top}$ be independent Gaussian unit variance input random variables. Define the output Gaussian random variables as

$$
\begin{aligned}
Y_1 &= \theta_1 \delta_1 \\
Y_2 &= \theta_2 \delta_1 + \theta_3 \delta_2 \\
Y_3 &= \theta_4 \delta_1 + \theta_5 \delta_3,
\end{aligned}
\tag{1.8}
$$

It is easy to see that this implies the conditional independence of Y_2 and Y_3 given Y_1. The covariance matrix of the $\{Y_i\}$ is

$$
C = \begin{pmatrix} c_{11} & c_{12} & c_{13} \\ c_{21} & c_{22} & c_{23} \\ c_{31} & c_{32} & c_{33} \end{pmatrix} = \begin{pmatrix} \theta_1^2 & \theta_1 \theta_2 & \theta_1 \theta_4 \\ \theta_1 \theta_2 & \theta_2^2 + \theta_3^2 & \theta_2 \theta_4 \\ \theta_1 \theta_4 & \theta_2 \theta_4 & \theta_4^2 + \theta_5^2 \end{pmatrix}.
$$

This is invertible (and positive definite) if and only if $\theta_1 \theta_3 \theta_5 \neq 0$. If we adjoin the saturation condition $\theta_1 \theta_3 \theta_5 v - 1 = 0$ and eliminate the θ_j and we obtain the symmetry conditions $c_{12} = c_{21}$ etc. plus the single equation $c_{11}c_{23} - c_{12}c_{13} = 0$. This is equivalent to the $(2,3)$ entry of C^{-1} being zero. The linear representation (1.8) can be derived from a graphical simple model: $2 - 1 - 3$, and points to a strong relationship between graphical models and conditions on covariance structures. The representation is also familiar in time series as the moving average representation. See (Drton *et al.* 2007) for some of the first work on the algebraic method for Gaussian models.

In practical statistics one does not rest with a single model, at least not until after a considerable effort on diagnostics, testing and so on. It is better to think in terms of hierarchies of models. At the bottom of the hierarchy may be simple models. In regression or log-linear models these may typically be additive models. More complex models may involve interactions, which for log-linear models may be representations of conditional independence. One can think of models of higher polynomial degree in the algebraic sense. The advent of very large data sets has stimulated work on model choice criteria and methods. The statistical kit-bag includes AIC, BIC, CART, BART, Lasso and many other methods. There are also close links to methods in data-mining and machine learning. The hope is that the algebra and algebraic and differential geometry will point to natural model structures be they rings, complexes, lattices, graphs, networks, trees and so on and also to suitable algorithms for climbing around such structures using model choice criteria.

In latent, or hidden, variable methods we extended the model top 'layer' with another layer which endows parameters from the first layer with distributions, that is to say *mixing*. This is also, of course, a main feature of Bayesian models and classical random effect models. Another generic term is hierarchical models, especially when we have many layers. This brings us naturally to *secant varieties* and we can push our climbing analogy one step further. A secant variety is a bridge which walks us from one first-level parameter value to another, that is it provides a support for the mixing. In its simplest form secant variety takes the form

$$\{r : r = (1 - \lambda)p + \lambda q, \ 0 \leq \lambda \leq 1\}$$

where p and q lie in varieties P and G respectively (which may be the same). See (Sturmfels and Sullivant 2006) for a useful study.

In probability models distinction should be made between a zero in a cell in data table, a zero *count*, and a structural zero in the sense that the model assigns zero probability to the cell. This distinction becomes a little cloudy when it is a cell which has a count but which, for whatever reason, could not be observed. One could refer to the latter as censoring which, historically, is when an observation is not observed because it has not happened yet, like the time of death or failure. In some fields it is referred to as having *partial information*.

As an example consider the toric idea for a simple balanced incomplete block design (BIBD). There are two factors, 'blocks' and 'treatments', and the arrangement of treatment in blocks is given by the scheme

$$\binom{1}{2} \binom{1}{3} \binom{1}{4} \binom{2}{3} \binom{2}{4} \binom{3}{4}$$

e.g. $\binom{1}{2}$ is the event that treatment 1 and 2 are in the first block. This corresponds to the following two-factor table where we have inserted the probabilities for observed cells, e.g. p_{11} and p_{21} are the probabilities that treatments one and two are in the first block,

p_{11}	p_{12}	p_{13}			
p_{21}			p_{24}	p_{25}	
	p_{32}		p_{34}		p_{36}
		p_{43}		p_{45}	p_{46}

The additive model $\log p_{ij} = \mu_0 + \alpha_i + \beta_j$ (ignoring the $\sum p_{ij} = 1$ constraint) has nine degrees of freedom (the rank of the X-matrix) and the kernel has rank 3 and one solution yields the terms:

$$p_{12}p_{21}p_{34} - p_{11}p_{24}p_{32} = 0$$
$$p_{24}p_{36}p_{45} - p_{25}p_{34}p_{46} = 0$$
$$p_{11}p_{25}p_{43} - p_{13}p_{21}p_{45} = 0.$$

A Gröbner basis and a Markov basis can also be found. For work on Markov bases for incomplete tables see (Aoki and Takemura 2008) and (Consonni and Pistone 2007).

1.3.2 Inference

If we condition on the sufficient statistics in a log-linear model for contingency tables, or its power-product form, the conditional distribution of the table does not depend on the parameters. If we take a classical test statistic for independence such as a χ^2 or likelihood ratio (deviance) statistics, then its conditional distribution, given the sufficient statistics T, will also not depend on the parameters, being a function of T. If we are able to find the conditional distribution and perform a conditional test, e.g. for independence, then (Type I) error rates will be the same as for the unconditional test. This follows simply by taking expectations. This technique is called an *exact conditional test*. For (very) small samples we can find the exact conditional distribution using combinatorial methods.

However, for tables which are small but too large for the combinatorics and not large enough for asymptotic methods to be accurate, algebraic Markov chain methods were introduced by (Diaconis and Sturmfels 1998). In the tradition of Markov Chain Monte Carlo (MCMC) methods we can simulate from the true conditional distribution of the tables by running a Markov chain whose steps preserve the appropriate margins. The collection of steps forms a *Markov basis* for the table. For example for a complete $I \times J$ table, under independence, the row and column sums (margins) are sufficient. A table is now a state of the Markov chain and a typical move is represented by a table with all zeros except values 1 at entry (i, i') and (j, j') and entry -1 at entries (j, i') and (i, j'). Adding this to or subtracting this from a current table (state) keeps the margins fixed, although one has to add the condition of non-negativity of the tables and adopt appropriate transition probabilities. In fact, as in MCMC practice, derived chains such as in the Metropolis–Hastings algorithm are used in the simulation.

It is not difficult to see that if we set up the X-matrix for the problem then a move corresponds to a column orthogonal to all the columns of X i.e. the kernel space. If we restrict to all probabilities being positive then the toric variety, the variety arising from a kernel basis and the Markov basis are all the same. In general the kernel basis is smaller than the Markov basis which is smaller than the associated Gröbner basis. In the terminology of ideals:

$$I_K \subset I_M \subset I_G,$$

with reverse inclusion for the varieties, where the sub-indices K, M, G stands for Kernel, Markov and Gröbner, respectively.

Given that one can carry out a single test, it should be possible to do multiple testing, close in spirit to the model-order choice problem mentioned above. There are several outstanding problems such as (i) finding the Markov basis for large problems and incomplete designs, (ii) decreasing the cost of simulation itself for example by repeat use of simulation, and (iii) alternatives to, or hybrids, simulation, using linear, integer programming, integer lattice theory (see e.g. Chapter 4).

The algebra can give insight into the solutions of the Maximum Likelihood Equations. In the Poisson/multinomial GLM case and when $p(\theta)$ is the vector of probabilities, the likelihood equations are

$$\frac{1}{n}X^\top Y = \frac{1}{n}T = X^\top p(\theta),$$

where $n = \sum_{x_i} Y(x_i)$ and T is the vector of sufficient statistics or generalised margins. We have emphasised the non-linear nature of these equations by showing that p depends on θ. Since $m = X^\top p$ are the moments with respect to the columns of X and $\frac{1}{n}X^\top Y$ are their sample counterpart, the equations simply equate the sample non-central moments to the population non-central moments. For the example in (1.1) the population non-central moments are $m_0 = 1$, $m_1 = \sum_i p_i x_i$, $m_2 = \sum_i p_i x_i^2$. Two types of result have been studied using algebra: (i) conditions for when the solution have closed form, meaning a rational form in the data Y and (ii) methods for counting the number of solutions. It is important to note that unrestricted solutions, $\hat{\theta}$, to these equations are not guaranteed to place the probabilities $p(\hat{\theta})$ in the region $\sum_i p_i = 1, p_i > 0, i = 1, \ldots, n$. Neither need they be real. Considerable progress has been made such as showing that decomposable graphical models have a simple form for the toric ideals and closed form of the maximum likelihood estimators: see (Geiger *et al.* 2006). But many problems remain such as in the study of non-decomposable models, models defined via various kinds of marginal independence and marginal conditional independence, and distinguishing real from complex solutions of the maximum likelihood equations.

As is well known, an advantage of the GLM formulation is that quantities which are useful in the asymptotics can be readily obtained, once the maximum likelihood estimators have been obtained. Two key quantities are the score statistic and the Fisher information for the parameters. The score (vector) is

$$U = \frac{\partial l}{\partial \theta} = X^\top Y - X^\top \mu,$$

where $j = (1, \ldots, n)^\top$ and we recall $\mu = \mathrm{E}[Y]$. The (Fisher) information is

$$\mathcal{I} = -\mathrm{E}\left[\frac{\partial^2 l}{\partial \theta_i \partial \theta_j}\right] = X^\top \mathrm{diag}(\mu)X,$$

which does not depend on the data.

As a simple exercise let us take the 2×2 contingency table, with the additive Poisson log-linear model (independence in the multinomial case representation) so that, after reparametrising to $\log \mu_{00} = \theta_0$, $\log \mu_{10} = \theta_0 + \theta_1$, $\log \mu_{01} = \theta_0 + \theta_2$ and

$\log \mu_{11} = \theta_0 + \theta_1 + \theta_2$, we have the rank 3 X-matrix:

$$X = \begin{pmatrix} 1 & 0 & 0 \\ 1 & 1 & 0 \\ 1 & 0 & 1 \\ 1 & 1 & 1 \end{pmatrix}.$$

In the power product formulation it becomes $\mu_{00} = \xi_0$, $\mu_{10} = \xi_0\xi_1$, $\mu_{01} = \xi_0\xi_2$, and $\mu_{11} = \xi_0\xi_1\xi_2$, and if we algebraically eliminate the ξ_i we obtain the following variety for the entries of $\mathcal{I} = \{\mathcal{I}_{ij}\}$, the information matrix for the θ

$$\mathcal{I}_{13} - \mathcal{I}_{33} = 0, \; \mathcal{I}_{12} - \mathcal{I}_{22} = 0, \; \mathcal{I}_{11}\mathcal{I}_{23} - \mathcal{I}_{22}\mathcal{I}_{33} = 0.$$

This implies that the $(2,3)$ entry in $\mathcal{I}^{-1}$, the asymptotic covariance of the maximum likelihood estimation of the parameters, is zero, as expected from the orthogonality of the problem.

1.3.3 Cumulants and moments

A key quantity in the development of the exponential model and associated asymptotics is the cumulant generating function. This is embedded in the Poisson/multinomial development as is perhaps most easily seen by writing the multinomial version in terms of repeated sampling from a given discrete distribution whose support is what we have been calling the 'design'. Let us return to Example 1.1 one more time. We can think of this as arising from a distribution with support $\{0, 1, 2, 3\}$ and probability mass function:

$$p(x; \theta_1, \theta_2) = \exp(\theta_1 x + \theta_2 x^2 - K(\theta_1, \theta_2)),$$

where we have suppressed θ_0 and incorporated it into $K(\theta_1, \theta_2)$. We clearly have

$$K(\theta_1, \theta_2) = \log(1 + e^{\theta_1 + \theta_2} + e^{2\theta_1 + 4\theta_2} + e^{3\theta_1 + 9\theta_2}).$$

The moment generating function is

$$M_X(s) = \mathrm{E}_X[e^{sX}] = e^{K(\theta_1 + s, \theta_2)} e^{-K(\theta_1, \theta_2)},$$

and the cumulant generating function is

$$K_X(s) = \log M_X(s) = K(\theta_1 + s, \theta_2) - K(\theta_1, \theta_2).$$

The expression for $K''(s)$ in terms of $K'(s)$ is sometime called the *variance function* in GLM theory and we note that $\mu = K'(0)$ and $\sigma^2 = K''(0)$ give the first two cumulants, which are respectively the mean and variance. If we make the power parametrisation $\xi_1 = e^{\theta_1}$, $\xi_2 = e^{\theta_2}$, $t = e^s$ and eliminate t from the expressions for K' and K'' (suppressing s), which are now rational, we obtain, after some algebra, the implicit representation

$$-8K'^2 + 24K' + (-12 - 12K' + 4K'^2 - 12K'\xi_2^2 + 36K'\xi_2^2)H$$
$$+ (8 - 24\xi_2^2)H^2 + (-9\xi_2^6 - 3\xi_2^4 + 5\xi_2^2 - 1)H^3$$

where $H = 3K' - K'^2 - K''$. Only at the value $\xi_2 = 1/\sqrt{3}$ the last term is zero and there is then an explicit quadratic variance function:

$$K'' = \frac{1}{3}K'(3 - K').$$

All discrete models of the log-linear type with integer support/design have an implicit polynomial relationship between K' and K'' where, in the multivariate case these are respectively a $(p-1)$-vector and a $(p-1) \times (p-1)$ matrix, and as in this example, we may obtain a polynomial variance function for special parameter values. Another interesting fact is that because of the finiteness of the support higher order moments can be expressed in terms of lower order moments. For our example we write the design variety $x(x-1)(x-2)(x-3) = 0$ as

$$x^4 = 6x^3 - 11x^2 + 6x$$

multiplying by x^r and taking expectation we have for the moments $m_r = \mathrm{E}[X^r]$ the recurrence relationship

$$m_{4+r} = 6m_{3+r} - 11m_{2+r} + 6m_{r+1}.$$

See (Pistone and Wynn 2006) and (Pistone and Wynn 1999) for work on cumulants.

This analysis generalises to the multivariate case and we have intricate relations between the defining Gröbner basis for the design, recurrence relationships and generating functions for the moments and cumulants, the implicit relationship between K and K' and implicit relation for raw probabilities and moments, arising from the kernel/toric representations. There is much work to be done to unravel all these relationships.

1.4 Information geometry on the simplex

In information geometry a statistical model is a family of probability densities (on the same sample space) and is viewed as a differential manifold. In the last twenty years there has been a development of information geometry in the non-parametric (infinite-dimensional) case and non-commutative (quantum) case. Here we consider the finite-dimensional case of a probability vector $p = (p_1, \ldots, p_n) \in \mathbb{R}^n$. Thus we may take the sample space to be $\Omega = \{1, ..., n\}$ and the manifold to be the interior of the standard simplex:

$$\mathcal{P}_n^1 = \{p : p_i > 0, \sum p_i = 1\}$$

(other authors use the notation $\mathcal{M}_>$). Each probability vector $p \in \mathcal{P}_n^1$ is a function from Ω to $\mathbb{R}^n$ and $f(p)$ is well defined for any reasonable real function f, e.g. any bounded function.

The tangent space of the simplex can be represented as

$$T_p(\mathcal{P}_n^1) = \{u \in \mathbb{R}^n : \sum_i u_i = 0\} \tag{1.9}$$

because the simplex is embedded naturally in $\mathbb{R}^n$. The tangent space at a given p can be also identified with the p-centered random variables, namely random variables with zero mean with respect to the density p

$$T_p(\mathcal{P}_n^1) = \{u \in \mathbb{R}^n : \mathrm{E}_p[u] = \sum_i u_i p_i = 0\}. \tag{1.10}$$

With a little abuse of language we use the same symbol for the two different representations (both will be useful in the sequel).

1.4.1 Maximum entropy and minimum relative entropy

Let p and q be elements of the simplex. Entropy and relative (Kullback–Leibler) entropy are defined by the following formulas

$$\mathrm{S}(p) = -\sum_i p_i \log p_i, \tag{1.11}$$

$$\mathrm{K}(p,q) = \sum_i p_i(\log p_i - \log q_i), \tag{1.12}$$

which for $q_0 = \left(\frac{1}{n}, \ldots, \frac{1}{n}\right)$ simplifies to $\mathrm{K}(p, q_0) = \sum_i p_i \log p_i - \sum_i p_i \log \frac{1}{n} = -\mathrm{S}(p) + \log n$.

In many applications, e.g. large deviations and maximum likelihood estimation, it is required to minimise the relative entropy, namely to determine a probability p on a manifold M that minimises $\mathrm{K}(p, q_0)$, equivalently that maximises the entropy $\mathrm{S}(p)$. Here Pythagorean-like theorems can be very useful. But the relative entropy is not the square of a distance between densities. For example, it is asymmetric and the triangle inequality does not hold. In Section 1.4.2 we illustrate some geometries on the simplex to bypass these difficulties.

In (Dukkipati 2008) the constrained maximum entropy and minimum relative entropy optimisation problems are translated in terms of toric ideals, following an idea introduced in (Hoşten *et al.* 2005) for maximum likelihood estimation. The key point is that the solution is an exponential model, hence a toric model, under the assumption of positive integer valued sufficient statistics. This assumption is embedded in the constraints of the optimisation, see e.g. (Cover and Thomas 2006). Ad hoc algorithms are to be developed to make this approach effective.

1.4.2 Paths on the simplex

To understand a geometry on a manifold we need to describe its geodesics in an appropriate context. The following are examples of curves that join the probability vectors p and q in $\mathcal{P}_n^1$:

$$(1 - \lambda)p + \lambda q, \tag{1.13}$$

$$\frac{p^{1-\lambda} q^{\lambda}}{C}, \tag{1.14}$$

$$\frac{((1 - \lambda)\sqrt{p} + \lambda\sqrt{q})^2}{B}, \tag{1.15}$$

where $C = \sum_i p_i^{1-\lambda} q_i^{\lambda}$ and $B = 2\sum_i [(1-\lambda)\sqrt{p_i}+\lambda\sqrt{q_i}]^2$ are suitable normalisation constants. We may ask which is the most 'natural' curve joining p and q. In the case (1.15) the answer is that the curve is a geodesic with respect to the metric defined by the Fisher information. Indeed, all the three curves above play important roles in this geometric approach to statistics.

1.5 Exponential–mixture duality

We consider the simplex and the localised representation of the tangent space. Define a parallel transport as

$$U_{pq}^m(u) = \frac{p}{q}u$$

for $u \in T_p(\mathcal{P}_n^1)$. This shorthand notation must be taken to mean $\left(\frac{p_1}{q_1}u_1, \ldots, \frac{p_n}{q_n}u_n\right)$. Then $\frac{p}{q}u$ is q-centred and composing the transports $U_{pq}^m U_{qr}^m$ gives U_{pr}^m. The geodesics associated to this parallel transport are the mixture curves in (1.13).

The parallel transport defined as

$$U_{pq}^e(u) = u - \mathrm{E}_q[u]$$

leads to a geometry whose geodesics are the exponential models as in (1.14). In the parametric case this can be considered arising from local representation of the models via their differentiated log-density or *score*.

There is an important and general duality between the mixture and exponential forms. Assume that v is p-centred and define

$$\langle u, v \rangle_p = \mathrm{E}_p[uv] = \mathrm{Cov}_p(u, v).$$

Then we have

$$\langle U_{pq}^e(u), U_{pq}^m(v) \rangle_q = \mathrm{E}_q\left[(u - \mathrm{E}_q[u.])\frac{p}{q}v\right] =$$

$$\mathrm{E}_p[uv] - \mathrm{E}_q[u]\,\mathrm{E}_p[v] = \mathrm{E}_p[uv] = \langle u, v \rangle_p. \quad (1.16)$$

1.6 Fisher information

Let us develop the exponential model in more detail. The exponential model is given in the general case by

$$p_\theta = \exp(u_\theta - K(u_\theta))p$$

where we have set $p = p_0$ and u_θ is a parametrised class of functions. In the simplex case we can write the one-parameter exponential model as

$$p_{\lambda,i} = \exp(\lambda(\log q_i - \log p_i) - \log(C))p_i.$$

Thus with θ replaced by λ, the ith component of u_θ by $\lambda(\log q_i - \log p_i)$ and $K = \log C$, we have the familiar exponential model. After an elementary calculation the

Fisher information at p in terms of the centred variable $\bar{u} = u - \mathrm{E}_p[u]$ is

$$\mathcal{I}_p = \sum_{i=1}^{n} \bar{u}_i^2 p_i$$

where $\bar{u} \in T_p(\mathcal{P}_n^1)$ as in Equation (1.10). Analogously, the Fisher metric is $\langle u, v \rangle_p = \sum_{i=1}^{n} \bar{u}_i \bar{v}_i p_i$. In the representation (1.9) of the tangent space the Fisher matrix is

$$\langle \bar{\bar{u}}, \bar{\bar{v}} \rangle_{p,FR} = \sum_i \frac{\bar{\bar{u}}_i \bar{\bar{v}}_i}{p_i}$$

with $\bar{\bar{u}}_i = u_i - \sum_i u_i / n$ where n is the total sample size.

The duality in (1.16) applies to the simplex case and exhibits a relationship endowed with the Fisher information. Let $u = \log \frac{q}{p}$ so that for the exponential model

$$\dot{p}_\lambda = \frac{\partial p_\lambda}{\partial \lambda} = u - \mathrm{E}_\lambda[u].$$

Now the mixture representative of the models is $\frac{p_\lambda}{p} - 1$, whose differential (in the tangent space) is $\frac{u}{p_\lambda} = \frac{p}{q} v$, say. Then putting $\lambda = 1$ the duality in (1.16) becomes

$$\langle \bar{u}, \bar{v} \rangle_p = \langle \bar{\bar{u}}, \bar{\bar{v}} \rangle_{p,FR} = \mathrm{Cov}_p(u, v).$$

Note that the manifold $\mathcal{P}_n^1$ with the Fisher metric is isometric with an open subset of the sphere of radius 2 in $\mathbb{R}^n$. Indeed, if we consider the map $\varphi : \mathcal{P}_n^1 \to S_2^{n-1}$ defined by

$$\varphi(p) = 2(\sqrt{p_1}, ..., \sqrt{p_n})$$

then the differential on the tangent space is given by

$$D_p\varphi(u) = \left(\frac{u_1}{\sqrt{p_1}}, ..., \frac{u_n}{\sqrt{p_n}} \right).$$

(Gibilisco and Isola 2001) shows that the Fisher information metric is the pull-back of the natural metric on the sphere.

This identification allows us to describe geometric objects of the Riemannian manifold, namely $(\mathcal{P}_n^1, \langle \cdot, \cdot \rangle_{p,FR})$, using properties of the sphere S_2^{n-1}. For example, as in (1.15), we obtain that the geodesics for the Fisher metric on the simplex are

$$\frac{\left(\lambda\sqrt{p} + (1-\lambda)\sqrt{q} \right)^2}{B}.$$

As shown above, the geometric approach to Fisher information demonstrates in which sense mixture and exponential models are dual of each other. This can be considered as a fundamental paradigm of information geometry and from this an abstract theory of statistical manifolds has been developed which generalises Riemannian geometry, see (Amari and Nagaoka 2000).

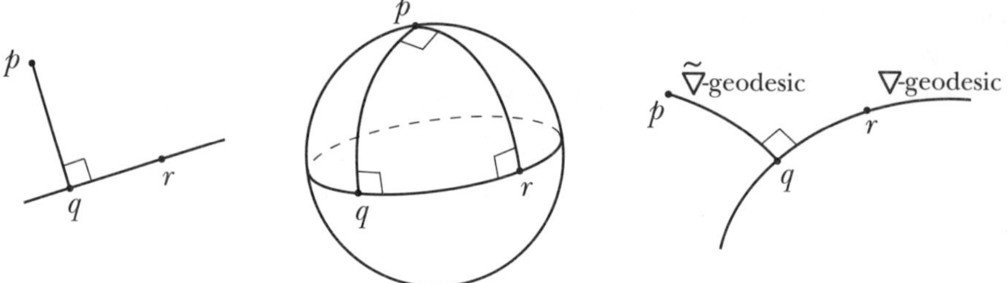

Fig. 1.1 Pythagora theorem: standard (left), geodesic triangle on the sphere (centre) and generalised (right).

1.6.1 The generalised Pythagorean theorem

We formulate the Pythagorean theorem in a form suitable to be generalised to a Riemannian manifold. Let p, q, r be points of the real plane and let $D(p|q)$ be the square of the distance between p and q. If γ is a geodesic connecting p and q, and δ is a geodesic connecting q with r, and furthermore if γ and δ intersect at q orthogonally, then $D(p|q) + D(q|r) = D(p|r)$, see Figure 1.1 (left). Figure 1.1 (centre) shows that on a general Riemannian manifold, like the sphere, $D(p|q) + D(q|r) \neq D(p|r)$, usually. This is due to the curvature of the manifold and a flatness assumption is required. The flatness assumption allows the formulation of the Pythagorean theorem in a context broader than the Riemannian one.

A *divergence* on a differential manifold M is a non-negative smooth function $D(\cdot|\cdot)\colon M \times M \to \mathbb{R}$ such that $D(p|q) = 0$ if, and only if, $p = q$ (note that here D stands for divergence and not derivative). A typical example is the Kullback-Leibler divergence, which we already observed is not symmetric hence it is not a distance.

It is a fundamental result of Information Geometry, see (Eguchi 1983, Eguchi 1992, Amari and Nagaoka 2000), that to any divergence D one may associate three geometries, namely a triple $\left(\langle \cdot, \cdot \rangle^D, \nabla^D, \tilde{\nabla}^D \right)$ where $\langle \cdot, \cdot \rangle^D$ is a Riemannian metric while $\nabla^D, \tilde{\nabla}^D$ are two linear connections in duality with respect to the Riemannian metric.

A statistical structure $\left(\langle \cdot, \cdot \rangle^D, \nabla^D, \tilde{\nabla}^D \right)$ is *dually flat* if both ∇ and $\tilde{\nabla}$ are flat. This means that curvature and torsion are (locally) zero for both connections. This is equivalent to the existence of an affine coordinate system. The triple given by the Fisher information metric, the mixture–exponential connection pair, whose geodesics are given in Equations (1.13) and (1.14), is an example of a dually flat statistical structure. The generalised Pythagorean theorem can be stated as follows.

Let $D(\cdot|\cdot)$ be a divergence on M such that the induced statistical structure is dually flat. Let $p, q, r \in M$, let γ be a ∇^D-geodesic connecting p and q, let δ be a $\tilde{\nabla}^D$-geodesic connecting q with r, and suppose that γ and δ intersect at q orthogonally with respect to the Riemannian metric $\langle \cdot, \cdot \rangle^D$. Then, as shown in Figure 1.1 (right),

$$D(p|q) + D(q|r) = D(p|r).$$

Summarising, if the divergence is the squared Euclidean distance, this is the usual Pythagorean theorem and if the divergence is the Kullback–Leibler relative entropy, this is the differential geometric version of the result proved in (Csiszár 1975), see also (Grünwald and Dawid 2004). In a quantum setting, (Petz 1998) proved a Pythagorean-like theorem with the Umegaki relative entropy instead of Kullback–Leibler relative entropy. Here as well the flatness assumption is essential.

1.6.2 General finite-dimensional models

In the above we really only considered the one-parameter exponential model, even in the finite-dimensional case. But as is clear from the early part of this introduction more complex exponential models of the form

$$p_\theta = \exp\left(\sum \theta_i u_i - K(\theta)\right) p$$

are studied. Here the u_i are the columns of the X-matrix, and we can easily compute the cumulant generating functions, as explained for the running example. More such examples are given in Chapter 21. A log-linear model becomes a flat manifold in the information geometry terminology. There remain problems, even in this case, for example when we wish to compute quantities of interest such as $K(\theta)$ at a maximum likelihood estimator and this does not have a closed form, there will be no closed form for K either.

More serious is when we depart from the log-linear formulation. To repeat: this is when u_θ is not linear. We may use the term *curved* exponential model (Efron 1975). As we have seen, the dual (kernel) space to the model is computable in the linear case and, with the help of algebra, we can obtain implicit representation of the model. But in the non-linear finite-dimensional case there will be often severe computational problems. Understanding the curvature and construction of geodesics may help both with the statistical analysis and also the computation e.g. those relying on gradients. The infinite-dimensional case requires special care as some obvious properties of submanifolds and, hence, tangent spaces could be missing. Concrete and useful examples of infinite-dimensional models do exists e.g. in the framework of Wiener spaces, see Chapter 21.

One way to think of a finite-dimensional mixture model is that it provides a special curved, but still finite-dimensional, exponential family, but with some attractive duality properties. As mentioned, mixture models are the basis of latent variable models (Pachter and Sturmfels 2005) and is to be hoped that the methods of secant varieties will be useful. See Chapter 2 and the on-line Chapter 22 by Yi Zhou. See also Chapter 4 in (Drton *et al.* 2009) for an algebraic exposition on the role of secant varieties for hidden variable models.

1.7 Appendix: a summary of commutative algebra (with Roberto Notari)

We briefly recall the basic results from commutative algebra we need to develop the subject. Without any further reference, we mention that the sources for the material in the present section are (Atiyah and Macdonald 1969) and (Eisenbud 2004).

Let $\mathcal{K}$ be a ground field, and let $R = \mathcal{K}[x_1,\ldots,x_k]$ be the polynomial ring over $\mathcal{K}$ in the indeterminates (or variables) $x_1,\ldots,x_k$. The ring operations in R are the usual sum and product of polynomials.

Definition 1.1 A subset $I \subset R$ is an *ideal* if $f + g \in I$ for all $f, g \in I$ and $fg \in I$ for all $f \in I$ and all $g \in R$.

Polynomial ideals

Proposition 1.1 *Let* $f_1,\ldots,f_r \in R$. *The set* $\langle f_1,\ldots,f_r \rangle = \{f_1 g_1 + \cdots + f_r g_r : g_1,\ldots,g_r \in R\}$ *is the smallest ideal in R with respect to the inclusion that contains* $f_1,\ldots,f_r$.

The ideal $\langle f_1,\ldots,f_r \rangle$ is called the *ideal generated by* $f_1,\ldots,f_r$. A central result in the theory of ideals in polynomial ring is the following Hilbert's basis theorem.

Theorem 1.1 *Given an ideal $I \subset R$, there exist $f_1,\ldots,f_r \in I$ such that $I = \langle f_1,\ldots,f_r \rangle$.*

The Hilbert's basis theorem states that R is a Noetherian ring, where a ring is Noetherian if every ideal is finitely generated.

As in the theory of $\mathcal{K}$-vector spaces, the intersection of ideals is an ideal, while the union is not an ideal, in general. However, the following proposition holds.

Proposition 1.2 *Let $I, J \subset R$ be ideals. Then,*

$$I + J = \{f + g : f \in I, g \in J\}$$

is the smallest ideal in R with respect to inclusion that contains both I and J, and it is called the sum of I and J.

Quotient rings

Definition 1.2 Let $I \subset R$ be an ideal. We write $f \sim_I g$ if $f - g \in I$ for $f, g \in R$.

Proposition 1.3 *The relation $\sim_I$ is an equivalence relation in R. Moreover, if $f_1 \sim_I f_2, g_1 \sim_I g_2$ then $f_1 + g_1 \sim_I f_2 + g_2$ and $f_1 g_1 \sim_I f_2 g_2$.*

Definition 1.3 The set of equivalence classes, the cosets, of elements of R with respect to $\sim_I$ is denoted as R/I and called the *quotient space (modulo I)*.

Proposition 1.3 shows that R/I is a ring with respect to the sum and product it inherits from R. Explicitly, if $[f], [g] \in R/I$ then $[f]+[g] = [f+g]$ and $[f][g] = [fg]$. Moreover, the ideals of R/I are in one-to-one correspondence with the ideals of R containing I.

Definition 1.4 If J is ideal in R, then I/J is the ideal of R/J given by $I \supseteq J$ where I is ideal in R.

Ring morphisms

Definition 1.5 Let R, S be two commutative rings with identity. A map $\varphi : R \to S$ is a *morphism of rings* if (i) $\varphi(f + g) = \varphi(f) + \varphi(g)$ for every $f, g \in R$; (ii) $\varphi(fg) = \varphi(f)\varphi(g)$ for every $f, g \in R$; (iii) $\varphi(1_R) = 1_S$ where $1_R, 1_S$ are the identities of R and S, respectively.

Theorem 1.2 *Let $I \subset R$ be an ideal. Then, the map $\varphi : R \to R/I$ defined as $\varphi(f) = [f]$ is a surjective (or onto) morphism of commutative rings with identity.*

An isomorphism of rings is a morphism that is both injective and surjective.

Theorem 1.3 *Let I, J be ideals in R. Then, $(I + J)/I$ is isomorphic to $J/(I \cap J)$.*

Direct sum of rings

Definition 1.6 Let R, S be commutative rings with identity. Then the set

$$R \oplus S = \{(r, s) : r \in R, s \in S\}$$

with component-wise sum and product is a commutative ring with $(1_R, 1_S)$ as identity.

Theorem 1.4 *Let I, J be ideals in R such that $I + J = R$. Let*

$$\phi : R \to R/I \oplus R/J$$

be defined as $\phi(f) = ([f]_I, [f]_J)$. It is an onto morphism, whose kernel is $I \cap J$. Hence, $R/(I \cap J)$ is isomorphic to $R/I \oplus R/J$.

Localisation of a ring

Let $f \in R, f \neq 0$, and let $S = \{f^n : n \in \mathbb{N}\}$. In $R \times S$ consider the equivalence relation $(g, f^m) \sim (h, f^n)$ if $gf^n = hf^m$. Denote with $\frac{g}{f^n}$ the cosets of $R \times S$, and R_f the quotient set.

Definition 1.7 The set R_f is called the localisation of R with respect to f.

With the usual sum and product of ratios, R_f is a commutative ring with identity.

Proposition 1.4 *The map $\varphi : R \to R_f$ defined as $\varphi(g) = \frac{g}{1}$ is an injective morphism of commutative rings with identity.*

Maximal ideals and prime ideals

Definition 1.8 An ideal $I \subset R$, $I \neq R$, is a *maximal ideal* if I is not properly included in any ideal J with $J \neq R$.

Of course, if $a_1, \ldots, a_k \in \mathcal{K}$ then the ideal $I = \langle x_1 - a_1, \ldots, x_k - a_k \rangle$ is a maximal ideal. The converse of this remark is called Weak Hilbert's Nullstellensatz, and it needs a non-trivial hypothesis.

Theorem 1.5 *Let $\mathcal{K}$ be an algebraically closed field. Then, I is a maximal ideal if, and only if, there exist $a_1, \ldots, a_k \in \mathcal{K}$ such that $I = \langle x_1 - a_1, \ldots, x_k - a_k \rangle$.*

Definition 1.9 An ideal $I \subset R, I \neq R$, is a prime ideal if $xy \in I, x \notin I$ implies that $y \in I$, where $x, y \in \{x_1, \ldots, x_k\}$.

Proposition 1.5 *Every maximal ideal is a prime ideal.*

Radical ideals and primary ideals

Definition 1.10 Let $I \subset R$ be an ideal. Then,

$$\sqrt{I} = \{f \in R : f^n \in I, \text{ for some } n \in \mathbb{N}\}$$

is the *radical* ideal in I.

Of course, I is a radical ideal if $\sqrt{I} = I$.

Definition 1.11 Let $I \subset R, I \neq R$, be an ideal. Then I is a *primary ideal* if $xy \in I, x \notin I$ implies that $y^n \in I$ for some integer n, with $x, y \in \{x_1, \ldots, x_k\}$.

Proposition 1.6 *Let I be a primary ideal. Then, $\sqrt{I}$ is a prime ideal.*

Often, the primary ideal I is called $\sqrt{I}$-primary.

Primary decomposition of an ideal

Theorem 1.6 *Let $I \subset R, I \neq R$, be an ideal. Then, there exist $I_1, \ldots, I_t$ primary ideals with different radical ideals such that $I = I_1 \cap \cdots \cap I_t$.*

Theorem 1.6 provides the so-called primary decomposition of I.

Corollary 1.1 *If I is a radical ideal, then it is the intersection of prime ideals.*

Proposition 1.7 links morphisms and primary decomposition, in a special case that is of interest in algebraic statistics.

Proposition 1.7 *Let $I = I_1 \cap \cdots \cap I_t$ be a primary decomposition of I, and assume that $I_i + I_j = R$ for every $i \neq j$. Then the natural morphism*

$$\varphi : R/I \to R/I_1 \oplus \cdots \oplus R/I_t$$

is an isomorphism.

Hilbert function and Hilbert polynomial

The Hilbert function is a numerical function that 'gives a size' to the quotient ring R/I.

Definition 1.12 Let $I \subset R$ be an ideal. The *Hilbert function* of R/I is the function

$$h_{R/I} : \mathbb{Z} \to \mathbb{Z}$$

defined as $h_{R/I}(j) = \dim_{\mathcal{K}}(R/I)_{\leq j}$, where $(R/I)_{\leq j}$ is the subset of cosets that contain a polynomial of degree less than or equal to j, and $\dim_{\mathcal{K}}$ is the dimension as $\mathcal{K}$-vector space.

The following (in)equalities follow directly from Definition 1.12.

Proposition 1.8 *For every ideal $I \subset R, I \neq R$, it holds: (i) $h_{R/I}(j) = 0$ for every $j < 0$; (ii) $h_{R/I}(0) = 1$; (iii) $h_{R/I}(j) \leq h_{R/I}(j+1)$.*

Theorem 1.7 *There exists a polynomial $p_{R/I}(t) \in \mathbb{Q}[t]$ such that $p_{R/I}(j) = h_{R/I}(j)$ for j much larger than zero, $j \in \mathbb{Z}$.*

Definition 1.13 (i) The polynomial $p_{R/I}$ is called the *Hilbert polynomial* of R/I. (ii) Let $I \subset R$ be an ideal. The *dimension of R/I* is the degree of the Hilbert polynomial $p_{R/I}$ of R/I.

If the ring R/I has dimension 0 then the Hilbert polynomial of R/I is a non-negative constant called the degree of the ring R/I and indicated as $\deg(R/I)$. The meaning of the degree is that $\deg(R/I) = \dim_{\mathcal{K}}(R/I)_{\leq j}$ for j large enough. Moreover, the following proposition holds.

Proposition 1.9 *Let $I \subset R$ be an ideal. The following are equivalent: (i) R/I is $0-$dimensional; (ii) $\dim_{\mathcal{K}}(R/I)$ is finite. Moreover, in this case, $\deg(R/I) = \dim_{\mathcal{K}}(R/I)$.*

Term-orderings and Gröbner bases

Next, we describe some tools that make effective computations with ideals in polynomial rings.

Definition 1.14 A *term* in R is $x^a = x_1^{a_1} \ldots x_k^{a_k}$ for $a = (a_1, \ldots, a_k) \in (\mathbb{Z}_{\geq 0})^k$. The set of terms is indicated as $\mathbb{T}^k$.

The operation in $\mathbb{T}^k$, of interest, is the product of terms.

Definition 1.15 A term-ordering is a *well ordering* $\preccurlyeq$ on $\mathbb{T}^k$ such that $1 \preccurlyeq x^a$ for every $x^a \in \mathbb{T}^k$ and $x^a \preccurlyeq x^b$ implies $x^a x^c \preccurlyeq x^b x^c$ for every $x^c \in \mathbb{T}^k$.

A polynomial in R is a linear combination of a finite set of terms in $\mathbb{T}^k$: $f = \sum_{a \in A} c_a x^a$ where A is a finite subset of $\mathbb{Z}^k_{\geq 0}$.

Definition 1.16 Let $f \in R$ be a polynomial, A the finite set formed by the terms in f and $x^b = \max_{\preccurlyeq}\{x^a : a \in A\}$. Let $I \subset R$ be an ideal.

(i) The term $\mathrm{LT}(f) = c_b x^b$ is called the *leading term* of f.
(ii) The ideal generated by $\mathrm{LT}(f)$ for every $f \in I$ is called the *order ideal* of I and is indicated as $\mathrm{LT}(\mathrm{I})$.

Definition 1.17 Let $I \subset R$ be an ideal and let $f_1, \ldots, f_t \in I$. The set $\{f_1, \ldots, f_t\}$ is a *Gröbner basis* of I with respect to $\preccurlyeq$ if $\mathrm{LT}(I) = \langle \mathrm{LT}(f_1), \ldots, \mathrm{LT}(f_t) \rangle$.

Gröbner bases are special sets of generators for ideals in R. Among the many results concerning Gröbner bases, we list a few, to stress their role in the theory of ideals in polynomial rings.

Proposition 1.10 *Let $I \subseteq R$ be an ideal. Then, $I = R$ if, and only if, $1 \in \mathcal{F}$, where $\mathcal{F}$ is a Gröbner basis of I, with respect to any term-ordering $\preccurlyeq$.*

Proposition 1.11 *Let $I \subset R$ be an ideal. The ring R/I is 0–dimensional if, and only if, $x_i^{a_i} \in \mathrm{LT}(I)$ for every $i = 1, \ldots, k$.*

Proposition 1.11, known as Buchberger's criterion for 0–dimensionality of quotient rings, states that for every $i = 1, \ldots k$, there exists $f_{j(i)} \in \mathcal{F}$, Gröbner basis of I, such that $\mathrm{LT}(f_{j(i)}) = x_i^{a_i}$ ..

Definition 1.18 Let $I \subset R$ be an ideal. A polynomial $f = \sum_{a \in A} c_a x^a$ is in *normal form* with respect to $\preccurlyeq$ and I if $x^a \notin \mathrm{LT}(I)$ for each $a \in A$.

Proposition 1.12 *Let $I \subset R$ be an ideal. For every $f \in R$ there exists a unique polynomial, indicated as $\mathrm{NF}(f) \in R$, in normal form with respect to $\preccurlyeq$ and I such that $f - \mathrm{NF}(f) \in I$. Moreover, $\mathrm{NF}(f)$ can be computed from f and a Gröbner basis of I with respect to $\preccurlyeq$.*

Gröbner bases allow us to compute in the quotient ring R/I, with respect to a term-ordering, because they provide canonical forms for the cosets. This computation is implemented in much software for symbolic computation.

As last result, we recall that Gröbner bases simplify the computation of Hilbert functions.

Proposition 1.13 *Let $I \subset R$ be an ideal. Then R/I and $R/\mathrm{LT}(I)$ have the same Hilbert function. Furthermore, a basis of the $\mathcal{K}$–vector space $(R/\mathrm{LT}(I))_{\leq j}$ is given by the cosets of the terms of degree $\leq j$ not in $\mathrm{LT}(I)$.*

References

4ti2 Team (2006). *4ti2 – A software package for algebraic, geometric and combinatorial problems on linear spaces* (available at www.4ti2.de).

Amari, S. and Nagaoka, H. (2000). *Methods of Information Geometry* (American Mathematical Society/Oxford University Press).

Aoki, S. and Takemura, A. (2008). The largest group of invariance for Markov bases and toric ideals, *Journal of Symbolic Computing* **43**(5), 342–58.

Atiyah, M. F. and Macdonald, I. G. (1969). *Introduction to Commutative Algebra* (Addison-Wesley Publishing Company).

Berstein, Y., Maruri-Aguilar, H., Onn, S., Riccomagno, E. and Wynn, H. P. (2007). Minimal average degree aberration and the state polytope for experimental design (available at arXiv:stat.me/0808.3055).

Casanellas, M. and Fernández-Sánchez, J. (2007). Performance of a new invariants method on homogeneous and nonhomogeneous quartet trees, *Molecular Biology and Evolution* **24**(1), 288–93.

Čencov, N. N. (1982). *Statistical decision rules and optimal inference* (Providence, RI, American Mathematical Society). Translation from the Russian edited by Lev J. Leifman.

Consonni, G. and Pistone, G. (2007). Algebraic Bayesian analysis of contingency tables with possibly zero-probability cells, *Statistica Sinica* **17**(4), 1355–70.

Cover, T. M. and Thomas, J. A. (2006). *Elements of Information Theory* 2nd edn (Hoboken, NJ, John Wiley & Sons).

Csiszár, I. (1975). *I*-divergence geometry of probability distributions and minimization problems, *Annals of Probability* **3**, 146–58.

Cox, D., Little, J. and O'Shea, D. (2008). *Ideals, Varieties, and Algorithms* 3rd edn (New York, Springer-Verlag).

Diaconis, P. and Sturmfels, B. (1998). Algebraic algorithms for sampling from conditional distributions, *Annals of Statistics* **26**(1), 363–97.

Drton, M., Sturmfels, B. and Sullivant, S. (2007). Algebraic factor analysis: tetrads pentads and beyond, *Probability Theory and Related Fields* **138**, 463–93.

Drton, M., Sturmfels, B. and Sullivant, S. (2009). *Lectures on Algebraic Statistics* (Vol. 40, Oberwolfach Seminars, Basel, Birkhäuser).

Dukkipati, A. (2008). Towards algebraic methods for maximum entropy estimation (available at arXiv:0804.1083v1).

Efron, B. (1975). Defining the curvature of a statistical problem (with applications to second–order efficiency) (with discussion), *Annals of Statistics* **3**, 1189–242.

Eisenbud, D. (2004). *Commutative Algebra*, GTM 150, (New York, Springer-Verlag).

Eguchi, S. (1983). Second order efficiency of minimum contrast estimators in a curved exponential family, *Annals of Statistics* **11**, 793–803.

Eguchi, S. (1992). Geometry of minimum contrast, *Hiroshima Mathematical Journal* **22**(3), 631–47.

Evans, S. N. and Speed, T. P. (1993). Invariants of some probability models used in phylogenetic inference, *Annals of Statistics* **21**(1), 355–77.

Fienberg, S. E. (1980). *The analysis of cross-classified categorical data* 2nd edn (Cambridge, MA, MIT Press).

Grayson, D. and Stillman, M. (2006). *Macaulay 2, a software system for research in algebraic geometry* (available at www.math.uiuc.edu/Macaulay2/).

Geiger, D., Meek, C. and Sturmfels, B. (2006). On the toric algebra of graphical models, *Annals of Statistics* **34**, 1463–92.

Gibilisco, P. and Isola, T. (2001). A characterisation of Wigner-Yanase skew information among statistically monotone metrics, *Infinite Dimensional Analysis Quantum Probability and Related Topics* **4**(4), 553–7.

Greuel, G.-M., Pfister, G. and Schönemann, H. (2005). Singular *3.0. A Computer Algebra System for Polynomial Computations. Centre for Computer Algebra* (available at www.singular.uni-kl.de).

Grünwald, P. D. and Dawid, P. (2004). Game theory, maximum entropy, minimum discrepancy and robust Bayesian decision theory, *Annals of Statistics* **32**(4), 1367–433.

Hoşten, S., Khetan, A. and Sturmfels, B. (2005). Solving the likelihood equations, *Foundations of Computational Mathematics* **5**(4), 389–407.

Pachter, L. and Sturmfels, B. eds. (2005). *Algebraic Statistics for Computational Biology* (New York, Cambridge University Press).

Petz, D. (1998). Information geometry of quantum states. In *Quantum Probability Communications*, vol. X, Hudson, R. L. and Lindsay, J. M. eds. (Singapore, World Scientific) 135–58.

Pistone, G., Riccomagno, E. and Wynn, H. P. (2001). *Algebraic Statistics* (Boca Raton, Chapman & Hall/CRC).

Pistone, G. and Wynn, H. P. (1996). Generalised confounding with Gröbner bases, *Biometrika* **83**(3), 653–66.

Pistone, G., and Wynn, H. P. (1999). Finitely generated cumulants, *Statistica Sinica* **9**(4), 1029–52.

Pistone, G., and Wynn, H. P. (2006). Cumulant varieties, *Journal of Symbolic Computing* **41**, 210–21.

Riccomagno, E. (2009). A short history of Algebraic Statisitcs, *Metrika* **69**, 397–418.

Sturmfels, B. and Sullivant, S. (2006). Combinatorial secant varieties, *Pure and Appl Mathematics Quarterly* **3**, 867–91.

Part I

Contingency tables

2

Maximum likelihood estimation in latent class models for contingency table data

Stephen E. Fienberg

Patricia Hersh

Alessandro Rinaldo

Yi Zhou

Abstract

Statistical models with latent structure have a history going back to the 1950s and have seen widespread use in the social sciences and, more recently, in computational biology and in machine learning. Here we study the basic latent class model proposed originally by the sociologist Paul F. Lazarfeld for categorical variables, and we explain its geometric structure. We draw parallels between the statistical and geometric properties of latent class models and we illustrate geometrically the causes of many problems associated with maximum likelihood estimation and related statistical inference. In particular, we focus on issues of non-identifiability and determination of the model dimension, of maximisation of the likelihood function and on the effect of symmetric data. We illustrate these phenomena with a variety of synthetic and real-life tables, of different dimension and complexity. Much of the motivation for this work stems from the '100 Swiss Francs' problem, which we introduce and describe in detail.

2.1 Introduction

Latent class (LC) or latent structure analysis models were introduced in the 1950s in the social science literature to model the distribution of dichotomous attributes based on a survey sample from a populations of individuals organised into distinct homogeneous classes on the basis of an unobservable attitudinal feature. See (Anderson 1954, Gibson 1955, Madansky 1960) and, in particular, (Henry and Lazarfeld 1968). These models were later generalised in (Goodman 1974, Haberman 1974, Clogg and Goodman 1984) as models for the joint marginal distribution of a set of manifest categorical variables, assumed to be conditionally independent given an unobservable or latent categorical variable, building upon the then recently developed literature on log-linear models for contingency tables. More recently, latent class models have been described and studied as a special case of a larger class of directed acyclic graphical models with hidden

Algebraic and Geometric Methods in Statistics, ed. Paolo Gibilisco, Eva Riccomagno, Maria Piera Rogantin and Henry P. Wynn. Published by Cambridge University Press. © Cambridge University Press 2010.

nodes, sometimes referred to as Bayes nets, Bayesian networks, or causal models, e.g., see (Lauritzen 1996, Cowell *et al.* 1999, Humphreys and Titterington 2003) and, in particular, (Geiger *et al.* 2001). A number of recent papers have established fundamental connections between the statistical properties of latent class models and their algebraic and geometric features, e.g., see (Settimi and Smith 1998, Settimi and Smith 2005, Smith and Croft 2003, Rusakov and Geiger 2005, Watanabe 2001) and (Garcia *et al.* 2005).

Despite these recent important theoretical advances, the basic statistical tasks of estimation, hypothesis testing and model selection remain surprisingly difficult and, in some cases, infeasible tasks, even for small latent class models. Nonetheless, LC models are widely used and there is a 'folklore' associated with estimation in various computer packages implementing algorithms such as Expectation Maximisation (EM) for estimation purposes, e.g., see (Uebersax 2006).

The goal of this chapter is two-fold. First, we offer a simplified geometric and algebraic description of LC models and draw parallels between their statistical and geometric properties. The geometric framework enjoys notable advantages over the traditional statistical representation and, in particular, offers natural ways of representing singularities and non-identifiability problems. Furthermore, we argue that the many statistical issues encountered in fitting and interpreting LC models are a reflection of complex geometric attributes of the associated set of probability distributions. Second, we illustrate with examples, most of which quite small and seemingly trivial, some of the computational, statistical and geometric challenges that LC models pose. In particular, we focus on issues of non-identifiability and determination of the model dimension, of maximisation of the likelihood function and on the effect of symmetric data. We also show how to use symbolic software from computational algebra to obtain a more convenient and simpler parametrisation and for unravelling the geometric features of LC models. These strategies and methods should carry over to more complex latent structure models, such as in (Bandeen-Roche *et al.* 1997).

In the next section, we describe the basic latent class model and introduce its statistical properties and issues, and we follow that, in Section 2.3, with a discussion of the geometry of the models. In Section 2.4, we turn to our examples exemplifying identifiability issues and the complexity of the likelihood function, with a novel focus on the problems arising from symmetries in the data. Finally, we present some computational results for two real-life examples, of small and very large dimension, and remark on the occurrence of singularities in the observed Fisher information matrix.

2.2 Latent class models for contingency tables

Consider k categorical variables, $X_1, \ldots, X_k$, where each X_i takes value on the finite set $[d_i] \equiv \{1, \ldots, d_i\}$. Letting $\mathcal{D} = \bigotimes_{i=1}^{k}[d_i]$, $\mathbb{R}^{\mathcal{D}}$ is the vector space of k-dimensional arrays of the format $d_1 \times \ldots \times d_k$, with a total of $d = \prod_i d_i$ entries. The cross-classification of N independent and identically distributed realisations of $(X_1, \ldots, X_k)$ produces a random integer-valued vector $\mathbf{n} \in \mathbb{R}^{\mathcal{D}}$, whose coordinate entry $\mathbf{n}_{i_i,\dots,i_k}$ corresponds to the number of times the label combination $(i_1,\dots,i_k)$ was observed in the sample, for each $(i_1,\dots,i_k) \in \mathcal{D}$. The table $\mathbf{n}$ has a Multinomial$_d(N,\mathbf{p})$ distribution, where $\mathbf{p}$ is a point in the $(d-1)$-dimensional probability simplex Δ_{d-1} with coordinates

$$p_{i_1,\dots,i_k} = Pr\left\{(X_1,\dots,X_k) = (i_1,\dots,i_k)\right\}, \qquad (i_1,\dots,i_k) \in \mathcal{D}.$$

Let H be an unobservable latent variable, defined on the set $[r] = \{1,\dots,r\}$. In its most basic version, also known as the *naive Bayes model*, the LC model postulates that, conditional on H, the variables $X_1,\dots,X_k$ are mutually independent. Specifically, the joint distributions of $X_1,\dots,X_k$ and H form the subset $\mathcal{V}$ of the probability simplex Δ_{dr-1} consisting of points with coordinates

$$p_{i_1,\dots,i_k,h} = p_1^{(h)}(i_1)\dots p_k^{(h)}(i_k)\lambda_h, \qquad (i_1,\dots,i_k,h) \in \mathcal{D} \times [r], \qquad (2.1)$$

where λ_h is the marginal probability $Pr\{H = h\}$ and $p_l^{(h)}(i_l)$ is the conditional marginal probability $Pr\{X_l = i_l | H = h\}$, which we assume to be strictly positive for each $h \in [r]$ and $(i_1,\dots,i_k) \in \mathcal{D}$.

The log-linear model specified by the polynomial mapping (2.1) is a decomposable graphical model, see e.g. (Lauritzen 1996), and $\mathcal{V}$ is the image set of a homomorphism from the parameter space

$$\Theta \equiv \left\{\theta : \theta = (p_1^{(h)}(i_1)\dots p_k^{(h)}(i_k),\lambda_h), (i_1,\dots,i_k,h) \in \mathcal{D} \times [r]\right\} = \bigotimes_i \Delta_{d_i-1} \times \Delta_{r-1},$$

so that global identifiability is guaranteed. The remarkable statistical properties of this type of model and the geometric features of the set $\mathcal{V}$ are well understood. Statistically, Equation (2.1) defines a linear exponential family of distributions, though not in its natural parametrisation. The maximum likelihood estimates of λ_h and $p_l^{(h)}(i_l)$ exist if and only if the minimal sufficient statistics, i.e., the empirical joint distributions of (X_i, H) for $i = 1, 2,\dots,k$, are strictly positive and are given in closed form as rational functions of the observed two-way marginal distributions between X_i and H for $i = 1, 2,\dots,k$. The log-likelihood function is strictly concave and the global maximum is always attainable, possibly on the boundary of the parameter space. Furthermore, the asymptotic theory of goodness-of-fit testing is fully developed. The statistical problem arises because H is latent and unobservable.

Geometrically, we can obtain the set $\mathcal{V}$ as the intersection of Δ_{dr-1} with an affine variety (see, e.g., (Cox *et al.* 1992)) consisting of the solution set of a system of $r \prod_i \binom{d_i}{2}$ homogeneous square-free polynomials. For example, when $k = 2$, each of these polynomials take the form of quadric equations of the type

$$p_{i_1,i_2,h}\, p_{i_1',i_2',h} = p_{i_1',i_2,h}\, p_{i_1,i_2',h}, \qquad (2.2)$$

with $i_1 \neq i_1'$, $i_2 \neq i_2'$ and for each fixed h. Equations of the form (2.2) are nothing more than conditional odds ratio of 1 for every pair $(X_i, X_{i'})$ given $H = h$ and, for each given h, the coordinate projections of the first two coordinates of the points satisfying (2.2) trace the surface of independence inside the simplex Δ_{d-1}. The strictly positive points in $\mathcal{V}$ form a smooth manifold whose dimension is $r \prod_i (d_i - 1) + (r-1)$ and whose co-dimension corresponds to the number of degrees of freedom.

The singular points in $\mathcal{V}$ all lie on the boundary of the simplex Δ_{dr-1} and identify distributions with degenerate probabilities along some coordinates. The singular locus of $\mathcal{V}$ can be described similarly in terms of stratified components of $\mathcal{V}$, whose dimensions and co-dimensions can also be computed explicitly.

Under the LC model, the variable H is unobservable and the new model $\mathcal{H}$ is a r-class mixture over the exponential family of distributions prescribing mutual independence among the manifest variables $X_1, \ldots, X_k$. Geometrically, $\mathcal{H}$ is the set of probability vectors in Δ_{d-1} obtained as the image of the marginalisation map from Δ_{dr-1} onto Δ_{d-1} which consists of taking the sum over the coordinate corresponding to the latent variable. Formally, $\mathcal{H}$ is made up of all probability vectors in Δ_{d-1} with coordinates satisfying the *accounting equations*, see, e.g., (Henry and Lazarfeld 1968)

$$p_{i_1,\ldots,i_k} = \sum_{h \in [r]} p_{i_1,\ldots,i_k,h} = \sum_{h \in [r]} p_1^{(h)}(i_1) \ldots p_k^{(h)}(i_k)\lambda_h, \qquad (2.3)$$

where $(i_1, \ldots, i_k, h) \in \mathcal{D} \times [r]$.

Despite being expressible as a convex combination of very well-behaved models, even the simplest form of the LC model (2.3) is far from well-behaved and, in fact, shares virtually none of the properties of the standard log-linear models (2.1) described above. In particular, latent class models described by Equations (2.3) do not define exponential families, but instead belong to a broader class of models called stratified exponential families, see (Geiger *et al.* 2001), whose properties are much weaker and less well understood. *The minimal sufficient statistics for an observed table* **n** *are the observed counts themselves and we can achieve no data reduction via sufficiency.* The model may not be identifiable, because for a given $\mathbf{p} \in \Delta_{d-1}$ defined by (2.3), there may be a subset of Θ, known as the *non-identifiable space*, consisting of parameter points all satisfying the same accounting equations. The non-identifiability issue has in turn considerable repercussions for the determination of the correct number of degrees of freedom for assessing model fit and, more importantly, on the asymptotic properties of standard model selection criteria (e.g. likelihood ratio statistic and other goodness-of-fit criteria such as BIC, AIC, etc.), whose applicability and correctness may no longer hold.

Computationally, maximising the log-likelihood can be a rather laborious and difficult task, particularly for high-dimensional tables, due to lack of concavity, the presence of local maxima and saddle points, and singularities in the observed Fisher information matrix. Geometrically, $\mathcal{H}$ is no longer a smooth manifold on the relative interior of Δ_{d-1}, with singularities even at probability vectors with strictly positive coordinates, as we show in the next section. The problem of characterising the singular locus of $\mathcal{H}$ and of computing the dimensions of its stratified components (and of the tangent spaces and tangent cones of its singular points) is of statistical importance: singularity points of $\mathcal{H}$ are probability distributions of lower complexity, in the sense that they are specified by lower-dimensional subsets of Θ, or, loosely speaking, by less parameters. Because the sample space is discrete, although the singular locus of $\mathcal{H}$ has typically Lebesgue measure zero, there is nonetheless a positive probability that the maximum likelihood estimates end up being either a

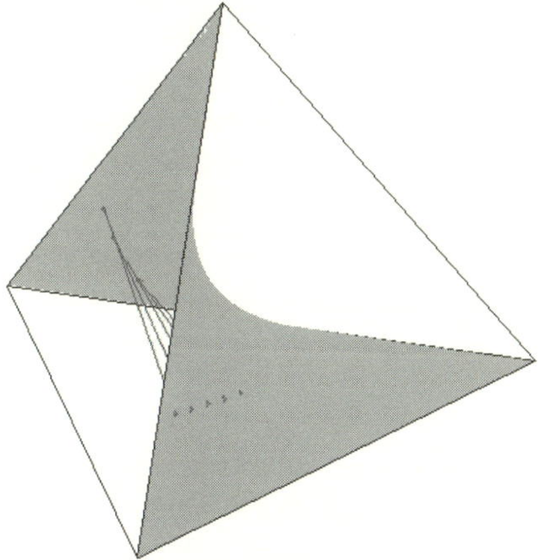

Fig. 2.1 Surface of independence for the 2×2 table with five secant lines.

singular point in the relative interior of the simplex Δ_{d-1} or a point on the boundary. In both cases, standard asymptotics for hypothesis testing and model selection fall short.

2.3 Geometric description of latent class models

In this section, we give a geometric representation of latent class models, summarise existing results and point to some of the relevant mathematical literature. For more details, see (Garcia *et al.* 2005) and (Garcia 2004).

The latent class model defined by (2.3) can be described as the set of all convex combinations of all r-tuple of points lying on the surface of independence inside Δ_{d-1}. Formally, let

$$\sigma: \quad \Delta^{d_1-1} \times \ldots \times \Delta^{d_k-1} \quad \to \quad \Delta_{d-1}$$
$$(p_1(i_1), \ldots, p_k(i_k)) \quad \mapsto \quad \textstyle\prod_j p_j(i_j)$$

be the map that sends the vectors of marginal probabilities into the k-dimensional array of joint probabilities for the model of complete independence. The set $\mathcal{S} \equiv \sigma(\Delta^{d_1-1} \times \ldots \times \Delta^{d_k-1})$ is a manifold in Δ_{d-1} known in statistics as the surface of independence and in algebraic geometry as (the intersection of Δ_{d-1} with) the Segre embedding of $\mathbb{P}^{d_1-1} \times \ldots \times \mathbb{P}^{d_k-1}$ into $\mathbb{P}^{d-1}$ see, e.g., (Harris 1992). The dimension of $\mathcal{S}$ is $\prod_i (d_i - 1)$, i.e., the dimension of the corresponding decomposable model of mutual independence. The set $\mathcal{H}$ can then be constructed geometrically as follows. Pick any combination of r points along the hyper-surface $\mathcal{S}$, say $\mathbf{p}^{(1)}, \ldots, \mathbf{p}^{(r)}$, and determine their convex hull, i.e. the convex subset of Δ_{d-1} consisting of all points of the form $\sum_h \mathbf{p}^{(h)} \lambda_h$, for some choice of $(\lambda_1, \ldots, \lambda_r) \in \Delta_{r-1}$. The coordinates of any point in this new subset satisfy, by construction, the accounting equations (2.3). In fact, the closure of the union of all such convex hulls is precisely the latent class model $\mathcal{H}$. In algebraic geometry, $\mathcal{H}$ would be described as the intersection of Δ_{d-1} with the r-th secant variety of the Segre embedding mentioned above.

Example 2.1 The simplest example of a latent class model is for a 2×2 table with $r = 2$ latent classes. The surface of independence, i.e. the intersection of the simplex Δ_3 with the Segre variety, is shown in Figure 2.1. The secant variety for this latent class models is the union of all the *secant lines*, i.e. the lines connecting any two distinct points lying on the surface of independence. Figure 2.1 displays five such secant lines. It is not to hard to picture that the union of all such secant lines is the enveloping simplex Δ_3 and, therefore, $\mathcal{H}$ fills up all the available space. For formal arguments, see Proposition 2.3 in (Catalisano *et al.* 2002).

The model $\mathcal{H}$, thought of as a portion of the r-th secant variety to the Segre embedding, is not a smooth manifold. Instead, it is a semi-algebraic set, see, e.g., (Benedetti 1990), clearly singular on the boundary of the simplex, but also at strictly positive points along the $(r-1)$st secant variety (both of Lebesgue measure zero). This means that the model is singular at all points in $\mathcal{H}$ which satisfy the accounting equations with one or more of the λ_h's equal to zero. In Example 2.1 above, the surface of independence is a singular locus for the latent class model. From the statistical viewpoint, singular points of $\mathcal{H}$ correspond to simpler models for which the number of latent classes is less than r (possibly 0). As usual, for these points one needs to adjust the number of degrees of freedom to account for the larger tangent space.

Unfortunately, we have no general closed-form expression for computing the dimension of $\mathcal{H}$ and the existing results only deal with specific cases. Simple considerations allow us to compute an upper bound for the dimension of $\mathcal{H}$, as follows. As Example 2.1 shows, there may be instances for which $\mathcal{H}$ fills up the entire simplex Δ_{d-1}, so that $d-1$ is an attainable upper bound. Counting the number of free parameters in (2.3), we can see that this dimension cannot exceed $r \sum_i (d_i - 1) + r - 1$, cf. (Goodman 1974, p. 219). This number, the *standard dimension*, is the dimension of the fully observable model of conditional independence. Incidentally, this value can be determined mirroring the geometric construction of $\mathcal{H}$ as follows, cf. (Garcia 2004). The number $r \sum_i (d_i - 1)$ arises from the choice of r points along the $\sum_i (d_i - 1)$-dimensional surface of independence, while the term $r - 1$ accounts for the number of free parameters for a generic choice of $(\lambda_1, \ldots, \lambda_r) \in \Delta_{r-1}$. Therefore, we conclude that the dimension of $\mathcal{H}$ is bounded by

$$\min \left\{ d - 1, r \sum_i (d_i - 1) + r - 1 \right\}, \tag{2.4}$$

a value known in algebraic geometry as the *expected dimension* of the variety $\mathcal{H}$.

Cases of latent class models with dimension strictly smaller than the expected dimension have been known for a long time, however. In the statistical literature, (Goodman 1974) noticed that the latent class models for 4 binary observable variables and a 3-level latent variable, whose expected dimension is 14, has dimension 13. In algebraic geometry, secant varieties with dimension smaller than the expected dimension (2.4) are called *deficient*, e.g. see (Harris 1992). In particular, Exercise 11.26 in (Harris 1992) gives an example of deficient secant variety, which corresponds to a latent class model for a two-way table with a latent variable taking on 2 values. In this case, the deficiency is 2, as is demonstrated in Equation (2.5) below. The true or *effective* dimension of a latent class model, i.e. the dimension of the semi-algebraic set $\mathcal{H}$ representing it, is crucial for establishing identifiability and for computing correctly the number of degrees of freedom. In fact, if a model is deficient, then the pre-image of each probability array in $\mathcal{H}$ arising from the accounting equations is a subset (in fact, a variety) of Θ called the *non-identifiable subspace*, with dimension exactly equal to the deficiency itself. Therefore, a deficient model is non-identifiable, with adjusted degrees of freedom equal to the number of degrees of freedom for the observable graphical model plus the value of the deficiency.

Theoretically, it is possible to determine the effective dimension of $\mathcal{H}$ by computing the maximal rank of the Jacobian matrix for the polynomial mapping from Θ into $\mathcal{H}$ given coordinatewise by (2.3). In fact, (Geiger *et al.* 2001) showed that this value is equal to the dimension of $\mathcal{H}$ almost everywhere with respect to the Lebesgue measure, provided the Jacobian is evaluated at strictly positive parameter points. These symbolic evaluations, however, require the use of symbolic software which can only handle small tables and models, so that, in practice, computing the effective dimension of a latent class model is computationally difficult and often infeasible.

Recently, in the algebraic-geometry literature, (Catalisano *et al.* 2002) have obtained explicit formulas for the effective dimensions of some secant varieties which are of statistical interest. In particular, they show that for $k = 3$ and $r \leq \min\{d_1, d_2, d_3\}$, the latent class model has the expected dimension and is identifiable. On the other hand, assuming $d_1 \leq d_2 \leq \ldots \leq d_k$, $\mathcal{H}$ is deficient when $\prod_{i=1}^{k-1} d_i - \sum_{i=1}^{k-1}(d_i - 1) \leq r \leq \min\left\{d_k, \prod_{i=1}^{k-1} d_i - 1\right\}$. Finally, under the same conditions, $\mathcal{H}$ is identifiable when $\frac{1}{2}\sum_i (d_i - 1) + 1 \geq \max\{d_k, r\}$. Obtaining bounds and results of this type is highly non-trivial and is an open area of research.

In the remainder of the chapter, we will focus on simpler latent class models for tables of dimension $k = 2$ and illustrate with examples the results mentioned above. For latent class models on two-way tables, there is an alternative, quite convenient way of describing $\mathcal{H}$ by representing each $\mathbf{p}$ in Δ_{d-1} as a $d_1 \times d_2$ matrix and by interpreting the map σ as a vector product. In fact, each point $\mathbf{p}$ in $\mathcal{S}$ is a rank one matrix obtained as $\mathbf{p}_1 \mathbf{p}_2^\top$, where $\mathbf{p}_1 \in \Delta_{d_1 - 1}$ and $\mathbf{p}_2 \in \Delta_{d_1 - 2}$ are the appropriate marginal distributions of X_1 and X_2 and $^\top$ stands for transpose. Then, the accounting equations for latent class models with r-level become

$$\mathbf{p} = \sum_h \mathbf{p}_1^{(h)}(\mathbf{p}_2^{(h)})^\top \lambda_h, \quad (\mathbf{p}_1, \mathbf{p}_2, (\lambda_1, \ldots, \lambda_r)) \in \Delta_{d_1 - 1} \times \Delta_{d_2 - 1} \times \Delta_{r - 1}$$

i.e. the matrix $\mathbf{p}$ is a convex combination of r rank 1 matrices lying on the surface of independence. Therefore all points in $\mathcal{H}$ are non-negative matrices with entries summing to one and with rank at most r. This simple observation allows one to compute the effective dimension of $\mathcal{H}$ for the two-way table as follows. In general, a real-valued $d_1 \times d_2$ matrix has rank r or less if and only if the homogeneous polynomial equations corresponding to all of its $(r + 1) \times (r + 1)$ minors vanish. Provided $k < \min\{d_1, d_2\}$, on $\mathbb{R}^{d_1} \times \mathbb{R}^{d_2}$, the zero locus of all such equations form a *determinantal* variety of co-dimension $(d_1 - r)(d_2 - r)$, see (Harris 1992, Proposition

12.2), and hence has dimension $r(d_1 + d_2) - r^2$. Subtracting this value from the expected dimension computed above, and taking into account the fact that all the points lie inside the simplex, we obtain

$$r(d_1 + d_2 - 2) + r - 1 - \left(r(d_1 + d_2) - r^2 - 1\right) = r(r - 1). \qquad (2.5)$$

This number is also the difference between the dimension of the fully identifiable (i.e., of expected dimension) graphical model of conditional independence X_1 and X_2 given H, and the deficient dimension of the latent class model obtained by marginalising over the variable H.

The study of higher-dimensional tables is still an open area of research. The mathematical machinery required to handle larger dimensions is considerably more complicated and relies on the notions of higher-dimensional tensors, rank tensors and non-negative rank tensors, for which only partial results exist. See (Kruskal 1975, Cohen and Rothblum 1993) and (Strassen 1983) for details. Alternatively, (Mond *et al.* 2003) conduct an algebraic-topological investigation of the topological properties of stochastic factorisation of stochastic matrices representing models of conditional independence with one hidden variable and (Allman and Rhodes 2006, Allman and Rhodes 2008) explore an overlapping set of problems framed in the context of trees with latent nodes and branches.

The specific case of k-way tables with two-level latent variables is a fortunate exception, for which the results for two-way tables just described apply. In fact, (Landsberg and Manivel 2004) show that these models are the same as the corresponding models for any two-dimensional table obtained by any 'flattening' of the $d_1 \times \ldots \times d_k$-dimensional array of probabilities $\mathbf{p}$ into a two-dimensional matrix. Flattening simply means collapsing the k variables into two new variables with f_1 and f_2 levels, and re-organising the entries of the k-dimensional tensor $\mathbf{p} \in \Delta_{d-1}$ into a $f_1 \times f_2$ matrix accordingly, where, necessarily, $f_1 + f_2 = \sum_i d_i$. Then, $\mathcal{H}$ is the determinantal variety which is the zero set of all 3×3 sub-determinants of the matrix obtained by any such flattening. The second example in Section 2.4.1 below illustrates this result.

2.4 Examples involving synthetic data

We further elucidate the non-identifiability phenomenon from the algebraic and geometric point of view, and the multi-modality of the log-likelihood function issue using few, small synthetic examples. In particular, in the '100 Swiss Francs' problem we embark on a exhaustive study of a table with symmetric data and describe the effects of such symmetries on both the parameter space and the log-likelihood function. Although this example involves one of the simplest cases of LC models, it already exhibits considerable statistical and geometric complexity.

2.4.1 Effective dimension and polynomials

We show how it is possible to take advantage of the polynomial nature of Equations (2.3) to gain further insights into the algebraic properties of distributions obeying latent class models. All the computations that follow were made in SIN-GULAR (Greuel *et al.* 2005) and are described in detail, along with more examples in the on-line supplement. Although in principle symbolic algebraic software allows one to compute the set of polynomial equations that fully characterise LC models and their properties, this is still a rather difficult and costly task that can be accomplished only for smaller models.

The accounting equations (2.3) determine a polynomial mapping f from Θ to Δ^{d-1} given by

$$(p_1(i_1)\ldots p_k(i_k), \lambda_h) \mapsto \sum_{h \in [r]} p_1(i_1)\ldots p_k(i_k)\lambda_h, \qquad (2.6)$$

so that the latent class model can be analytically defined as the image of this map, i.e. $\mathcal{H} = f(\Theta)$. Then, following the geometry–algebra dictionary principle, see e.g., (Cox *et al.* 1992), the problem of computing the effective dimension of $\mathcal{H}$ can in turn be geometrically cast as a problem of computing the dimension of the image of a polynomial map. We illustrate how this representation offers considerable advantages with some small examples.

Example 2.2 Consider a $2 \times 2 \times 2$ table with $r = 2$ latent classes. From Proposition 2.3 in (Catalisano *et al.* 2002), the latent class models with 2 classes and 3 manifest variables are identifiable. The standard dimension, i.e. the dimension of the parameter space Θ is $r\sum_i(d_i - 1) + r - 1 = 7$, which coincides with the dimension of the enveloping simplex Δ_7. Although this condition implies that the number of parameters to estimate is no larger than the number of cells in the table, a case which, if violated, would entail non-identifiability, it does not guarantee that the effective dimension is also 7. This can be verified by checking that the symbolic rank of the Jacobian matrix of the map (2.6) is indeed 7, almost everywhere with respect to the Lebesgue measure. Alternatively, one can determine the dimension of the non-identifiable subspace using computational symbolic algebra. First, we define the ideal of polynomials determined by the eight equations in (2.6) in the polynomial ring in which the (redundant) 16 indeterminates are the 8 joint probabilities in Δ_7 and the 3 pairs of marginal probabilities in Δ_1 for the observable variables, and the marginal probabilities in Δ_1 for the latent variable. Then we use implicitisation, e.g. (Cox *et al.* 1992, Ch. 3), to eliminate all the marginal probabilities and to study the Gröbner basis of the resulting ideal in which the indeterminates are the joint probabilities only. There is only one element in the basis, namely $p_{111} + p_{112} + p_{121} + p_{122} + p_{211} + p_{212} + p_{221} + p_{222} = 1$, which gives the trivial condition for probability vectors. This implies the map (2.6) is surjective, so that $\mathcal{H} = \Delta_7$ and the effective dimension is also 7, showing identifiability, at least for positive distributions.

Example 2.3 We consider the $2 \times 2 \times 3$ table with $r = 2$. For this model Θ has dimension 9 and the image of the mappings (2.6) is Δ_9. The symbolic rank of the associated Jacobian matrix is 9 as well and the model is identifiable. The image of the polynomial mapping determined by (2.6) is the variety associated to the ideal for which a Gröbner basis consists of the trivial equation $p_{111} + p_{112} + p_{113} + p_{121} + p_{122} + p_{123} + p_{211} + p_{212} + p_{213} + p_{221} + p_{222} + p_{223} = 1$, and four polynomials corresponding to the determinants

$$\begin{vmatrix} p_{121} & p_{211} & p_{221} \\ p_{122} & p_{212} & p_{222} \\ p_{123} & p_{213} & p_{223} \end{vmatrix} \qquad \begin{vmatrix} p_{1+1} & p_{211} & p_{221} \\ p_{1+2} & p_{212} & p_{222} \\ p_{1+3} & p_{213} & p_{223} \end{vmatrix}$$

$$\begin{vmatrix} p_{+11} & p_{121} & p_{221} \\ p_{+12} & p_{122} & p_{222} \\ p_{+13} & p_{123} & p_{223} \end{vmatrix} \qquad \begin{vmatrix} p_{111} & p_{121}+p_{211} & p_{221} \\ p_{112} & p_{122}+p_{212} & p_{222} \\ p_{113} & p_{123}+p_{213} & p_{223} \end{vmatrix}$$

where the subscript symbol '+' indicates summation over that coordinate. In turn, the zero set of the above determinants coincide with the determinantal variety specified by the zero set of all 3×3 minors of the 3×4 matrix

$$\begin{pmatrix} p_{111} & p_{121} & p_{211} & p_{221} \\ p_{112} & p_{122} & p_{212} & p_{222} \\ p_{113} & p_{123} & p_{213} & p_{223} \end{pmatrix} \qquad (2.7)$$

which is a flattening of the $2 \times 2 \times 3$ array of probabilities describing the joint distribution for the latent class model under study. This is in accordance with the result in (Landsberg and Manivel 2004) mentioned above. Now, the determinantal variety given by the vanishing locus of all the 3×3 minors of the matrix (2.7) is the latent class model for a 3×4 table with 2 latent classes, which, according to (2.5), has deficiency equal to 2. The effective dimension of this variety is 9, computed as the standard dimension, 11, minus the deficiency. Then, the effective dimension of the model we are interested in is also 9 and we conclude that the model is identifiable.

Table 2.1 summarises some of our numerical evaluations of the different notions of dimension for a different LC models. We computed the effective dimensions by evaluating with MATLAB the numerical rank of the Jacobian matrix, based on the simple algorithm suggested in (Geiger *et al.* 2001) and also using SINGULAR, for which only computations involving small models were feasible.

2.4.2 The 100 Swiss Franc problem

Introduction

Now we study the problem of fitting a non-identifiable two-level latent class model to a two-way table with symmetry counts. This problem was suggested by Bernd Sturmfels to the participants of his postgraduate lectures on Algebraic Statistics held at ETH Zurich in the summer semester of 2005 (where he offered 100 Swiss Francs for a rigorous solution), and is described in detail as Example 1.16 in (Pachter

Table 2.1 *Different dimensions of some latent class models. The Complete Dimension is the dimension $d-1$ of the enveloping probability simplex Δ_{d-1}. See also Table 1 in (Kocka and Zhang, 2002).*

Latent Class Model		Effective Dimension	Standard Dimension	Complete Dimension	Deficiency
Δ_{d-1}	r				
2×2	2	3	5	3	0
3×3	2	7	9	8	1
4×5	3	17	23	19	2
$2 \times 2 \times 2$	2	7	7	7	0
$2 \times 2 \times 2$	3	7	11	7	0
$2 \times 2 \times 2$	4	7	15	7	0
$3 \times 3 \times 3$	2	13	13	26	0
$3 \times 3 \times 3$	3	20	20	26	0
$3 \times 3 \times 3$	4	25	27	26	1
$3 \times 3 \times 3$	5	26	34	26	0
$3 \times 3 \times 3$	6	26	41	26	0
$5 \times 2 \times 2$	3	17	20	19	2
$4 \times 2 \times 2$	3	14	17	15	1
$3 \times 3 \times 2$	5	17	29	17	0
$6 \times 3 \times 2$	5	34	44	35	1
$10 \times 3 \times 2$	5	54	64	59	5
$2 \times 2 \times 2 \times 2$	2	9	9	15	0
$2 \times 2 \times 2 \times 2$	3	13	14	15	1
$2 \times 2 \times 2 \times 2$	4	15	19	15	0
$2 \times 2 \times 2 \times 2$	5	15	24	15	0
$2 \times 2 \times 2 \times 2$	6	15	29	15	0

and Sturmfels 2005). The observed table is

$$
\mathbf{n} = \begin{pmatrix} 4 & 2 & 2 & 2 \\ 2 & 4 & 2 & 2 \\ 2 & 2 & 4 & 2 \\ 2 & 2 & 2 & 4 \end{pmatrix}.
\tag{2.8}
$$

For the basic latent class model, the standard dimension of $\Theta = \Delta_3 \times \Delta_3 \times \Delta_1$ is $2(3+3)+1 = 13$ and, by (2.5), the deficiency is 2. Thus, the model is not identifiable and the pre-image of each point $\mathbf{p} \in \mathcal{H}$ by the map (2.6) is a two-dimensional surface in Θ. To keep the notation light, we write α_{ih} for $p_1^{(h)}(i)$ and β_{jh} for $p_2^{(h)}(j)$, where $i, j = 1, \ldots, 4$ and $\alpha^{(h)}$ and $\beta^{(h)}$ for the conditional marginal distribution of X_1 and X_2 given $H = h$, respectively. The accounting equations for the points in $\mathcal{H}$ become

$$
p_{ij} = \sum_{h \in \{1,2\}} \lambda_h \alpha_{ih} \beta_{jh}, \qquad i, j \in [4]
\tag{2.9}
$$

and the log-likelihood function, ignoring an irrelevant additive constant, is

$$\ell(\theta) = \sum_{i,j} n_{ij} \log \left(\sum_{h \in \{1,2\}} \lambda_h \alpha_{ih} \beta_{jh} \right), \qquad \theta \in \Delta_3 \times \Delta_3 \times \Delta_1.$$

Again we emphasise that the observed counts are minimal sufficient statistics.

Alternatively, we can re-parametrize the log-likelihood function using directly points in $\mathcal{H}$ rather the points in the parameter space Θ. Recall from Section 2.3 that the 4×4 array $\mathbf{p}$ is in $\mathcal{H}$ if and only if each 3×3 minor vanishes. Then, we can write the log-likelihood function as

$$\ell(\mathbf{p}) = \sum_{i,j} n_{ij} \log p_{ij}, \qquad \mathbf{p} \in \Delta_{15}, \ \det(\mathbf{p}_{ij}^*) = 0 \text{ for all } i, j \in [4], \tag{2.10}$$

where $\mathbf{p}_{ij}^*$ is the 3×3 sub-matrix of $\mathbf{p}$ obtained by erasing the i-th row and the j-th column.

Although the first order optimality conditions for the Lagrangian corresponding to the parametrisation (2.10) are algebraically simpler and can be given the form of a system of a polynomial equations, in practice, the classical parametrisation (2.9) is used in both the EM and the Newton–Raphson implementations in order to compute the maximum likelihood estimate of $\mathbf{p}$. See (Goodman 1979, Haberman 1988) and (Redner and Walker 1984) for more details about these numerical procedures.

Global and local maxima

Using both the EM and Newton–Raphson algorithms with several different starting points, we found seven local maxima of the log-likelihood function, reported in Table 2.2. The global maximum was found experimentally to be $-20.8074 + const.$, where $const.$ denotes the additive constant stemming from the multinomial coefficient. The maximum is achieved by the three tables of fitted values in Table 2.2 **a)**. The remaining four tables are local maximum of $-20.8616 + const.$, close in value to the actual global maximum. Using SINGULAR, we checked that the found tables satisfy the first-order optimality conditions (2.10). After verifying numerically the second-order optimality conditions, we conclude that those points are indeed local maxima. Furthermore, as indicated in (Pachter and Sturmfels 2005), the log-likelihood function also has a few saddle points.

A striking feature of the global maxima in Table 2.2 is their invariance under the action of the symmetric group on four elements acting simultaneously on the row and columns. Different symmetries arise for the local maxima. We will give an explicit representation of these symmetries under the classical parametrisation (2.9) in the next section.

Despite the simplicity and low-dimensionality of the LC model for the Swiss Francs problem and the strong symmetric features of the data, we have yet to provide a purely mathematical proof that the three top arrays in Table 2.2 correspond to a global maximum of the likelihood function.[1] We view the difficulty and

---

[1] The 100 Swiss Francs were awarded to Mingfu Zhu at Clemson University on 14 September 2008 for a mathematical proof based on the present chapter (editors' note).

Table 2.2 *Tables of fitted values corresponding to the seven maxima of the likelihood equation for the observed table (2.8). (a): global maximum (log-likelihood value -20.8079). (b): local maxima (log-likelihood value -20.8616).*

(a)

$$
\begin{pmatrix} 3 & 3 & 2 & 2 \\ 3 & 3 & 2 & 2 \\ 2 & 2 & 3 & 3 \\ 2 & 2 & 3 & 3 \end{pmatrix}
\quad
\begin{pmatrix} 3 & 2 & 3 & 2 \\ 2 & 3 & 2 & 3 \\ 3 & 2 & 3 & 2 \\ 2 & 3 & 2 & 3 \end{pmatrix}
\quad
\begin{pmatrix} 3 & 2 & 2 & 3 \\ 2 & 3 & 3 & 2 \\ 2 & 3 & 3 & 2 \\ 3 & 2 & 2 & 3 \end{pmatrix}
$$

(b)

$$
\begin{pmatrix} 8/3 & 8/3 & 8/3 & 2 \\ 8/3 & 8/3 & 8/3 & 2 \\ 8/3 & 8/3 & 8/3 & 2 \\ 2 & 2 & 2 & 4 \end{pmatrix}
\quad
\begin{pmatrix} 8/3 & 8/3 & 2 & 8/3 \\ 8/3 & 8/3 & 2 & 8/3 \\ 2 & 2 & 4 & 2 \\ 8/3 & 8/3 & 2 & 8/3 \end{pmatrix}
$$

$$
\begin{pmatrix} 8/3 & 2 & 8/3 & 8/3 \\ 2 & 4 & 2 & 2 \\ 8/3 & 2 & 8/3 & 8/3 \\ 8/3 & 2 & 8/3 & 8/3 \end{pmatrix}
\quad
\begin{pmatrix} 4 & 2 & 2 & 2 \\ 2 & 8/3 & 8/3 & 8/3 \\ 2 & 8/3 & 8/3 & 8/3 \\ 2 & 8/3 & 8/3 & 8/3 \end{pmatrix}
$$

complexity of the 100 Swiss Francs problem as a consequence of the inherent difficulty of even small LC models and perhaps an indication that the current theory has still many open, unanswered problems. In Section 2.6, we present partial results towards the completion of the proof.

Unidentifiable space

It follows from Equation (2.5) that the non-identifiable subspace is a two-dimensional subset of Θ. We give an explicit algebraic description of this space, which we will then use to obtain interpretable plots of the profile likelihood. For a coloured version of our figures see the on-line version.

Firstly, we focus on the three global maxima in Table 2.2 (a). By the well-known properties of the EM algorithm, if θ is a stationary point in the maximisation step of the EM algorithm, then θ is a critical point and hence a good candidate for a local maximum. It follows that any point in Θ satisfying the equations

$$
\begin{aligned}
\alpha_{1h} &= \alpha_{2h}, \; \alpha_{3h} = \alpha_{4h} \quad h = 1,2 \\
\beta_{1h} &= \beta_{2h}, \; \beta_{3h} = \beta_{4h} \quad h = 1,2 \\
\textstyle\sum_h \lambda_h \alpha_{1h}\beta_{1h} &= \textstyle\sum_h \lambda_h \alpha_{3h}\beta_{3t} = 3/40 \\
\textstyle\sum_h \lambda_h \alpha_{1h}\beta_{3h} &= \textstyle\sum_h \lambda_h \alpha_{3h}\beta_{1t} = 2/40
\end{aligned}
\tag{2.11}
$$

is a stationary point. The first four equations in (2.11) require $\alpha^{(h)}$ and $\beta^{(h)}$ to each have the first and second pairs of coordinates identical, for $h = 1, 2$. Equation (2.11) defines a two-dimensional surface in Θ. Using SINGULAR, we can verify that, holding, for example, α_{11} and β_{11} fixed, determines all of the other parameters

Fig. 2.2 The two-dimensional surface defined by Equation (2.12), when evaluated over the ball in $\mathbb{R}^3$ of radius 3, centred at the origin. The inner box is the unit cube $[0,1]^3$.

according to the equations

$$
\begin{cases}
\lambda_1 = \frac{1}{80\alpha_{11}\beta_{11} - 20\alpha_{11} - 20*\beta_{11} + 6} \\
\lambda_2 = 1 - \lambda_1 \\
\alpha_{21} = \alpha_{11} \\
\alpha_{31} = \alpha_{41} = 0.5 - \alpha_{11} \\
\alpha_{12} = \alpha_{22} = \frac{10\beta_{11} - 3}{10(4\beta_{11} - 1)} \\
\alpha_{32} = \alpha_{42} = 0.5 - \alpha_{12} \\
\beta_{21} = \beta_{11} \\
\beta_{31} = \beta_{41} = 0.5 - \beta_{11} \\
\beta_{12} = \beta_{22} = \frac{10\alpha_{11} - 3}{10(4\alpha_{11} - 1)} \\
\beta_{32} = \beta_{42} = 0.5 - \beta_{12}.
\end{cases}
$$

Using *elimination* to remove all the variables in the system except for λ_1, we are left with one equation

$$80\lambda_1\alpha_{11}\beta_{11} - 20\lambda_1\alpha_{11} - 20\lambda_1\beta_{11} + 6\lambda_1 - 1 = 0. \tag{2.12}$$

Without the constraints for the coordinates of α_{11}, β_{11} and λ_1 to be probabilities, (2.12) defines a two-dimensional object in $\mathbb{R}^3$, depicted in Figure 2.2. Notice that the axes do not intersect this surface, so that zero is not a possible value for α_{11}, β_{11} and λ_1. Because the non-identifiable space in Θ is two dimensional, Equation (2.12) actually defines a bijection between α_{11}, β_{11} and λ_1 and the rest of the parameters. Then, the intersection of the surface (2.12) with the unit cube $[0,1]^3$, given as a red box in Figure 2.2, is the projection of the whole non-identifiable subspace into the three-dimensional unit cube. Figure 2.3 displays two different views of this projection.

The preceding arguments hold unchanged if we replace the symmetry conditions in the first two lines of Equation (2.11) with either of these other two conditions, requiring different pairs of coordinates to be identical, namely

$$\alpha_{1h} = \alpha_{3h}, \ \alpha_{2h} = \alpha_{4h}, \ \beta_{1h} = \beta_{3h}, \ \beta_{2h} = \beta_{4h} \tag{2.13}$$

(a)   (b)

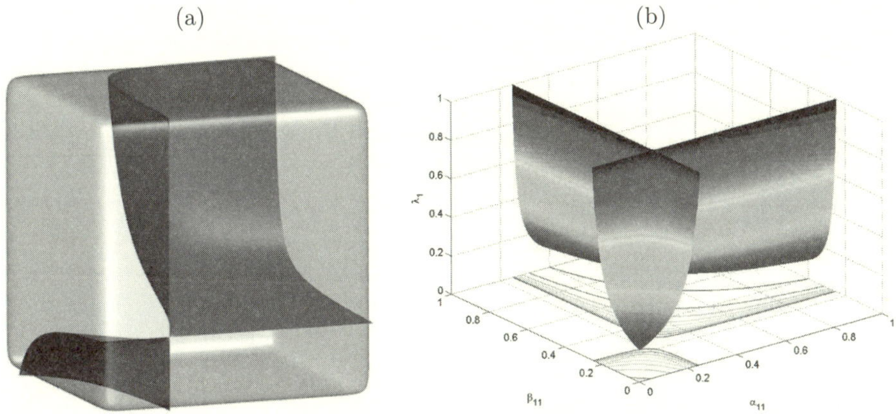

Fig. 2.3 Intersection of the surface defined by Equation (2.12) with the unit cube $[0,1]^3$, different views obtained using `surf` in (a) and `MATLAB` in (b).

Fig. 2.4 Projection of the non-identifiable spaces corresponding to the first and second and third MLE from Table 2.2 (a) into the three-dimensional unit cube where λ_1, α_{11} and β_{21} take values.

and

$$\alpha_{1h} = \alpha_{4h}, \quad \alpha_{2h} = \alpha_{3h}, \quad \beta_{1h} = \beta_{4h}, \quad \beta_{2h} = \beta_{3h}, \tag{2.14}$$

where $h = 1, 2$.

By our computations, the non-identifiable surfaces inside Θ corresponding each to one of the three pairs of coordinates held fixed in Equations (2.11), (2.13) and (2.14), produce the three distinct tables of maximum likelihood estimates reported in Table 2.2 (a). Figure 2.3 shows the projection of the non-identifiable subspaces for the three MLEs in Table 2.2 (a) into the three-dimensional unit cube for λ_1, α_{11} and β_{11}. Although each of these three subspaces are disjoint subsets of Θ, their lower-dimensional projections comes out as unique. By projecting onto the different coordinates λ_1, α_{11} and β_{21} instead, we obtain two disjoint surfaces for the first, and second and third MLE, shown in Figure 2.4.

Table 2.3 *Estimated parameters by the EM algorithm for the three global maxima in Table 2.2 (a).*

Estimated Means	Estimated Parameters		

$$
\begin{pmatrix} 3 & 3 & 2 & 2 \\ 3 & 3 & 2 & 2 \\ 2 & 2 & 3 & 3 \\ 2 & 2 & 3 & 3 \end{pmatrix}
\qquad
\widehat{\alpha}^{(1)} = \widehat{\beta}^{(1)} = \begin{pmatrix} 0.3474 \\ 0.3474 \\ 0.1526 \\ 0.1526 \end{pmatrix}
\qquad
\widehat{\alpha}^{(2)} = \widehat{\beta}^{(2)} = \begin{pmatrix} 0.1217 \\ 0.1217 \\ 0.3783 \\ 0.3783 \end{pmatrix}
\qquad
\widehat{\lambda} = \begin{pmatrix} 0.5683 \\ 0.4317 \end{pmatrix}
$$

$$
\begin{pmatrix} 3 & 2 & 3 & 2 \\ 2 & 3 & 2 & 3 \\ 3 & 2 & 3 & 2 \\ 2 & 3 & 2 & 3 \end{pmatrix}
\qquad
\widehat{\alpha}^{(1)} = \widehat{\beta}^{(1)} = \begin{pmatrix} 0.3474 \\ 0.1526 \\ 0.3474 \\ 0.1526 \end{pmatrix}
\qquad
\widehat{\alpha}^{(2)} = \widehat{\beta}^{(2)} = \begin{pmatrix} 0.1217 \\ 0.3783 \\ 0.1217 \\ 0.3783 \end{pmatrix}
\qquad
\widehat{\lambda} = \begin{pmatrix} 0.5683 \\ 0.4317 \end{pmatrix}
$$

$$
\begin{pmatrix} 3 & 2 & 2 & 3 \\ 2 & 3 & 3 & 2 \\ 2 & 3 & 3 & 2 \\ 3 & 2 & 2 & 3 \end{pmatrix}
\qquad
\widehat{\alpha}^{(1)} = \widehat{\beta}^{(1)} = \begin{pmatrix} 0.3474 \\ 0.1526 \\ 0.1526 \\ 0.3474 \end{pmatrix}
\qquad
\widehat{\alpha}^{(2)} = \widehat{\beta}^{(2)} = \begin{pmatrix} 0.1217 \\ 0.3783 \\ 0.3783 \\ 0.1217 \end{pmatrix}
\qquad
\widehat{\lambda} = \begin{pmatrix} 0.5683 \\ 0.4317 \end{pmatrix}
$$

Table 2.3 presents some estimated parameters using the EM algorithm. Though these estimates are hardly meaningful, because of the non-identifiability issue, they show the symmetry properties we pointed out above and implicit in Equations (2.11), (2.13) and (2.14), and they explain the invariance under simultaneous permutation of the fitted tables. In fact, the number of global maxima is the number of different configurations of the four-dimensional vectors of estimated marginal probabilities with two identical coordinates, namely three. This phenomenon, entirely due to the strong symmetry in the observed table (2.8), is completely separate from the non-identifiability issues, but just as problematic.

By the same token, we can show that vectors of marginal probabilities with three identical coordinates also produce stationary points for the EM algorithms. This type of stationary points trace surfaces inside Θ which determine the local maxima of Table 2.2 (b). The number of these local maxima corresponds, in fact, to the number of possible configurations of four-dimensional vectors with three identical coordinates, namely four. Figure 2.5 depicts the lower-dimensional projections into λ_1, α_{11} and β_{11} of the non-identifiable subspaces for the first MLE in Table 2.2 (a), the first three local maxima and the last local maxima in Table 2.2 (b).

We can summarise our finding as follows: the maxima in Table 2.2 define disjoint two-dimensional surfaces inside the parameter space Θ and the projection of one of them is depicted in Figure 2.3. While non-identifiability is a structural feature of these models which is independent of the observed data, the multiplicity and invariance properties of the maximum likelihood estimates and the other local maxima is a phenomenon caused by the symmetry in the observed table of counts.

Fig. 2.5 Projection of the non-identifiable spaces the first MLE in Table 2.2 (a), the first three local maxima and the last local maxima in Table 2.2 (b) into the three-dimensional unit cube where λ_1, α_{11} and β_{11} take values. In this coordinate system, the projection of non-identifiable subspaces for the first three local maxima in Table 2.2 (b) results in the same surface; in order to obtain distinct surfaces, it would be necessary to change the coordinates over which the projections are made.

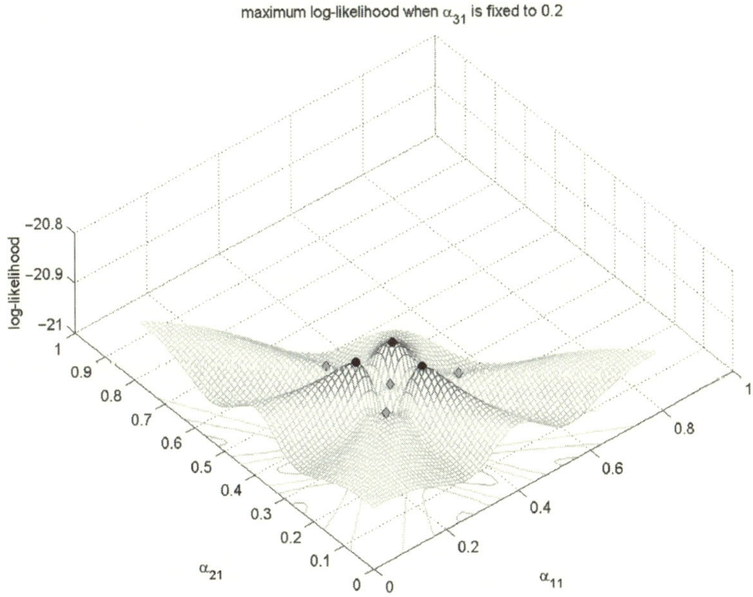

Fig. 2.6 The plot of the profile likelihood as a function of α_{11} and α_{21} when α_{31} is fixed to 0.2. There are seven peaks: the three black points are the MLEs and the four grey diamonds are the other local maxima.

Plotting the log-likelihood function

Having determined that the non-identifiable space is two-dimensional and that there are multiple maxima, we proceed with some plots of the profile log-likelihood function. To obtain a non-trivial surface, we need to consider three parameters.

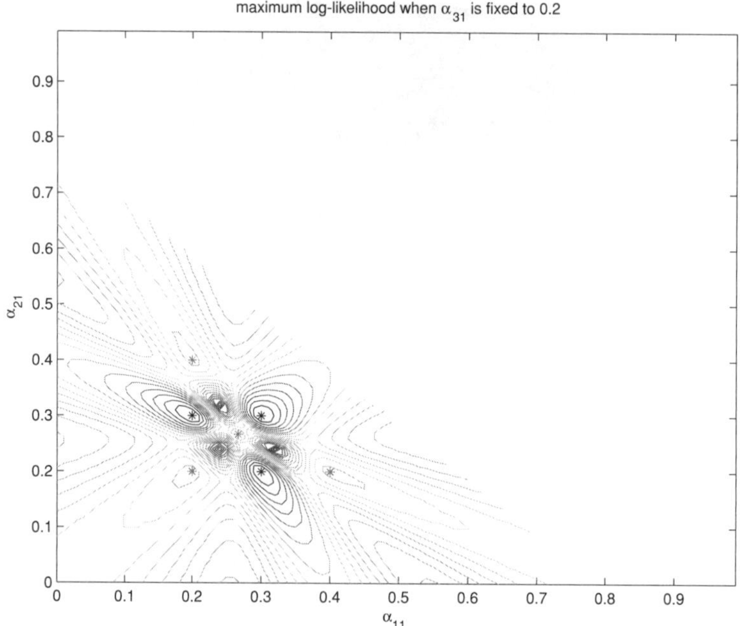

Fig. 2.7 The contour plot of the profile likelihood as a function of α_{11} and α_{21} when α_{31} is fixed. There are seven peaks: the three black points are the MLEs and the four grey points are the other local maxima.

Figures 2.6 and 2.7 display the surface and contour plot of the profile log-likelihood function for α_{11} and α_{21} when α_{31} is one of the fixed parameters. Both figures show clearly the different maxima of the log-likelihood function, each lying on the top of 'ridges' of the log-likelihood surface which are placed symmetrically with respect to each others. The position and shapes of these ridges reflect, once again, the invariance properties of the estimated probabilities and parameters.

Further remarks and open problems

An interesting aspect we came across while fitting the table (2.8) was the proximity of the values of the local and global maxima of the log-likelihood function. Although these values are very close, the fitted tables corresponding to global and local maxima are remarkably different. Even though the data (2.8) are not sparse, we wonder about the effect of cell sizes. Figure 2.8 shows the same profile log-likelihood for the table (2.8) multiplied by 10 000. While the number of global and local maxima, the contour plot and the basic symmetric shape of the profile log-likelihood surface remain unchanged after this rescaling, the peaks around the global maxima have become much more pronounced and so has the difference between the values of the global and local maxima.

We studied a number of variations of table (2.8), focusing in particular on symmetric data. We report only some of our results and refer to the on-line supplement for a more extensive study. Table 2.4 shows the values and number of local and

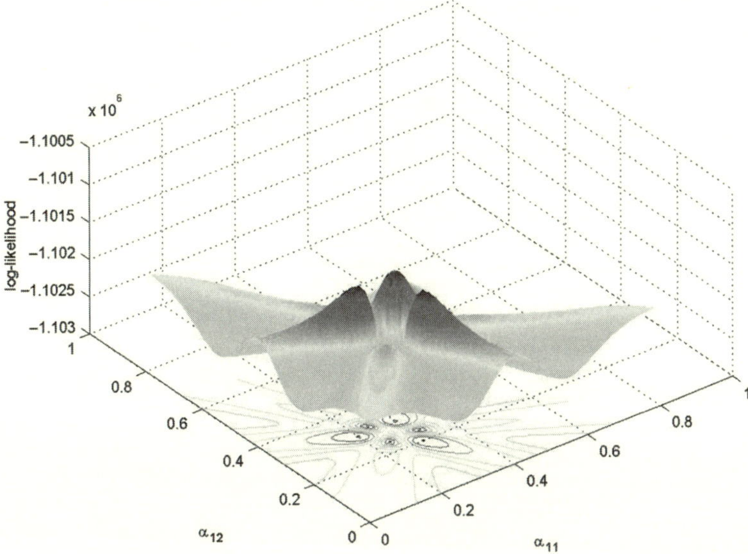

Fig. 2.8 The contour plot of the profile likelihood as a function of α_{11} and α_{21} when α_{31} is fixed for the data (2.8) multiplied by 10 000. As before, there are seven peaks: three global maxima and four identical local maxima.

global maxima for the 6×6 version of (2.8). As for the 4×4 case, we notice strong invariance features of the various maxima of the likelihood function and a very small difference between the value of the global and local maxima.

Fitting the same model to the table

$$\begin{pmatrix} 1 & 2 & 2 & 2 \\ 2 & 1 & 2 & 2 \\ 2 & 2 & 1 & 2 \\ 2 & 2 & 2 & 1 \end{pmatrix}$$

we found six global maxima of the likelihood function, which give as many maximum likelihood estimates, all obtainable via simultaneous permutation of rows and columns of the table below

$$\begin{pmatrix} 7/4 & 7/4 & 7/4 & 7/4 \\ 7/4 & 7/4 & 7/4 & 7/4 \\ 7/4 & 7/4 & 7/6 & 7/3 \\ 7/4 & 7/4 & 7/3 & 7/6 \end{pmatrix}, \quad \text{log-likelihood} = -77.2927 + const.$$

Based on our investigations, we formulate the following conjecture, which we verified computationally up to dimension $k = 50$. For the $n \times n$ table with values x along the diagonal and values $y \le x$ off the diagonal elements, the maximum likelihood estimates for the latent class model with two latent classes are the 2×2 block diagonal matrices of the form $\begin{pmatrix} A & B \\ B' & C \end{pmatrix}$ and the permuted versions of it,

Table 2.4 *Stationary points for the 6 × 6 version of the table (2.8). All the maxima are invariant under simultaneous permutations of the rows and columns of the corresponding fitted tables.*

Fitted counts	Log-likelihood
$\begin{pmatrix} 4 & 2 & 2 & 2 & 2 & 2 \\ 2 & 12/5 & 12/5 & 12/5 & 12/5 & 12/5 \\ 2 & 12/5 & 12/5 & 12/5 & 12/5 & 12/5 \\ 2 & 12/5 & 12/5 & 12/5 & 12/5 & 12/5 \\ 2 & 12/5 & 12/5 & 12/5 & 12/5 & 12/5 \\ 2 & 12/5 & 12/5 & 12/5 & 12/5 & 12/5 \end{pmatrix}$	$-300.2524 + const.$
$\begin{pmatrix} 7/3 & 7/3 & 7/3 & 7/3 & 7/3 & 7/3 \\ 7/3 & 13/5 & 13/5 & 13/5 & 29/15 & 29/15 \\ 7/3 & 13/5 & 13/5 & 13/5 & 29/15 & 29/15 \\ 7/3 & 13/5 & 13/5 & 13/5 & 29/15 & 29/15 \\ 7/3 & 29/15 & 29/15 & 29/15 & 44/15 & 44/15 \\ 7/3 & 29/15 & 29/15 & 29/15 & 44/15 & 44/15 \end{pmatrix}$	$-300.1856 + const.$
$\begin{pmatrix} 3 & 3 & 2 & 2 & 2 & 2 \\ 3 & 3 & 2 & 2 & 2 & 2 \\ 2 & 2 & 5/2 & 5/2 & 5/2 & 5/2 \\ 2 & 2 & 5/2 & 5/2 & 5/2 & 5/2 \\ 2 & 2 & 5/2 & 5/2 & 5/2 & 5/2 \\ 2 & 2 & 5/2 & 5/2 & 5/2 & 5/2 \end{pmatrix}$	$-300.1729 + const.$
$\begin{pmatrix} 8/3 & 8/3 & 8/3 & 2 & 2 & 2 \\ 8/3 & 8/3 & 8/3 & 2 & 2 & 2 \\ 8/3 & 8/3 & 8/3 & 2 & 2 & 2 \\ 2 & 2 & 2 & 8/3 & 8/3 & 8/3 \\ 2 & 2 & 2 & 8/3 & 8/3 & 8/3 \\ 2 & 2 & 2 & 8/3 & 8/3 & 8/3 \end{pmatrix}$	$-300.1555 + const.$ (MLE)
$\begin{pmatrix} 7/3 & 7/3 & 7/3 & 7/3 & 7/3 & 7/3 \\ 7/3 & 7/3 & 7/3 & 7/3 & 7/3 & 7/3 \\ 7/3 & 7/3 & 7/3 & 7/3 & 7/3 & 7/3 \\ 7/3 & 7/3 & 7/3 & 7/3 & 7/3 & 7/3 \\ 7/3 & 7/3 & 7/3 & 7/3 & 7/3 & 7/3 \\ 7/3 & 7/3 & 7/3 & 7/3 & 7/3 & 7/3 \end{pmatrix}$	$-301.0156 + const.$
$\begin{pmatrix} 7/3 & 7/3 & 7/3 & 7/3 & 7/3 & 7/3 \\ 7/3 & 35/9 & 35/18 & 35/18 & 35/18 & 35/18 \\ 7/3 & 35/18 & 175/72 & 175/72 & 175/72 & 175/72 \\ 7/3 & 35/18 & 175/72 & 175/72 & 175/72 & 175/72 \\ 7/3 & 35/18 & 175/72 & 175/72 & 175/72 & 175/72 \\ 7/3 & 35/18 & 175/72 & 175/72 & 175/72 & 175/72 \end{pmatrix}$	$-300.2554 + const.$

where A, B, and C are

$$A = \left(y + \tfrac{x-y}{p}\right) \cdot \mathbf{1}_{p \times p},$$
$$B = y \cdot \mathbf{1}_{p \times q},$$
$$C = \left(y + \tfrac{x-y}{q}\right) \cdot \mathbf{1}_{q \times q},$$

and $p = \left\lfloor \tfrac{n}{2} \right\rfloor$, $q = n - p$.

We also noticed other interesting phenomena, which suggest the need for further geometric analysis. For example, consider fitting the (non-identifiable) latent class model with two classes to the table of counts (B. Sturmfels: private communication)

$$\begin{pmatrix} 5 & 1 & 1 \\ 1 & 6 & 2 \\ 1 & 2 & 6 \end{pmatrix}.$$

Based on numerical computations, the maximum likelihood estimates appear to be unique, namely the table of fitted values

$$\begin{pmatrix} 5 & 1 & 1 \\ 1 & 4 & 4 \\ 1 & 4 & 4 \end{pmatrix}. \tag{2.15}$$

Looking at the non-identifiable subspace for this model, we found that the MLEs (2.15) can arise from combinations of parameters some of which can be 0, such as $\alpha^{(1)} = \beta^{(1)}$, $\alpha^{(2)} = \beta^{(2)}$ and

$$\alpha^{(1)} = \begin{pmatrix} 0.7143 \\ 0.1429 \\ 0.1429 \end{pmatrix}, \quad \alpha^{(2)} = \begin{pmatrix} 0 \\ 0.5 \\ 0.5 \end{pmatrix}, \quad \lambda = \begin{pmatrix} 0.3920 \\ 0.6080 \end{pmatrix}.$$

This might indicate the presence of singularities besides the obvious ones given by marginal probabilities for H containing 0 coordinates (which have the geometric interpretation as lower order secant varieties) and by points $\mathbf{p}$ along the boundary of the simplex Δ_{d-1}.

2.5 Two applications

2.5.1 Example: Michigan influenza

(Monto *et al.* 1985) present data for 263 individuals on the outbreak of influenza in Tecumseh, Michigan during the four winters of 1977–1981: (1) Influenza type A (H3N2), December 1977–March 1978; (2) Influenza type A (H1N1), January 1979–March 1979; (3) Influenza type B, January 1980–April 1980 and (4) Influenza type A (H3N2), December 1980–March 1981. The data have been analysed by others including (Haber 1986) and we reproduce them here as Table 2.5. The table is characterised by a large count for the cell corresponding to lack of infection from any type of influenza.

The LC model with one binary latent variable (identifiable by Theorem 3.5 in (Settimi and Smith 2005)) fits the data extremely well, as shown in Table 2.5. We also conducted a log-linear model analysis of this dataset and concluded that there is no indication of second- or higher-order interaction among the four types of influenza. The best log-linear model selected via both Pearson's chi-squared and the likelihood ratio statistics was the model of conditional independence of influenza of type (2), (3) and (4) given influenza of type (1) and was outperformed by the LC model.

Table 2.5 *Infection profiles and frequency of infection for four influenza outbreaks for a sample of 263 individuals in Tecumseh, Michigan during the winters of 1977–1981. A value of of 0 in the first four columns indicates Source: Monto et al. (1985). The last column is the values fitted by the naive Bayes model with $r = 2$.*

Type of Influenza				Observed Counts	Fitted Values
(1)	(2)	(3)	(4)		
0	0	0	0	140	139.5135
0	0	0	1	31	31.3213
0	0	1	0	16	16.6316
0	0	1	1	3	2.7168
0	1	0	0	17	17.1582
0	1	0	1	2	2.1122
0	1	1	0	5	5.1172
0	1	1	1	1	0.4292
1	0	0	0	20	20.8160
1	0	0	1	2	1.6975
1	0	1	0	9	7.7354
1	0	1	1	0	0.5679
1	1	0	0	12	11.5472
1	1	0	1	1	0.8341
1	1	1	0	4	4.4809
1	1	1	1	0	0.3209

Despite the reduced dimensionality of this problem and the large sample size, we report on the instability of the Fisher scoring algorithm implemented in the R package `gllm`, e.g., see (Espeland 1986). As the algorithm cycles through, the evaluations of Fisher information matrix become increasing ill-conditioned and eventually produce instabilities in the estimated coefficients and in the standard errors. These problems disappear in the modified Newton–Raphson implementation, originally suggested by (Haberman 1988), based on an inexact line search method known in the convex optimization literature as the Wolfe conditions.

2.5.2 Data from the National Long Term Care Survey

(Erosheva 2002) and (Erosheva *et al.* 2007) analyse an extract from the National Long Term Care Survey in the form of a 2^{16} contingency table that contains data on six activities of daily living (ADL) and ten instrumental activities of daily living (IADL) for community-dwelling elderly from 1982, 1984, 1989, and 1994 survey waves. The six ADL items include basic activities of hygiene and personal care (eating, getting in/out of bed, getting around inside, dressing, bathing, and getting to the bathroom or using toilet). The ten IADL items include basic activities necessary to reside in the community (doing heavy housework, doing light housework, doing laundry, cooking, grocery shopping, getting about outside, travelling, managing money, taking medicine and telephoning). Of the 65 536 cells in the table, 62 384 (95.19%) contain zero counts, 1729 (2.64%) contain counts of 1, 499 (0.76%)

Table 2.6 *BIC and log-likelihood values for various values of r for the NLTCS dataset.*

r	Dimension	Maximal log-likelihood	BIC
2	33	−152527.32796	305383.97098
3	50	−141277.14700	283053.25621
4	67	−137464.19759	275597.00455
5	84	−135272.97928	271384.21508
6	101	−133643.77822	268295.46011
7	118	−132659.70775	266496.96630
8	135	−131767.71900	264882.63595
9	152	−131367.70355	264252.25220
10	169	−131033.79967	263754.09160
11	186	−130835.55275	263527.24492
12	203	−130546.33679	263118.46015
13	220	−130406.83312	263009.09996
14	237	−130173.98208	262713.04502
15	254	−129953.32247	262441.37296
16	271	−129858.83550	262422.04617
17	288	−129721.02032	262316.06296
18	305	−129563.98159	262171.63265
19	322	−129475.87848	262165.07359
20	339	−129413.69215	262210.34807

contain counts of 2. The largest cell count, corresponding to the $(1, 1, \ldots, 1)$ cell, is 3853.

(Erosheva 2002) and (Erosheva *et al.* 2007) use an individual-level latent mixture model that bears a striking resemblance to the LC model. Here we report on analyses with the latter.

We use both the EM and Newton–Raphson algorithms to fit a number of LC models with up to 20 classes, which can be shown to be all identifiable in virtue of Proposition 2.3 in (Catalisano *et al.* 2002). Table 2.6 reports the maximal value of log-likelihood function and the value of BIC (the Bayesian Information Criterion), which seem to indicate that larger LC models with many levels are to be preferred. To provide a better sense of how well these LC models fit the data, we show in Table 2.7 the fitted values for the six largest cells, which, as mentioned, deviates considerably from most of the cell entries. We have also considered alternative model selection criteria such as AIC and modifications of it. AIC (with and without a second-order correction) points to $k > 20$! An ad-hoc modification of AIC due to (Anderson *et al.* 1994) for overdispersed data gives rather bizarre results. The dimensionality of a suitable LC model for these data appears to be much greater than for the individual level mixture model in (Erosheva *et al.* 2007).

Because of its high dimensionality and remarkable degree of sparsity, this example offers an ideal setting in which to test the relative strengths and disadvantages of the EM and Newton–Raphson algorithms. In general, the EM algorithm, as a hill-climbing method, moves steadily towards solutions with higher value of the log-likelihood, but converges only linearly. On the other hand, despite its faster quadratic rate of convergence, the Newton–Raphson method tends to be very time

Table 2.7 *Fitted values for the largest six cells for the NLTCS dataset for various r.*

r	Fitted values					
2	826.78	872.07	6.7	506.61	534.36	237.41
3	2760.93	1395.32	152.85	691.59	358.95	363.18
4	2839.46	1426.07	145.13	688.54	350.58	383.19
5	3303.09	1436.95	341.67	422.24	240.66	337.63
6	3585.98	1294.25	327.67	425.37	221.55	324.71
7	3659.80	1258.53	498.76	404.57	224.22	299.52
8	3663.02	1226.81	497.59	411.82	227.92	291.99
9	3671.29	1221.61	526.63	395.08	236.95	294.54
10	3665.49	1233.16	544.95	390.92	237.69	297.72
11	3659.20	1242.27	542.72	393.12	244.37	299.26
12	3764.62	1161.53	615.99	384.81	235.32	260.04
13	3801.73	1116.40	564.11	374.97	261.83	240.64
14	3796.38	1163.62	590.33	387.73	219.89	220.34
15	3831.09	1135.39	660.46	361.30	261.92	210.31
16	3813.80	1145.54	589.27	370.48	245.92	219.06
17	3816.45	1145.45	626.85	372.89	236.16	213.25
18	3799.62	1164.10	641.02	387.98	219.65	221.77
19	3822.68	1138.24	655.40	365.49	246.28	213.44
20	3836.01	1111.51	646.39	360.52	285.27	220.47
Observed	3853	1107	660	351	303	216

and space consuming when the number of variables is large, and may be numerically unstable if the Hessian matrices are poorly conditioned around critical points, which again occurs more frequently in large problems (but also in small ones, such as the Michigan Influenza examples above).

For the class of basic LC models considered here, the time complexity for one single step of the EM algorithm is $\mathcal{O}\left(d \cdot r \cdot \sum_i d_i\right)$, while the space complexity is $\mathcal{O}\left(d \cdot r\right)$. In contrast, for the Newton–Raphson algorithm, both the time and space complexity are $\mathcal{O}\left(d \cdot r^2 \cdot \sum_i d_i\right)$. Consequently, for the NLTCS dataset, when r is bigger than 4, Newton–Raphson is sensibly slower than EM, and when r goes up to 7, Newton–Raphson needs more than 1G of memory. Another significant drawback of the Newton–Raphson method we experienced while fitting both the Michigan influenza and the NLTCS datasets is its potential numerical instability, due to the large condition numbers of the Hessian matrices. As already remarked, following (Haberman 1988), a numerically convenient solution is to modify the Hessian matrices so that they remain negative definite and then approximate locally the log-likelihood by a quadratic function. However, since the log-likelihood is neither concave nor quadratic, these modifications do not necessarily guarantee an increase of the log-likelihood at each iteration step. As a result, the algorithm may experience a considerable slowdown in the rate of convergence, which we in fact observed with the NLTCS data. Table 2.8 shows the condition numbers for the true Hessian matrices evaluated at the numerical maxima, for various values of r. This table

Table 2.8 *Condition numbers of Hessian matrices at the maxima for the NLTCS data.*

r	Condition number
2	$2.1843e + 03$
3	$1.9758e + 04$
4	$2.1269e + 04$
5	$4.1266e + 04$
6	$1.1720e + 08$
7	$2.1870e + 08$
8	$4.2237e + 08$
9	$8.7595e + 08$
10	$8.5536e + 07$
11	$1.2347e + 19$
12	$3.9824e + 08$
13	$1.0605e + 20$
14	$3.4026e + 18$
15	$3.9783e + 20$
16	$3.2873e + 09$
17	$1.0390e + 19$
18	$2.1018e + 09$
19	$2.0082e + 09$
20	$2.5133e + 16$

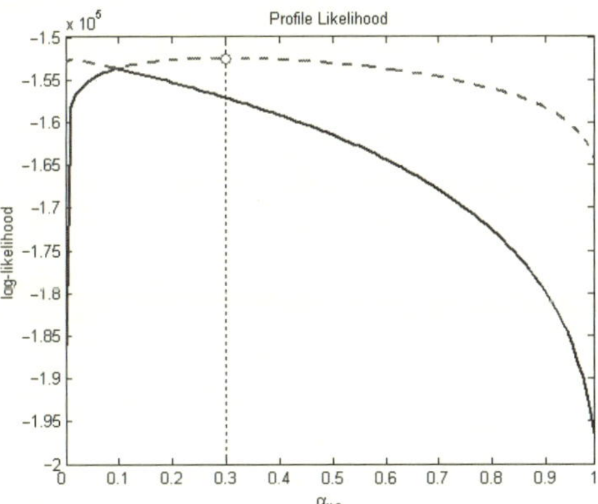

Fig. 2.9 The plot of the profile likelihood for the NLCST dataset, as a function of α_{12}. The vertical line indicates the location of the maximizer.

suggests that, despite full identifiability, the log-likelihood has a very low curvature around the maxima and that the log-likelihood may, in fact, look quite flat.

To elucidate this point and some of the many difficulties in fitting LC models, we show in Figure 2.9 the profile likelihood plot for the parameter α_{12} in simplest LC model with $r = 2$. The actual profile log-likelihood is shown in dashed and is obtained as the upper envelope of two distinct, smooth curves, each corresponding to a local maxima of the log-likelihood. The location of the optimal value of α_{12} is displayed with a vertical line. Besides illustrating multimodality, the log-likelihood function in this example is notable for its relative flatness around its global maximum.

2.6 On symmetric tables and the MLE

In this section, we show how symmetry in data allows one to symmetrise via averaging local maxima of the likelihood function and to obtain critical points that are more symmetric. In various examples we looked at, these have larger likelihood than the tables from which they are obtained. We also prove that if the aforementioned averaging process always causes likelihood to go up, then among the 4×4 matrices of rank 2, the ones maximizing the log-likelihood function for the 100 Swiss Francs problem (2.16) are given in Table 2.9. We will further simplify the notation and will write L for the matrix of observed counts and M for the matrix of MLEs.

2.6.1 Introduction and motivation

A main theme in this section is to understand in what ways symmetry in data forces symmetry in the global maxima of the likelihood function. One question is whether our ideas can be extended at all to non-symmetric data by suitable scaling. We prove that non-symmetric local maxima will imply the existence of more symmetric points which are critical points at least within a key subspace and are related in a very explicit way to the non-symmetric ones. Thus, if the EM algorithm leads to a local maximum which lacks certain symmetries, then one may deduce that certain other, more symmetric points are also critical points (at least within certain subspaces), and so check these to see if they give larger likelihood. There is numerical evidence that they do, and also a close look at our proofs shows that for 'many' data points this symmetrisation process is guaranteed to increase maximum likelihood, by virtue of a certain single-variable polynomial encoding of the likelihood function often being real-rooted.

Here is an example of our symmetrisation process. Given the data

$$
\begin{matrix}
4 & 2 & 2 & 2 & 2 & 2 \\
2 & 4 & 2 & 2 & 2 & 2 \\
2 & 2 & 4 & 2 & 2 & 2 \\
2 & 2 & 2 & 4 & 2 & 2 \\
2 & 2 & 2 & 2 & 4 & 2 \\
2 & 2 & 2 & 2 & 2 & 4
\end{matrix}
$$

one of the critical points located by the EM algorithm is

$$
\begin{matrix}
7/3 & 7/3 & 7/3 & 7/3 & 7/3 & 7/3 \\
7/3 & 13/5 & 13/5 & 13/5 & 29/15 & 29/15 \\
7/3 & 13/5 & 13/5 & 13/5 & 29/15 & 29/15 \\
7/3 & 13/5 & 13/5 & 13/5 & 29/15 & 29/15 \\
7/3 & 29/15 & 29/15 & 29/15 & 44/15 & 44/15 \\
7/3 & 29/15 & 29/15 & 29/15 & 44/15 & 44/15
\end{matrix}
$$

One way to interpret this matrix is that $M_{i,j} = 7/3 + e_i f_j$ where

$$\mathbf{e} = \mathbf{f} = (0, 2/\sqrt{15}, 2/\sqrt{15}, 2/\sqrt{15}, -3/\sqrt{15}, -3/\sqrt{15}).$$

Our symmetrisation process suggests replacing the vectors $\mathbf{e}$ and $\mathbf{f}$ each by the vector

$$(1/\sqrt{15}, 1/\sqrt{15}, 2/\sqrt{15}, 2/\sqrt{15}, -3/\sqrt{15}, -3/\sqrt{15})$$

in which two coordinates are averaged; however, since one of the values being averaged is zero, it is not so clear whether this should increase likelihood. Repeatedly applying such symmetrisation steps to this example, does converge to a local maximum. More generally, let M be an n by n matrix of rank at most two which has row and column sums all equalling kn, implying (by results of Section 2.6.2) that we may write $M_{i,j}$ as $k + e_i f_j$ where e, f are each vectors whose coordinates sum to 0.

We are interested in the following general questions.

Question 2.1 Suppose a data matrix is fixed under simultaneously swapping rows and columns i, j. Consider any M as above, i.e. with $M_{i,j} = k + e_i f_j$. Does $e_i > e_j > 0, f_i > f_j > 0$ (or similarly $e_i < e_j < 0, f_i < f_j < 0$) imply that replacing e_i, e_j by $\frac{e_i + e_j}{2}$ and f_i, f_j by $\frac{f_i + f_j}{2}$ always increases the likelihood?

Remark 2.1 The weaker conditions $e_i > e_j = 0$ and $f_i > f_j = 0$ (resp. $e_i < e_j = 0, f_i < f_j = 0$) do not always imply that this replacement will increase likelihood. However, one may consider the finite list of possibilities for how many zeros the vectors $\mathbf{e}$ and $\mathbf{f}$ may have; an affirmative answer to Question 2.1 would give a way to find the matrix maximizing likelihood in each case, and then we could compare this finite list of maxima to find the global maximum.

Question 2.2 Are all real-valued critical points of the likelihood function obtained by setting some number of coordinates in the $\mathbf{e}$ and $\mathbf{f}$ vectors to zero and then averaging by the above process so that the eventual vectors $\mathbf{e}$ and $\mathbf{f}$ have all positive coordinates equal to each other and all negative coordinates equal to each other? This seems to be true in many examples.

One may check that the example discussed in Chapter 1 of (Pachter and Sturmfels 2005) gives another instance where this averaging approach leads quickly to a global maximum. Namely, given the data matrix

$$\begin{array}{cccc} 4 & 2 & 2 & 2 \\ 2 & 4 & 2 & 2 \\ 2 & 2 & 4 & 2 \\ 2 & 2 & 2 & 4 \end{array}$$

and a particular starting point, the EM algorithm converges to the saddle point

$$\frac{1}{48}\begin{pmatrix} 4 & 2 & 3 & 3 \\ 2 & 4 & 3 & 3 \\ 3 & 3 & 3 & 3 \\ 3 & 3 & 3 & 3 \end{pmatrix}$$

which we may write as $M_{i,j} = 1/48(3 + a_i b_j)$ for $\mathbf{a} = (-1, 1, 0, 0)$ and $\mathbf{b} = (-1, 1, 0, 0)$. Averaging -1 with 0 and 1 with the other 0 simultaneously in $\mathbf{a}$ and $\mathbf{b}$ immediately yields the global maximum directly by symmetrising the saddle point, i.e. rather than finding it by running the EM algorithm repeatedly from various starting points.

An affirmative answer to Question 2.1 would imply several things. It would yield a (positive) solution to the 100 Swiss Francs problem, as discussed in Section 2.6.3. More generally, it would explain in a rather precise way how certain symmetries in data seem to impose symmetry on the global maxima of the maximum likelihood function. Moreover it would suggest good ways to look for global maxima, as well as constraining them enough that in some cases they can be characterised, as we demonstrate for the 100 Swiss Francs problem. To make this concrete, for an n by n data matrix which is fixed by the S_n action simultaneously permuting rows and columns in the same way, it would follow that any probability matrix maximising likelihood for such a data matrix will have at most two distinct types of rows.

We do not know the answer to this question, but we do prove that this type of averaging will at least give a critical point within the subspace in which e_i, e_j, f_i, f_j may vary freely but all other parameters are held fixed. Data also provides evidence that the answer to the question may very well be yes. At the very least, this type of averaging appears to be a good heuristic for seeking local maxima, or at least finding a way to continue to increase maximum likelihood beyond what it is at a critical point one reaches. Moreover, while real data is unlikely to have these symmetries, perhaps it could come close, and this could still be a good heuristic to use in conjunction with the EM algorithm.

2.6.2 Preservation of marginals and some consequences

Proposition 2.1 *Given data in which all row and column sums (i.e. marginals) are equal, then for M to maximise the likelihood function for this data among matrices of a fixed rank, row and column sums of M all must be equal.*

We prove the case mentioned in the abstract, which should generalise by adjusting exponents and ratios in the proof. It may very well also generalise to distinct marginals and tables with more rows and columns.

Proof Let R_1, R_2, R_3, R_4 be the row sums of M. Suppose $R_1 \geq R_2 \geq R_3 > R_4$; other cases will be similar. Choose δ so that $R_3 = (1 + \delta)R_4$. We will show that multiplying row 4 by $1 + \epsilon$ with $0 < \epsilon < \min(1/4, \delta/2)$ will strictly increase L, giving a contradiction to M maximising L. The result for column sums follows by symmetry. We write $L(M')$ for the new matrix M' in terms of the variables $x_{i,j}$ for the original matrix M, so as to show that $L(M') > L(M)$. The first inequality below is proven in Lemma 2.1. Then

$$
\begin{aligned}
L(M') &= \frac{(1+\epsilon)^{10}(\prod_{i=1}^{4} x_{i,i})^4 (\prod_{i\neq j} x_{i,j})^2}{R_1 + R_2 + R_3 + (1+\epsilon)R_4)^{40}} \\[2mm]
&> \frac{(1+\epsilon)^{10}(\prod_{i=1}^{4} x_{i,i})^4 (\prod_{i\neq j} x_{i,j})^2}{[(1+1/4(\epsilon-\epsilon^2))(R_1 + R_2 + R_3 + R_4)]^{40}} \\[2mm]
&= \frac{(1+\epsilon)^{10}(\prod_{i=1}^{4} x_{i,i})^4 (\prod_{i\neq j} x_{i,j})^2}{[(1+1/4(\epsilon-\epsilon^2))^4]^{10}[R_1 + R_2 + R_3 + R_4]^{40}} \\[2mm]
&= \frac{(1+\epsilon)^{10}(\prod_{i=1}^{4} x_{i,i})^4 (\prod_{i\neq j} x_{i,j})^2}{A} \geq \frac{(1+\epsilon)^{10}}{(1+\epsilon)^{10}} \cdot L(M)
\end{aligned}
$$

where $A = [1+4(1/4)(\epsilon-\epsilon^2)+6(1/4)^2(\epsilon-\epsilon^2)^2+\cdots+(1/4)^4(\epsilon-\epsilon^2)^4]^{10}[\sum_{i=1}^{4} R_i]^{40}$. $\qquad\square$

Lemma 2.1 *If* $\epsilon < \min(1/4, \delta/2)$ *and* $R_1 \geq R_2 \geq R_3 = (1+\delta)R_4$, *then* $R_1 + R_2 + R_3 + (1+\epsilon)R_4 < (1+1/4(\epsilon-\epsilon^2))(R_1 + R_2 + R_3 + R_4)$.

Proof It is equivalent to show $\epsilon R_4 < (1/4)(\epsilon)(1-\epsilon)\sum_{i=1}^{4} R_i$. However,

$$
(1/4)(\epsilon)(1-\epsilon)\Big(\sum_{i=1}^{4} R_i\Big) \geq (3/4)(\epsilon)(1-\epsilon)(1+\delta)R_4 + (1/4)(\epsilon)(1-\epsilon)R_4
$$

$$
> (3/4)(\epsilon)(1-\epsilon)(1+2\epsilon)R_4 + (1/4)(\epsilon)(1-\epsilon)R_4
$$

$$
= (3/4)(\epsilon)(1+\epsilon-2\epsilon^2)R_4 + (1/4)(\epsilon-\epsilon^2)R_4
$$

$$
= \epsilon R_4 + [(3/4)(\epsilon^2)-(6/4)(\epsilon^3)]R_4 - (1/4)(\epsilon^2)R_4
$$

$$
= \epsilon R_4 + [(1/2)(\epsilon^2)-(3/2)(\epsilon^3)]R_4 \geq \epsilon R_4 + [(1/2)(\epsilon^2)-(3/2)(\epsilon^2)(1/4)]R_4 > \epsilon R_4.
$$

$\qquad\square$

Corollary 2.1 *There exist two vectors* (e_1, e_2, e_3, e_4) *and* (f_1, f_2, f_3, f_4) *such that* $\sum_{i=1}^{4} e_i = \sum_{i=1}^{4} f_i = 0$ *and* $M_{i,j} = K + e_i f_j$. *Moreover,* K *equals the average entry size.*

In particular, it follows that L may be maximised by treating it as a function of just six variables, namely $e_1, e_2, e_3, f_1, f_2, f_3$, since e_4, f_4 are also determined by these; changing K before solving this maximisation problem simply has the impact of multiplying the entire matrix M that maximises likelihood by a scalar.

Let E be the *deviation matrix* associated to M, where $E_{i,j} = e_i f_j$.

Question 2.3 Another natural question to ask, in light of this corollary, is whether the matrix of rank at most r maximising L is expressible as the sum of a rank one matrix and a matrix of rank at most $r-1$ maximising L.

Remark 2.2 When we consider matrices with fixed row and column sums, then we may ignore the denominator in the likelihood function and simply maximise the numerator.

Corollary 2.2 *If M which maximises L has $e_i = e_j$, then it also has $f_i = f_j$. Consequently, if it has $e_i \neq e_j$, then it also has $f_i \neq f_j$.*

Proof One consequence of having equal row and column sums is that it allows the likelihood function to be split into a product of four functions, one for each row, or else one for each column; this is because the sum of all table entries equals the sum of those in any row or column multiplied by four, allowing the denominator to be written just using variables from any one row or column. Thus, once the vector e is chosen, we find the best possible f for this given e by solving four separate maximisation problems, one for each f_i, i.e. one for each column. Setting $e_i = e_j$ causes the likelihood function for column i to coincide with the likelihood function for column j, so both are maximised at the same value, implying $f_i = f_j$. $\square$

Next we prove a slightly stronger general fact for matrices in which rows and columns i, j may simultaneously be swapped without changing the data matrix.

Proposition 2.2 *If a matrix M maximising likelihood has $e_i > e_j > 0$, then it also has $f_i > f_j > 0$.*

Proof Without loss of generality, set $i = 1, j = 3$. We will show that if $e_1 > e_3$ and $f_1 < f_3$, then swapping columns one and three will increase likelihood, yielding a contradiction. Let

$$L_1(e_1) = (1/4 + e_1 f_1)^4 (1/4 + e_1 f_2)^2 (1/4 + e_1 f_3)^2 (1/4 + e_1 f_4)^2$$

and

$$L_3(e_3) = (1/4 + e_2 f_1)^2 (1/4 + e_2 f_2)^2 (1/4 + e_3 f_3)^4 (1/4 + e_3 f_4)^2,$$

namely the contributions of rows 1 and 3 to the likelihood function. Let

$$K_1(e_1) = (1/4 + e_1 f_3)^4 (1/4 + e_1 f_2)^2 (1/4 + e_1 f_1)^2 (1/4 + e_1 f_4)^2$$

and

$$K_3(e_3) = (1/4 + e_3 f_3)^2 (1/4 + e_3 f_2)^2 (1/4 + e_3 f_1)^4 (1/4 + e_3 f_4)^2,$$

so that after swapping the first and third columns, the new contribution to the likelihood function from rows 1 and 3 is $K_1(e_1)K_3(e_3)$. Since the column swap does not impact that contributions from rows 2 and 4, the point is to show $K_1(e_1)K_3(e_3) > L_1(e_1)L_3(e_3)$. Ignoring common factors, this reduces to showing

$$(1/4 + e_1 f_3)^2 (1/4 + e_3 f_1)^2 > (1/4 + e_1 f_1)^2 (1/4 + e_3 f_3)^2,$$

in other words $(1/16 + 1/4(e_1 f_3 + e_3 f_1) + e_1 e_3 f_1 f_3)^2$ is greater than $(1/16 + 1/4(e_1 f_1 + e_3 f_3) + e_1 e_3 f_1 f_3)^2$, namely $e_1 f_3 + e_3 f_1 > e_1 f_1 + e_3 f_3$. But since $e_3 < e_1, f_1 < f_3$, we have $0 < (e_1 - e_3)(f_3 - f_1) = (e_1 f_3 + e_3 f_1) - (e_1 f_1 + e_3 f_3)$, just as needed. $\square$

Table 2.9 *Tables of fitted values corresponding to the global maximum of the likelihood equation for the observed table (2.16) (log-likelihood value* -20.8079*).*

$$
\begin{pmatrix} 3 & 3 & 2 & 2 \\ 3 & 3 & 2 & 2 \\ 2 & 2 & 3 & 3 \\ 2 & 2 & 3 & 3 \end{pmatrix}
\begin{pmatrix} 3 & 2 & 3 & 2 \\ 2 & 3 & 2 & 3 \\ 3 & 2 & 3 & 2 \\ 2 & 3 & 2 & 3 \end{pmatrix}
\begin{pmatrix} 3 & 2 & 2 & 3 \\ 2 & 3 & 3 & 2 \\ 2 & 3 & 3 & 2 \\ 3 & 2 & 2 & 3 \end{pmatrix}
$$

Question 2.4 Does having a data matrix which is symmetric with respect to transpose imply that matrices maximising likelihood will also be symmetric with respect to transpose?

Perhaps this could also be verified again by averaging, similarly to what we suggest for involutions swapping a pair of rows and columns simultaneously.

2.6.3 The 100 Swiss Francs problem

We use the results derived so far to solve the '100 Swiss Francs' problem discussed in detail in the on-line supplement. Here we provide a mathematical proof that the three tables in Table 2.9 are global maxima of the log-likelihood function for the basic LC model with $r = 2$ and data given in (2.16)

$$
\mathbf{n} = \begin{pmatrix} 4 & 2 & 2 & 2 \\ 2 & 4 & 2 & 2 \\ 2 & 2 & 4 & 2 \\ 2 & 2 & 2 & 4 \end{pmatrix}. \tag{2.16}
$$

Theorem 2.1 *If the answer to Question 2.1 is yes, then the 100 Swiss Francs problem is solved.*

Proof Proposition 2.1 will show that for M to maximise L, M must have row and column sums which are all equal to the quantity which we call $R_1, R_2, R_3, R_4, C_1,$ $C_2, C_3,$ or C_4 at our convenience. The denominator of L may therefore be expressed as $(4C_1)^{10}(4C_2)^{10}(4C_3)^{10}(4C_4)^{10}$ or as $(4R_1)^{10}(4R_2)^{10}(4R_3)^{10}(4R_4)^{10}$, enabling us to rewrite L as a product of four smaller functions using distinct sets of variables.

Note that letting S_4 simultaneously permute rows and columns will not change L, so we assume the first two rows of M are linearly independent. Moreover, we may choose the first two rows in such a way that the next two rows are each non-negative combinations of the first two. Since row and column sums are all equal, the third row, denoted v_3, is expressible as $xv_1 + (1-x)v_2$ for v_1, v_2 the first and second rows and $x \in [0, 1]$. One may check that M does not have any row or column with values all equal to each other, because if it has one, then it has the other, reducing to a three by three problem which one may solve, and one may check that the answer does not have as high a likelihood as

$$
\begin{array}{cccc}
3 & 3 & 2 & 2 \\
3 & 3 & 2 & 2 \\
2 & 2 & 3 & 3 \\
2 & 2 & 3 & 3
\end{array}
$$

Proposition 2.3 shows that if the answer to Question 2.1 is yes, then for M to maximize L, we must have $x = 0$ or $x = 1$, implying row 3 equals either row 1 or row 2, and likewise row 4 equals one of the first two rows. Proposition 2.4 below shows M does not have three rows all equal to each other, and therefore must have two pairs of equal rows. Thus, the first column takes the form $(a, a, b, b)^\top$, so it is simply a matter of optimising a and b, then noting that the optimal choice will likewise optimise the other columns (by virtue of the way we broke L into a product of four expressions which are essentially the same, one for each column). Thus, M takes the form

$$
\begin{array}{cccc}
a & a & b & b \\
a & a & b & b \\
b & b & a & a \\
b & b & a & a
\end{array}
$$

since this matrix does indeed have rank two. Proposition 2.5 shows that to maximise L one needs $2a = 3b$, finishing the proof. $\qquad\square$

Proposition 2.3 *If the answer to Question 2.1 is yes, then row 3 equals either row 1 or row 2 in any matrix M which maximises likelihood. Similarly, each row i with $i > 2$ equals either row 1 or row 2.*

Proof $M_{3,3} = xM_{1,3} + (1-x)M_{2,3}$ for some $x \in [0, 1]$, so $M_{3,3} \le \max(M_{1,3}, M_{2,3})$. If $M_{1,3} = M_{2,3}$, then all entries of this column are equal, and one may use calculus to eliminate this possibility as follows: either M has rank 1, and then we may replace column 3 by $(c, c, 2c, c)^\top$ for suitable constant c to increase likelihood, since this only increases rank to at most two, or else the column space of M is spanned by $(1, 1, 1, 1)^\top$ and some (a_1, a_2, a_3, a_4) with $\sum a_i = 0$; specifically, column 3 equals $(1/4, 1/4, 1/4, 1/4) + x(a_1, a_2, a_3, a_4)$ for some x, allowing its contribution to the likelihood function to be expressed as a function of x whose derivative at $x = 0$ is non-zero, provided that $a_3 \ne 0$, implying that adding or subtracting some small multiple of $(a_1, a_2, a_3, a_4)^\top$ to the column will make the likelihood increase. If $a_3 = 0$, then row 3 is also constant, i.e. $e_3 = f_3 = 0$. But then, an affirmative answer to the second part of Question 2.1 will imply that this matrix does not maximise likelihood.

Suppose, on the other hand, $M_{1,3} > M_{2,3}$. Our goal then is to show $x = 1$. By Proposition 2.1 applied to columns rather than rows, we know that $(1, 1, 1, 1)$ is in the span of the rows, so each row may be written as $1/4(1, 1, 1, 1) + cv$ for some fixed vector v whose coordinates sum to 0. Say row 1 equals $1/4(1, 1, 1, 1) + kv$ for $k = 1$. Writing row 3 as $1/4(1, 1, 1, 1) + lv$, what remains is to rule out the possibility $l < k$. However, Proposition 2.2 shows that $l < k$ and $a_1 < a_3$ together imply that swapping columns 1 and 3 will yield a new matrix of the same rank with larger likelihood.

Now we turn to the case of $l < k$ and $a_1 \geq a_3$. If $a_1 = a_3$ then swapping rows 1 and 3 will increase likelihood. Assume $a_1 > a_3$. By Corollary 2.1, we have (e_1, e_2, e_3, e_4) with $e_1 > e_3$ and (f_1, f_2, f_3, f_4) with $f_1 > f_3$. Therefore, if the answer to Question 2.1 is yes, then replacing e_1, e_3 each by $(e_1 + e_3)/2$ and f_1, f_3 each by $(f_1 + f_3)/2$ yields a matrix with larger likelihood, completing the proof. $\square$

Proposition 2.4 *In any matrix M maximising L among rank 2 matrices, no three rows of M are equal to each other.*

Proof Without loss of generality, if M had three equal rows, then M would take the form

$$\begin{matrix} a & c & e & g \\ b & d & f & h \\ b & d & f & h \\ b & d & f & h \end{matrix}$$

but then the fact that M maximises L ensures $d = f = h$ and $c = e = g$ since L is a product of four expressions, one for each column, so that the second, third and fourth columns will all maximise their contribution to L in the same way. Since all row and column sums are equal, simple algebra may be used to show that all entries must be equal. However, we have already shown that such matrices do not maximise L. $\square$

Proposition 2.5 *To maximise M requires a, b related by $2a = 3b$.*

Proof We must maximise $a^6 b^4 (8a + 8b)^{-10}$. We may assume $a + b = 1$ since multiplying the entire matrix by a constant does not change L, so we maximise $(1/8)^{10} a^6 b^4$ with $b = 1 - a$; in other words, we maximise $f(a) = a^6 (1 - a)^4$. But solving $f'(a) = 0 = 6a^5 (1 - a)^4 + a^6 (4)(1 - a)^3 (-1) = a^5 (1 - a)^3 [6(1 - a) - 4a]$ yields $6(1 - a) - 4a = 0$, so $a = 6/10$ and $b = 4/10$ as desired. $\square$

2.7 Conclusions

In this chapter we have reconsidered the classical latent class model for contingency table data and studied its geometric and statistical properties. For the former we have exploited tools from algebraic geometry and computation tools that have allowed us to display the complexities of the latent class model. We have focused on the problem of maximum likelihood estimation under LC models and have studied the singularities arising from symmetries in the contingency table data and the multiple maxima that appear to result from these. We have given an informal characterisation of this problem, but a strict mathematical proof of the existence of identical multiple maxima has eluded us; we describe elements of a proof in a separate section in the on-line supplement.

We have also applied LC models data arising in two applications. In one, the models and maximum likelihood estimation are well behaved whereas in the other high-dimensional example various computational and other difficulties arise. The EM algorithm is especially vulnerable to problems of multimodality and it provides little in the way of clues regarding the dimensionality difficulties associated with the underlying structure of LC models.

Based on our work, we would advise practitioners to exercise caution in applying LC models. They have a tremendous heuristic appeal and in some examples provide a clear and convincing description of the data. But in many situations, the kind of complex behaviour explored in this chapter may lead to erroneous inferences.

Acknowledgement

This research was supported in part by the National Institutes of Health under Grant No. R01 AG023141-01, by NSF Grant DMS-0631589, and by a grant from the Pennsylvania Department of Health through the Commonwealth Universal Research Enhancement Program, all to the Department of Statistics to Carnegie Mellon University, and by NSF Grant DMS-0439734 to the Institute for Mathematics and Its Application at the University of Minnesota. We thank Bernd Sturmfels for introducing us to the 100 Swiss Francs problem, which motivated much of this work, and for his valuable comments and feedback.

References

Allman, E. S. and Rhodes, J. A. (2006). Phylogenetic invariants for stationary base composition, *Journal of Symbolic Computation* **41**, 138–50.

Allman, E. S. and Rhodes, J. A. (2008). Phylogenetic ideals and varieties for the general Markov model, *Advances in Applied Mathematics* **40**(2), 127–48.

Anderson, D. R., Burham, K. P. and White, G. C. (1994). AIC model selection in overdispersed capture-recapture data, *Ecology* **75**, 1780–93.

Anderson, T. W. (1954). On estimation of parameters in latent structure analysis, *Psychometrika* **19**, 1–10.

Bandeen-Roche, K., Miglioretti, D. L., Zeger, S. and Rathouz, P. J. (1997). Latent variable regression for multiple discrete outcomes, *Journal of the American Statistical Association* **92**, 1375–86.

Benedetti, R. (1990). *Real Algebraic and Semi-algebraic Sets* (Paris, Hermann).

Catalisano, M. V., Geramita, A. V. and Gimigliano, A. (2002). Ranks of tensors, secant varieties of Segre varieties and fat points, *Linear Algebra and Its Applications* **355**, 263–85. Corrigendum (2003). **367**, 347–8.

Clogg, C. and Goodman, L. (1984). Latent structure analysis of a set of multidimensional contingency tables, *Journal of the American Statistical Association* **79**, 762–771.

Cohen, J. E. and Rothblum, U. G. (1993). Nonnegative rank, decompositions and factorisations of nonnegative matrices, *Linear Algebra and Its Applications* **190**, 149–68.

Cox, D., Little, J. and O'Shea, D. (1992). *Ideals, Varieties, and Algorithms* (New York, Springer-Verlag).

Cowell, R. G., Dawid, P. A., Lauritzen, S. L. and Spiegelhalter, D. J. (1999). *Probabilistic Networks and Expert Systems* (New York, Springer-Verlag).

Erosheva, E. A. (2002). Grade of membership and latent structure models with application to disability survey data. PhD thesis, Department of Statistics, Carnegie Mellon University.

Erosheva, E. A., Fienberg, S. E. and Joutard, C. (2007). Describing disability through individual-level mixture models for multivariate binary data, *Annals of Applied Statistics* **1**(2) 502–37.

Espeland, M. A. (1986). A general class of models for discrete multivariate data, *Communications in Statistics: Simulation and Computation* **15**, 405–24.

Garcia, L. D. (2004). Algebraic statistics in model selection. In *Proc. UAI–04* (San Mateo, CA, Morgan Kaufmann) 177–84.

Garcia, L., Stillman, M. and Sturmfels, B. (2005). Algebraic Geometry of Bayesian Networks, *Journal of Symbolic Computation* **39**, 331–55.

Geiger, D., Heckerman, D., King, H. and Meek, C. (2001). Stratified exponential families: graphical models and model selection, *Annals of Statistics* **29**(2), 505–29.

Gibson, W. A. (1995). An extension of Anderson's solution for the latent structure equations, *Psychometrika* **20**, 69–73.

Goodman, L. (1974). Exploratory latent structure analysis using both identifiable and unidentifiable models, *Biometrika* **61**, 215–31.

Goodman, L. (1979). On the estimation of parameters in latent structure analysis, *Psychometrika* **44**(1), 123–8.

Greuel, G.-M., Pfister, G. and Schönemann, H. (2005). Singular *3.0. A Computer Algebra System for Polynomial Computations. Centre for Computer Algebra* (available at `www.singular.uni-kl.de`).

Haber, M. (1986). Testing for pairwise independence, *Biometrics* **42**, 429–35.

Haberman, S. J. (1974). Log-linear models for frequency tables derived by indirect observations: maximum likelihood equations, *Annals of Statistics* **2**, 911–24.

Haberman, S. J. (1988). A stabilized Newton-Raphson algorithm for log-linear models for frequency tables derived by indirect observation, *Sociological Methodology* **18**, 193–211.

Harris, J. (1992). *Algebraic Geometry: A First Course* (New York, Springer-Verlag).

Henry, N. W. and Lazarfeld, P.F. (1968). *Latent Structure Analysis* (Boston, Houghton Mufflin Company).

Humphreys, K. and Titterington, D. M. (2003). Variational approximations for categorical causal modeling with latent variables, *Psychometrika* **68**, 391–412.

Kocka, T. and Zhang, N. L. (2002). Dimension correction for hierarchical latent class models, In *Proc. UAI–02* (San Mateo, CA, Morgan Kaufmann) 267–74.

Kruskal, J. B. (1975). More factors than subjects, tests and treatments: An indeterminacy theorem for canonical decomposition and individual differences scaling, *Psychometrica* **41**, 281–93.

Landsberg, J. M. and Manivel, L. (2004). On the ideals of secant varieties of Segre varieties, *Foundations of Computational Mathematics* **4**, 397–422.

Lauritzen, S. L. (1996). *Graphical Models* (New York, Oxford University Press).

Madansky, A. (1960). Determinantal methods in latent class analysis, *Psychometrika* **25**, 183–98.

Mond, D. M. Q., Smith, J. Q. and Van Straten, D. (2003). Stochastic factorisations, sandwiched simplices and the topology of the space of explanations. In *Proceedings of the Royal Society of London, Series A* **459**, 2821–45.

Monto, A. S., Koopman, J. S. and Longini, I. M. (1985). Tecumseh study of illness. XIII. Influenza infection and disease. *American Journal of Epidemiology* **121**, 811–22.

Pachter, L. and Sturmfels, B. eds. (2005). *Algebraic Statistics for Computational Biology* (New York, Cambridge University Press).

Redner, R. A. and Walker, H. F. (1984). Mixture densities, maximum likelihood and the EM algorithm, *SIAM Review* **26**, 195–239.

Rusakov, D. and Geigerm, D. (2005). Asymptotic model selection for naive Bayesian networks, *Journal of Machine Learning Research* **6**, 1–35.

Settimi, R. and Smith, J. Q. (1998). On the geometry of Bayesian graphical models with hidden variables. In *Proc. UAI–98* (San Mateo, CA, Morgan Kaufmann) 479–2.

Settimi, R. and Smith, J. Q. (2005). Geometry, moments and conditional independence trees with hidden variables, *Annals of Statistics* **28**, 1179–205.

Smith, J. Q. and Croft, J. (2003). Bayesian networks for discrete multivariate data: an algebraic approach to inference, *Journal of Multivariate Analysis* **84**, 387–402.

Strassen, V. (1983). Rank and optimal computation of generic tensors, *Linear Algebra and Its Applications* **52/53**, 654–85.

Uebersax, J. (2006). Latent Class Analysis, A web-site with bibliography, software, links and FAQ for latent class analysis (available at `http://ourworld.compuserve.com/homepages/jsuebersax/index.htm`).

Watanabe, S. (2001). Algebraic analysis for non-identifiable learning machines, *Neural Computation* **13**, 899–933.

3
Algebraic geometry of 2×2 contingency tables

Aleksandra B. Slavković

Stephen E. Fienberg

Abstract

Contingency tables represent the joint distribution of categorical variables. In this chapter we use modern algebraic geometry to update the geometric representation of 2×2 contingency tables first explored in (Fienberg 1968) and (Fienberg and Gilbert 1970). Then we use this geometry for a series of new ends including various characterizations of the joint distribution in terms of combinations of margins, conditionals, and odds ratios. We also consider incomplete characterisations of the joint distribution and the link to latent class models and to the phenomenon known as Simpson's paradox. Many of the ideas explored here generalise rather naturally to $I \times J$ and higher-way tables. We end with a brief discussion of generalisations and open problems.

3.1 Introduction

(Pearson 1956) in his presidential address to the Royal Statistical Society was one of the earliest statistical authors to write explicitly about the role of geometric thinking for the theory of statistics, although many authors previously, such as (Edgeworth 1914) and (Fisher 1921), had relied heuristically upon geometric characterisations.

For contingency tables, beginning with (Fienberg 1968) and (Fienberg and Gilbert 1970), several authors have exploited the geometric representation of contingency table models, in terms of quantities such as margins and odds ratios, both for the proof of statistical results and to gain deeper understanding of models used for contingency table representation. For example, see (Fienberg 1970) for the convergence of iterative proportional fitting procedure, (Diaconis 1977) for the geometric representation of exchangeability, and (Kenett 1983) for uses in exploratory data analysis. More recently, (Nelsen 1995, Nelsen 2006) in a discussion of copulas for binary variables points out that two faces of the tetrahedron form the Fréchet upper bound, the other two the lower bound, and the surface of independence is the independence copula.

There has also been considerable recent interest in geometric descriptions of contingency tables models and analytical tools, from highly varying perspectives.

Algebraic and Geometric Methods in Statistics, ed. Paolo Gibilisco, Eva Riccomagno, Maria Piera Rogantin and Henry P. Wynn. Published by Cambridge University Press. © Cambridge University Press 2010.

(Erosheva 2005) employed a geometric approach to compare the potential value of using the Grade of Membership, latent class, and Rasch models in representing population heterogeneity for 2^J tables. Similarly, (Heiser 2004, De Rooij and Anderson 2007, De Rooij and Heiser 2005) have given geometric characterisations linked to odds ratios and related models for $I \times J$ tables, (Greenacre and Hastie 1987) focus on the geometric interpretation of correspondence analysis for contingency tables, (Carlini and Rapallo 2005) described some of the links to (Fienberg and Gilbert 1970) as well as the geometric structure of statistical models for case-control studies, and (Flach 2003) linked the geometry to Receiver Operating Characteristic space.

In this chapter we return to the original geometric representation of (Fienberg and Gilbert 1970) and link the geometry to some modern notions from algebraic geometry, e.g., as introduced to statistical audiences in (Diaconis and Sturmfels 1998) and (Pistone *et al.* 2001), to provide a variety of characterisations of the joint distribution of two binary variables, some old and some new. There are numerous ways we can characterise bivariate distributions, e.g., see (Arnold *et al.* 1999, Ramachandran and Lau 1991, Kagan *et al.* 1973). In related work, (Slavkovic and Sullivant 2006) give an algebraic characterisation of compatibility of full conditionals for discrete random variables. In this chapter, however, we are interested in the 'feasibility' question; that is, when do compatible conditionals and/or marginals correspond to an actual table. *Under the assumption that given sets of marginal and conditional binary distributions are compatible, we want to check whether or not they are sufficient to uniquely identify the existing joint distribution.* We are under the assumptions of the uniqueness theorem of (Gelman and Speed 1993) as redefined by (Arnold *et al.* 1999). More specifically, we allow cell entries to be zero as long as we do not condition on an event of zero probability. We draw on a more technical discussion in (Slavkovic 2004), and we note the related discussion in (Luo *et al.* 2004) and in (Carlini and Rapallo 2005).

3.2 Definitions and notation

Contingency tables are arrays of non-negative integers that arise from the cross-classification of a sample or a population of N objects based on a set of categorical variables of interest, see (Bishop *et al.* 1975) and (Lauritzen 1996). We represent the contingency table **n** as a vector of non-negative integers, each indicating the number of times a given configuration of classifying criteria has been observed in the sample. We also use the contingency table representation for probabilities **p** for the joint occurrence of the set of categorical variables.

We let X and Y be binary random variables and denote by n_{ij} the observed cell counts in a 2×2 table **n**. When we sum over a subscript we replace it by a '+'. Thus n_{i+} and n_{+j} denote the row and column totals, respectively, and these in turn sum to the grand total n_{++}. See the left-hand panel of Table 3.1. Similarly, we represent the *joint probability distribution* for X and Y as a 2×2 table of cell probabilities $\mathbf{p} = (p_{ij})$, where $p_{ij} = P(X = i, Y = j), i, j = 1, 2$, are non-negative and sum to one. See the right-hand panel of Table 3.1.

Table 3.1 *Notation for* 2 × 2 *tables: Sample point on the left and parameter value on the right.*

	Y_1	Y_2	Total		Y_1	Y_2	Total
X_1	n_{11}	n_{12}	n_{1+}	X_1	p_{11}	p_{12}	p_{1+}
X_2	n_{21}	n_{22}	n_{2+}	X_2	p_{21}	p_{22}	p_{2+}
Total	n_{+1}	n_{+2}	n_{++}	Total	p_{+1}	p_{+2}	1

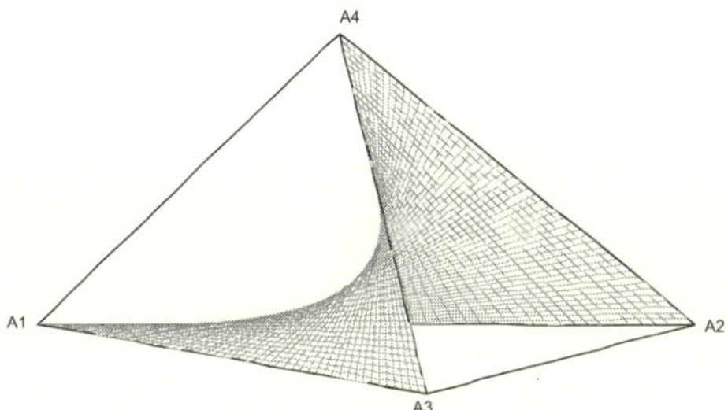

Fig. 3.1 Surface of independence for the 2 × 2 table. The tetrahedron represents the set of all probability distributions $\mathbf{p} = (p_{11}, p_{12}, p_{21}, p_{22})$ for the 2 × 2 tables, while the enclosed surface identifies the probability distributions satisfying the equation $p_{11}p_{22} = p_{12}p_{21}$, i.e., the toric variety for the model of independence.

Denote by $\mathbb{R}^4_p$ the four-dimensional real space with coordinates $\mathbf{p} = (p_{11}, p_{12}, p_{21}, p_{22})$. Geometrically, $\mathbf{p}$ is a point lying in a three-dimensional simplex (tetrahedron):

$$\mathbf{p} \in \Delta_3 = \{(p_{11}, p_{12}, p_{21}, p_{22}) : p_{ij} \geq 0, \sum_{i,j} p_{ij} = 1\}.$$

In barycentric coordinates, this tetrahedron of reference has vertices $A_1 = (1, 0, 0, 0)$, $A_2 = (0, 1, 0, 0)$, $A_3 = (0, 0, 1, 0)$, and $A_4 = (0, 0, 0, 1)$; see Figure 3.1. When the observed counts, $\mathbf{n} = \{n_{ij}\}$, come from a multinomial distribution, $Multi(N, \mathbf{p})$, we refer to Δ_3 as a *full parameter space*. If we consider a different parametrisation, the parameter space Θ parametrises a related surface.

The *marginal probability distributions* for X and Y are $\mathbf{p_X} = (p_{1+}, p_{2+}) = (s, 1 - s)$ and $\mathbf{p_Y} = (p_{+1}, p_{+2}) = (t, 1 - t)$. The lines $A_1 A_3$ and $A_2 A_4$ in the tetrahedron represent the set of all probability distributions, $\mathbf{p} = (s, 0, 1 - s, 0)$ and $\mathbf{p} = (0, s, 0, 1 - s)$ whose joint distributions are equivalent to the marginal distribution of $\mathbf{p_X} = (s, 1 - s)$. Similarly, the lines $A_1 A_2$ and $A_3 A_4$ represent the set of all probability distributions, $\mathbf{p} = (t, 1 - t, 0, 0)$ and $\mathbf{p} = (0, 0, t, 1 - t)$, whose joint distributions are equivalent to the marginal distribution of $\mathbf{p_Y} = (t, 1 - t)$.

We represent the *conditional probability distributions*, $\mathbf{p_{X|Y}}$ and $\mathbf{p_{Y|X}}$, by 2 × 2 *conditional probability matrices* $C = (c_{ij})$ and $R = (r_{ij})$, and denote by $\mathbb{R}^4_c$ and $\mathbb{R}^4_r$ the four-dimensional real spaces with coordinates $\mathbf{c} = (c_{11}, c_{12}, c_{21}, c_{22})$ and $\mathbf{r} = (r_{11}, r_{12}, r_{21}, r_{22})$, respectively. Given that we have observed $Y = j$, the conditional probability values are $c_{ij} = P(X = i|Y = j) = p_{ij}/p_{+j}$, such that $\sum_{i=1}^{2} c_{ij} = 1, j = 1, 2$, and

$$C = \begin{pmatrix} c_{11} & c_{12} \\ c_{21} & c_{22} \end{pmatrix}.$$

Given that we have observed $X = i$, the conditional probability values are $r_{ij} = P(Y = j|X = i) = p_{ij}/p_{i+}$ such that $\sum_{j=1}^{2} r_{ij} = 1, i = 1, 2$, and

$$R = \begin{pmatrix} r_{11} & r_{12} \\ r_{21} & r_{22} \end{pmatrix}.$$

Defined as such, the conditional probabilities can be considered as two-dimensional linear fractional transformations of either the cell counts or the cell probabilities. Recall that two-dimensional linear fractional transformations take the form $g(x, y) = (axy + cx + ey + g)/(bxy + dx + fy + h)$, e.g., $r_{11} = g(n_{11}, n_{12}) = n_{11}/(n_{11} + n_{12})$. The joint distribution $\mathbf{p}$ has the columns of C and rows of R as its conditional distributions. In the next section we provide a more careful geometric description of these conditionals.

We can now write the *odds ratio* or *cross-product ratio* for a 2×2 table

$$\alpha = \frac{p_{11}p_{22}}{p_{12}p_{21}} = \frac{c_{11}c_{22}}{c_{12}c_{21}} = \frac{r_{11}r_{22}}{r_{12}r_{21}}. \tag{3.1}$$

The odds ratio α is the fundamental quantity that measures the association in the 2×2 table whether we think in terms of probabilities that add to 1 across the entire table or conditional probabilities for rows, or conditional probabilities for columns. We can define two other odds ratios as follows:

$$\alpha^* = \frac{p_{11}p_{12}}{p_{22}p_{21}} = \frac{c_{11}c_{12}}{c_{22}c_{21}}, \tag{3.2}$$

$$\alpha^{**} = \frac{p_{11}p_{21}}{p_{12}p_{22}} = \frac{r_{11}r_{21}}{r_{12}r_{22}}. \tag{3.3}$$

Here α^* is characterised by the column conditionals and α^{**} by the row conditionals.

If we use the usual saturated log-linear model parametrization for the cell probabilities, e.g., see (Bishop *et al.* 1975) or (Fienberg 1980):

$$\log p_{ij} = u + u_{1(i)} + u_{2(j)} + u_{12(ij)}$$

where $\sum_{i=1}^{2} u_{1(i)} = \sum_{j=1}^{2} u_{2(j)} = \sum_{i=1}^{2} u_{12(ij)} = \sum_{j=1}^{2} u_{12(ij)} = 0$, then it turns out that $u_{1(1)} = \frac{1}{4}\log\alpha^*$, $u_{2(1)} = \frac{1}{4}\log\alpha^{**}$, and $u_{12(11)} = \frac{1}{4}\log\alpha$. Thus we can use the three odds ratios in Equations (3.1), (3.2), and (3.3) to completely characterise the standard saturated log-linear model, and thus the joint distribution $\mathbf{p}$.

3.3 Parameter surfaces and other loci for 2×2 tables

(Fienberg and Gilbert 1970) show that (a) the locus of all points corresponding to tables with independent margins is a hyperbolic paraboloid (Figure 3.1), (b) the locus of all points corresponding to tables with constant degree of association, α, is a hyperboloid of one sheet (Figure 3.2), and (c) the locus of all points corresponding to tables with fixed both margins is a line. Clearly, the other odds ratios in Equations (3.2) and (3.3) correspond to tables with constant column and row 'effects', respectively, and their surfaces are also hyperboloids of one sheet. All of these surfaces lie within the simplex Δ_3.

Fixing marginals implies imposing sets of linear constraints on the cell counts or the cell probabilities. We can fully specify log-linear models for the vector $\mathbf{p}$ of cell probabilities by a 0-1 *design matrix* $\mathbf{A}$, in the sense that, for each $\mathbf{p}$ in the model, $\log \mathbf{p}$ belongs to the row span of $\mathbf{A}$. The surface of independence, which geometrically represents the *independence model*, corresponds to the *Segre* variety in algebraic geometry (Figure 3.1). If we consider a knowledge of a single marginal, then the vector $\mathbf{p}$ is geometrically described by an intersection of a plane with the simplex, Δ_3. For example, fix the marginal $\mathbf{p_X}$. Then the plane, π_X, is defined by

$$A = \begin{pmatrix} 1 & 1 & 0 & 0 \\ 0 & 0 & 1 & 1 \end{pmatrix}, \; \mathbf{t} = \begin{pmatrix} s \\ 1-s \end{pmatrix}. \tag{3.4}$$

Similarly, we can define the plane π_Y for the fixed marginal $\mathbf{p_Y}$.

Now consider a set of linear constraints on the cell probabilities imposed by fixing conditional probabilities and clearing the denominators for the values from the matrix R (analogously from C). Then the vector $\mathbf{p}$ can be specified by a *constraint matrix* $\mathbf{A}$ and a vector $\mathbf{t}$ of the following form:

$$A = \begin{pmatrix} 1 & 1 & 1 & 1 \\ r_{12} & -r_{11} & 0 & 0 \\ 0 & 0 & r_{22} & -r_{21} \end{pmatrix}, \; \mathbf{t} = \begin{pmatrix} 1 \\ 0 \\ 0 \end{pmatrix}.$$

In the related sample space of integer-valued tables, the constraint matrix $\mathbf{A}$ can also be constructed by using the observed conditional frequencies, or relevant observed cell counts, but adding the parameter N for the sample size as follows:

$$A = \begin{pmatrix} 1 & 1 & 1 & 1 \\ n_{12} & -n_{11} & 0 & 0 \\ 0 & 0 & n_{22} & -n_{21} \end{pmatrix}, \; \mathbf{t} = \begin{pmatrix} N \\ 0 \\ 0 \end{pmatrix}.$$

Hence, any contingency table with fixed marginals and/or conditional probability values is a point in a convex polytope defined by a linear system of equations induced by observed marginals and conditionals. An affine algebraic variety is the common zero set of finitely many polynomials. Thus our problem of finding the loci of all possible tables given an arbitrary set of conditionals and marginals for 2×2 tables translates into an algebraic problem of studying zero sets in $\mathbb{R}^4_p$.

In the next section we derive the geometric description of the parameter space of $\mathbf{p}$ for fixed values of conditional probabilities defined by matrices C and R.

3.3.1 Space of tables for fixed conditional probabilities

Consider a system of linear equations for four unknowns, $p_{11}, p_{12}, p_{21}, p_{22}$, imposed by observing or fixing conditional probabilities defined by the matrix R.

Proposition 3.1 *The locus of probability distributions* $\mathbf{p}$ *for a* 2×2 *table satisfying a set of conditional probability distributions defined by* R *is a ruling of two surfaces of constant associations,* α *and* α^{**}.

Proof Let $f_{p,r} : \mathbb{R}_p^4 \setminus W \to \pi_r$ be the map given by $r_{ij} = p_{ij}/p_{i+}$, where W is a union of two varieties, $W = V(\langle p_{11}+p_{12}\rangle) \cup V(\langle p_{21}+p_{22}\rangle)$. Since $\sum_{j=1}^{2} p_{ij}/p_{i+} = 1$, $i = 1,2$, the image of f is contained in the plane $\pi_r \subset \mathbb{R}_r^4$ of equations $r_{11} + r_{12} = 1, r_{21} + r_{22} = 1$, and we can represent a point $\mathbf{r}$ in this plane by the coordinates $\mathbf{r} = (r_{11}, r_{22})$. Then the preimage of a point $\mathbf{r} \in \pi_r, f^{-1}(\mathbf{r})$, is the plane in $\mathbb{R}_p^4$ of equations $(1 - r_{11})p_{11} - r_{11}p_{12} = 0$ and $-r_{22}p_{21} - (1 - r_{22})p_{22} = 0$.

Since we are interested in $\mathbf{p}$, we restrict the function $f_{p,r}$ on the simplex Δ_3. The intersection $\Delta_3 \cap V(\langle p_{11} + p_{12}\rangle)$ is the face $\overline{12}$, that is the line $A_1 A_2$ consisting of the points of the form $\mathbf{p} = (s, 0, 1 - s, 0)$. Similarly, $\Delta_3 \cap V(\langle p_{21} + p_{22}\rangle)$ is the face $\overline{34}$ consisting of the points of the form $\mathbf{p} = (0, s, 0, 1 - s)$. With $\tilde{W} = \overline{12} \cup \overline{34}$, the map becomes $\tilde{f}_{p,r} : \Delta_3 \setminus \tilde{W} \to \pi_r$. Observe that the condition for the $\mathbf{p}$ to lie in $\Delta_3 \setminus \tilde{W}$ forces $0 \leq r_{11} \leq 1$ and $0 \leq r_{22} \leq 1$ such that $\tilde{f}_{p,r} : \Delta_3 \setminus (\tilde{W}) \to \Delta_1 \times \Delta_1$. Thus the preimage of a point $\mathbf{r} \in \pi_r, \tilde{f}^{-1}(\mathbf{r})$, is the segment in Δ_3 of equations

$$V_{\Delta_3} := \{(r_{11}s, (1 - r_{11})s, (1 - r_{22})(1 - s), r_{22}(1 - s)) : 0 < s < 1\}.$$

Finally take the closure of V for a given $\mathbf{r}$,

$$\overline{V}_{\Delta_3,r} := \{(r_{11}s, (1 - r_{11})s, (1 - r_{22})(1 - s), r_{22}(1 - s)) : 0 \leq s \leq 1, \text{fixed } \mathbf{r}\}, \quad (3.5)$$

and parametrise the probability variety by the probability of the margin s we condition upon. $\qquad \square$

By taking the closure of V we can understand what is happening with points $\mathbf{p}$ in the closure of the parameter space; that is, the points of $\tilde{W}$. If $s = 0$ we obtain a point $T^* = (0, 0, (1 - r_{22}), r_{22})$ on the line $A_3 A_4$, while if $s = 1$ we obtain a point $T = (r_{11}, 1 - r_{11}, 0, 0)$ on the line $A_1 A_2$. The point T^* is in the closure of the preimage of every point in $\Delta_1 \times \Delta_1$ of the form (t, r_{22}), $0 \leq t \leq 1$. As t varies, the preimage of (t, r_{22}), that is the segment TT^*, represents a ruling of the surface with different odds ratio; see Figure 3.2. All these rulings pass through the same point (t, r_{22}). Recall from Equations (3.1) and (3.3) that the conditional distributions from R define the association coefficients α and α^{**}. For a fixed value of $\mathbf{r}$-parameter, as we vary the values of s, the segment defined in Equation (3.5) belongs to a family of lines that determine the surface of constant association α, which we denote as S_α. They are also rulings for the surface of constant association defined by α^{**}, that is of $S_{\alpha^{**}}$.

In a similar way, we define the map $f_{p,c} : \mathbb{R}_p^4 \setminus W' \to \pi_c$ given by $c_{ij} = p_{ij}/p_{+i}$, where $W' = V(\langle p_{11} + p_{21}\rangle) \cup V(\langle p_{12} + p_{22}\rangle)$ and π_c the plane $\pi_c \subset \mathbb{R}_c^4$ of equations $c_{11} + c_{21} = 1, c_{12} + c_{22} = 1$. The segment with coordinates

$$\overline{V}_{\Delta_3,c} = \{(c_{11}t, (1 - c_{22})(1 - t), (1 - c_{11})t, c_{22}(1 - t)) : 0 \leq t \leq 1, \text{fixed } \mathbf{c}\}, \quad (3.6)$$

represents an equivalence class with fixed value of the matrix C that is the $\mathbf{c}$-parameter. Thus the lines SS^* are the second set of rulings for the *surface of*

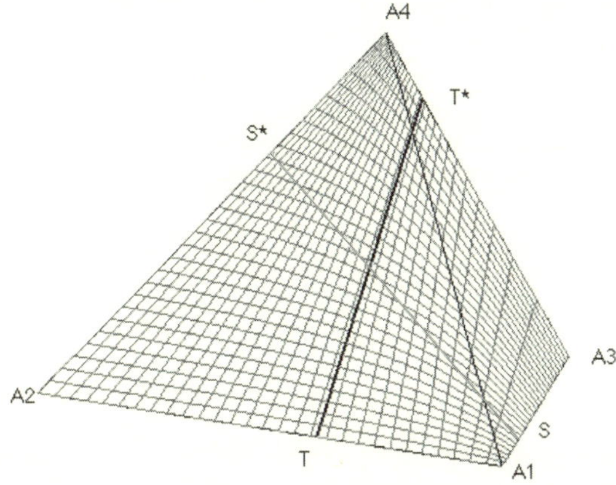

Fig. 3.2 Surface of constant association $\alpha = 6$. The line SS^* represents all probability distributions $\mathbf{p} = (p_{11}, p_{12}, p_{21}, p_{22})$ satisfying fixed **c**-conditional parameter. The line TT^* represent all probability distributions $\mathbf{p} = (p_{11}, p_{12}, p_{21}, p_{22})$ satisfying fixed **r**-conditional parameter.

constant association, α, and also rulings for the surface of association defined by α^*.

If X and Y are independent, then $\mathbf{p_{Y|X}} = \mathbf{p_Y}$ and $\mathbf{p_{X|Y}} = \mathbf{p_X}$. Thus, we confirm the result of (Fienberg and Gilbert 1970), who state that for *surface of independence* ($\alpha = 1$, see Figure 3.1), the rulings are two families of straight lines corresponding to constant column and row margins.

In the following sections we use the above described measures and their geometry, and consider the geometric interpretation of the Uniqueness Theorem, see (Gelman and Speed 1993, Arnold *et al.* 1996, Arnold *et al.* 1999), and complete specification of joint distribution via log-linear models. A geometric interpretation of incomplete specification of the joint distribution $\mathbf{p}$ is also considered.

3.4 Complete specification of the joint distribution

When we examine observed 2×2 tables, our statistical goal is usually to make inferences about the joint distribution of the underlying categorical variables, e.g., finding estimates of and models for $\mathbf{p}$. In this section, we discuss possible complete specifications of the joint distribution and give their geometric interpretations. In Section 3.5, we turn to incomplete specifications, i.e., reduced models.

3.4.1 Specification I

From the definition of conditional probability, we know that the joint distribution for any 2×2 table is uniquely identified by one marginal and the related conditional:

$$P(X, Y) = P(X)P(Y|X) = P(Y)P(X|Y),$$

or equivalently $p_{ij} = p_{i+} r_{ij} = p_{j+} c_{ij}$.

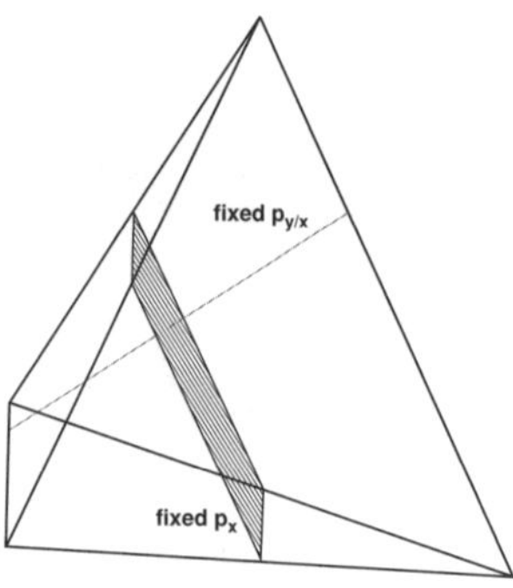

Fig. 3.3 Specification I. The intersection of the simplex Δ_3, the line for fixed $\mathbf{r}$, and the plane π_X, is a fully specified joint distribution $\mathbf{p}$.

We can use the geometric representations in Section 3.3 to demonstrate this uniqueness. For example, consider the locus of points $\mathbf{p}$ for fixed $\mathbf{r}$ as described by $\overline{V}_{\Delta_3,r}$ in Equation (3.5); see the line segment in Figure 3.3. The other locus of points $\mathbf{p}$ is a plane π_X defined by (3.4) observing a specific value of s corresponding to p_{1+}. The intersection of Δ_3 with these two varieties is a unique point representing the joint distribution $\mathbf{p}$. This is a geometric description of the basic factorisation theorem in statistics.

3.4.2 Specification II

The joint distribution for a 2×2 table is also fully specified by knowing two sets of conditionals: $\mathbf{p}_{X|Y}$ and $\mathbf{p}_{Y|X}$, equivalent to Specification I under independence of X and Y. Note that this is the simplest version of the Hammersley–Clifford theorem, see (Besag 1974).

Its geometric representation is the intersection of lines representing $\mathbf{p}$ for fixed $\mathbf{p}_{Y|X}$ and $\mathbf{p}_{X|Y}$ (Figure 3.2). It is an intersection of two varieties defined by Equations (3.5) and (3.6), $\overline{V}_{\Delta_3,r} \cap \overline{V}_{\Delta_3,c}$. Specifically, it is a point on the surface of the constant association, α, identifying the unique table given these conditional distributions.

Lemma 3.1 *The specification of joint distribution* $\mathbf{p}$ *by two sets of conditional parameters,* $\mathbf{r}$ *and* $\mathbf{c}$, *is equivalent to its specification by a saturated log-linear model.*

Proof Based on Proposition 3.1, each conditional includes full information on two out of three odds ratios; $\mathbf{r}$ has full information on α and α^{**}, while $\mathbf{c}$ has information on α and α^*. As seen at the end of Section 3.2 all three odds ratios together represent the key parameters of the saturated log-linear model and thus they fully characterise the joint distribution for a 2×2 table. $\square$

This specification is clearly implicit in many treatments of log-linear models and 2×2 tables, e.g., see (Fienberg 1980), but to our knowledge has never been made explicit. We discuss further related specifications with odds ratios in Section 1.4.4.

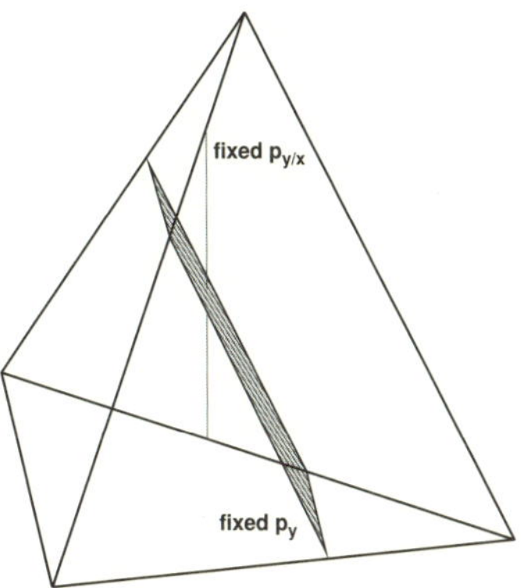

Fig. 3.4 Specification III. The intersection of the simplex Δ_3 with the line segment and the plane is a fully specified joint distribution **p**.

3.4.3 Specification III

(Arnold *et al.* 1996, Arnold *et al.* 1999) show that sometimes a conditional and the 'wrong' marginal (e.g., $\mathbf{p_{Y|X}}$ and $\mathbf{p_Y}$) also uniquely identify the joint distribution, provided Arnold's *positivity condition*. Here the geometric representation of **p** lies in the intersection of simplex Δ_3 with $\overline{V}_{\Delta_3, r}$, see Equation (3.5) and Figure 3.4, and the plane π_Y, see Section 3.3. For 2×2 tables, this result *always* holds and states that for two *dependent* binary random variables, X and Y, either the collection $\{\mathbf{p_{X|Y}}, \mathbf{p_X}\}$ or $\{\mathbf{p_{Y|X}}, \mathbf{p_Y}\}$ uniquely identifies the joint distribution.

If the matrix $\mathbf{p} = (p_{ij})$ has rank 1, X and Y are independent and this implies that common odds ratio $\alpha = 1$. Since conditional distributions also preserve α, this implies that the ranks of matrices $C = (c_{ij})$ and $R = (r_{ij})$ are also both 1. Thus any rank greater than 1 implies a dependence between X and Y. Specifically for 2×2 tables, when the conditional matrices have full rank, X and Y are dependent random variables. We redefine the result on the uniqueness of the joint distribution.

Proposition 3.2 *For two binary discrete random variables, X and Y, either collection* $\{\mathbf{p_{X|Y}}, \mathbf{p_X}\}$ *or* $\{\mathbf{p_{Y|X}}, \mathbf{p_Y}\}$ *uniquely identifies the joint distribution if the conditional matrices* $C = (c_{ij})$ *and* $R = (r_{ij})$ *have full rank.*

Proof Consider $\mathbf{p_X} = (p_{1+}, p_{2+}) = (s, 1 - s)$ and $\mathbf{p_{X|Y}} = (c_{11} = p_{11}/p_{+1}, c_{21} = p_{21}/p_{+1}, c_{12} = p_{12}/p_{+2}, c_{22} = p_{22}/p_{+2})$. Recall that we are assuming that there exists a joint probability distribution **p** from which $\mathbf{p_{X|Y}}$ and $\mathbf{p_X}$ are derived, and thus they are compatible. Imposing $p_{ij} \in [0, 1]$ requires that either $0 \leq c_{11} \leq s \leq c_{12} \leq 1$ or $0 \leq c_{12} \leq s \leq c_{11}$. If the conditional matrix C has a full rank there are two linearly independent equations from observing $\mathbf{p_{X|Y}}$ that describe relationships on the cell probabilities (p_{ij}). If C has a full rank this implies that the marginal array $\mathbf{p_X}$ also has a full rank, and there are two additional linearly independent constraints describing relationships among the (p_{ij}).

Consider the ideal I generated by the four polynomials obtained after clearing the denominators in the ratios defining relationships between the conditionals c_{ij}'s

Table 3.2 *Representation of the joint distribution* $\mathbf{p}$ *as a function of the* $\mathbf{p_X} = (s, 1-s)$ *and the conditional* $\mathbf{p_{X|Y}} = (c_{11}, c_{12}, c_{21}, c_{22})$.

	Y_1	Y_2
X_1	$\dfrac{c_{11}(c_{12}-s)}{c_{12}-c_{11}}$	$\dfrac{-c_{12}(c_{11}-s)}{c_{12}-c_{11}}$
X_2	$\dfrac{c_{12}+sc_{11}-s-c_{11}c_{12}}{c_{12}-c_{11}}$	$\dfrac{(c_{11}-s)((c_{12}-1)}{c_{12}-c_{11}}$

and cell probabilities p_{ij}'s, namely $p_{11} + p_{12} - s, p_{21} + p_{22} - 1 + s, (1 - c_{11})p_{11} - c_{11}p_{21}, c_{12}p_{22} - (1 - c_{12})p_{12}$. Then a Gröbner basis of I using lexicographic order is $\{p_{21}+p_{22}+s-1, p_{11}+p_{12}-s, p_{12}c_{12}+p_{22}c_{12}-p_{12}, p_{12}c_{11}+p_{22}c_{11}-p_{12}+sc_{11}, p_{22}c_{11}- p_{22}c_{12} - sc_{12} + c_{11}c_{12} + s - c_{11}\}$. Set these polynomials equal to zero. Then, (1) if $c_{11} \neq c_{12}$, matrix C has a full rank, and the equivalent unique solution is given in Table 3.2; and (2) if $c_{11} = c_{12}$, then $c_{11} = 1$ or $c_{11} = s$. When $c_{11} = c_{12} = s$, we have independence of X and Y. However, if $c_{11} = c_{12} = s = 1$ then $\mathbf{p}$ is not identifiable. In this case the matrix C does not have a full rank and conditions of the proposition are not satisfied. Furthermore, $\mathbf{p} = \mathbf{p_Y}$ and solutions would lie on the face $A_1 A_2$ or $A_3 A_4$ of the simplex Δ_3 (see Figure 3.1). $\square$

(Slavkovic 2004) derived a result similar to that in Theorem 4.2. but for $I \times 2$ tables. This characterisation is far more subtle than the previous two and we have not found it in any other setting.

3.4.4 Odds-ratio specification

In Section 3.2 we showed that all three odds ratios, α, α^*, and α^{**} together represent the key parameters of the saturated log-linear model: $\log p_{ij} = u + u_{1(i)} + u_{2(j)} + u_{12(ij)}$. That is $u_{12(11)} = \frac{1}{4}\log\alpha$, $u_{1(1)} = \frac{1}{4}\log\alpha^*$, and $u_{2(1)} = \frac{1}{4}\log\alpha^{**}$, and thus they too specify the joint distribution for 2×2 tables. If we add a representation for the 'constant' term, i.e., $u = \frac{1}{4}\log(p_{11}p_{12}p_{21}p_{22})$, then the implicit representation of the joint distribution is defined by simultaneously solving the equations from

$$V_{\Delta_3} = (p_{11}p_{22} - \alpha p_{12}p_{21}, p_{11}p_{12} - \alpha^* p_{21}p_{22}, p_{11}p_{21} - \alpha^{**} p_{12}p_{22}). \tag{3.7}$$

Let $r_1 = p_{11}/p_{12} = r_{11}/r_{12}$ and $r_2 = p_{21}/p_{22} = r_{21}/r_{22}$ be the row odds. The column odds are $c_1 = p_{11}/p_{21} = c_{11}/c_{21}$ and $c_2 = p_{12}/p_{22} = c_{12}/c_{22}$. (Kadane *et al.* 1999) gave an alternative parametrisation to the one given by Equation (3.7), and showed in the context of capture–recapture type problems that it is sufficient to have α and the odds, r_1 and c_1 to identify the joint distribution. In this setting, r_1 are the odds of a unit being counted twice given that it was counted in the first sample, and c_1 is the odds of a unit being counted twice given that the same unit was counted in the second sample.

Geometrically, the intersection of the probability simplex, Δ_3, with two surfaces of constant associations is a line segment that would be defined by a fixed set of

Table 3.3 *Representation of the joint distribution* $\mathbf{p}$ *as a function of the margins* $\mathbf{p_X} = (s, 1 - s)$ *and* $\mathbf{p_Y} = (t, 1 - t)$, *and the odds ratios,* α, α^* *and* α^{**}.

	Y_1	Y_2
X_1	$\dfrac{\sqrt{\alpha\alpha^{**}}}{1+\sqrt{\alpha\alpha^{**}}}s = \dfrac{\sqrt{\alpha\alpha^{*}}}{1+\sqrt{\alpha\alpha^{*}}}t$	$\dfrac{1}{1+\sqrt{\alpha\alpha^{**}}}s = \dfrac{\alpha^{*}}{\alpha^{*}+\sqrt{\alpha\alpha^{*}}}(1-t)$
X_2	$\dfrac{\alpha^{**}}{\alpha^{**}+\sqrt{\alpha\alpha^{**}}}(1-s) = \dfrac{1}{1+\sqrt{\alpha\alpha^{*}}}t$	$\dfrac{\sqrt{\alpha\alpha^{**}}}{\alpha^{**}+\sqrt{\alpha\alpha^{**}}}(1-s) = \dfrac{\sqrt{\alpha\alpha^{*}}}{\alpha^{*}+\sqrt{\alpha\alpha^{*}}}(-t)$

conditional probabilities as we saw in Section 3.3.1. This line is one of the rulings for each of the respective hyperbolic surfaces for joint distributions $\mathbf{p}$ with constant associations. The observation naturally leads to an equivalence statement about Specification I and the following two sets of parameters: (1) $\{\mathbf{p_X}, \alpha, \alpha^{**}\}$ and (2) $\{\mathbf{p_Y}, \alpha, \alpha^{*}\}$. Let $\{\mathbf{p_X}, \mathbf{p_{Y|X}}\}$ and $\{\mathbf{p_Y}, \mathbf{p_{X|Y}}\}$ uniquely identify the joint distribution $\mathbf{p}$. Then the following lemma holds:

Lemma 3.2 *For a 2×2 table, the specification of $\mathbf{p}$ by $\{\mathbf{p_X}, \mathbf{p_{Y|X}}\}$ is equivalent to characterisation by $\{\mathbf{p_X}, \alpha, \alpha^{**}\}$, and $\{\mathbf{p_Y}, \mathbf{p_{X|Y}}\}$ is equivalent to characterisation by $\{\mathbf{p_Y}, \alpha, \alpha^{*}\}$.*

Proof The two odds ratios will completely specify the missing conditional distributions on the probability simplex (cf. Section 3.4), and thus completely specify the joint distribution. Consider the two ideals generated by

$$p_{11} + p_{12} - s, p_{21} + p_{22} - 1 + s, p_{11}p_{22} - \alpha p_{12}p_{21}, p_{11}p_{12} - \alpha^{*}p_{21}p_{22}$$

and

$$p_{11} + p_{21} - t, p_{12} + p_{22} - 1 + t, p_{11}p_{22} - \alpha p_{12}p_{21}, p_{11}p_{21} - \alpha^{**}p_{12}p_{22}.$$

Finding the Gröbner basis, and setting the defining polynomials equal to zero results in the solution in Table 3.3. More specifically, the probabilities $p_{ij} = g(\alpha, \alpha^{**})\mathbf{p_X} = h(\alpha, \alpha^{*})\mathbf{p_Y}$ where g, and h are functions of the three odds ratios given in Table 3.3. □

If $\alpha = 1$, $\mathbf{p} = \{\frac{\sqrt{\alpha^{**}}}{1+\sqrt{\alpha^{**}}}s, \frac{1}{1+\sqrt{\alpha^{**}}}s, \frac{\alpha^{**}}{\alpha^{**}+\sqrt{\alpha^{**}}}(1-s), \frac{\sqrt{\alpha^{**}}}{\alpha^{**}+\sqrt{\alpha^{**}}}(1-s)\}$. Clearly $\mathbf{p_{X|Y}} = \mathbf{p_X}$, and $\mathbf{p_Y} = \{\frac{\sqrt{\alpha^{**}}}{1+\sqrt{\alpha^{**}}}, \frac{1}{1+\sqrt{\alpha^{**}}}\}$ and we have independence of X and Y. If $\alpha = \alpha^{**} = 1$ then the joint distribution $\mathbf{p}$ is identified as $\{\frac{1}{2}s, \frac{1}{2}s, \frac{1}{2}(1-s), \frac{1}{2}(1-s)\}$. Notice that if $s = 1$ then $c_{11} = c_{12} = s = 1$ and $\mathbf{p}$ is not identifiable. Furthermore, $\mathbf{p} = \mathbf{p_Y}$ and potential solutions would lie on the face A_1A_2 or A_3A_4 of the simplex Δ_3. Similar considerations can be made for t, α, and α^{*}.

This specification is related to the parametrisation given by (Kadane *et al.* 1999). Then the following sets of parameters will also uniquely identify the joint distribution: (3) $\{\mathbf{p_X}, \alpha, r_1\}$ and (4) $\{\mathbf{p_Y}, \alpha, c_1\}$. These characterisations are different from any previously described in the literature and may be of special interest to those attempting to elicit joint distributions via components in a Bayesian context.

3.4.5 Specification via the non-central hypergeometric distribution

Finally we point out a well-established fact in statistical literature that both sets of one-way marginals, $\mathbf{p_X}$ and $\mathbf{p_Y}$, and the odds-ratio, α give a complete specification of the joint probability distribution $\mathbf{p}$ via the non-central hypergeometric distribution. Within Δ_3, as shown in (Fienberg and Gilbert 1970), the locus of joint probability distributions $\mathbf{p}$ given $\{\mathbf{p_X}, \mathbf{p_Y}\}$ is a line segment. This line segment intersects the hyperboloid specified by α in a unique point $V_{\Delta_3,s,t,\alpha}$ with coordinates

$$\left\{ \left(st, s(1-t), \frac{(1-s)t}{\alpha(1-t)+t}, \frac{\alpha(1-s)(1-t)}{\alpha(1-t)+t} \right) : \text{fixed } s,t,\alpha \right\}.$$

3.5 Incomplete specification of the joint distribution

Statistical models come from restricting values of one or more parameters and focusing on subspaces. A natural question arises as to the specification of the joint distribution if one of the parameters from the complete specification is set to zero or missing. For example, setting $\alpha = 1$ in Equation (3.7) defines the model of independence which corresponds to a hyperbolic paraboloid surface and the *Segre variety* in Figure 3.1.

3.5.1 Space of tables for a fixed marginal and odds-ratio

As noted in Section 3.4.5, both sets of one-way marginals and the odds-ratio, $\{\mathbf{p_X}, \mathbf{p_Y}, \alpha\}$ give a complete specification of $\mathbf{p}$ via the non-central hypergeometric distribution. In this section we consider the specification if one of the margins is missing.

Partial specification of the joint probability distribution $\mathbf{p}$ based solely on one odds-ratio, e.g., α, is an intersection of a hyperbolic surface with the probability simplex Δ_3, see (Fienberg and Gilbert 1970); knowledge of odds-ratio also specifies the locus of conditional distributions (see Section 1.5.2). Partial specification via one margin and α yields points lying on the intersection of a hyperbola and the probability simplex Δ_3:

$$V_{\Delta_3,s,\alpha} =$$
$$\left\{ \left(st, s(1-t), \frac{(1-s)t}{\alpha(1-t)+t}, \frac{\alpha(1-s)(1-t)}{\alpha(1-t)+t} \right) : 0 \leq t \leq 1, \text{fixed } s,\alpha \right\} \quad (3.8)$$

as shown in Figure 3.5. This is a *rational parametric representation* requiring that $\alpha(1-t)+t \neq 0$ and it implies not conditioning on the event of probability zero.

3.5.2 Space of conditional tables

Proposition 3.3 *The locus of conditional distributions* $\mathbf{r}$ *or* $\mathbf{c}$, *given a fixed odds-ratio lies in the intersection of a quadric with the plane* π_r *or* π_c, *respectively.*

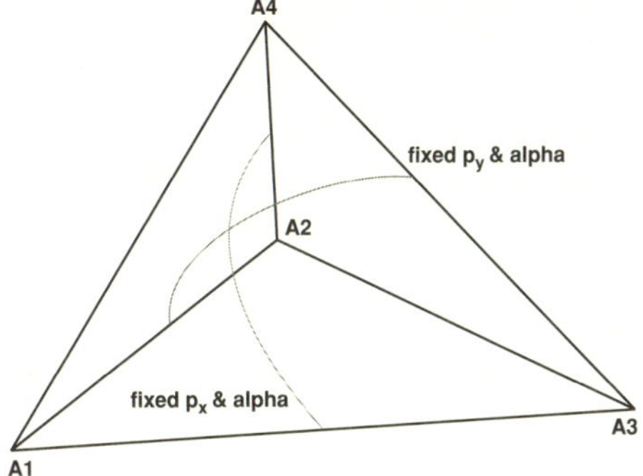

Fig. 3.5 Incomplete specification of the joint distribution **p** is given by the intersection of the simplex Δ_3 with the curve defined by one marginal and odds-ratio.

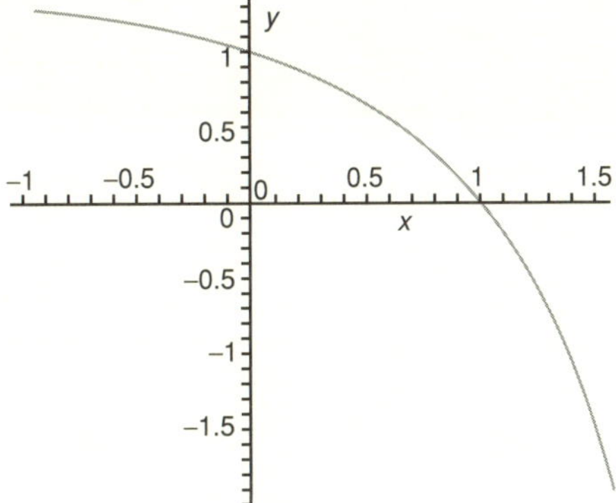

Fig. 3.6 Specification of the conditional distribution $\mathbf{p}_{Y|X}$ lies in the intersection of a quadric and π_r.

We treat the case of α and $\mathbf{r}$ and $\mathbf{c}$, but the α^{**} or α^* with either $\mathbf{r}$ or $\mathbf{c}$ would work in a similar way.

Proof Fix the odds-ratio α. Recall that the joint probabilities **p** satisfying the odds-ratio lie on the intersection of the hyperbolic surface S_α and Δ_3 where $S_\alpha :=$ $V(\langle p_{11}p_{22} - \alpha p_{12}p_{21} \rangle)$ and $\alpha = p_{11}p_{22}/p_{12}p_{21} = r_{11}r_{22}/r_{12}r_{21}$. Restrict our attention on the plane π_X. A bijection $\tilde{f}_{\pi_X} : \pi_X \to \pi_r$ given by

$$\begin{pmatrix} r_{11} \\ r_{22} \end{pmatrix} = \begin{pmatrix} \frac{1}{s} & 0 \\ 0 & \frac{1}{1-s} \end{pmatrix} \begin{pmatrix} p_{11} \\ p_{22} \end{pmatrix}$$

is the restriction of $\tilde{f}$ to the plane π_X. The image of surface S_α under the map $\tilde{f}$ is the curve

$$C_{r,\alpha} := V(\langle \alpha(1 - r_{11})(1 - r_{22}) - r_{11}r_{22} \rangle)$$

which is clearly the intersection of a quadric with the plane π_r. Similar derivation can be done for the intersection of a quadric and the plane π_c defined by the equation $\alpha(1 - c_{11})(1 - c_{22}) = c_{11}c_{22}$. $\qquad\square$

Once we fix a plane π_X, the curve $C_{r,\alpha}$ is in the bijection with the curve $S_\alpha \cap \pi_X$. Note that this bijection exists only when you fixed a specific plane π_X which is needed to define a conditional distribution. In fact, a point $\mathbf{r}$ on the curve $C_{r,\alpha}$ has as preimage the segment $\overline{V}_{\Delta_3,r}$ defined by Equation (3.5). Once we fix a plane π_X, the preimage of $\mathbf{r}$ is exactly the point determined by the intersection $\overline{V}_{\Delta_3,r} \cap \pi_X$. If we fix another plane π'_X, the preimage of $\mathbf{r}$ will be another point in $\overline{V}_{\Delta_3,r}$ but given by the intersection $\overline{V}_{\Delta_3,r} \cap \pi'_X$. This corresponds with the fact that, given a conditional distribution $\mathbf{p_{Y|X}}$ (i.e., a point $\mathbf{r}$) and a marginal $\mathbf{p_X}$ (i.e., a plane π_X) the probabilities of $\mathbf{p}$ are uniquely determined (the point in the intersection $\overline{V}_{\Delta_3,r} \cap \pi_X$).

From the above we directly derived the corresponding probability variety given in Equation (3.8).

3.5.3 Margins

If we are given the row and column totals, then the well-known Fréchet bounds for the individual cell counts are:

$$\min(n_{i+}, n_{+j}) \geq n_{ij} \geq \max(n_{i+} + n_{+j} - n, 0) \text{ for } i = 1, 2, j = 1, 2.$$

The extra lower bound component comes from the upper bounds on the cells complementary to (i, j). These bounds have been widely exploited in the disclosure limitation literature and have served as the basis for the development of statistical theory on copulas (Nelsen 2006). The link to statistical theory comes from recognizing that the minimum component $n_{i+} + n_{+j} - n$ corresponds to the MLE of the expected cell value under independence, $n_{i+}n_{+j}/n$. For further details see (Dobra 2001, Dobra 2003) and Chapter 8 in this volume.

Geometric interpretation corresponds to fixing $\mathbf{p_X}$ and $\mathbf{p_Y}$, that is restricting the parameter space to the intersection of Δ_3 with π_X and π_Y, respectively (see Section 1.3). The points $\mathbf{p}$ then lie in intersection of Δ_3 with the segment $\pi_X \cap \pi_Y$ given by $C_{s,t} := V(\langle p_{11} - p_{22} - (s + t - 1)\rangle)$.

3.5.4 Two odds-ratios

In this section we address the question of specification of the joint probability distribution $\mathbf{p}$ when we have two odds ratios, e.g. α and α^*. This is the case when we are missing the marginal from the log-linear model specification, e.g., non-hierarchical log-linear model. We treat the case with α and α^{**}, but α^* would work in a similar way. This characterisation is related to the specifications of $\mathbf{p}$ discussed in Section 1.4.4, and results in Table 1.2. (Carlini and Rapallo 2005) describe an analogous question but with application to case-control studies.

Lemma 3.3 *The points* $\mathbf{p}$ *with given* α *and* α^{**} *lie in the intersection of* Δ_3 *with the line segment defined by*

$$V_{\alpha,\alpha^{**}} :=$$

$$\left\{ \frac{s\sqrt{\alpha\alpha^{**}}}{\sqrt{\alpha\alpha^{**}}+1}, \frac{s}{\sqrt{\alpha\alpha^{**}}+1}, \frac{\sqrt{\alpha^{**}}(1-s)}{\sqrt{\alpha_1}+\sqrt{\alpha^{**}}}, \frac{\sqrt{\alpha}(1-s)}{\sqrt{\alpha}+\sqrt{\alpha^{**}}} \Big| 0 < s < 1 \right\}. \quad (3.9)$$

We first note that the partial specification based solely on two odds ratios uniquely specifies the missing conditional. We used this result in the proof of Lemma 2 in Section 1.4.4.

Proof The points in the plane π_r with the given odds ratio lie on two curves, $C_{r,\alpha} := V(\langle \alpha(1-r_{11})(1-r_{22}) - r_{11}r_{22}\rangle)$ and $C_{r,\alpha^{**}} := V(\langle \alpha^{**}(1-r_{11})r_{22} - r_{11}(1-r_{22})\rangle)$ (see Section 1.5.2), whose intersection, $C_{r,\alpha} \cap C_{r,\alpha^{**}}$, consists of two points:

$$r_{11} = \frac{\sqrt{\alpha\alpha^{**}}}{1+\sqrt{\alpha\alpha^{**}}} \qquad r_{12} = \frac{1}{1+\sqrt{\alpha\alpha^{**}}}$$
$$r_{21} = \frac{\sqrt{\alpha^{**}}}{\sqrt{\alpha}+\sqrt{\alpha^{**}}} \qquad r_{22} = \frac{\sqrt{\alpha}}{\sqrt{\alpha}+\sqrt{\alpha^{**}}}$$

or

$$r_{11} = \frac{\sqrt{\alpha\alpha^{**}}}{-1+\sqrt{\alpha\alpha^{**}}} \qquad r_{12} = -\frac{1}{-1+\sqrt{\alpha\alpha^{**}}}$$
$$r_{21} = -\frac{\sqrt{\alpha^{**}}}{\sqrt{\alpha}-\sqrt{\alpha^{**}}} \qquad r_{22} = \frac{\sqrt{\alpha}}{\sqrt{\alpha}-\sqrt{\alpha^{**}}}$$

The second point does not represent conditional probabilities since it has two negative coordinates. The preimage of the other point is the segment given by Equation (3.9) which consists of points $\mathbf{p}$ in the intersection of the surfaces (in Δ_3) $S_\alpha := V(\langle p_{11}p_{22} - \alpha p_{12}p_{21}\rangle)$ and $S_{\alpha^{**}} := V(\langle p_{11}p_{21} - \alpha^{**}p_{12}p_{22}\rangle)$; that is, points $\mathbf{p}$ with given odds ratios α and α^{**}. The set $V_{\alpha,\alpha^{**}}$ corresponds to points on a ruling for each surface S_i. $\qquad\square$

These line segments are the rulings discussed in Section 3.3.1, and thus describe the equivalent segments as when we fix the conditional, in this case, the $\mathbf{r}$-conditional (see Figure 3.2).

3.6 Extensions and discussion

The geometric representation described in Section 1.3.1 about the space of tables given fixed conditionals extend to $I \times J$ tables via linear manifolds. The specification results on $\mathbf{p}$ also generalise, in part (e.g., using $\mathbf{p}_{Y|X}$ and $\mathbf{p}_X$), but when we are given margins we need to define multiple odds ratios. The bounds are also directly applicable to $I \times J$ tables and essentially a related argument can be used to derive exact sharp bounds for multi-way tables whenever the marginal totals that are fixed correspond to the minimal sufficient statistics of a log-linear model that is *decomposable.*

The natural extension to k-way tables is via log-linear models and understanding the specifications via fixed margins and combinations of margins and odds ratios, and ratios of odds ratios. For $I \times J \times K$ tables, we use a triple subscript notation and we model the logarithms of the cell probabilities as

$$\log(p_{ijk}) = u + u_{1(i)} + u_{2(j)} + u_{3(k)} + u_{12(ij)} + u_{13(ik)} + u_{23(jk)} + u_{123(ijk)} \quad (3.10)$$

where we set the summation of a u-term over any subscript equal to 0 for identification. There is a one-to-one correspondence between the u terms and odds ratio. For example, for $2 \times 2 \times 2$ tables, we can rewrite the parameters as a function of the logarithm of the cell probabilities

$$u_{123(111)} = \frac{1}{8} \log \left(\frac{\alpha^{(1)}}{\alpha^{(2)}} \right) \quad (3.11)$$

where $\alpha^{(k)} = p_{11k}p_{22k}/p_{12k}p_{21k}$. See (Bishop *et al.* 1975, Chapter 2) for further details. The toric variety corresponding to the model of no second-order interaction, i.e., $u_{123(ijk)} = 0$ for $i, j, k = 1, 2$, is a hyper-surface with three sets of generators corresponding to the first-order interactions, $p_{11k}p_{22k} - \alpha^{(k)}p_{12k}p_{21k}$, $p_{1j1}p_{2j2} - \alpha^{(j)}p_{1j2}p_{2j1}$, $p_{i11}p_{i22} - \alpha^{(i)}p_{i12}p_{i21}$, such that $\alpha^{(i=1)} = \alpha^{(i=2)}$, $\alpha^{(j=1)} = \alpha^{(j=2)}$, $\alpha^{(k=1)} = \alpha^{(k=2)}$. Each of the other subscripted u-terms in the log-linear model of Equation (3.10) can also be represented in terms of a ratio of odds ratios of the form of Equation (3.11).

3.6.1 Simpson's paradox

For three events A, B, and C, (Simpson 1951) observed that it was possible that $P(A|B) < P(A|\bar{B})$ (where $\bar{B}$ is the complementary set of B) but that $P(A|BC) > P(A|\bar{B}C)$ and $P(A|B\bar{C}) > P(A|\bar{B}\bar{C})$. This became known as Simpson's paradox although (Yule 1903) had made a similar observation 50 years earlier. For an extensive discussion of related aggregation phenomena, see (Good and Mittal 1987) and for an early geometrical treatment see (Shapiro 1982). As many authors have observed, another way to think about Simpson's paradox is as the reversal of the direction of an association when data from several groups are combined to form a single group. Thus for a $2 \times 2 \times 2$ table we are looking at three sets of 2×2 tables, one for each level of the third variable and another for the marginal table, and we can display all three within the same simplex Δ_3.

Consider the model of complete independence for a $2 \times 2 \times 2$ table:

$$\log p_{ijk} = u + u_{1(i)} + u_{2(j)} + u_{3(k)}$$

where $u_{12(ij)} = u_{13(ik)} = u_{23(jk)} = u_{123(ijk)} = 0$, for $i, j, k = 1, 2$ that is the corresponding odds ratios and ratios of odds ratios are all equal to 1. Now consider the marginal 2×2 table with vector of probabilities $\mathbf{p} = (p_{ij+})$. The complete independence model implies marginal independence, i.e., $\log p_{ij+} = v + v_{1(i)} + v_{2(j)}$, so that the marginal odds ratios $\alpha_{12}=1$, and $\mathbf{p}$ would be a point on the surface of independence.

Next suppose that variables 1 and 2 are conditionally independent given 3, i.e., $\log p_{ijk} = u + u_{1(i)} + u_{2(j)} + u_{3(k)} + u_{13(ik)} + u_{23(jk)}$. The marginal odds ratio $\alpha_{12} \neq 1$, but the two conditional odds ratios for each level of the third variable equal one,

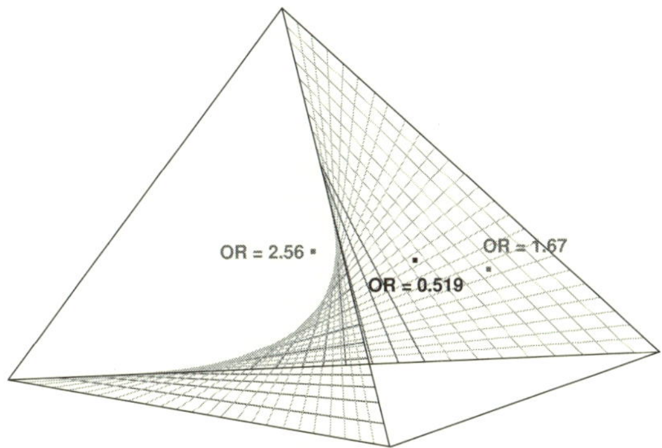

Fig. 3.7 An example of Simpson's paradox. Two dots with odds ratios (OR) > 1 are conditional 2 × 2 tables and on the same side of surface of independence. The **p** with odds-ratio (OR) < 1 is the marginal 2 × 2 table.

that is $\alpha_{12(3)} = 1$, and $\mathbf{p_{12|3}}$ would be two points on the surface of independence. When we connect such two points on the surface of independence, the line segment corresponds to tables with either positive association or negative association. This is the boundary for the occurrence of Simpson's paradox.

Simpson's paradox occurs when we have two tables corresponding to points lying on one side of the surface of independence, but the line segment connecting them cuts the surface and includes points on the 'other side'. Figure 3.7 gives one such example. If we put a probability measure over the simplex, we could begin to discuss 'the probability of the occurrence of Simpson's paradox,' cf. (Hadjicostas 1998).

When we connect two points lying on the surface of independence then we are combining two *different* independence models and the line connecting them will either consists of all weighted combinations of the two tables, or in the sense described above, all possible marginal tables. These will either all have values of $\alpha > 1$ or values of $\alpha < 1$ unless the two original tables being combined share either row or column margins, in which case $\alpha = 1$. The locus of all possible such lines corresponds to the $k = 2$ latent class model described in Chapter 2 in this volume and it consists of the entire simplex Δ_3.

3.7 Generalisations and questions

In this chapter we have employed an algebraic geometric approach to describe a variety of characterisations, both complete and incomplete, of bivariate distributions for two categorical variables. We have updated some older geometric representations of 2 × 2 contingency tables, e.g., from (Fienberg and Gilbert 1970), and we have described a series of new characterisations of the joint distribution using arbitrary sets of margins, conditionals, and odds ratios. We also considered incomplete characterisations of the joint distribution, and their links to latent class models and to Simpson's paradox. Many of the ideas explored here generalise rather naturally to $I \times J$ and higher-way tables. For higher-way tables, the usual characterisations corresponding to log-linear models come in terms of specifying marginal totals (minimal sufficient statistics) and setting various sets of generalised odds ratios equal to zero. The number of such specifications grows dramatically with the dimensionality of the table.

Many questions remain to be explored; e.g. (i) What are the partial specifications arising from subset of ratio of odds ratios? (ii) When are subsets of odds ratios implied by conditionals? (iii) When do combinations of margins and conditionals reduce to higher-order margins? (iv) What are the implications of such results for bounds in contingency tables? About question (iv), see also Chapter 8 in this volume.

Acknowledgements

We thank Cristiano Bocci and Eva Riccomagno for helpful suggestions regarding some proofs. This research was supported in part by NSF Grant SES-0532407 to the Department of Statistics, Penn State University, NSF grants EIA9876619 and IIS0131884 to the National Institute of Statistical Sciences, NSF Grant DMS-0439734 to the Institute for Mathematics and Its Application at the University of Minnesota, and NSF Grant DMS-0631589 to Carnegie Mellon University.

References

Arnold, B., Castillo, E. and Sarabia, J. M. (1996). Specification of distributions by combinations of marginal and conditional distributions, *Statistics and Probability Letters* **26**, 153–57.

Arnold, B., Castillo, E. and Sarabia, J. M. (1999). *Conditional Specification of Statistical Models*, (New York, Springer-Verlag).

Besag, J. (1974). Spatial interaction and the statistical analysis of lattice systems (with discussion), *Journal of the Royal Statistical Society, Series B* **36**, 192–236.

Bishop, Y. M. M., Fienberg, S. E. and Holland, P. W. (1975). *Discrete Multivariate Analysis: Theory and Practice* (Cambridge, MA, MIT Press). Reprinted (2007) (New York, Springer-Verlag).

Carlini, E. and Rapallo, F. (2005). The geometry of statistical models for two-way contingency tables with fixed odds ratios, *Rendiconti dell'Istituto di Matematica dell'Università di Trieste* **37**, 71–84.

De Rooij, M. and Anderson, C.J. (2007). Visualizing, summarizing, and comparing odds ratio structures, *Methodology* **3**, 139–48.

De Rooij, M., and Heiser, W. J. (2005). Graphical representations and odds ratios in a distance-association model for the analysis of cross-classified data, *Psychometrika* **70**, 99–123.

Diaconis, P. (1977). Finite forms of de Finetti's theorem on exchangeability, *Synthese* **36**, 271–81.

Diaconis, P. and Sturmfels, B. (1998). Algebraic algorithms for sampling from conditional distributions, *Annals of Statistics* **26**(1), 363–97.

Dobra, A. (2001). Statistical tools for disclosure limitation in multi-way contingency tables. PhD thesis, Department of Statistics, Carnegie Mellon University.

Dobra, A. (2003). Markov bases for decomposable graphical models, *Bernoulli* **9**(6), 1–16.

Edgeworth, F. Y. (1914). On the use of analytical geometry to represent certain kinds of statistics, *Journal of the Royal Statistical Society* **77**, 838–52.

Erosheva, E. A. (2005). Comparing latent structures of the grade of membership, Rasch, and latent class models, *Psychometrika* **70**, 619–28.

Fienberg, S. E. (1968). The geometry of an $r \times c$ contingency table, *Annals of Mathematical Statistics* **39**, 1186–90.

Fienberg, S. E. (1970). An iterative procedure for estimation in contingency tables, *Annals of Mathematical Statistics* **41**, 907–17. Corrigenda **42**, 1778.

Fienberg, S. E. and Gilbert, J. P. (1970). The geometry of a two by two contingency table, *Journal of the American Statistical Association* **65**, 694–701.

Fienberg, S. E. (1980). *The Analysis of Cross-Classified Categorical Data* 2nd edn (Cambridge, MA, MIT Press). Reprinted (2007) (New York, Springer-Verlag).

Fisher, R. A. (1921). On the interpretation of χ^2 from contingency tables, and the calculation of P, *Journal of the Royal Statistical Society* **85**, 87–94.

Flach, P. A. (2003). The geometry of ROC space: understanding machine learning metrics through ROC isometrics, In *Proc. ICML-2003*, Washington DC, 194–201.

Gelman, A. and Speed, T. P. (1993). Characterizing a joint probability distribution by conditionals, *Journal of the Royal Statistical Society. Series B* **55**, 185–8. Corrigendum **6**, 483 (1993).

Good, I. J. and Mittal, Y. (1987). The amalgamation and geometry of two-by-two contingency tables, *Annals of Statistics* **15**, 694–711. Addendum **17**, 947 (1989).

Greenacre, M. and Hastie, T. (1987). The geometric interpretation of correspondence analysis, *Journal of the American Statistical Association* **82**, 437–47.

Hadjicostas, P. (1998). The asymptotic proportion of subdivisions of a 2×2 table that result in Simpson's paradox, *Combinatorics, Probability and Computing* **7**, 387–96.

Heiser, W. J. (2004). Geometric representation of association between categories, *Psychometrika* **69**, 513–45.

Kadane, J. B., Meyer, M. M. and Tukey, J. W. (1999). Yule's association paradox and ignored stratum heterogeneity in capture-recapture studies, *Journal of the American Statistical Association* **94**, 855–9.

Kagan, A. M., Linnik, Y. V. and Rao, C. R. (1973). *Characterization Problems in Mathematical Statistics* (New York, John Wiley & Sons).

Kenett, R. S. (1983). On an exploratory analysis of contingency tables, *The Statistician* **32**, 395–403.

Lauritzen, S. L. (1996). *Graphical Models* (New York, Oxford University Press).

Luo, D., Wood, G. and Jones, G. (2004). Visualising contingency table data, *Australian Mathematical Society Gazette* **31**, 258–62.

Nelsen, R. B. (2006). *An Introduction to Copulas* 2nd edn (New York, Springer-Verlag).

Nelsen, R. B. (1995). Copulas, characterization, correlation, and counterexamples, *Mathematics Magazine* **68**, 193–8.

Pearson, E. S. (1956). Some aspects of the geometry of statistics, *Journal of the Royal Statistical Society. Series A* **119**, 125–46.

Pistone, G., Riccomagno, E. and Wynn, H. P. (2001). *Algebraic Statistics* (Boca Raton, Chapman & Hall/CRC).

Ramachandran, B. and Lau, K. S. (1991). *Functional Equations in Probability Theory* (New York, Academic Press).

Shapiro, S. H. (1982). Collapsing contingency tables – A geometric approach, *American Statistician* **36**, 43–6.

Simpson, E. H. (1951). The interpretation of interaction in contingency tables, *Journal of the Royal Statistical Society. Series B* **13**, 238–41.

Slavkovic, A. B. (2004). Statistical disclosure limitation beyond the margins: characterization of joint distributions for contingency tables. PhD thesis, Department of Statistics, Carnegie Mellon University.

Slavkovic, A. B. and Sullivant, S. (2004). The space of compatible full conditionals is a unimodular toric variety, *Journal of Symbolic Computing* **46**, 196–209.

Yule, G. U. (1903). Notes on the theory of association of attributes in statistics, *Biometrika* **2**, 121–34.

4

Model selection for contingency tables with algebraic statistics

Anne Krampe

Sonja Kuhnt

Abstract

Goodness-of-fit tests based on chi-square approximations are commonly used in the analysis of contingency tables. Results from algebraic statistics combined with MCMC methods provide alternatives to the chi-square approximation. However, within a model selection procedure usually a large number of models is considered and extensive simulations would be necessary. We show how the simulation effort can be reduced by an appropriate analysis of the involved Gröbner bases.

4.1 Introduction

Categorical data occur in many different areas of statistical applications. The analysis usually concentrates on the detection of the dependence structure between the involved random variables. Log-linear models are adopted to describe such association patterns, see (Bishop *et al.* 1995, Agresti 2002) and model selection methods are used to find the model from this class, which fits the data best in a given sense. Often, goodness-of-fit tests for log-linear models are applied, which involve chi-square approximations for the distribution of the test statistic. If the table is sparse such an approximation might fail. By combining methods from compu tational commutative algebra and from statistics, (Diaconis and Sturmfels 1998) provide the background for alternative tests. They use the MCMC approach to get a sample from a conditional distribution of a discrete exponential family with given sufficient statistic. In particular Gröbner bases are used for the construction of the Markov chain. This approach has been applied to a number of tests for the analysis of contingency tables (Rapallo 2003, Rapallo 2005, Krampe and Kuhnt 2007). Such tests have turned out to be a valuable addition to traditional exact and asymptotic tests.

However, if applied within a model selection procedure, goodness-of-fit tests have to be conducted with respect to a number of considered models. The algebraic approach involves the computation of an individual Gröbner basis for an ideal constructed from the sufficient statistics of each model. This also means that a new simulation of a Markov chain has to be conducted for each tested model. Thus, the

Algebraic and Geometric Methods in Statistics, ed. Paolo Gibilisco, Eva Riccomagno, Maria Piera Rogantin and Henry P. Wynn. Published by Cambridge University Press. © Cambridge University Press 2010.

83

selection of a model based on the current algebraic approaches is time consuming and computationally extensive. Based on an analysis of properties of log-linear models and Gröbner bases we propose a new model selection approach. It is shown that it suffices to compute the Gröbner basis and to simulate a Markov chain for the model of mutual independence. All other test decisions can then be derived from this chain.

The outline of this chapter is as follows: Section 4.2 introduces the treated model selection problem and the traditional Pearson goodness-of-fit test. In Section 4.3 we recall the algebraic approach by Diaconis–Sturmfels, leading to the proposal of a new alternative model selection procedure in Section 4.4. We compare the performance of the new model selection approach with classical model selection procedures by simulation studies in Section 4.5.

4.2 Model selection

In the analysis of categorical data the main interest lies in identifying the dependence structure between variables. In so-called graphical models a mathematical graph represents the random variables and independence properties of a statistical model, which vice versa fulfils the independence properties described by the graph. Graphical models with undirected graphs and joint multinomial distribution belong to the well-known class of hierarchical log-linear models (Bishop *et al.* 1995, Agresti 2002) . Model building strategies which aim at finding a most suitable model in a set of candidate models, can also be applied to find an appropriate graphical model (Edwards 2000, Chapter 6), (Borgelt and Kruse 2002, Madigan and Raftery 1994). We focus on p-values for strategies based on goodness-of-fit tests.

To fix ideas and to introduce some notation, consider the case of three categorical variables X_1, X_2, X_3 with I, J and K possible outcomes. The number of observations in a sample of size n with outcome i for the first, outcome j for second and k for the third variable is denoted by $n_{i,j,k}$ or n_{ijk} for short. This defines a mapping $z : \mathcal{H} \rightarrow \mathbb{N}$ of the finite sample space $\mathcal{H}$ into the set of non-negative integers $\mathbb{N}$, where $\mathcal{H} = \{(i,j,k) \mid i = 1, \ldots, I, j = 1, \ldots, J, k = 1, \ldots, K\}$. Each frequency count n_{ijk} is seen as the possible outcome of a random variable N_{ijk} with expected value m_{ijk}. The vector of cell counts $(N_x)_{x \in \mathcal{H}}$ follows a multinomial distribution. The class of graphical models for three variables is characterised by the set of undirected graphs on three vertices as depicted in Figure 4.1. A missing edge between two vertices means that the two random variables are conditionally independent given the remaining variables. The corresponding log-linear models are described in Table 4.1: logarithms of the cell probabilities p_{ijk} are expressed by linear functions in unknown real u-parameters. Each of these functions contains a u-term associated with each individual variable and interaction terms depending on the considered graphical model. Note that we omitted the saturated model as p-values for the considered tests always equal one.

The well-known Pearson test statistic to evaluate the adequacy of a model is given by the standardised sum of the squared difference between the observed and

Table 4.1 *Log-linear models of a three-dimensional table.*

Model 1	$\log(p_{ijk}) = u + u_{i(X_1)} + u_{j(X_2)} + u_{k(X_3)} + u_{ij(X_1X_2)} + u_{jk(X_2X_3)}$
Model 2	$\log(p_{ijk}) = u + u_{i(X_1)} + u_{j(X_2)} + u_{k(X_3)} + u_{ij(X_1X_2)} + u_{ik(X_1X_3)}$
Model 3	$\log(p_{ijk}) = u + u_{i(X_1)} + u_{j(X_2)} + u_{k(X_3)} + u_{ik(X_1X_3)} + u_{jk(X_2X_3)}$
Model 4	$\log(p_{ijk}) = u + u_{i(X_1)} + u_{j(X_2)} + u_{k(X_3)} + u_{ij(X_1X_2)}$
Model 5	$\log(p_{ijk}) = u + u_{i(X_1)} + u_{j(X_2)} + u_{k(X_3)} + u_{jk(X_2X_3)}$
Model 6	$\log(p_{ijk}) = u + u_{i(X_1)} + u_{j(X_2)} + u_{k(X_3)} + u_{ik(X_1X_3)}$
Model 7	$\log(p_{ijk}) = u + u_{i(X_1)} + u_{j(X_2)} + u_{k(X_3)}$
	$i = 1,\ldots,I,\ j = 1,\ldots,J,\ k = 1,\ldots,K$

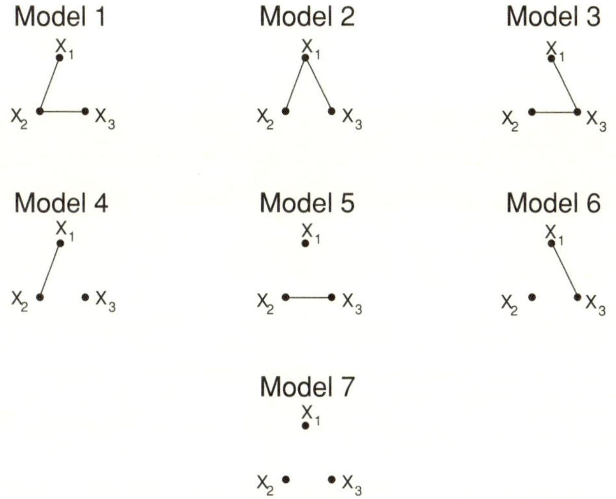

Fig. 4.1 Graphical models for three-dimensional tables.

the estimated expected cell counts $\hat{m}_{ijk}$. In the example this is $\sum_{ijk}(N_{ijk} - \hat{m}_{ijk})^2/ \hat{m}_{ijk}$, which is approximately chi-square distributed under the tested model with degrees of freedom given by the difference in dimensionality between the parameter space of the saturated and of the tested model. The resulting χ^2 goodness-of-fit test has the disadvantage that the approximation may not be good for tables with small cell counts, (Cochran 1954, Conover 1971).

Alternatively, an exact conditional goodness-of-fit test can be derived using the fact that multinomial distributions constitute an exponential family, see (Lehmann 1986). Let the sufficient statistic of a model be denoted by T with observed outcome t. Let further the set of all data sets with value t of the sufficient statistic be denoted by $\mathscr{Z}_t$. Hence, if $z \in \mathscr{Z}_t$, the set $\{z(x),\ x \in \mathscr{H}\}$ gives a data set for which the sufficient statistic takes on the value t. The sufficient statistics for the case of three variables are given in Table 4.2. The set $\mathscr{Z}_t$ is finite, non-empty and the probability function on $\mathscr{Z}_t$ is hypergeometric. The p-value of the exact conditional test is the probability under the null hypothesis of getting a more extreme value of the test statistic in the direction of the alternative than the observed value. Hence, an exact p-value can be derived by evaluating all elements in $\mathscr{Z}_t$ using the

Table 4.2 *Sufficient statistics for three-dimensional tables.*

M 1	$T^{(1)} = (N_{+jk}, j = 1, \ldots, J,\ k = 1, \ldots, K,\ N_{ij+},\ i = 1, \ldots, I,\ j = 1, \ldots, J)'$
M 2	$T^{(2)} = (N_{ij+},\ i = 1, \ldots, I,\ j = 1, \ldots, J\ N_{i+k},\ i = 1, \ldots, I,\ k = 1, \ldots, K)'$
M 3	$T^{(3)} = (N_{i+k},\ i = 1, \ldots, I,\ k = 1, \ldots, K,\ N_{+jk}, j = 1, \ldots, J,\ k = 1, \ldots, K)'$
M 4	$T^{(4)} = (N_{ij+},\ i = 1, \ldots, I,\ j = 1, \ldots, J,\ N_{++k},\ k = 1, \ldots, K)'$
M 5	$T^{(5)} = (N_{i++},\ i = 1, \ldots, I,\ N_{+jk}, j = 1, \ldots, J,\ k = 1, \ldots, K)'$
M 6	$T^{(6)} = (N_{+j+},\ j = 1, \ldots, J,\ N_{i+k},\ i = 1, \ldots, I,\ k = 1, \ldots, K)'$
M 7	$T^{(7)} = (N_{i++},\ i = 1, \ldots, I,\ N_{+j+},\ j = 1, \ldots, J,\ N_{++k},\ k = 1, \ldots, K)'$

hypergeometric probability function. This, however, is only possible for tables with very few observations.

(Diaconis and Sturmfels 1998) show how computational commutative algebra can be combined with Markov Chain Monte Carlo methods to sample from the hypergeometric distribution on $\mathscr{Z}_t$ for a specific model. Corresponding p-values can then be determined from the simulated distribution of the Pearson goodness-of-fit statistic. We will discuss this in more detail and show how just one simulation can be used to derive p-values for all considered models.

4.3 MCMC and algebra

Markov Chain Monte Carlo (MCMC) methods are used to sample from the distribution of interest. In the first part of this section we describe the Metropolis–Hastings algorithm thus showing how MCMC works to sample from the distribution on $\mathscr{Z}_t$. We then demonstrate how the Diaconis–Sturmfels algorithm combines the MCMC approach with computational commutative algebra in order to derive the Markov basis.

4.3.1 Metropolis–Hastings algorithm

The Metropolis–Hastings algorithm is a very powerful MCMC method (Chib and Greenberg 1995, Ewens and Grant 2001, Sørensen and Gianola 2002). A Markov chain is generated whose stationary density π equals a target density, here the hypergeometric density on $\mathscr{Z}_t$ denoted by H. This is done in two steps: First a potential new state of the Markov chain is generated. Let E denote the state space of the Markov chain and assume that the chain is currently in state r. State s is then proposed with probability $q(r, s)$, $r, s \in E$. We refer to $q(\cdot, \cdot)$ as the proposal probability function. To ensure that the generated Markov chain is reversible an acceptance probability

$$\alpha(r, s) = \begin{cases} \min(\frac{\pi(s)q(s,r)}{\pi(r)q(r,s)}, 1), & \text{if } \pi(r)q(r, s) > 0, \\ 1, & \text{otherwise,} \end{cases}$$

$r, s \in E$, is introduced in the second step of the algorithm. The term π is the invariant density of the Markov chain if also some mild regularity conditions (aperiodicity and irreducibility) hold. The objective now is to find an adequate proposal probability function. (Diaconis and Sturmfels 1998) use the notion of a Markov basis. It is essential here that $\mathscr{Z}_t$ can be written as $\mathscr{Z}_t := \{z : \mathscr{H} \to \mathbb{N} \mid \sum_{x \in \mathscr{H}} z(x)T^*(x) = t\}$, with a mapping $T^* : \mathscr{H} \to \mathbb{N}^d$.

The mapping T^* is determined by the sufficient statistic T. For example $T^{(7)*}$ has the same length as $T^{(7)}$, $I \cdot J \cdot K$, and can be divided into three parts. The first part has length I, the second has length J and the last one consists of K entries. The i-th, the $I+j$-th, and the $I+J+k$-th entry are all one, all others are zero.

A Markov basis is a set of functions $m_1, m_2, \ldots, m_L : \mathscr{H} \to \mathbb{Z}$, called moves, such that

(i) $\sum\limits_{x \in \mathscr{H}} m_i(x)T^*(x) = 0$ for all $1 \le i \le L$ and

(ii) for any t and $z, z' \in \mathscr{Z}_t$ there is a sequence of moves $(m_{i_1}, \ldots, m_{i_A})$ as well as a sequence of directions $(\epsilon_1, \ldots, \epsilon_A)$ with $\epsilon_j = \pm 1$, such that $z' = z + \sum\limits_{j=1}^{A} \epsilon_j m_{i_j}$ and $z + \sum\limits_{j=1}^{a} \epsilon_j m_{i_j} \ge 0$, $1 \le a \le A$.

These conditions ensure the irreducibility of the simulated Markov chain and also that the value t of the sufficient statistic T is the same for each state z of the Markov chain. A Markov chain on $\mathscr{Z}_t$ can now be generated. Its stationary probability function equals the hypergeometric probability function H.

Assume that a Markov basis $m_1, \ldots, m_L$ is given. We select a move m_U uniformly in $\{1, \ldots, L\}$. We also choose a direction of the move $\epsilon = \pm 1$ with probability $1/2$ independently of U. Suppose that the chain is currently in state $z \in \mathscr{Z}_t$. Since $q(\cdot, \cdot)$ is symmetric the chain moves to $z' = z + \epsilon m_U \in \mathscr{Z}_t$ with probability

$$\alpha = \min\left(\frac{H(z')}{H(z)}, 1\right) = \min\left(\frac{\prod\limits_{x \in \mathscr{H}} z(x)!}{\prod\limits_{x \in \mathscr{H}} (z(x) + \epsilon m_U(x))!}, 1\right),$$

see e.g. (Rapallo 2003). If an entry of the proposed new state z' is negative, then z' is not defined as a contingency table and thus not an element of $\mathscr{Z}_t$. In this case, the hypergeometric density $H(z')$ and hence α are zero and the new state is again z. As a consequence, the problem to identify a suitable proposal distribution can be restated in terms of finding a Markov basis.

4.3.2 Diaconis–Sturmfels algorithm

(Diaconis and Sturmfels 1998) apply results from computational commutative algebra to identify an appropriate Markov basis. An introduction to computational commutative algebra can be found in (Cox *et al.* 1997, Pistone *et al.* 2001).

Diaconis and Sturmfels define for each $x \in \mathscr{H}$ an indeterminate also denoted by x and identify a function $f : \mathscr{H} \to \mathbb{N}$ by a monomial $\prod_{x \in \mathscr{H}} x^{f(x)}$, where $\mathbb{N}$ denotes the natural numbers. Using our notation for three-dimensional tables we represent each $x = (i, j, k) \in \mathscr{H}$ by an indeterminate x_{ijk}. Then a table $\begin{pmatrix} 0 & 1 \\ 0 & 0 \end{pmatrix}\begin{pmatrix} 0 & 0 \\ 1 & 0 \end{pmatrix}$

is represented by $x_{111}^0 x_{121}^1 x_{211}^0 x_{221}^0 x_{112}^0 x_{122}^0 x_{212}^1 x_{222}^0$. Let $\mathcal{T} = \{T_1, \ldots, T_d\}$ be the set of all entries of the sufficient statistic T. Assume that $T^* : \mathcal{H} \to \mathbb{N}^d$ with $T^* = (T_1^*, \ldots, T_d^*)'$ is given as well as a monomial ordering $\succ$ for $\mathcal{H}$. This ordering will be extended to $\mathcal{H} \cup \mathcal{T}$ such that $T_i \succ x$ for all $x \in \mathcal{H}$ and $T_i \in \mathcal{T}$, $i = 1, \ldots, d$, in the polynomial ring $k[\mathcal{H}, \mathcal{T}]$. Following the implicitation algorithm, Diaconis and Sturmfels construct the ideal $\mathscr{I}^* = \{x - \mathcal{T}^{T^*(x)}, x \in \mathcal{H}\}$ with $\mathcal{T}^{T^*(x)} := T_1^{T_1^*(x)} \cdot T_2^{T_2^*(x)} \cdot \ldots \cdot T_d^{T_d^*(x)}$, where $T_i^*(x)$ is the i-th entry of $T^*(x)$, $i = 1, \ldots, d$. Using e.g. the free software CoCoA (CoCoA Team 2007) we can compute the reduced Gröbner basis G^* for $\mathscr{I}^*$. In the next step we set $\mathscr{I}_T := \mathscr{I}^* \cap k[\mathcal{H}]$. The reduced Gröbner basis G for $\mathscr{I}_T$ contains only the polynomials of G^* involving elements of $\mathcal{H}$. It can be shown that G equals the Markov basis needed for the Metropolis–Hastings algorithm, see (Diaconis and Sturmfels 1998, Theorems 3.1, 3.2) and (Cox *et al.* 1997, § 3.1, § 3.3). A Gröbner basis and hence the derived Markov basis is described as a set of polynomials. Thereby each element m of the Markov basis represents a function $m : \mathcal{H} \to \mathbb{Z}$, which can be written as $m(x) = m^+(x) - m^-(x)$ with $m^+(x), m^-(x) : \mathcal{H} \to \mathbb{N}$, $m^+(x) := \max(m(x), 0)$ and $m^-(x) := \max(-m(x), 0)$.

Now, consider a move m given by adding $\begin{pmatrix} 0 & 1 \\ -1 & 0 \end{pmatrix} \begin{pmatrix} 0 & -1 \\ 1 & 0 \end{pmatrix}$ to a $2 \times 2 \times 2$ table. Using $m^+ = \begin{pmatrix} 0 & 1 \\ 0 & 0 \end{pmatrix} \begin{pmatrix} 0 & 0 \\ 1 & 0 \end{pmatrix}$ and $m^- = \begin{pmatrix} 0 & 0 \\ 1 & 0 \end{pmatrix} \begin{pmatrix} 0 & 1 \\ 0 & 0 \end{pmatrix}$ we can convert this move into a polynomial $x_{121} x_{212} - x_{211} x_{122}$.

To exemplify the Diaconis–Sturmfels algorithm we apply it to a three-dimensional table. In particular, we will use model 7 described in Section 4.2. We take the graded lexicographic monomial ordering but other ordering such as graded reverse lexicographic ordering give the same results. The procedure for the other models is similar and therefore not presented here.

Example 4.1 We consider data from a study on the effect of an antiretroviral drug (azidothymidine, AZT) on the development of AIDS symptoms, published in the New York Times in 1991. A total of 338 probands whose immune systems exhibit first symptoms after the infection with the AIDS virus were randomly assigned to receive medication immediately or to wait until the immune systems were affected by the virus, see (Agresti 2002). The medication with AZT is represented by $X_1 = i$ ($i = 1$: take AZT immediately, $i = 2$: otherwise), the probands' status of disease by $X_2 = j$ ($j = 1$: AIDS symptoms developed, $j = 2$: no AIDS symptoms developed), and their race by $X_3 = k$ ($k = 1$: white, $k = 2$: black).

Assuming model 7 given in Table 4.2 we get the sufficient statistic $T^{(7)} = (N_{1++}, N_{2++}, N_{+1+}, N_{+2+}, N_{++1}, N_{++2})'$ and the mapping $T^{(7)*}$ is given by $T^{(7)*}((1,1,1)) = (1,0,1,0,1,0)'$, $T^{(7)*}((1,1,2)) = (1,0,1,0,0,1)'$, $\ldots$, $T^{(7)*}((2,2,2)) = (0,1,0,1,0,1)'$.

Table 4.3 *Data set 1 (Agresti 2002).*

	$j = 1$	$j = 2$		$j = 1$	$j = 2$
$i = 1$	14	93		11	52
$i = 2$	32	81		12	43
		$k = 1$			$k = 2$

Table 4.4 *Reduced Gröbner basis for model 7 for a $2 \times 2 \times 2$-table.*

Model 7: $\mathscr{G}^{(7)} = \{g_1^{(7)}, g_2^{(7)}, g_3^{(7)}, g_4^{(7)}, g_5^{(7)}, g_6^{(7)}, g_7^{(7)}, g_8^{(7)}, g_9^{(7)}\}$

$$g_1^{(7)} = x_{121}\, x_{222} - x_{122}\, x_{221}, \qquad g_2^{(7)} = x_{112}\, x_{222} - x_{122}\, x_{212},$$
$$g_3^{(7)} = x_{111}\, x_{222} - x_{122}\, x_{211}, \qquad g_4^{(7)} = x_{211}\, x_{222} - x_{212}\, x_{221},$$
$$g_5^{(7)} = x_{111}\, x_{222} - x_{121}\, x_{212}, \qquad g_6^{(7)} = x_{111}\, x_{222} - x_{112}\, x_{221},$$
$$g_7^{(7)} = x_{111}\, x_{212} - x_{112}\, x_{211}, \qquad g_8^{(7)} = x_{111}\, x_{122} - x_{112}\, x_{121},$$
$$g_9^{(7)} = x_{111}\, x_{221} - x_{121}\, x_{211}.$$

Applying the Diaconis–Sturmfels procedure, in the following procedure abbreviated by DS, we consider the ideal

$$\mathscr{I}^* = \langle x_{111} - T_1^{(7)} \cdot T_3^{(7)} \cdot T_5^{(7)},\ x_{121} - T_1^{(7)} \cdot T_4^{(7)} \cdot T_5^{(7)},$$
$$x_{211} - T_2^{(7)} \cdot T_3^{(7)} \cdot T_5^{(7)},\ x_{221} - T_2^{(7)} \cdot T_4^{(7)} \cdot T_5^{(7)},$$
$$x_{112} - T_1^{(7)} \cdot T_3^{(7)} \cdot T_6^{(7)},\ x_{122} - T_1^{(7)} \cdot T_4^{(7)} \cdot T_6^{(7)},$$
$$x_{212} - T_2^{(7)} \cdot T_3^{(7)} \cdot T_6^{(7)},\ x_{222} - T_2^{(7)} \cdot T_4^{(7)} \cdot T_6^{(7)} \rangle.$$

CoCoA gives the reduced Gröbner basis for $\mathscr{I}^*$ and we obtained the Gröbner basis for $\mathscr{I}_T$: $\mathscr{G}^{(7)} = \{g_1^{(7)}, \ldots, g_9^{(7)}\}$ as given in Table 4.4.

In the above example we treated only one possible model. Since the analysis is conditioned on the set of sufficient statistics $\mathscr{Z}_t$ we get a different Gröbner basis for each model. In Table 4.5, we list the Gröbner bases of the models 1-6 introduced in Section 4.2 for a $2 \times 2 \times 2$-table. For each model a p-value for the Pearson goodness-of-fit test can be simulated as follows. First a Markov chain with chain length l is simulated based on the Gröbner basis for the considered model and the Metropolis–Hastings algorithm. According to the usual MCMC procedures we disregard the first b data sets and sample each sth table. The values of the Pearson goodness-of-fit test are calculated for each of the sampled tables. The simulated p-value is given by $p = \frac{1}{\lfloor \frac{l-b}{s} \rfloor} \sum_{i=1}^{\lfloor \frac{l-b}{s} \rfloor} \mathbf{1}_{\{\chi^2_{\text{obs}} \geq \chi^2_i\}}(i)$, where χ^2_{obs} denotes the observed value of the Pearson χ^2-test, and χ^2_i the values for the simulated data sets.

Table 4.5 *Reduced Gröbner basis for the models 1-6 for a $2 \times 2 \times 2$-table.*

$$\text{Model 1: } \mathscr{G}^{(1)} = \{g_1^{(1)}, g_2^{(1)}\}$$
$$g_1^{(1)} = x_{121}\, x_{222} - x_{122}\, x_{221}, \qquad g_2^{(1)} = x_{111}\, x_{212} - x_{112}\, x_{211}$$

$$\text{Model 2: } \mathscr{G}^{(2)} = \{g_1^{(2)}, g_2^{(2)}\}$$
$$g_1^{(2)} = x_{211}\, x_{222} - x_{212}\, x_{221}, \qquad g_2^{(2)} = x_{111}\, x_{122} - x_{112}\, x_{121}$$

$$\text{Model 3: } \mathscr{G}^{(3)} = \{g_1^{(3)}, g_2^{(3)}\}$$
$$g_1^{(3)} = x_{112}\, x_{222} - x_{122}\, x_{212}, \qquad g_2^{(3)} = x_{111}\, x_{221} - x_{121}\, x_{211}$$

$$\text{Model 4: } \mathscr{G}^{(4)} = \{g_1^{(4)}, g_2^{(4)}, g_3^{(4)}, g_4^{(4)}, g_5^{(4)}, g_6^{(4)}\}$$
$$g_1^{(4)} = x_{211}\, x_{222} - x_{212}\, x_{221}, \qquad g_2^{(4)} = x_{121}\, x_{222} - x_{122}\, x_{221},$$
$$g_3^{(4)} = x_{111}\, x_{222} - x_{112}\, x_{221}, \qquad g_4^{(4)} = x_{121}\, x_{212} - x_{122}\, x_{211},$$
$$g_5^{(4)} = x_{111}\, x_{212} - x_{112}\, x_{211}, \qquad g_6^{(4)} = x_{111}\, x_{122} - x_{112}\, x_{121}$$

$$\text{Model 5: } \mathscr{G}^{(5)} = \{g_1^{(5)}, g_2^{(5)}, g_3^{(5)}, g_4^{(5)}, g_5^{(5)}, g_6^{(5)}\}$$
$$g_1^{(5)} = x_{121}\, x_{222} - x_{122}\, x_{221}, \qquad g_2^{(5)} = x_{112}\, x_{222} - x_{122}\, x_{212},$$
$$g_3^{(5)} = x_{112}\, x_{221} - x_{121}\, x_{212}, \qquad g_4^{(5)} = x_{111}\, x_{222} - x_{122}\, x_{211},$$
$$g_5^{(5)} = x_{111}\, x_{221} - x_{121}\, x_{211}, \qquad g_6^{(5)} = x_{111}\, x_{212} - x_{112}\, x_{211}$$

$$\text{Model 6: } \mathscr{G}^{(6)} = \{g_1^{(6)}, g_2^{(6)}, g_3^{(6)}, g_4^{(6)}, g_5^{(6)}, g_6^{(6)}\}$$
$$g_1^{(6)} = x_{211}\, x_{222} - x_{212}\, x_{221}, \qquad g_2^{(5)} = x_{112}\, x_{222} - x_{122}\, x_{212},$$
$$g_3^{(6)} = x_{112}\, x_{221} - x_{122}\, x_{211}, \qquad g_4^{(5)} = x_{111}\, x_{222} - x_{121}\, x_{212},$$
$$g_5^{(6)} = x_{111}\, x_{221} - x_{121}\, x_{211}, \qquad g_6^{(6)} = x_{111}\, x_{122} - x_{112}\, x_{121}.$$

4.4 Reduction of computational costs

In the following we present a way to reduce the computational expenses for the above model selection procedure. So far we have derived p-values for a model selection procedure by simulating an individual Markov chain of length l for each model. This yields a large computational effort. We will now discuss, how the structure of the considered models can be used to let the model selection be based on a single simulation.

Graphical models for contingency tables are hierarchical log-linear models, see (Edwards 2000). The most general model is the saturated model with no independence constraints and a complete graph. In this case the sufficient statistic equals the cell counts and the set $\mathscr{Z}_t$ always has only one element, namely the observed table. Each additional missing edge from the graph mirrors an additional conditional independence constraint on the model, which is also reflected in the sufficient statistics. Let us compare two distinct graphical models, $M1$ and $M2$, for the same data set. Let $M1$ be a sub-model of $M2$, which means that the set of edges of the graph of $M1$ is a subset of the set of edges of $M2$. From this it follows that the set $\mathscr{Z}_{t(M2)}$ for $M2$ is a subset of the set $\mathscr{Z}_{t(M1)}$ for $M1$. The model with the largest number of conditional independence constraints is the model of complete independence, corresponding to a graph with no edges at all. Hence the model of complete independence is a sub-model of all other models. For any given data set, the set $\mathscr{Z}_{t(7)}$

of the complete independence model contains the corresponding sets for all other models.

These structural properties of graphical models are also found in the derived Gröbner bases.

Theorem 4.1 *Let $M1$ and $M2$ be two log-linear models with $M1 \subset M2$, i.e. $M1$ is a sub-model of $M2$. Following the Diaconis–Sturmfels approach we get the corresponding elimination ideals $\mathscr{I}^{(M1)}$ and $\mathscr{I}^{(M2)}$ with $\mathscr{I}^{(M1)} \supset \mathscr{I}^{(M2)}$.*

Proof According to Section 4.3.2 we denote the sets of entries of the sufficient statistics for the parameters of $M1$ and $M2$ by $\mathscr{T}^{(M1)}$ and $\mathscr{T}^{(M2)}$, respectively. Because of the hierarchical structure of the models it holds that $\mathscr{T}^{(M1)}$ is entirely determined by $\mathscr{T}^{(M2)}$. We abbreviate this relationship by $\mathscr{T}^{(M1)} \subset \mathscr{T}^{(M2)}$.

From $\mathscr{T}^{(M1)} \subset \mathscr{T}^{(M2)}$ it follows for the respective varieties that $V^{(M1)} \supset V^{(M2)}$, where $V^{(M1)}$ and $V^{(M2)}$ are defined by the power product representation in the Diaconis–Sturmfels algorithm for the models $M1$ and $M2$. This implies that $\mathscr{I}(V^{(M1)}) \supset \mathscr{I}(V^{(M2)})$, see (Cox *et al.* 1997, Proposition 8, p. 34). $\qquad\square$

In the context of model selection described in Section 4.2, we observe that $\{\mathscr{T}^{(1)}, \mathscr{T}^{(2)}, \mathscr{T}^{(3)}\} \supset \{\mathscr{T}^{(3)}, \mathscr{T}^{(4)}, \mathscr{T}^{(5)}\} \supset \mathscr{T}^{(7)}$. Using the results of Theorem 4.1 we get $\mathscr{I}^{(7)} \supset \{\mathscr{I}^{(6)}, \mathscr{I}^{(5)}, \mathscr{I}^{(4)}\} \supset \{\mathscr{I}^{(3)}, \mathscr{I}^{(2)}, \mathscr{I}^{(1)}\}$. Now we will focus again on the simple case of $2 \times 2 \times 2$-tables. Gröbner bases for all graphical models in this situation, except for the saturated model, are given in Table 4.5. The reduced Gröbner bases for the models 1, 2, and 3, which are characterised by one missing edge in the graph, consist of two elements. For models 4, 5 and 6 with two missing edges the reduced Gröbner bases have six elements. Each Gröbner basis of model 1, 2, 3 can be found directly in the Gröbner basis $\mathscr{G}^{(7)}$ of model 7. For models 4, 5, 6 we observe that there is one basis polynomial of each model that is not an element of $\mathscr{G}^{(7)}$, which are $g_4^{(4)}$, $g_3^{(5)}$, $g_3^{(6)}$. However, these polynomials can be described by linear combinations of basis polynomials of $\mathscr{G}^{(7)}$:

$$
\begin{aligned}
g_4^{(4)} &= x_{121}\, x_{212} - x_{122}\, x_{211} \\
&= x_{111}\, x_{222} - x_{122}\, x_{211} - (x_{111}\, x_{222} - x_{121}\, x_{212}) = g_3^{(7)} - g_5^{(7)}, \\
g_3^{(5)} &= x_{112}\, x_{221} - x_{121}\, x_{212} \\
&= x_{111}\, x_{222} - x_{121}\, x_{212} - (x_{211}\, x_{222} - x_{212}\, x_{221}) = g_5^{(7)} - g_4^{(7)}
\end{aligned}
$$

and

$$
\begin{aligned}
g_3^{(6)} &= x_{112}\, x_{221} - x_{122}\, x_{211} \\
&= x_{111}\, x_{222} - x_{122}\, x_{211} - (x_{211}\, x_{222} - x_{212}\, x_{221}) = g_3^{(7)} - g_4^{(7)}.
\end{aligned}
$$

This ensures that each possible state for models 1–7 is attainable when using the proposal distribution constructed by $\mathscr{G}^{(7)}$ in the Metropolis–Hastings algorithm. Therefore all possible data sets with the same values of the sufficient statistic for models 1–6 can be extracted from the Markov chain generated assuming model 7. We will denote these six new chains as 'selected chains'. Assuming that the simulation for model 7 gives an adequate approximation of the hypergeometric distribution on $\mathscr{L}_{t^{(7)}}$ the selected chains give an adequate approximation of the respective conditional distributions: The simulation for the independence model with sufficient statistic $T^{(7)}$ leads to the approximation

$$P((N_x)_{x\in\mathscr{H}} = (n_x)_{x\in\mathscr{H}}|T^{(7)} = t^{(7)}) \approx \frac{|\text{simulated states equal to } (n_x)_{x\in\mathscr{H}}|}{|\text{simulated states}|}$$

for all $(n_x)_{x\in\mathscr{H}} \in \{(n_x)_{x\in\mathscr{H}}|n_x \geq 0, \sum_{x\in\mathscr{H}} n_x = n\}$. For all models 1–6 it holds that $\{(n_x)_{x\in\mathscr{H}}|T^{(i)} = t^{(i)}\} \subseteq \{(n_x)_{x\in\mathscr{H}}|T^{(7)} = t^{(7)}\}$, $i = 1,\ldots,6$, where $t^{(7)}$ and $t^{(i)}$ are calculated from the same observed table. This implies for all $i = 1,\ldots,6$:

$$P((N_x)_{x\in\mathscr{H}} = (n_x)_{x\in\mathscr{H}}|T^{(i)} = t^{(i)})$$

$$= P((N_x)_{x\in\mathscr{H}} = (n_x)_{x\in\mathscr{H}}|T^{(i)} = t^{(i)} \cap T^{(7)} = t^{(7)})$$

$$= \frac{P((N_x)_{x\in\mathscr{H}} = (n_x)_{x\in\mathscr{H}} \cap T^{(i)} = t^{(i)}|T^{(7)} = t^{(7)})}{P(T^{(i)} = t^{(i)}|T^{(7)} = t^{(7)})}$$

$$\approx \frac{|\text{simulated states equal to } (n_x)_{x\in\mathscr{H}} \text{ and with } T^{(i)} = t^{(i)}|}{|\text{simulated states with } T^{(i)} = t^{(i)}|}.$$

We conjecture that the simulation of only one Markov chain is sufficient for the analysis of the dependence structure in graphical models for contingency tables. This is the Markov chain for the model of mutual independence of all variables considered in the graphical model selection problem, depicted by a graph without any edges. Irrespective of the considered set of models the model of mutual independence is always a sub-model of all other models. Hence, its set $\mathscr{L}_{t^{(7)}}$ of all tables with the same values of the sufficient statistic includes the respective sets of the other models. Thereby 'selected chains' for all other models can be derived from its simulated Markov chain.

However, the approximation above and hence the benefit of the new procedure depends on the amount of simulated states with $T^{(i)} = t^{(i)}$. For example, if the sample size is large this amount decreases. To ensure that the new procedure works well, the chain length of the simulated Markov chain needs to be adjusted. Hence, if the appropriate chain length for the new approach is large, distinct computations might be more efficient.

Overall, we suggest a new procedure for the model selection: In the first step we generate a Markov chain as described in Section 4.3 using the Gröbner basis of the complete independence model. The second step is the 'selection' step. For the models 1–6 we extract the data sets from the simulated Markov chain according to their observed sufficient statistics. Thus, we obtain six 'selected chains'. Finally, these chains are used to calculate the different p-values for all considered models.

4.5 Simulation results

In this section we focus on the simulation of Markov chains in order to compare the new approach with theoretical results as well as with the Diaconis–Sturmfels

Table 4.6 *Data set 2.*

	$j=1$	$j=2$		$j=1$	$j=2$
$i=1$	1	0		0	1
$i=2$	2	1		0	2
		$k=1$			$k=2$

procedure. The simulation design chosen is as follows. For the DS procedure we generate a Markov chain with chain length 500 000 for each model. We delete the first 50 000 tables in the burn-in-phase and sample each 100th table (step length). For the new procedure we construct a Markov chain with 1 000 000 states for model 7 and extract the 'selected chains' for models 1–6. As the selected data sets are typically not arranged in the originally simulated Markov chain, we disregard only the first 10 tables in the burn-in-phase and sample each 10th table. Of course the resulting chain length of the selected chains vary randomly, which has to be kept in mind when comparing with the DS approach. The chain length, the burn-in-phase and the step length can be regarded as parameters of the simulation process which influence the rate of convergence of the Markov chain. We consider only the parameter values as given above, a discussion of rates of convergence can be found in (Diaconis and Sturmfels 1998, Section 2.3).

4.5.1 Comparison of theoretical and simulated probabilities

In Section 4.4 we discussed the use of 'selected chains' to derive values for the probabilities on the sets $\mathscr{Z}_{t(i)}, i = 1, \ldots, 6$. We compare such derived simulated probabilities with the true hypergeometric probabilities, the computation of which is only feasible, however, for very small data sets. To this purpose we employ a sparse $2 \times 2 \times 2$-table with only seven observations, which we call data set 2, in Table 4.6. We find that the theoretical hypergeometric and the simulated probabilities from the new approach almost coincide for all considered models and for all possible tables from the sets $\mathscr{Z}_{t(i)}, i = 1, \ldots, 6$, i.e. the maximal absolute difference is in the third decimal place. As an example, we give the results for the six possible data sets with the same value of the sufficient statistic as in data set 2 with respect to model 2 in Table 4.7 ($|\mathscr{Z}_{t(2)}| = 6$).

4.5.2 A simulation study of p-values

As the main aim of our new approach lies within model selection we are particularly interested in the resulting p-values. For 100 randomly generated $2 \times 2 \times 2$-tables we examine the p-values of the DS procedure and of the new approach. We simulate the data sets from a log-linear model with expected values as given in Table 4.8, hence from model 4.

Using the new approach we obtain 'selected chains' of different lengths for the 100 simulated tables. The number of extracted data sets for models 1–6 is displayed

Table 4.7 *Exact hypergeometric and simulated probabilities using the new approach for all elements in $\mathscr{Z}_{t(2)}$.*

exact probability	simulated probability
0.050	0.051
0.150	0.150
0.300	0.301
0.300	0.295
0.150	0.152
0.050	0.051

Table 4.8 *Expected values of the simulation model.*

	$j = 1$	$j = 2$		$j = 1$	$j = 2$
$i = 1$	4	8		3	6
$i = 2$	4	4		3	3
	$k = 1$			$k = 2$	

by boxplots in Figure 4.2. The variations of the number of extracted data sets for models 1–3 is much smaller than for models 4–6, due to the missing basis polynomials $g_4^{(4)}, g_3^{(5)}$ and $g_3^{(6)}$ in $\mathscr{G}^{(7)}$.

For each of the 100 tables, p-values for the DS procedure and for the new approach are calculated and plotted against each other (Figure 4.3).

The p-values lie very close to or on the bisecting line for each model, indicating nearly identical test results for the two simulation based procedures. In particular we come always to the same test decisions at level $\alpha = 0.05$ for both tests.

4.5.3 Results for AZT data set

We now return to data set 1 (Table 4.3) and compute p-values for the Diaconis–Sturmfels procedure, the new approach and the chi-square approximation.

The sample size of $n = 338$ probands is relatively large compared to the sample sizes of the data sets considered above. As a consequence, $\mathscr{Z}_{t(i)}$, $i = 1, \ldots, 7$, becomes very large for each of the seven considered models. To ensure that the Markov chains converge, we increase the chain length to $800\,000$ states for the DS procedure and to $12\,000\,000$ for the new approach. We keep the length of the burn-in-phase and the step length as before. In Table 4.9 we give the χ^2 test results for all seven models.

The two simulated p-values almost coincide and we conclude that the new approach is a reasonable alternative. Here also the chi-square approximation seems to

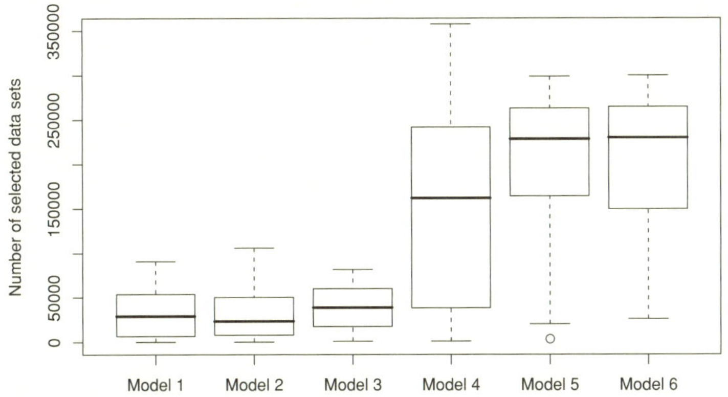

Fig. 4.2 Boxplot of the number of the selected data sets for the models 1–6.

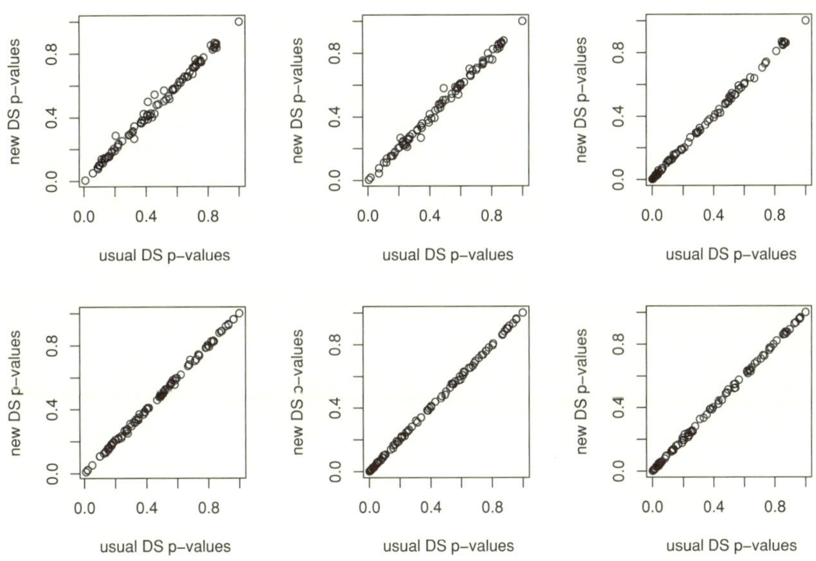

Fig. 4.3 p-values of the Diaconis–Sturmfels method (xlab) and of the new procedure (ylab) for models 1–6 (starting top from left to right).

work well as it returns nearly identical *p*-values. Examples of data sets where this is not the case can be found in (Rapallo 2003, Krampe and Kuhnt 2007).

We note that models without the conditional independence assumption between the medication with AZT and the probands status of disease exhibit relatively large *p*-values, whereas models that incorporate this independence constraint possess much smaller *p*-values.

Table 4.9 *Pearson goodness-of-fit test results for data set 1.*

		p-values	
	approximate	D-S procedure	new procedure
Model 1	0.359	0.365	0.361
Model 2	0.493	0.496	0.490
Model 3	0.018	0.021	0.017
Model 4	0.552	0.497	0.550
Model 5	0.033	0.033	0.031
Model 6	0.040	0.037	0.039
Model 7	0.060	0.058	0.059

4.6 Discussion

Using the Diaconis–Sturmfels algorithm to derive p-values within a model selection procedure so far required the simulation of an individual Markov chain for each model. The involved generation of a Gröbner basis yields high computational costs, especially when the dimension of the considered table gets larger. By concentrating on p-values for goodness-of-fit tests for graphical log-linear models we showed that it suffices to generate a Markov chain according to the independence model. For all other models approximate p-values can then be calculated by selecting those data sets from the Markov chain with the same value of the sufficient statistic as the observed data set. Further research is needed concerning the convergence rate of p-values from the simulated 'selected chains'. The choice of parameters for the simulation of the Markov chain should also be optimised. We believe, however, that the additional computational effort for the selection is out-weighed by the fact that only the Gröbner basis for the independence model is required. For graphical models with decomposable and reducible graphs as the graph of the complete independence model, (Dobra 2003, Dobra and Sullivant 2004) derived efficient procedures for the construction of Gröbner bases. Future research should also cover the comparison with other 'exact' methods as given by (Booth and Butler 1999) and extended by (Caffo and Booth 2001).

Acknowledgements

The financial support of the Deutsche Forschungsgemeinschaft (SFB 475: 'Reduction of Complexity for Multivariate Data Structures' and Graduiertenkolleg 'Statistical modelling') as well as the helpful comments of Ursula Gather and the referees are gratefully acknowledged.

References

Agresti, A. (2002). *Categorical Data Analysis*, 2nd edn (New York, John Wiley & Sons).

Bishop, Y. M. M., Fienberg, S. E. and Holland, P. W. (1995). *Discrete Multivariate Analysis* (Cambridge, MA, MIT Press).

Booth, J. G., and Butler, J. W. (1999). An importance sampling algorithm for exact conditional tests in loglinear models, *Biometrika* **86**, 321–2.

Borgelt, Ch. and Kruse, R. (2002). *Graphical Models* (Chichester, John Wiley & Sons).

Caffo, B. S. and Booth, J. G. (2001). A Markov Chain Monte Carlo algorithm for approximating exact conditional probabilities, *Journal of Computational and Graphical Statistics* **10**, 730–45.

CoCoATeam (2007). *CoCoA, a system for doing Computations in Commutative Algebra*, 4.7 edn (available at `http://cocoa.dima.unige.it`).

Chib, S. and Greenberg, E. (1995). Understanding the Metropolis-Hastings-Algorithm, *American Statistician* **49**, 327–35.

Cochran, W. G. (1954.). Some methods for strengthening the common χ^2 tests, *Biometrics* **10**, 417–51.

Conover W. J. (1971). *Practical Nonparametric Statistics* (New York, John Wiley & Sons).

Cox, D., Little, J. and O'Shea, D. (1997). *Ideals, Varieties, and Algorithms* 2nd edn (New York, Springer-Verlag).

Diaconis, P. and Sturmfels, B. (1998). Algebraic algorithms for sampling from conditional distributions, *Annals of Statistics* **26**(1), 363–97.

Dobra, A. (2003). Markov bases for decomposable graphical models, *Bernoulli* **9**, 1093–108.

Dobra, A. and Sullivant, S. (2004). A divide-and-conquer algorithm for generating Markov bases of multi-way tables, *Computational Statistics* **19**, 347–66.

Edwards, D. (2000). *Introduction to Graphical Modelling* 2nd edn (New York, Springer-Verlag).

Ewens, W. J. and Grant, G. R. (2001). *Statistical Methods in Bioinformatics. An Introduction* (New York, Springer-Verlag).

Krampe, A. and Kuhnt, S. (2007). Bowker's test for symmetry and modifications within the algebraic framework, *Computational Statistics and Data Analysis* **51**, 4124–42.

Lehmann, E. L. (1986). *Testing Statistical Hypotheses* 2nd edn (New York, John Wiley & Sons).

Madigan, D. and Raftery, A. (1994). Model selection and accounting for model uncertainty in graphical models using Occam's window, *Journal of the American Statistical Association* **89**, 1535–46.

Pistone, G., Riccomagno, E. and Wynn, H. P. (2001). *Algebraic Statistics* (Boca Raton, Chapman & Hall/CRC).

Rapallo, F. (2003). Algebraic Markov bases and MCMC for two-way contingency tables, *Scandinavian Journal of Statistics* **30**, 358–97.

Rapallo, F. (2005). Algebraic exact inference for rater agreement models, *Statistical Methods and Applications* **14**, 45–66.

Sørensen, D. and Gianola, D. (2002). *Likelihood, Bayesian, and MCMC Methods in Qualitative Genetics* (New York, Springer-Verlag).

5

Markov chains, quotient ideals and connectivity with positive margins

Yuguo Chen

Ian H. Dinwoodie

Ruriko Yoshida

Abstract

We present algebraic methods for studying connectivity of Markov moves with margin positivity. The purpose is to develop Markov sampling methods for exact conditional inference in statistical models where a Markov basis is hard to compute. In some cases positive margins are shown to allow a set of Markov connecting moves that are much simpler than the full Markov basis.

5.1 Introduction

Advances in algebra have impacted in a fundamental way the study of exponential families of probability distributions. In the 1990s, computational methods of commutative algebra were brought into statistics to solve both classical and new problems in the framework of exponential family models. In some cases, the computations are of an algebraic nature or could be made algebraic with some work, as in the cumulant methods of (Pistone and Wynn 1999). In other cases, the computations are ultimately Monte Carlo averages and the algebra plays a secondary role in designing algorithms. This is the nature of the work of (Diaconis and Sturmfels 1998). Commutative algebra is also used in statistics for experimental design (Pistone *et al.* 2001) where exponential families are not the focus.

(Diaconis and Sturmfels 1998) showed how computing a generating set for a toric ideal is fundamental to irreducibility of a Markov chain on a set of constrained tables. This theory gives a method for obtaining Markov chain moves, such as the genotype sampling method of (Guo and Thompson 1992), extensions to graphical models (Geiger *et al.* 2006) and beyond (Hosten and Sullivant 2004).

It has been argued that irreducibility is not essential (Besag and Clifford 1989), but that view is not conventional. Sparse tables in high dimensions can be very difficult to study.

Algorithms and software have been developed for toric calculations that are much faster than early methods. The volumes (Sturmfels 1996) and (Kreuzer and Robbiano 2000) are good introductions to toric ideals and some algorithms for computation. In addition, the software 4ti2 (4ti2 Team 2006) is essential to research on statistics and algebra. It is easy to use and very fast (Hemmecke and Malkin 2005).

Algebraic and Geometric Methods in Statistics, ed. Paolo Gibilisco, Eva Riccomagno, Maria Piera Rogantin and Henry P. Wynn. Published by Cambridge University Press. © Cambridge University Press 2010.

99

Despite these significant computational advances, there are applied problems where one may never be able to compute a Markov basis. Recall that a Markov basis is a collection of vector increments that preserve the table constraints and connect all tables with the same constraints, see Section 5.2. Models of no-3-way interaction and constraint matrices of Lawrence type seem to be arbitrarily difficult, in that the degree and support of elements of a minimal Markov basis can be arbitrarily large (De Loera and Onn 2005). Thus, it is useful to compute a smaller number of moves which connect tables with given constraints rather than all constraints. The purpose of this chapter is to develop algebraic tools for understanding sets of Markov moves that connect tables with positive margins, because sets of Markov moves that work with certain margins may be much simpler than a full Markov basis. Such connecting sets were formalised in (Chen *et al.* 2006) with the terminology Markov sub-basis.

Connectivity of a set of Markov moves is traditionally studied through primary decomposition (Diaconis *et al.* 1998). As a practical tool, this is problematic because the primary decomposition is very difficult to compute and also can be hard to interpret in a useful way. In our experience, the computation is very slow or impossible with 20 or more cells in the table (giving 20 or more indeterminates). Theoretical results on primary decomposition of lattice ideals are relevant, for example (Hosten and Shapiro 2000), but are generally not sufficient to determine connecting properties of sets of Markov moves. Therefore we believe that developing algebraic tools based on quotient operations and radical ideals may be more practical in large problems.

A motivating example is the following, see also Example 5.4. In logistic regression at 10 levels of an integer covariate, one has a table of counts that gives the number of 'yes' responses and the number of 'no' responses at each covariate level $i = 1, 2, \ldots, 10$. The sufficient statistics for logistic regression are (1) the total number of 'yes' responses over all levels, (2) the quantity which is the sum over i of the 'yes' count at level i multiplied by the covariate level i, and (3) the total counts of 'yes' and 'no' responses at each level i. Conditional inference requires that one works with all tables that fix these 12 values and which have non-negative entries. A Markov chain with 2465 moves from 'primitive partition identities' (Sturmfels 1996, p. 47) is irreducible in this collection of constrained tables, no matter what the 12 constraint values are. However, when each of the 10 sums over 'yes' and 'no' counts at the 10 levels of i is positive, a Markov chain with only 36 moves is irreducible (Chen *et al.* 2005). Therefore the property of positive margins can greatly simplify computations.

5.2 Arbitrary margins and toric ideals

A *contingency table* records counts of events at combinations of factors and is used to study the relationship between the factors. All possible combinations of factor labels or 'levels' make 'cells' in an array and the count in each cell may be viewed as the outcome of a multinomial probability distribution.

In this section a contingency table is written as a vector of length c and this representation comes from numbering the cells in a multi-way table. Let A be an

$r \times c$ matrix of non-negative integers with columns $\mathbf{a}_1, \ldots, \mathbf{a}_c$ in Z_+^r. The matrix A is the *design matrix* or *constraint matrix*, and the r rows are the vectors for computing sufficient statistics. The total number of constraints when sufficient statistics are fixed is r, which is also the number of parameters in a log-linear representation of the cell probabilities p_i:

$$p_i = \frac{e^{\theta' \mathbf{a}_i}}{z_\theta}$$

where z_θ is the normalising constant, and θ is a column vector of parameters in $\mathbb{R}^r$. Then the points $(p_1, \ldots, p_c)$ are in the toric variety defined by the matrix A, while also being non-negative and summing to 1.

For example, for 2×3 tables under the independence model, A is the 5×6 matrix given by

$$A = \begin{pmatrix} 1 & 1 & 1 & 0 & 0 & 0 \\ 0 & 0 & 0 & 1 & 1 & 1 \\ 1 & 0 & 0 & 1 & 0 & 0 \\ 0 & 1 & 0 & 0 & 1 & 0 \\ 0 & 0 & 1 & 0 & 0 & 1 \end{pmatrix}$$

and the rows of A compute row and column sums of the contingency table.

Assume that a strictly positive vector is in the row space of A. The toric ideal I_A in the ring $\mathbb{Q}[\mathbf{x}] = \mathbb{Q}[x_1, x_2, \ldots, x_c]$ is defined by

$$I_A = \langle \mathbf{x}^{\mathbf{a}} - \mathbf{x}^{\mathbf{b}} : A\mathbf{a} = A\mathbf{b} \rangle$$

where $\mathbf{x}^{\mathbf{a}} = x_1^{a_1} x_2^{a_2} \cdots x_c^{a_c}$ is the usual monomial notation. Define the fiber $\Omega_{\mathbf{t}} := \{\mathbf{n} \in Z_+^c : A\mathbf{n} = \mathbf{t}\}$ (non-negative integer lattice points) for $\mathbf{t} = (t_1, \ldots, t_r) \in Z_+^r$. That is, the fiber is the set of all contingency tables satisfying the given constraints. It is known that a generating set of binomials $\{\mathbf{x}^{\mathbf{a}_i^+} - \mathbf{x}^{\mathbf{a}_i^-}\}$ for I_A provide increments $\{\pm(\mathbf{a}_i^+ - \mathbf{a}_i^-)\}$ that make an irreducible Markov chain in $\Omega_{\mathbf{t}}$, whatever the value of $\mathbf{t}$ (Diaconis and Sturmfels 1998). Here $\mathbf{a}_i^+ = \max\{\mathbf{a}_i, 0\}$ and $\mathbf{a}_i^- = \max\{-\mathbf{a}_i, 0\}$. Such a generating set is called a Markov basis. The Markov chain is run by randomly choosing one of the increments $\mathbf{a}_i^+ - \mathbf{a}_i^-$ and randomly choosing a sign, then adding the increment to the current state if the result is non-negative. Irreducible means that for any two non-negative integer vectors $\mathbf{m}, \mathbf{n}$ that satisfy $A\mathbf{m} = A\mathbf{n} = \mathbf{t}$, there is a sequence of signed vectors $\sigma_j(\mathbf{a}_{i_j}^+ - \mathbf{a}_{i_j}^-)$, $j = 1, 2, \ldots, J$ ($\sigma_j = \pm 1$), that connects $\mathbf{m}$ and $\mathbf{n}$. That is, $\mathbf{n} = \mathbf{m} + \sum_{j=1}^{J} \sigma_j(\mathbf{a}_{i_j}^+ - \mathbf{a}_{i_j}^-)$ and furthermore every intermediate point in the path remains in the domain:

$$\mathbf{m} + \sum_{j=1}^{I} \sigma_j(\mathbf{a}_{i_j}^+ - \mathbf{a}_{i_j}^-) \in \Omega_{\mathbf{t}}, \quad 1 \leq I \leq J.$$

In particular, intermediate points on the path are non-negative.

When one allows entries in the table to go negative, connecting Markov chains are easier to find. Proposition 5.1 below uses some standard terminology. Let $M := \{\pm \mathbf{a}_i \in Z^c : i = 1, \ldots, g\} \subset \ker(A)$ be signed Markov moves (that is, integer vectors in $\ker(A)$ that are added or subtracted randomly from the current state), not necessarily a Markov basis. Let $I_M := \langle \mathbf{x}^{\mathbf{a}_i^+} - \mathbf{x}^{\mathbf{a}_i^-} , \ i = 1, \ldots, g \rangle$ be the corresponding ideal, which satisfies $I_M \subset I_A$. For the definition of radical ideals see Appendix 1.7.

A set of integer vectors $M \subset Z^c$ is called a *lattice basis* for A if every integer vector in $\ker(A)$ can be written as an integral linear combination of the vectors (or moves) in M. Computing a lattice basis is very simple and does not require symbolic computation.

Proposition 5.1 *Suppose I_M is a radical ideal and suppose the moves in M form a lattice basis. Then the Markov chain using the moves in M that allow entries to drop down to -1 connects a set that includes $\Omega_{\mathbf{t}}$.*

Proof Let $\mathbf{m}, \mathbf{n}$ be two elements in $\Omega_{\mathbf{t}}$. By allowing entries to drop down to -1 in the Markov chain, it is enough to show that $\mathbf{m} + \mathbf{1}$ and $\mathbf{n} + \mathbf{1}$ are connected with a non-negative path using moves in M. By (Sturmfels 2002, Theorem 8.14) $\mathbf{m} + \mathbf{1}$ and $\mathbf{n} + \mathbf{1}$ are connected in this way if $\mathbf{x}^{\mathbf{m}+\mathbf{1}} - \mathbf{x}^{\mathbf{n}+\mathbf{1}}$ are in the ideal $I_M \subset \mathbf{Q}[\mathbf{x}]$. Let $p = x_1 \cdot x_2 \cdot \ldots \cdot x_c$. Since the moves are a lattice basis, it follows that $I_M : p^n = I_A$ for some integer $n > 0$ (Sturmfels 1996, Lemma 12.2). Thus $p^n (\mathbf{x}^{\mathbf{m}} - \mathbf{x}^{\mathbf{n}}) \in I_M$ by the definition of the quotient ideal. Hence $p^n (\mathbf{x}^{\mathbf{m}} - \mathbf{x}^{\mathbf{n}})^n \in I_M$, and since I_M is radical it follows that $\mathbf{x}^{\mathbf{m}+\mathbf{1}} - \mathbf{x}^{\mathbf{n}+\mathbf{1}} = p (\mathbf{x}^{\mathbf{m}} - \mathbf{x}^{\mathbf{n}}) \in I_M$. $\qquad\square$

The idea of allowing some entries to drop down to -1 appears in (Bunea and Besag 2000) and (Chen *et al.* 2005). In high-dimensional tables (c large), the enlarged state space that allows entries to drop down to -1 may be much larger than the set of interest $\Omega_{\mathbf{t}}$, even though each dimension is only slightly extended. Nevertheless, Proposition 5.1 makes it possible to use the following approach on large tables: compute a lattice basis, compute the radical of the ideal of binomials from the lattice basis, run the Markov chain in the larger state space, and do computations on $\Omega_{\mathbf{t}}$ by conditioning. More precisely, suppose $\Omega_{\mathbf{t}} \subset \Omega_0$ where the set Ω_0 is the connected component of the Markov chain that is allowed to drop down to -1. Suppose the desired sampling distribution μ on $\Omega_{\mathbf{t}}$ is uniform. If one runs a symmetric Markov chain $X_1, X_2, X_3, \ldots, X_n$ in Ω_0, then a Monte Carlo estimate of $\mu(A)$ for any subset $A \subset \Omega_{\mathbf{t}}$ is

$$\mu(A) \approx \frac{\sum_{i=1}^n I_A(X_i)}{\sum_{i=1}^n I_{\Omega_{\mathbf{t}}}(X_i)}$$

where I_A is the indicator function of the set A.

5.3 Survey of computational methods

A log-linear model for a multi-way table of counts can be fitted and evaluated many ways. Maximum likelihood fitting and asymptotic measures of goodness-of-fit are available from Poisson regression on a data frame, part of any generalised linear model package such as the one in R (R Development Core Team 2004). The R command `loglin` also does table fitting, using iterative proportional fitting and this is more convenient than Poisson regression when the data is in a multidimensional array. Both methods rely on χ^2 asymptotics on either the Pearson χ^2 statistic or likelihood ratio statistics for goodness-of-fit. For sparse tables, one often wants exact conditional methods to avoid asymptotic doubts. The basic command `chisq.test` in R has an option for the exact method on two-way tables, usually called Fisher's exact test.

For higher-way tables, the package `exactLoglinTest` is maintained by Brian Caffo (Caffo 2006). This implements an importance sampling method of (Booth and Butler 1999). There are certain examples where it has difficulty generating valid tables, but user expertise can help.

Markov chains can be run with a set of Markov moves that come from generators of a toric ideal. Computing these generators can be done in many algebra software packages, including CoCoA (CoCoATeam 2007), Macaulay 2 (Grayson and Stillman 2006) and SINGULAR (Greuel *et al.* 2005) which implement several algorithms. Finally, 4ti2 (4ti2 Team 2006) was used for computing Markov bases in this chapter. It is very fast, it has a natural coding language for statistical problems and it has utilities for filtering output.

A Monte Carlo method that is extremely flexible and does not require algebraic computations in advance is sequential importance sampling (Chen *et al.* 2006). This method uses linear programming to generate tables that in practice satisfy constraints with very high probability. Efficient implementation requires a good proposal distribution.

5.4 Margin positivity

The Markov basis described in Section 5.2 is a very powerful construction. It can be used to construct an irreducible Markov chain for any margin values $\mathbf{t}$. It is possible that a smaller set of moves may connect tables when $\mathbf{t}$ is strictly positive. The notion of Markov sub-basis was introduced in (Chen *et al.* 2006) to study connecting sets of moves in $\Omega_{\mathbf{t}}$ for certain values of $\mathbf{t}$.

Now a lattice basis for $\ker(A)$ has the property that any two tables can be connected by its vector increments if one is allowed to swing negative in the connecting path. See (Schrijver 1989, p. 47) and (Sturmfels 1996, Chapter 12) for definitions and properties of a lattice basis. One may expect that if the margin values $\mathbf{t}$ are sufficiently large positive numbers, then the paths can be drawn out of negative territory and one may get non-negative connecting paths and so remain in $\Omega_{\mathbf{t}}$. However, in general, large positive margin values do not make every lattice basis a connecting set, as illustrated below.

Example 5.1 This example is from (Sturmfels, 2002, p. 112). With moves of adjacent minors (meaning the nine adjacent $\begin{smallmatrix} + & - \\ - & + \end{smallmatrix}$ sign pattern vector increments in the matrix), it is clear that one cannot connect the following tables, no matter how large the margins $3n$ may be:

$$
\begin{array}{|cccc|}
\hline
n & n & 0 & n \\
0 & 0 & 0 & n \\
n & 0 & 0 & n \\
n & 0 & n & n \\
\hline
\end{array}
\, , \quad
\begin{array}{|cccc|}
\hline
n & n & 0 & n \\
n & 0 & 0 & n \\
0 & 0 & 0 & n \\
n & n & 0 & n \\
\hline
\end{array}
$$

Adjacent minors have been studied in depth, see e.g. (Hosten and Sullivant 2002).

Proposition 5.2 *Let A be a 0-1 matrix. Suppose there is an integer lower bound $b > 0$ on all the constraint values: $t_m \geq b$, $m = 1, 2, \ldots, r$. Let $I_m = \langle x_k \rangle_{A_{m,k} > 0}$ be the monomial ideal generated by all the indeterminates for the cells that contribute to margin m. If*

$$
I_A \cap \bigcap_{m=1}^{r} I_m^b \subset I_M
$$

where $I_m^b = \langle x_{i_1} x_{i_2} \cdots x_{i_b} \rangle_{A_{m,i_k} > 0}$, then the moves in M connect all tables in Ω_t.

Proof Let $\mathbf{m}$ and $\mathbf{n}$ be two tables in Ω_t. It is sufficient to show that $\mathbf{x^m} - \mathbf{x^n} \in I_M$, by (Sturmfels, 2002, Theorem 8.14). Now clearly $\mathbf{x^m} - \mathbf{x^n} \in I_A$. Since all the constraint values t_m are positive and A has 0-1 entries, it follows that each monomial $\mathbf{x^m}$ and $\mathbf{x^n}$ belongs to $I_m^b = \langle x_{i_1} x_{i_2} \cdots x_{i_b} \rangle_{A_{m,i_k} > 0}$. Thus the binomial $\mathbf{x^m} - \mathbf{x^n} \in I_A \cap \bigcap_{m=1}^{r} I_m^b$.

Thus it is sufficient to show that

$$
I_A \cap \bigcap_{m} I_m^b \subset I_M
$$

which is the condition of the proposition. $\qquad\square$

This result can establish connectivity in examples where the primary decomposition is hard to compute. It does not require I_M to be radical.

Let $p = x_1 x_2 \cdots x_c$ and let $I_M : p^\infty$ be the *saturation* of I_M by p, namely,

$$
I_M : p^\infty := \{ g \in \mathbb{Q}[\mathbf{x}] : p^k \cdot g \in I_M \text{ for some } k \geq 0 \}.
$$

Then $I_A = I_M : p^\infty$ when the moves in M form a lattice basis (Sturmfels 1996, Lemma 12.2). One can show easily that

$$
I_A \cap \bigcap_{m=1}^{r} I_m \subset \left(I_M \cap \bigcap_{m=1}^{r} I_m \right) : p^\infty
$$

but the right-hand side seems hard to compute directly, so this way of computing moves for tables with positive margins does not seem efficient. The ideal $\bigcap_m I_m$ is a monomial ideal for the Stanley–Reisner complex given by subsets of sets of cell indices *not* in the margins. For example, for 2×3 tables with fixed row and column sums as in Example 5.3 and cells labelled left to right, the ideals are $\langle x_1, x_2, x_3 \rangle \cap \langle x_4, x_5, x_6 \rangle \cap \langle x_1, x_4 \rangle \cap \langle x_2, x_5 \rangle \cap \langle x_3, x_6 \rangle$ and the simplicial complex is all subsets of the sets $\{\{4, 5, 6\}, \{1, 2, 3\}, \{2, 3, 5, 6\}, \{1, 3, 4, 6\}, \{1, 2, 4, 5\}\}$.

Example 5.2 Consider the collection of 3×3 tables with fixed row and column sums. If the margin values are all positive, then the collection of four moves of adjacent minors is not necessarily a connecting set. Consider the two tables below:

<table>
<tr><td>1</td><td>0</td><td>0</td></tr>
<tr><td>0</td><td>0</td><td>1</td></tr>
<tr><td>0</td><td>1</td><td>0</td></tr>
</table>

,

<table>
<tr><td>0</td><td>1</td><td>0</td></tr>
<tr><td>0</td><td>0</td><td>1</td></tr>
<tr><td>1</td><td>0</td><td>0</td></tr>
</table>

.

However, if all the six margin values are at least $b = 2$, then one can apply Proposition 5.2 to the moves M of adjacent minors, which do not form a radical ideal. The toric ideal I_A can be computed and the containment required can be shown with $I_M : \left(I_A \cap \bigcap_{m=1}^{6} I_m^2 \right) = \langle 1 \rangle$.

Theorem 5.1 *Suppose I_M is a radical ideal and suppose M is a lattice basis. Let $p = x_1 \cdot x_2 \cdot \ldots \cdot x_c$. For each row index m with $t_m > 0$, let $I_m = \langle x_k \rangle_{A_{m,k} > 0}$ be the monomial ideal generated by indeterminates for cells that contribute to margin m. Let $\mathcal{M}$ be the collection of indices m with $t_m > 0$. Define*

$$I_{\mathcal{M}} = I_M : \prod_{m \in \mathcal{M}} I_m .$$

If $I_{\mathcal{M}} : (I_{\mathcal{M}} : p) = \langle 1 \rangle$, then the moves in M connect all tables in Ω_t.

Proof Let **m** and **n** be two tables in Ω_t with margins $\mathcal{M}$ positive. It is sufficient to show that $\mathbf{x^m} - \mathbf{x^n} \in I_M$, by (Sturmfels, 2002, Theorem 8.14). Now clearly $\mathbf{x^m} - \mathbf{x^n} \in I_A$ and since the margins $\mathcal{M}$ are positive it follows that $\mathbf{x^m} - \mathbf{x^n} \in \bigcap_{m \in \mathcal{M}} I_m$. Thus it is sufficient to show that

$$I_A \cap \bigcap_{m \in \mathcal{M}} I_m \subset I_M .$$

Since I_M is radical, this will follow if

$$I_A \cdot \prod_{m \in \mathcal{M}} I_m \subset I_M ,$$

which holds if $I_M : (\prod_{m \in \mathcal{M}} I_m \cdot I_A) = (I_M : \prod_{m \in \mathcal{M}} I_m) : I_A = \langle 1 \rangle$. This condition follows if $I_A \subset I_M : \prod_{m \in \mathcal{M}} I_m = I_{\mathcal{M}}$.

If $I_{\mathcal{M}} : (I_{\mathcal{M}} : p) = \langle 1 \rangle$, it follows that $I_{\mathcal{M}} = I_{\mathcal{M}} : p$. Then furthermore, $I_{\mathcal{M}} = I_{\mathcal{M}} : p^\infty$. Since M is a lattice basis, it follows (Sturmfels 1996, Lemma 12.2) that $I_A = I_M : p^\infty \subset I_{\mathcal{M}} : p^\infty = I_{\mathcal{M}} : p$. This shows that $I_A \subset I_{\mathcal{M}} : p = I_{\mathcal{M}}$ and the result is proven. $\qquad \square$

5.5 Additional examples

In this section we apply the results on further examples, starting with the simplest for illustration and clarification of notation. We also do an example of logistic regression where the results are useful and an example of no-3-way interaction where it is seen that the results are not useful.

Example 5.3 Consider the simplest example, the 2×3 table with fixed row and column sums, which are the constraints from fixing sufficient statistics in an independence model. If the second column sum is positive, then tables can be connected with adjacent minors. This is well known based on primary decomposition. Indeed, the two moves corresponding to increments

$$
\begin{array}{|c|c|c|}
\hline
+1 & -1 & 0 \\
\hline
-1 & +1 & 0 \\
\hline
\end{array}
,
\qquad
\begin{array}{|c|c|c|}
\hline
0 & +1 & -1 \\
\hline
0 & -1 & +1 \\
\hline
\end{array}
$$

make the radical ideal $I_M = \langle x_{11}x_{22} - x_{12}x_{21}, x_{12}x_{23} - x_{13}x_{22} \rangle$ in $\mathbb{Q}[x_{11}, x_{12}, x_{13}, x_{21}, x_{22}, x_{23}]$. Then I_M has primary decomposition equal to $I_A \cap \langle x_{12}, x_{22} \rangle$, which shows that the binomial $\mathbf{x}^{\mathbf{m}} - \mathbf{x}^{\mathbf{n}}$ for two tables $\mathbf{m}, \mathbf{n}$ with the same row and column sums can be connected by the two moves of adjacent minors if either x_{12} or x_{22} is present in $\mathbf{x}^{\mathbf{m}}$ and either is present in $\mathbf{x}^{\mathbf{n}}$, in other words, if the second column sum is positive.

Also, Theorem 5.1 applies. The set $\mathcal{M}$ has one index for the second column margin and $I_M = I_M : \langle x_{12}, x_{22} \rangle = I_A$. Hence $I_M : (I_M : x_{11}x_{12}x_{13}x_{21}x_{22}x_{23}) = I_A : (I_A : x_{11}x_{12}x_{13}x_{21}x_{22}x_{23}) = \langle 1 \rangle$.

Example 5.4 Consider the logistic regression problem with a 2×7 table and constraints of fixed row and column sums (9 constraints) in addition to fixed regression weighted sum $\sum_{i=1}^{7} i\, n_{1,i}$. The set-up and connection with exponential families is described in (Diaconis and Sturmfels 1998, p. 387). Consider the 15 moves like

$$
\begin{array}{|c|c|c|c|c|c|c|}
\hline
0 & +1 & -1 & 0 & -1 & +1 & 0 \\
\hline
0 & -1 & +1 & 0 & +1 & -1 & 0 \\
\hline
\end{array}
.
$$

The ideal I_M is radical, even though initial terms in a Gröbner basis are not square-free. It is known that such moves connect tables with positive column sums (Chen *et al.* 2005). This was not deduced from the primary decomposition, which we have not yet computed. Theorem 5.1 does apply and computing the radical ideal in order to verify the conditions of the theorem is not difficult. We have seven monomial ideals for the column sums given by $I_i = \langle x_{1,i}, x_{2,i} \rangle$ and the quotient ideal $I_M = I_M : (I_1 \cdot I_2 \cdots I_7)$ is the toric ideal I_A with 127 elements in the reduced Gröbner basis.

A widely used class of models in applications is the no-3-way interaction class. For example, if one has four factors A, B, C, D for categorical data, each with several levels, the no-3-way interaction model is the log-linear model described with the common notation [A, B], [A, C], [A, D], [B, C], [B, D], [C, D]; see (Christensen 1990) for notation and definitions. That is, the sufficient statistics are given by sums of counts that fix all pairs of factors at specified levels. The Markov basis calculations for these models are typically hard, even for the $4 \times 4 \times 4$ case. (Whittaker 1990) presents an 8-way binary table of this type, for which we have not yet computed the Markov basis but which can be approached with sequential importance sampling.

Given the difficulty of these models, it would be interesting and useful if positive margins lead to simpler Markov bases. The answer seems to be no. Consider the natural class of moves $M = \{(e_{i,j,\mathbf{k}} + e_{i',j',\mathbf{k}} - e_{i',j,\mathbf{k}} - e_{i,j',\mathbf{k}}) - (e_{i,j,\mathbf{k}'} + e_{i',j',\mathbf{k}'} - e_{i',j,\mathbf{k}'} - e_{i,j',\mathbf{k}'}), \ldots\}$. Also, permute the location of $i, j, \mathbf{k}$. That is, choose two different coordinates from the d coordinates (above it is the first two), and choose two different levels i, i' and j, j' from each. Choose two different vectors $\mathbf{k}, \mathbf{k}'$ for all the remaining coordinates. This collection is in $\ker(A)$. The example below shows that these moves do not connect tables with positive margins.

Example 5.5 Consider 4-way binary data and order the 2^4 cells 0000, 1000, 0100, 1100, ..., 1111. There are 20 moves M of degree 8 as described above which preserve sufficient statistics for the no-3-way interaction model. More precisely, the representation of moves M above $(e_{i,j,\mathbf{k}} + e_{i',j',\mathbf{k}} - e_{i',j,\mathbf{k}} - e_{i,j',\mathbf{k}}) - (e_{i,j,\mathbf{k}'} + e_{i',j',\mathbf{k}'} - e_{i',j,\mathbf{k}'} - e_{i,j',\mathbf{k}'})$ gives square-free degree-8 moves, including for example $(e_{1100} + e_{0000} - e_{0100} - e_{1000}) - (e_{1101} + e_{0001} - e_{0101} - e_{1001})$. The representation is redundant and only 20 of them are needed to connect the same set of tables. To see this, first compute a Gröbner basis using 4ti2 for the model. This gives 61 moves and 20 square-free moves of lowest total degree 8, under a graded term order. Each of the degree-8 moves in M reduces to 0 under long division by the Gröbner basis, and this division process can only use the degree-8 moves of the Gröbner basis, since the dividend has degree 8. Now the degree-8 moves in the Gröbner basis are the 20 degree-8 moves from M. Therefore these 20 moves connect everything that M connects.

Consider two tables given by

$$(0,0,1,0,1,0,0,2,0,1,0,0,0,0,1,0), (0,0,0,1,0,1,2,0,1,0,0,0,0,0,0,1).$$

These tables have the same positive margin vectors, but the 20 moves do not connect the two tables. This can be verified in Singular (Greuel *et al.* 2005) by division, long division, of the binomial $x_3 x_5 x_8^2 x_{10} x_{15} - x_4 x_6 x_7^2 x_9 x_{16}$ by a Gröbner basis for the ideal of 20 moves does not leave remainder 0.

Example 5.6 Consider $4 \times 4 \times 2$ tables with constraints [A, C], [B, C], [A, B] for factors A, B, C, which would arise for example in case-control data with two factors A and B at four levels each.

The constraint matrix that fixes row and column sums in a 4×4 table gives a toric ideal with a $\binom{4}{2} \times \binom{4}{2}$ element Gröbner basis. Each of these moves can be paired with its signed opposite to get 36 moves of $4 \times 4 \times 2$ tables that preserve sufficient statistics:

$$\begin{array}{|cccc|}
\hline
0 & 0 & 0 & 0 \\
+1 & 0 & -1 & 0 \\
0 & 0 & 0 & 0 \\
-1 & 0 & +1 & 0 \\
\hline
\end{array}, \quad \begin{array}{|cccc|}
\hline
0 & 0 & 0 & 0 \\
-1 & 0 & +1 & 0 \\
0 & 0 & 0 & 0 \\
+1 & 0 & -1 & 0 \\
\hline
\end{array}.$$

These elements make an ideal with a Gröbner basis that is square-free in the initial terms and hence the ideal is radical (Sturmfels, 2002, Proposition 5.3). Then applying Theorem 5.1 with 16 margins of case-control counts shows that these 36 moves do connect tables with positive case-control sums. The full Markov basis has

204 moves. This example should generalise to a useful proposition on extending Markov moves for simple models to an extra binary variable. The results of (Bayer *et al.* 2001) on Lawrence liftings may be useful for a more general result.

(Fallin *et al.* 2001) present case-control data with four binary factors, which are nucleotides at four loci related to Alzheimer's disease. The statistical question is whether the model of independence of nucleotides at these loci fits the data. One has five factors: L1, L2, L3, L4, for the four loci and C for the binary case-control variable. The constraint matrix for exact conditional analysis is the Lawrence lifting of the independence model on L1, L2, L3, L4, which is described in log-linear notation as [L1, C], [L2, C], [L3, C], [L4, C], [L1, L2, L3, L4]. The next example is an algebraic treatment of the situation with three loci L1, L2, L3. A general result for any number of binary factors would be interesting. Further examples of case-control data where such results could be applied are in (Chen *et al.* 2007).

Example 5.7 Consider the 4-way binary model [L1, C], [L2, C], [L3, C], [L1, L2, L3]. There is a natural set of 12 degree 8 moves that comes from putting the degree 4 moves from the independence model [L1], [L2], [L3] at level C=1 and matching them with the opposite signs at level C=0. This construction is very general for case-control data. The resulting ideal I_M is radical. Suppose the case-control sums are positive, or, in other words, suppose that the 2^3 constraints described by [L1, L2, L3] are positive. Then one can show that these 12 moves connect all tables.

5.6 Conclusions

We have presented algebraic methods for studying connectivity of moves with margin positivity. The motivation is that two kinds of constraint matrices lead to very difficult Markov basis calculations and they arise often in applied categorical data analysis. The first kind are the matrices of Lawrence type, which come up in case-control data. The second kind are the models of no-3-way interaction, which come up when three or more factors are present and one terminates the model interaction terms at 2-way interaction.

The examples that we have studied suggest that further research on connecting moves for tables with constraints of Lawrence type and with positive margins would have theoretical and applied interest. In this setting it does appear that there can be Markov connecting sets simpler than the full Markov basis. On the other hand, margin positivity does not seem to give much simplification of a Markov connecting set in problems of no-3-way interaction. Finally, radical ideals of Markov moves have valuable connectivity properties and efficient methods for computing radicals and verifying radicalness would be useful. When the full toric ideal is too complicated, working with a radical ideal may be possible.

Acknowledgements

Yuguo Chen was partly supported under NSF grant DMS-0503981.

References

4ti2 Team (2006). *4ti2 – A software package for algebraic, geometric and combinatorial problems on linear spaces* (available at www.4ti2.de).

Bayer, D., Popescu, S., and Sturmfels, B. (2001). Syzygies of unimodular Lawrence ideals, *Journal für die reine und angewandte Mathematik* **534**, 169–86.

Besag, J., and Clifford, P. (1989). Generalized Monte Carlo significance tests, *Biometrika* **76**, 633–42.

Booth, J. G., and Butler, J. W. (1999). An importance sampling algorithm for exact conditional tests in loglinear models, *Biometrika* **86**, 321–32.

Bunea, F., and Besag, J. (2000). MCMC in $I \times J \times K$ contingency tables, Fields Institute Communications **26**, 23–36.

Caffo, B. (2006). *exactLoglinTest: A Program for Monte Carlo Conditional Analysis of Log-linear Models* (available at www.cran.r-project.org).

Chen, Y., Dinwoodie, I. H., Dobra, A. and Huber, M. (2005). Lattice points, contingency tables and sampling. In *Contemporary Mathematics*. Barvinok, A., Beck, M., Haase, C., Reznick, B., and Welker, V. eds. (American Mathematical Society Vol. 374) 65–78.

Chen, Y., Dinwoodie, I. H., and MacGibbon, B. (2007). Sequential importance sampling for case-control data, *Biometrics* **63**(3), 845–55.

Chen, Y., Dinwoodie, I. H., and Sullivant, S. (2006). Sequential importance sampling for multiway tables, *Annals of Statistics* **34**, 523–45.

Christensen, R. (1990). *Log-Linear Models* (New York, Springer-Verlag).

CoCoA Team (2007). *CoCoA, a system for doing Computations in Commutative Algebra*, 4.7 edn (available at http://cocoa.dima.unige.it).

De Loera, J. and Onn, S. (2005). Markov bases of three-way tables are arbitrarily complicated, *Journal of Symbolic Computation* **41**, 173–81.

Diaconis, P. and Sturmfels, B. (1998). Algebraic methods for sampling from conditional distributions, *Annals of Statistics* **26**, 363–97.

Diaconis, P., Eisenbud, D., and Sturmfels, B. (1998). Lattice walks and primary decomposition. In *Mathematical Essays in Honor of Gian-Carlo Rota*, Sagan, B. E. and Stanley, R. P. eds. (Boston, Birkhauser) 173–93.

Fallin, D., Cohen, A., Essioux, L., Chumakov, I., Blumenfeld, M., Cohen, D., and Schork, N. J. (2001). Genetic analysis of case/control data using estimated haplotype frequencies: application to APOE locus variation and Alzheimer's disease, *Genome Research* **11**, 143–51.

Geiger, D., Meek, C., and Sturmfels, B. (2006). On the toric algebra of graphical models, *Annals of Statistics* **34**, 1463–92.

Grayson, D. and Stillman, M. (2006). *Macaulay 2, a software system for research in algebraic geometry* (available at www.math.uiuc.edu/Macaulay2/).

Greuel, G.-M., Pfister, G. and Schönemann, H. (2005). SINGULAR *3.0. A Computer Algebra System for Polynomial Computations. Centre for Computer Algebra* (available at www.singular.uni-kl.de).

Guo, S. W., and Thompson, E. A. (1992). Performing the exact test of Hardy-Weinberg proportion for multiple alleles, *Biometrics* **48**, 361–72.

Hemmecke, R., and Malkin, P. (2005). Computing generating sets of lattice ideals (available at arXiv:math.CO/0508359).

Hosten, S., and Shapiro, J. (2000). Primary decomposition of lattice basis ideals, *Journal of Symbolic Computation* **29**, 625–39.

Hosten, S., and Sullivant, S. (2002). Gröbner basis and polyhedral geometry of reducible and cyclic models, *Journal of Combinatorial Theory A* **100**, 277–301.

Hosten, S., and Sullivant, S. (2004). Ideals of adjacent minors, *Journal of Algebra* **277**, 615–42.

Kreuzer, M., and Robbiano, L. (2000). *Computational Commutative Algebra* (New York, Springer-Verlag).

Pistone, G., and Wynn, H. (1999). Finitely generated cumulants, *Statistica Sinica* **9**(4), 1029–52.

Pistone, G., Riccomagno, E. and Wynn, H. P. (2001). *Algebraic Statistics* (Boca Raton, Chapman & Hall/CRC).

R Development Core Team (2004). *R: A Language and Environment for Statistical Computing* (available at www.R-project.org).

Schrijver, A. (1989). *Theory of linear and integer programming* (Chichester, John Wiley & Sons).

Sturmfels, B. (1996). *Gröbner Bases and Convex Polytopes* (Providence, RI, American Mathematical Society).

Sturmfels, B. (2002). *Solving Systems of Polynomial Equations* (Providence, RI, American Mathematical Society).

Whittaker, J. (1990). *Graphical Models in Applied Mathematical Multivariate Statistics* (Chichester, John Wiley & Sons).

6

Algebraic modelling of category distinguishability

Enrico Carlini

Fabio Rapallo

Abstract

Algebraic Statistics techniques are used to define a new class of probability models which encode the notion of category distinguishability and refine the existing approaches. We study such models both from a geometric and statistical point of view. In particular, we provide an effective characterisation of the sufficient statistic.

6.1 Introduction

In this work we focus on a problem coming from rater agreement studies. We consider two independent raters. They classify n subjects using the same ordinal scale with I categories. The data are organised in a square contingency table which summarises the classifications. The cell (i, j) contains the number of items classified i by the first observer and j by the second observer.

Many applications deal with ordinal scales whose categories are partly subjective. In most cases, the ordinal scale is the discretisation of an underlying quantity continuous in nature. Classical examples in the field of medical applications are the classification of a disease in different grades through the reading of diagnostic images or the classification of the grade of a psychiatric disease based on the observation of some behavioural traits of the patients. An example of such problem is presented in detail in (Garrett-Mayer *et al.* 2004) and it is based on data about pancreatic neoplasia. Other relevant applications are, for instance, in lexical investigations, see e.g. (Bruce and Wiebe 1998) and (Bruce and Wiebe 1999). In their papers, category distinguishability is used as a tool to study when the definitions of the different meanings of a word in a dictionary can be considered as unambiguous. Table 6.1 presents a numerical example from (Agresti 1988). The data concern diagnoses of multiple sclerosis for two neurologists A and B classifying 149 patients on a scale with four levels from certain (1) to unlikely (4). In case of perfect distinguishability the table would be diagonal. But, in our situation, some non-diagonal cells seem to be non-negligible or, in our terminology, some categories seem to be confused.

Algebraic and Geometric Methods in Statistics, ed. Paolo Gibilisco, Eva Riccomagno, Maria Piera Rogantin and Henry P. Wynn. Published by Cambridge University Press. © Cambridge University Press 2010.

Table 6.1 *Contingency table concerning diagnoses of multiple sclerosis.*

		A			
		1	2	3	4
B	1	38	5	0	1
	2	33	11	3	0
	3	10	14	5	6
	4	3	7	3	0

A well-defined grading scale must have distinguishable categories. When two or more categories are confused, then the ordinal scale has to be redesigned following one of the strategies below:

- to reduce the number of the categories, by collapsing the confused categories;

- to improve the specifications of the 'boundaries' between the confused categories.

Therefore, a crucial problem concerning such tables is the one to check whether the categories are distinguishable or not.

To our knowledge, the first attempt to address this problem was based on the use of some techniques coming from rater agreement analysis, see e.g. (Landis and Koch 1975). Among these methods there are Cohen's κ, weighted κ and some particular log-linear models, such as quasi-independence and quasi-symmetry. Recent references for rater agreement techniques are (Agresti 2002) and (von Eye and Mun 2005).

(Darroch and McCloud 1986) showed that such methods are not suitable to solve our problem. The reasons will be discussed later in Section 6.2. They introduced the notion of category distinguishability through the analysis of some odds-ratios of the contingency table. New efforts in this direction can be found in the paper by (Agresti 1988) which considers a model of rater agreement as a sum of two components: a baseline association model plus an additional component concentrated on the main diagonal which represents the effect of the agreement. As a baseline association model one can simply use the independence model or more complex models, such as the linear-by-linear association model, see (Goodman 1979).

We use tools from Algebraic Statistics to define and analyse statistical models for the category distinguishability problem. Starting from (Diaconis and Sturmfels 1998) and (Pistone *et al.* 2001), the description of discrete probability models in terms of algebraic equations has received a great deal of attention.

The material is organised as follows. In Section 6.2 we recall some basic facts and we introduce the models we study, while in Section 6.3 we analyse the models, we show how to determine the sufficient statistic, and we present connections to estimation and goodness-of-fit testing. Section 6.4 is devoted to the description of the geometry related to these models. In Section 6.5 we add symmetry conditions and we compare our models with the classical quasi-independence and quasi-symmetry models. Finally, in Section 6.6 we present a real data example.

6.2 Background and definitions

We first review the basic ideas of category distinguishability as given in (Darroch and McCloud 1986). Let us consider an ordinal rating scale with I categories. The data are collected in an $I \times I$ contingency table and the corresponding probability distribution is a matrix of raw probabilities. We denote the probability of the cell (i, j) by $p_{i,j}$. We assume that the probabilities belong to the positive simplex

$$\Delta_{>} = \left\{ (p_{1,1}, \ldots, p_{I,I}) \in \mathbb{R}^{I \times I} \; : \; p_{i,j} > 0, \; \sum_{i,j} p_{i,j} = 1 \right\}.$$

The relevant quantities in the analysis are the odds-ratios

$$\tau_{i,j} = \frac{p_{i,i} p_{j,j}}{p_{i,j} p_{j,i}}$$

for $i, j = 1, \ldots, I$, $i \neq j$.

The categories i and j are indistinguishable if $\tau_{i,j} = 1$. We remark that the higher the agreement is the smaller the off-diagonal elements are. Therefore, large odds-ratios correspond to strong agreement. The degree of distinguishability for categories i and j is defined as

$$\delta_{i,j} = 1 - \tau_{i,j}^{-1}.$$

Notice that $\tau_{i,j} = 1$ if and only if

$$p_{i,i} p_{j,j} - p_{i,j} p_{j,i} = 0. \tag{6.1}$$

Thus, according to this first definition, two categories i and j are indistinguishable when the minor with the elements i and j of the main diagonal vanishes. We note that the binomial in Equation (6.1) is equivalent to the independence statement for the sub-table formed by the cells (i, i), (i, j), (j, i) and (j, j).

In this context, a number of authors have discussed the role of additional conditions such as marginal homogeneity and symmetry, see e.g. (Landis and Koch 1975). In (Agresti 1988) the connections between the $\tau_{i,j}$ and the quasi-independence and quasi-symmetry models are presented. Although it represents a relevant issue for applications, we leave aside this problem at the present stage. We will come back to that issue later in Section 6.5.

The approach to distinguishability in terms of the odds-ratios $\tau_{i,j}$ presents some difficulties when applied to large contingency tables as it implies the computation of $I(I-1)/2$ odds-ratios. Moreover, the $\tau_{i,j}$ approach is easily applied to pairwise comparisons of the categories, while multiple analyses are difficult to perform. On the other hand, the use of quasi-independence and quasi-symmetry models makes a global analysis easy but local properties of the table can be detected only through the analysis of the residuals.

In order to define a simple model to analyse the problem of category distinguishability, we introduce the patterns of indistinguishability in the contingency table by means of suitable subsets of $C = \{1, \ldots, I\}$, the set of the I categories. We use subsets $C_1, \ldots, C_k$ of C to determine the patterns of distinguishability. Two categories i and j are confused if they belong to one of the subsets C_r, while they are distinct if there is no subset C_r which contains both of them.

Definition 6.1 Let $C_1, \ldots, C_k$ be non-empty subsets of C of cardinality $n_1, \ldots, n_k$, possibly overlapping. We say that $C_1, \ldots, C_k$ define patterns of indistinguishability if they satisfy the following properties:

(i) C_r is a set of consecutive integers: $C_r = \{i_r, \ldots, i_r + n_r - 1\}$;

(ii) the sets $C_1, \ldots, C_k$ cover C:

$$\bigcup_{r=1}^{k} C_r = C \, ;$$

(iii) $C_{r'} \not\subseteq C_r$ for all r, r', $r \neq r'$.

The last condition in the definition of $C_1, \ldots, C_k$ prevents from trivialities and redundancies. In view of Definition 6.1, the perfect distinguishability of all categories corresponds to the partition $C_1 = \{1\}, \ldots, C_I = \{I\}$.

Some authors, see e.g. (Bernard 2003), use the notion of local independence for the analysis of local patterns of a contingency table. That notion rests on the following definition. The cell (i, j) is a cell of local independence if $p_{i,j} = p_{i,+} p_{+,j}$, where $p_{i,+}$ and $p_{+,j}$ are the marginal probabilities. Such a definition differs substantially from our models, as will be clear in the next section.

6.3 Analysis of the models and inference

We use the subsets $C_1, \ldots, C_k$ in Definition 6.1 to define constraints on the raw probabilities $p_{i,j}$ in terms of quadratic binomial equations. For all $r = 1, \ldots, k$, let n_r be the cardinality of C_r and let $C_r = \{i_r, \ldots, i_r + n_r - 1\}$. Then we define the constraints:

$$p_{i,j} p_{i+1,j+1} - p_{i,j+1} p_{i+1,j} = 0 \tag{6.2}$$

for all $i, j \in \{i_r, \ldots, i_r + n_r - 2\}$. If $n_r = 1$, then no equation is defined. In particular notice that, for each r, the constraints are equivalent to the independence model for the sub-table with rows and columns labelled $\{i_r, \ldots, i_r + n_r - 1\}$. For each subset C_r, Equation (6.2) states that $(n_r - 1)^2$ adjacent minors vanish.

Definition 6.2 The statistical model associated to $C_1, \ldots, C_k$ is defined through the set of binomials $\mathcal{B}$ *in Equation* (6.2). Therefore, the probability model assumes the form

$$M = \{p_{i,j} \ : \ \mathcal{B} = 0\} \cap \Delta_{>} \, .$$

We restrict our analysis to the open simplex $\Delta_{>}$. However, algebraic statistics allows us to consider structural zeros, i.e., statistical models in the closed simplex $\Delta_{\geq}$ with $p_{i,j} \geq 0$. In this setting, the statistical models become non-exponential and some of the properties we discuss below no longer hold. The interested reader

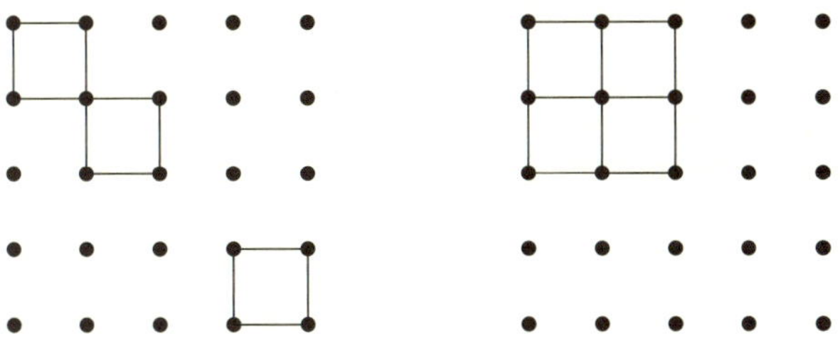

Fig. 6.1 2×2 minors for the first model (left) and for the second model (right) in Example 6.1.

can refer to (Rapallo 2007), where the behaviour of the statistical models on the boundary is studied.

In case of distinguishability of all categories, i.e.

$$C_1 = \{1\}, \ldots, C_I = \{I\},$$

we do not define any binomial equation and the corresponding probability model is saturated. Let us analyse some non-trivial examples.

Example 6.1 Suppose we have a set of five categories, $C = \{1, 2, 3, 4, 5\}$ and consider the following subsets: $C_1 = \{1, 2\}, C_2 = \{2, 3\}, C_3 = \{4, 5\}$. The corresponding probability model is defined through three binomial equations: $p_{1,1}p_{2,2} - p_{1,2}p_{2,1}$, $p_{2,2}p_{3,3} - p_{2,3}p_{3,2}$, $p_{4,4}p_{5,5} - p_{4,5}p_{5,4}$. On the other hand, if we consider the subsets $C_1 = \{1, 2, 3\}, C_2 = \{4\}, C_3 = \{5\}$, the binomials to define the model are: $p_{1,1}p_{2,2} - p_{1,2}p_{2,1}$, $p_{1,2}p_{2,3} - p_{1,3}p_{2,2}$, $p_{2,1}p_{3,2} - p_{2,2}p_{3,1}$, $p_{2,2}p_{3,3} - p_{2,3}p_{3,2}$. In Figure 6.1 the relevant 2×2 adjacent minors for these two models are illustrated.

One can also define binomial equations using the $\tau_{i,j}$. The most natural way to do this is to define

$$M_\tau = \{p_{i,j} : \tau_{h,k} = 1 \text{ for } (h, k) \in C_r \text{ for some } r\} \cap \Delta_> .$$

Notice that the equations of M_τ are not adjacent minors, but they are functions of some adjacent minors defining M. Hence, it is immediate to see that $M \subseteq M_\tau$. As M is defined only by adjacent minors, we can provide an elementary characterization of the sufficient statistic. The case of M_τ is more involved and its study is currently in progress.

Note that in our modelling the notion of indistinguishability is clearly symmetric and reflexive, but it fails to verify transitivity. As a counterexample, simply consider $I = 3$ and the subsets $C_1 = \{1, 2\}$ and $C_2 = \{2, 3\}$. The categories 1 and 2 are indistinguishable, as are the categories 2 and 3, but the categories 1 and 3 are not. In terms of the $\tau_{i,j}$ to add the transitivity property means to add more complicated binomial equations to the model. In our example, under the hypotheses $\tau_{1,2} = 1$ and $\tau_{2,3} = 1$ simple computations show that $\tau_{1,3} = 1$ is equivalent to the binomial constraint

$$p_{1,2}p_{1,3}p_{2,1}p_{3,1} - p_{1,1}^2 p_{2,3}p_{3,2} = 0\,.$$

This equation does not have an immediate meaning in terms of the probability model.

Now, we follow the theory in (Pistone *et al.* 2001) to compute the sufficient statistic for our models. As a reference in Polynomial Algebra, see (Cox *et al.* 1992). Using a vector notation, let

$$p = (p_{1,1},\ldots,p_{1,I},\ldots,p_{I,1},\ldots,p_{I,I})^t$$

be the column vector of the raw probabilities. Let $\mathbb{R}[p]$ be the polynomial ring in the indeterminates $p_{i,j}$ with real coefficients. Moreover, for any binomial $m = p^a - p^b \in \mathcal{B}$, we define its log-vector as $(a - b)$. The log-vectors of the binomials define a sub-vector space of $\mathbb{R}^{I \times I}$.

The sufficient statistic is a linear map T from the sample space $\mathcal{X} = \{1,\ldots,I\}^2$ to $\mathbb{R}^s$ for some integer s. The function T can be extended to a homomorphism from $\mathbb{R}^{I \times I}$ to $\mathbb{R}^s$ and we denote by A_T its matrix representation.

As we require the raw probabilities to be strictly positive, a binomial equation of the form $p^a - p^b = 0$ is equivalent to $\langle (a - b), \log(p) \rangle = 0$, where $\log(p) = (\log(p_{1,1}),\ldots,\log(p_{I,I}))^t$ and $\langle \cdot, \cdot \rangle$ is the inner product in $\mathbb{R}^{I \times I}$. Therefore, taking the log-probabilities, the binomials in $\mathcal{B}$ define a linear system of equations and we denote this system by

$$\log(p)^t Z_{\mathcal{B}} = 0\,. \tag{6.3}$$

The columns of $Z_{\mathcal{B}}$ are the log-vectors of the binomials in $\mathcal{B}$. If A_T is such that its columns are a basis of the orthogonal complement of the column space of $Z_{\mathcal{B}}$ in $\mathbb{R}^{I \times I}$, then the solutions of the system in Equation (6.3) are the column space of A_T, i.e.

$$\log(p) = A_T \zeta \tag{6.4}$$

for a vector ζ of unrestricted parameters.

Now, let $\#\mathcal{B}$ be the cardinality of $\mathcal{B}$. It is easy to show that the log-vectors of the elements in $\mathcal{B}$ are linearly independent, see (Haberman 1974), Chapter 5. Hence, to compute the sufficient statistic for our statistical models, we need to produce $(I^2 - \#\mathcal{B})$ linearly independent vectors.

In order to make it easier to find these vectors the following notion is useful: We say that a cell is a *free cell* if the corresponding indeterminate does not belong to any minor in $\mathcal{B}$. Now, a system of generators of the orthogonal to $Z_{\mathcal{B}}$ can be found using the following.

Proposition 6.1 *Let $C_1,\ldots,C_k \subset \{1,\ldots,I\}$ be as in Definition 6.1 and consider the corresponding set $\mathcal{B}$ of binomials defined in Equation (6.2). A system of generators of the orthogonal space to $Z_{\mathcal{B}}$ is given by the indicator vectors of the rows, of the columns and of the free cells.*

Proof Let $Z_\mathcal{B}$ be the column matrix of the log-vectors of minors in $\mathcal{B}$ and let $C_\mathcal{B}$ be its column space in $\mathbb{R}^{I \times I}$. We also let $\mathcal{L}$ be the vector space generated by the indicator functions of the rows, of the columns and of the free cells. In the case $\mathcal{B} = \mathcal{B}_0$ is the set of all adjacent minors, we have the following:

$$(C_\mathcal{B})^\perp = \mathcal{L}.$$

To build $\mathcal{B}$ from $\mathcal{B}_0$ we have to remove minors $m_1, \ldots, m_t$ and $n_1, \ldots, n_t$ which can be chosen in such a way that:

- m_i and n_i are symmetric with respect to the diagonal. (If m_i is on the main diagonal, then $m_i = n_i$);
- the monomials m_i are ordered in such a way that the difference of the indices of the topmost-rightmost variable is decreasing.

Now we proceed by induction. Let $\mathcal{B}_i$ be obtained by $\mathcal{B}_0$ removing the minors $m_1, \ldots, m_i$ and define as above $Z_{\mathcal{B}_i}$, $C_{\mathcal{B}_i}$ and $\mathcal{L}_i$. Now we assume that

$$(C_{\mathcal{B}_i})^\perp = \mathcal{L}_i.$$

When the minor m_{i+1} is removed we create at least a new free cell. Each new free cell has indicator vector not in $\mathcal{L}_i$ as it is not orthogonal to the log-vector of m_{i+1} but it is in $\left(C_{\mathcal{B}_{i+1}}\right)^\perp$. Pick one of the free cells and let v_{i+1} be its indicator vector. We conclude that

$$\left(C_{\mathcal{B}_{i+1}}\right)^\perp \supset \mathcal{L}_{i+1} = \mathcal{L}_i + \langle v_{i+1} \rangle$$

and as $\dim \mathcal{L}_{i+1} + \dim \left(C_{\mathcal{B}_{i+1}}\right)^\perp = I^2$ we have that $\left(C_{\mathcal{B}_{i+1}}\right)^\perp = \mathcal{L}_{i+1}$. Repeating this process we obtain the proof. $\qquad\square$

Equation (6.4) allows us to consider our models as log-linear models. Thus, maximum likelihood estimates of the cell probabilities can be found through numerical algorithms, such as the Fisher scoring or the Iterative Proportional Fitting. The R-package `gllm` (Generalized Log-Linear Models) is an easy tool to compute the maximum likelihood estimates of the cell probabilities. The input is formed by the observed cell counts and the design matrix A_T, see (Duffy 2006). Asymptotic chi-square p-values are then easy to compute. Non-asymptotic inference can be made through Algebraic Statistics, as extensively described for two-way tables in (Rapallo 2005). Moreover, Chapter 8 in (Sturmfels 2002) highlights connections between the maximum likelihood problem for contingency tables and the theory of systems of polynomial equations.

6.4 Geometric description of the models

The notions of distinguishability and of indistinguishability as modelled in Section 6.3 produce interesting varieties in the real affine space. In this section we introduce some properties of such varieties using a descriptive approach. To accomplish a thorough study of these objects one can follow the approach of (Hoşten and Sullivant 2004).

When the subsets $C_1, \ldots, C_k$ as in Definition 6.1 are given, we also have a partition of the set of binomials $\mathcal{B}$. Indeed, each C_r identifies a square matrix whose adjacent 2×2 minors we are considering. Hence, each C_r defines a variety V_r via the minors in Equation (6.2). The variety describing the model we study is the intersection

$$V_1 \cap \ldots \cap V_k \cap \Delta_> \, .$$

We begin with describing the variety V produced by an index set C which for the sake of simplicity we assume to be $C = \{1, \ldots, L\}$. We recall that a variety $\mathbb{X}$ is a cone of vertex $\mathbb{Y}$ if for all points $P \in \mathbb{X}$ the line joining P with any point of $\mathbb{Y}$ is contained in $\mathbb{X}$, see e.g. (Hodge and Pedoe 1994). With this definition in mind one sees that V is a cone with vertex the linear span of the coordinate points with non-zero coordinate $p_{i,j}$ with $i > L$ or $j > L$.

The cone V can also be described as the set of lines joining its vertex with a base variety B. In our case, the base variety naturally lies in a linear subspace of $\mathbb{R}^{I \times I}$

$$\{p_{i,j} = 0 \ : \ i > L \text{ or } j > L\} \supset B.$$

The base B is then defined in $\mathbb{R}^{L \times L}$ by all the adjacent minors of a general $L \times L$ matrix and it is well understood. The variety B can be described as an enlarged Segre variety, see (Harris 1995). The Segre variety describes matrices of rank 1, while our base B describes matrices having all 2×2 adjacent minors vanishing. Thus, B and the Segre variety coincide in the interior of the simplex as the vanishing of the adjacent minors implies the vanishing of all the minors if there is no zero row or column.

More precisely, B is a Segre variety unioned with some secant spaces. We recall that a secant space to B is a linear space spanned by points of B in generic linear position, e.g. a line spanned by two points of B, a plane spanned by three points of B and so on. In our case, the secant spaces to add are the ones lying on the linear spaces defined by the vanishing of a row or of a column. In other words, we have to consider the linear secant spaces spanned by points of B lying on the boundary of the simplex. Finally we remark that, as V is a cone and its vertex lies in $\Delta_>$, to describe $V \cap \Delta_>$ it is enough to describe $B \cap \Delta_>$.

As the simplex is convex and the V_r are cones, we conclude that $V_1 \cap \ldots \cap V_k \cap \Delta_>$ contains lines. To see why, notice that by the definition of the sets C_r, the vertices $\mathbb{Y}_r$ have points in common, e.g. the points corresponding to the free cells.

6.5 Adding symmetry

As mentioned in Section 6.1, in some cases the special adjacent minors in $\mathcal{B}$ are not sufficient to efficiently describe category distinguishability models. (Agresti 1988), following (Darroch and McCloud 1986), introduces further constraints to model a symmetry hypothesis. In this case, the indistinguishability of the categories i and j is defined as

$$\tau_{i,j} = 1 \quad \text{and} \quad \tau_{i,l} = \tau_{j,l} \quad \text{for all } l \neq i,j.$$

Writing down the second set of equations one obtains, for fixed i and j, $I - 2$ binomials of degree three of the form:

$$p_{i,i}p_{j,l}p_{l,j} - p_{j,j}p_{i,l}p_{l,i} = 0 \tag{6.5}$$

for $l \neq i, j$. In our construction, given any set C_r, that new condition adds to the model the constraints in Equation (6.5) for all i, j in C_r, with $i \neq j$.

Example 6.2 Consider the first model in Example 6.1, with subsets $C_1 = \{1, 2\}$, $C_2 = \{2, 3\}$ and $C_3 = \{4, 5\}$. Setting to zero the adjacent minors in Equation (6.2) and the binomials in Equation (6.5), we obtain a set of 12 binomial equations: 3 equations of degree 2 and 9 equations of degree 3. Their log-vectors define a sub-vector space with dimension 9. Standard linear algebra techniques show that a sufficient statistic for this model is given by the indicator vectors of the rows, of the columns plus 8 more vectors: $v_{i,j} = \mathbb{I}(i, j) - \mathbb{I}(j, i)$ for any free cell (i, j) and the indicator vector of the 3×3 sub-matrix obtained by deleting the last two rows and columns.

When the symmetry conditions in Equation (6.5) are assumed, the model is described by binomials of degree 3 and the computation of a sufficient statistic is more difficult than in the purely quadratic case. However, one can use symbolic software (or linear algebra software) to define the relevant binomials and to determine the matrix A_T. In our work we have used the free symbolic software CoCoA, see (CoCoA Team 2007). Below we present the pseudo-code for the model in Example 6.2.

(i) Define the list `ListS:=[[1,2],[2,3],[4,5]]` of the subsets;

(ii) Define an empty list of vectors Z. For each `C In ListS`, append to Z:

    – for each adjacent minor with rows and columns indices in C, append to Z the log-vector of the binomial of degree 2 as in Equation (6.2).

    – for each `[I,J] In C`, I<J and for each `L In 1..I`, L$\neq$I,J, append to Z the log-vector of the binomial of degree 3 as in Equation (6.5).

(iii) Define the matrix `ZMat:=Mat(Z)` and compute `AT:=LinKer(ZMat)`.

In order to compare our models with the quasi-independence and quasi-symmetry models we describe the case $I = 3$. In such a case quasi-symmetry and quasi-independence models have the same expression. We use here the classical notation for log-linear models as in (Agresti 2002). The classical quasi-independence model has the log-linear expression

$$\log p_{i,j} = \lambda + \lambda_i^X + \lambda_j^Y + \delta_i \mathbb{I}(i = j) \tag{6.6}$$

with the constraints $\sum_i \lambda_i^X = 0$, $\sum_j \lambda_j^Y = 0$, while the simplified quasi-independence model has the form

$$\log p_{i,j} = \lambda + \lambda_i^X + \lambda_j^Y + \delta \mathbb{I}(i = j) \tag{6.7}$$

where $\mathbb{I}(i = j)$ is equal to 1 when $i = j$ and 0 otherwise. The difference between Equations (6.6) and (6.7) is that the first model has one parameter for each diagonal cell, while the second one has one global parameter for all the diagonal cells. (Agresti 1988) argues that both of them have nice properties to detect category distinguishability.

In terms of binomials, the first model is described by one binomial equation, namely

$$M_{qi} = \{p_{1,2}p_{2,3}p_{3,1} - p_{1,3}p_{2,1}p_{3,2} = 0\} \cap \Delta_> \,,$$

while the second model is described by three binomial equations:

$$M_{sqi} = \{p_{1,2}p_{2,3}p_{3,1} - p_{1,3}p_{2,1}p_{3,2} = 0 \,, p_{1,1}p_{2,3}p_{3,2} - p_{1,2}p_{2,1}p_{3,3} = 0 \,,$$
$$p_{1,3}^2 p_{2,2} p_{3,2} - p_{1,2}^2 p_{2,3} p_{3,3} = 0\} \cap \Delta_> \,.$$

The models from Definition 6.1, apart from the trivial partition $C_1 = \{1\}$, $C_2 = \{2\}$, $C_3 = \{3\}$, allows three different configurations:

- Model M_1, with $C_1 = \{1, 2\}$, $C_2 = \{3\}$: C_1 and C_2 define two binomials and their log-vectors are linearly independent;
- Model M_2, with $C_1 = \{1, 2\}$, $C_2 = \{2, 3\}$: C_1 and C_2 define four binomials and their log-vectors define a sub-vector space with dimension 3;
- Model M_3, with $C_1 = \{1, 2, 3\}$: C_1 defines seven binomials and their log-vectors define a sub-vector space with dimension 4.

Simple computations based on rank of matrices show that:

- $M_{qis} \subset M_{qi}$, as is clear from their definitions in parametric form, see Equations (6.6) and (6.7);
- $M_3 \subset M_2 \subset M_1$, i.e., the models from our definition are embedded;
- M_3 is a subset of both M_{qi} and M_{qis}.

No other inclusion holds. Thus, modelling the category distinguishability through the subsets $C_1, \dots, C_k$ as in Definition 6.1, possibly adding the binomials in Equation (6.5), represents a more flexible tool with respect to log-linear models.

6.6 Final example

In Section 6.1 we presented a 4×4 contingency table. Considering the non-diagonal cell counts one can foresee indistinguishability between categories 1 and 2 and between categories 2 and 3. Therefore, we computed the maximum likelihood estimates of the cell counts for the model with three subsets $C_1 = \{1, 2\}$, $C_2 = \{2, 3\}$ and $C_3 = \{4\}$. These estimates are enclosed in parentheses in Table 6.2. The ML estimates show a good fit. In fact, the Pearson chi-square statistic is 2.5858. From the chi-square distribution with 2 df, we find an approximate p-value of 0.274. To compare this result with other common log-linear models used in rater agreement analyses, the quasi-independence model produces a chi-square statistic 21.2017 (p-value = 0.0007, based on 5 df), while the quasi-symmetry model leads to chi-square

Table 6.2 *Estimates for the example in Section 6.6.*

		1	2	3	4
				A	
B	1	38	5	0	1
		(35.07)	(7.93)	(0)	(1)
	2	33	11	3	0
		(35.93)	(8.12)	(2.95)	0
	3	10	14	5	6
		(10)	(13.95)	(5.05)	(6)
	4	3	7	3	0
		(3)	(7)	(3)	(0)

statistic 7.0985 (p-value $= 0.068$ based on 3 df). Thus, the model defined through the partition $\{\{1,2\},\{2,3\},\{4\}\}$ presents the best fit.

Acknowledgements

We are grateful to Professor Giovanni Pistone for the insight and knowledge in the field of Algebraic Statistics he shared with us. This contribution profited deeply from many fruitful conversations with him.

References

Agresti, A. (1988). A model for agreement between ratings on an ordinal scale, *Biometrics* **44**, 539–48.

Agresti, A. (2002). *Categorical Data Analysis*, 2nd edn (New York, John Wiley & Sons).

Bernard, J.-M. (2003). Analysis of local or asymmetric dependencies in contingency tables using the imprecise Dirichlct model, *Proc. ISIPTA 03*, Lugano, Switzerland, 46–61.

Bruce, R. and Wiebe, J. (1998). Word-sense distinguishability and inter-coder agreement, *Proc. EMNLP-98*, Granada, Spain, 1–8.

Bruce, R. and Wiebe, J. (1999). Recognizing subjectivity: A case study in manual tagging, *Natural Language Engineering* **5**, 187–205.

CoCoATeam (2007). *CoCoA, a system for doing Computations in Commutative Algebra*, 4.7 edn (available at `http://cocoa.dima.unige.it`).

Cox, D., Little, J. and O'Shea, D. (1992). *Ideals, Varieties, and Algorithms*, (New York, Springer Verlag).

Darroch, J. N. and McCloud, P. I. (1986). Category distinguishability and observer agreement, *Australian Journal of Statistics* **28**(3), 371–88.

Diaconis, P. and Sturmfels, B. (1998). Algebraic algorithms for sampling from conditional distributions, *Annals of Statistics* **26**(1), 363–97.

Duffy, D. (2006). *The gllm package*, 0.31 edn. (available from `http://cran.r-project.org`).

Garrett-Mayer, E., Goodman, S. N. and Hruban, R. H. (2004). The proportional odds model for assessing rater agreement with multiple modalities. Cobra Preprint #64.

Goodman, L. A. (1979). Simple models for the analysis of association in cross-classifications having ordered categories, *Journal of the American Statistical Association* **74**(367), 537–52.

Haberman, S. J. (1974). *The Analysis of Frequency Data* (Chicago and London, The University of Chicago Press).

Harris, J. (1995). *Algebraic Geometry: A First Course* (New York, Springer-Verlag).

Hodge, W. V. D. and Pedoe, D. (1994). *Methods of Algebraic Geometry, Vol. I* (Cambridge, Cambridge University Press). Reprint of the 1947 original.

Hoşten, S. and Sullivant, S. (2004). Ideals of adjacent minors, *Journal of Algebra* **277**, 615–42.

Landis, R. J. and Koch, G. G. (1975). A review of statistical methods in the analysis of data arising from observer reliability studies, Parts I and II, *Statistica Neerlandica* **29**, 101–23, 151–61.

Pistone, G., Riccomagno, E. and Wynn, H. P. (2001). *Algebraic Statistics* (Boca Raton, Chapman & Hall/CRC).

Rapallo, F. (2005). Algebraic exact inference for rater agreement models, *Statistical Methods and Applications* **14**(1), 45–66.

Rapallo, F. (2007). Toric statistical models: Binomial and parametric representations, *Annals of the Institute of Statistical Mathematics* **4**, 727–40.

Sturmfels, B. (2002). *Solving Systems of Polynomial Equations* (Providence, RI, American Mathematical Society).

von Eye, A. and Mun, E. Y. (2005). *Analyzing Rater Agreement. Manifest Variable Methods* (Mahway, NJ, Lawrence Erlbaum Associates).

7

The algebraic complexity of maximum likelihood estimation for bivariate missing data

Serkan Hoşten

Seth Sullivant

Abstract

We study the problem of maximum likelihood estimation for general patterns of bivariate missing data for normal and multinomial random variables, under the assumption that the data is missing at random (MAR). For normal data, the score equations have nine complex solutions, at least one of which is real and statistically relevant. Our computations suggest that the number of real solutions is related to whether or not the MAR assumption is satisfied. In the multinomial case, all solutions to the score equations are real and the number of real solutions grows exponentially in the number of states of the underlying random variables, though there is always precisely one statistically relevant local maxima.

7.1 Introduction

A common problem in statistical analysis is dealing with missing data in some of the repeated measures of response variables. A typical instance arises during longitudinal studies in the social and biological sciences, when participants may miss appointments or drop out of the study altogether. Over very long term studies nearly all measurements will involve some missing data, so it is usually impractical to throw out these incomplete cases. Furthermore, the underlying cause for the missing data (e.g. a subject dies) might play an important role in inference with the missing data that will lead to false conclusions in the complete case analysis. Thus, specialised techniques are needed in the setting where some of the data is missing. A useful reference for this material is (Little and Rubin 2002), from which we will draw notation and definitions. See also (Dempster *et al.* 1977) and (Little and Rubin 1983) for reviews, and (Rubin 1976) for an early reference.

In this chapter, we undertake an algebraic study of maximum likelihood estimation for general patterns of bivariate missing data, under the assumption that the data is *missing at random* (MAR) (Little and Rubin 2002). This implies, in particular, that the missing data mechanism does not affect the maximisation of the likelihood function with respect to the underlying parameters of the model, and thus the non-response is ignorable.

Algebraic and Geometric Methods in Statistics, ed. Paolo Gibilisco, Eva Riccomagno, Maria Piera Rogantin and Henry P. Wynn. Published by Cambridge University Press. © Cambridge University Press 2010.

123

Let $Y_1, \ldots, Y_n$ be i.i.d. repeated measures where $Y_j = (X_1, \ldots, X_d)$ with d response variables. We assume that the joint distribution of X_i's can be described by a parametric model. Let M be the $d \times n$ 0/1-matrix that is the indicator function for the missing entries of the Y_j; that is $M_{ij} = 1$ if and only if X_i in Y_j is missing. The missing data mechanism is determined by the conditional distribution of M given $Y = (Y_1, \ldots, Y_n)$. If we let this conditional distribution be $f(M|Y, \theta)$ where θ denotes the unknown parameters then the two missing data mechanisms, namely *missing completely at random* (MCAR) and *missing at random* (MAR) can be easily defined. The former is given by $f(M|Y, \theta) = f(M|\theta)$ for all Y and θ (i.e. the missingness does not depend on the data, missing or observed), and the latter is given by $f(M|Y, \theta) = f(M|Y_{\text{obs}}, \theta)$, for all Y_{mis} and θ (i.e. the missingness depends only on Y_{obs}, the observed components of Y, and not on the components Y_{mis} that are missing). Under MAR the log-likelihood function for the observed data is

$$\ell(\theta|Y, M) = \sum_{j=1}^{n} \log f(Y_j = y_j|\theta, M),$$

where $f(Y_j = y_j|\theta, M)$ denotes the marginal probability of observing $Y_j = y_j$ with appropriate entries of y_j missing

$$f(Y_j = y_j|\theta, M) = \int_{X_i|M_{ij}=1} f(X_{\text{obs}} = y_{\text{obs}}, X_{\text{mis}} = x_{\text{mis}}|\theta)dx_{\text{mis}}.$$

We wish to find the parameter values $\hat{\theta}$ that maximise this likelihood function.

Our focus in this chapter is on the case when $d = 2$. With a general pattern of missing data in the bivariate case, we assume that our data comes in the following form. There are n complete cases where we obtain a two-dimensional vector Y_j. There are r cases where we only obtain variable X_1, and s cases where we only obtain variable X_2. We denote these by Z_j and W_j, respectively. The log-likelihood function becomes

$$\ell(\theta; y, w, z) = \sum_{j=1}^{n} \log f(Y_j = y_j|\theta) + \sum_{j=1}^{r} \log f(Z_j = z_j|\theta) + \sum_{j=1}^{s} \log f(W_j = w_j|\theta)$$

and our goal is to maximise this function. Note that since we are assuming MAR missing data, we can ignore cases where neither variable is observed.

One approach to determining the maximum likelihood estimate uses computational algebraic geometry. The connections between maximum likelihood estimation and algebraic geometry was first extensively studied in (Catanese *et al.* 2004). These and similar approaches have been also used in (Buot and Richards 2006), (Buot *et al.* 2007) and (Hoşten *et al.* 2005). A basic fact is that, if the critical equations (score equations) are rational functions of the parameters and the data, then the number of complex solutions to the critical equations is constant for generic (i.e. almost all) data. This fixed number is called the *maximum likelihood degree* (ML-degree for short) of the model. The ML-degree is an intrinsic complexity measure of the score equations, and it is expected to give a hint about how difficult it would be to solve the maximum likelihood problem. In this chapter, we compute the

ML-degree in the bivariate missing data problem for Gaussian random variables and for multinomial random variables.

The outline of this chapter is as follows. In Section 7.2 we focus on the case where (X_1, X_2) have a jointly normal distribution. We show that the ML-degree in this case is nine. Our simulations show that if the data is indeed generated from bivariate normal distributions, and the censoring mechanism is MCAR or MAR, then there is a unique real solution to the score equations, which is a local maximum. On the other hand, we also present examples of data, where either the model or the missing data mechanism are misspecified, where there can be two statistically relevant local maxima. The possible existence of multiple maxima is important to take into account when using the EM-algorithm to find the maximum likelihood estimate. In Section 7.3 we focus on the discrete case, where (X_1, X_2) have a jointly multinomial distribution. In this setting, we give a combinatorial formula for the ML-degree.

7.2 Bivariate normal random variables

We assume that $X = (X_1, X_2) \sim \mathcal{N}(\mu, \Sigma)$ where $\mathrm{E}[X] = \mu = (\mu_1, \mu_2)$ and $\Sigma = \begin{bmatrix} \sigma_{11} & \sigma_{12} \\ \sigma_{12} & \sigma_{22} \end{bmatrix}$ is the covariance matrix. Then we have $Z_j \sim \mathcal{N}(\mu_1, \sigma_{11})$ for $j = 1, \ldots, r$ and $W_j \sim \mathcal{N}(\mu_2, \sigma_{22})$ for $j = 1, \ldots, s$. Up to scaling by a constant the log-likelihood function is equal to

$$\ell(\mu, \Gamma | y, w, z) = -\frac{1}{2} n \log(\det \Sigma) - \frac{1}{2} \Big(\sum_{j=1}^{n} (Y_j - \mu)^t \Sigma^{-1} (Y_j - \mu) \Big)$$

$$- \frac{1}{2} r \log(\sigma_{11}) - \frac{1}{2\sigma_{11}} \sum_{j=1}^{r} (Z_j - \mu_1)^2 - \frac{1}{2} s \log(\sigma_{22}) - \frac{1}{2\sigma_{22}} \sum_{j=1}^{s} (W_j - \mu_2)^2.$$

It is more convenient to use the entries of $\Gamma := \Sigma^{-1} = \begin{bmatrix} \gamma_{11} & \gamma_{12} \\ \gamma_{12} & \gamma_{22} \end{bmatrix}$ in our computations. With this substitution, we get the identities $\sigma_{11} = \gamma_{22}/\det \Gamma$, $\sigma_{22} = \gamma_{11}/\det \Gamma$, and $\sigma_{12} = -\gamma_{12}/\det \Gamma$. In the computations below we will also use a bar over a quantity to denote its average. The log-likelihood function becomes

$$\frac{1}{2}(n + r + s) \log(\det \Gamma) - \frac{1}{2} r \log \gamma_{22} - \frac{1}{2} s \log \gamma_{11} - \frac{n}{2} \Big[(\overline{Y_1^2} - 2\mu_1 \overline{Y_1} + \mu_1^2)\gamma_{11}$$

$$+ 2(\overline{Y_1 Y_2} - (\overline{Y_1}\mu_2 + \overline{Y_2}\mu_1) + \mu_1\mu_2)\gamma_{12} + (\overline{Y_2^2} - 2\mu_2 \overline{Y_2} + \mu_2^2)\gamma_{22} \Big]$$

$$- \frac{r}{2} \frac{\det \Gamma}{\gamma_{22}} (\overline{Z^2} - 2\mu_1 \overline{Z} + \mu_1^2) - \frac{s}{2} \frac{\det \Gamma}{\gamma_{11}} (\overline{W^2} - 2\mu_2 \overline{W} + \mu_2^2).$$

The critical equations for $\ell(\mu, \Gamma; y, z, w)$ are:

$$0 = \frac{\partial \ell}{\partial \mu_1} = n \big[(\overline{Y_1} - \mu_1)\gamma_{11} + (\overline{Y_2} - \mu_2)\gamma_{12} \big] + r \frac{\det \Gamma}{\gamma_{22}} (\overline{Z} - \mu_1)$$

$$0 = \frac{\partial \ell}{\partial \mu_2} = n \big[(\overline{Y_2} - \mu_2)\gamma_{22} + (\overline{Y_1} - \mu_1)\gamma_{12} \big] + s \frac{\det \Gamma}{\gamma_{11}} (\overline{W} - \mu_2)$$

$$0 = \frac{\partial \ell}{\partial \gamma_{11}} = \frac{1}{2}(n+r+s)\frac{\gamma_{22}}{\det \Gamma} - \frac{1}{2}\frac{s}{\gamma_{11}} - \frac{n}{2}(\overline{Y_1^2} - 2\mu_1 \overline{Y_1} + \mu_1^2)$$

$$- \frac{r}{2}(\overline{Z^2} - 2\mu_1 \overline{Z} + \mu_1^2) - \frac{s}{2}\frac{\gamma_{12}^2}{\gamma_{11}^2}(\overline{W^2} - 2\mu_2 \overline{W} + \mu_2^2)$$

$$0 = \frac{\partial \ell}{\partial \gamma_{22}} = \frac{1}{2}(n+r+s)\frac{\gamma_{11}}{\det \Gamma} - \frac{1}{2}\frac{r}{\gamma_{22}} - \frac{n}{2}(\overline{Y_2^2} - 2\mu_2 \overline{Y_2} + \mu_2^2)$$

$$- \frac{s}{2}(\overline{W^2} - 2\mu_2 \overline{W} + \mu_2^2) - \frac{r}{2}\frac{\gamma_{12}^2}{\gamma_{22}^2}(\overline{Z^2} - 2\mu_1 \overline{Z} + \mu_1^2)$$

$$0 = \frac{\partial \ell}{\partial \gamma_{12}} = (n+r+s)\frac{\gamma_{12}}{\det \Gamma} - n(\overline{Y_1 Y_2} - (\overline{Y_1}\mu_2 + \overline{Y_2}\mu_1) + \mu_1 \mu_2)$$

$$+ r\frac{\gamma_{12}}{\gamma_{22}}(\overline{Z^2} - 2\mu_1 \overline{Z} + \mu_1^2) + s\frac{\gamma_{12}}{\gamma_{11}}(\overline{W^2} - 2\mu_2 \overline{W} + \mu_2^2)$$

Theorem 7.1 *The ML-degree of the bivariate normal missing data problem is equal to nine, and at least one of the critical solutions to (7.1) is real. Moreover, for generic data at least one such real critical solution is a local maximum in the statistically relevant parameter space.*

Proof The theorem follows from a general principle about the number of complex solutions to a system of polynomial equations with parametric coefficients. Namely, if such a system has $N < \infty$ complex solutions (counted with multiplicity) for a 'random' choice of parameter values then other random choices of parameter values will also produce N complex solutions. Here we sketch a proof of this statement. Suppose I is an ideal in $\mathbb{C}(p_1, \ldots, p_k)[x_1, \ldots, x_t]$, the ring of polynomials in the indeterminates $x_1, \ldots, x_n$ with coefficients from the field of rational functions in $p_1, \ldots, p_k$ over $\mathbb{C}$. Pick any term order and compute a Gröbner basis G of I with respect to this term order. Now let U be the Zariski open set in $\mathbb{C}^k$ such that no denominator of the coefficients and no initial coefficient of the polynomials encountered during the Buchberger algorithm that produces G vanish on any point in U. If $\bar{p} \in U$ then both the initial ideal of I and that of $I(\bar{p})$ will have the same set of standard monomials: these are the monomials that no initial term in G and $G(\bar{p})$, respectively, divide. It is a well-known result that $I(\bar{p})$ has $N < \infty$ complex solutions (counted with multiplicity) if and only if the number of such standard monomials is N. This implies that for all $\bar{q} \in U$ the ideal $I(\bar{q})$ will have N complex solutions.

Now, in the setting of the critical Equations (7.1) let J be the ideal generated by the five polynomials obtained by clearing the denominators in (7.1). Furthermore, let K be the ideal generated by the product of these cleared denominators. Then the ML-degree we are after is the number of complex solution of $I = J : K$. A random choice of n, r, s and data vectors $y_1, \ldots, y_n, z_1, \ldots, z_r$, and $w_1, \ldots, w_s$, and a quick computation in SINGULAR shows that $I(n, r, s, y, w, z)$ has nine complex solutions. Our discussion above implies that the ML-degree of the bivariate normal missing data problem is nine. Since complex solutions to real polynomial equations come in complex conjugate pairs, at least one must be a real solution. Note that since we are taking the ideal quotient $J : K$, these nine solutions do not contain degenerate solutions where the covariance matrix is singular (i.e $\det \Gamma = 0$).

We can also see directly that there must be at least one real local maximum in the interior of the statistically relevant parameter space $\mathbb{R}^2 \times PD_2$ (where PD_2 denotes the space of 2×2 positive definite matrices). To see this, note that for generic data if any parameter has a large absolute value the log-likelihood function tends to $-\infty$. Similarly, if the Σ parameters approach the boundary of the positive definite cone the log-likelihood function tends to $-\infty$. Thus, the log-likelihood function must have a local maximum in the interior of $\mathbb{R}^2 \times PD_2$. $\qquad\square$

How many of the nine complex solutions in Theorem 7.1 can be real? We know that at least one is, but is it possible that there are three, five, seven, or nine? For various choices of the data parameters, we have observed that all of these values are possible. A more surprising fact is that the number of real solutions seems to be indicative of how well-specified the MAR assumption is. Here is a summary of the observations that emerge from our computations for which we have used Mathematica, Maple, and SINGULAR. We describe the separate cases in more details in the paragraphs following the list.

(i) When the data was generated from a Gaussian or uniform distribution and the missing data mechanism was MCAR (missing completely at random) or MAR, we consistently observed exactly one real critical point, which was necessarily a local maximum.

(ii) When the data was generated from a Gaussian distribution and the missing data mechanism was NMAR (not missing at random), we consistently observed three real critical points, all of which were in $\mathbb{R}^2 \times PD_2$ and two were local maxima.

(iii) When the joint distribution of Y and the marginal distributions of W and Z were unrelated to each other by a natural censoring mechanism, we observed seven real critical points, of which three were in the statistically relevant region, and two were statistically relevant local maxima.

(iv) When the twelve sufficient statistics $(n, r, s, \overline{Y_1}, \ldots)$ were generated randomly (without regard to an underlying distribution) we observed nine real critical points.

Of course, we could not test all possible scenarios for the above data types, and there will always be the possibility that data generated by one of the strategies will have a different number of real solutions than we observed.

When the missing data mechanism was MCAR, we generated data in an obvious way, by first generating data from a randomly chosen Gaussian distribution, and then deleting cell entries with the fixed probability $1/5$. For a more general MAR scenario, we generated data by taking a mixture of the MCAR scenario, with the missing data mechanism that covariate X_2 is not observed whenever $X_1 < -1$. Out of 1000 runs of the MAR scenario 985 cases produced a single real solution which is also a statistically relevant maximum. In fact, both of the above scenarios consistently had one real solution.

For the NMAR missing data mechanism, we generated data from a random, strongly negatively correlated Gaussian distribution, and censored covariate X_i when $X_i < -1$. Out of 1000 sample runs under this scenario 765 generated three real solutions, all statistically relevant, with two being local maxima.

For a family of 'wild' examples, we choose Y and Z to be generated from the same Gaussian distributions with mean $(0,0)$ but W to be generated from a uniform distribution on the interval $[5,6]$. We tested this scenario with 1000 sample runs as well, and we observed 831 of them having seven real solutions, three of them statistically relevant, with two local maxima.

For the case of randomly generated data without regard to an underlying distribution we also ran 1000 sample runs where we observed 134 cases with nine real critical solutions.

In summary, our computations suggest that the number of real solutions of the critical equations can be a gauge of how well the MAR assumption fits the data. For missing data sets with three or more covariates where direct computation of all critical points will not be possible, if the EM-algorithm produces more than one local maximum, this might suggest that one should pay more careful attention to whether or not the MAR assumption makes sense for the data.

7.3 Bivariate discrete random variables

In this section, we focus on the case where X_1 and X_2 are discrete multinomial random variables. We suppose that $X_1 \in \{1, 2, \ldots, m\}$ and $X_2 \in \{1, 2, \ldots, n\}$. We give a combinatorial formula of the ML-degree which shows that it grows exponentially as a function of m and n.

In the bivariate multinomial case, the data can be summarised by a table of counts $T = (t_{ij})$ which records the complete cases, and two vectors $R = (r_i)$ and $S = (s_j)$ which record the observations of only X_1 and only X_2, respectively. In this multinomial case, we want to estimate the raw probabilities $p_{ij} = P(X_1 = i, X_2 = j)$. The log-likelihood function becomes

$$\ell(p; R, S, T) = \sum_{i=1}^{m} \sum_{j=1}^{n} t_{ij} \log p_{ij} + \sum_{i=1}^{m} r_i \log p_{i+} + \sum_{j=1}^{n} s_j \log p_{+j}. \qquad (7.1)$$

We want to find p that maximises $\ell(p; R, S, T)$ subject to $p > 0$ and $p_{++} = 1$.

Theorem 7.2 *The ML-degree of the bivariate multinomial missing data problem is equal to the number of bounded regions in the arrangement of hyperplanes $\{p_{ij} = 0, p_{i+} = 0, p_{+j} = 0 : i \in [m], j \in [n]\}$ inside the hyperplane $p_{++} = 1$. Every solution to the score equations for (7.1) is real. For generic R, S, T there is exactly one non-negative critical point, and it is a local maximum.*

Proof Maximising the product of linear forms has a standard formula for the ML-degree as the number of bounded regions in the arrangement defined by these linear forms (Catanese *et al.* 2004). Each bounded region contains precisely one critical solution which is real. Furthermore, since all the coordinate probability functions are linear in the parameters, the objective function is convex so there is exactly one non-negative critical point that must be a local maximum. $\qquad\square$

From Theorem 7.2 we see that to calculate the ML-degree we need to count the number of bounded regions in a hyperplane arrangement. The remainder of this section is devoted to performing this count. First we provide some definitions which allow us to state Theorem 7.3. Then we proceed with the proof in a number of steps.

For integers k and l, the *Stirling numbers of the second kind* are the numbers

$$S(l,k) = \frac{1}{k!}\sum_{i=0}^{k}(-1)^{k-i}\binom{k}{i}i^{l}.$$

The *negative index poly-Bernoulli numbers* are the numbers:

$$B(l,k) = \sum_{i=0}^{l}(-1)^{l-i}i!S(l,i)(i+1)^{k}.$$

Theorem 7.3 *The ML-degree of the bivariate multinomial $m \times n$ missing data problem is*

$$ML(m,n) = \sum_{k=0}^{m}\sum_{l=0}^{n}(-1)^{m+n-k-l}\binom{m}{k}\binom{n}{l}B(m-k,n-l). \qquad (7.2)$$

For small values of m, we can explicitly work out formulas for this ML-degree. In particular, one can show that $ML(2,n) = 2^{n+1} - 3$. Since the ML-degree is monotone as a function of m and n, this shows that the ML-degree in the bivariate discrete case is exponential in the size of the problem. Let

$$S - \{p_{ij} : i \in [m] \cup \{+\}, j \in [n] \cup \{+\}\} \setminus \{p_{++}\}$$

be the set of all hyperplanes in the hyperplane arrangement that determines the ML-degree. Specifying a (possibly empty) region of the arrangement amounts to choosing a partition $S = N \cup P$. The resulting open region on the hyperplane $p_{++} = 1$ consists of all matrices p such that $p_{ij} < 0$ if $p_{ij} \in N$ and $p_{ij} > 0$ if $p_{ij} \in P$ and $\sum_{i,j} p_{ij} = 1$. We denote this set of matrices by $\mathcal{M}(N,P)$. Our goal is characterise and count the partitions $N \cup P$ such that $\mathcal{M}(N,P)$ is non-empty and bounded. We prove a sequence of results classifying the type of sub-configurations that can appear in N and P.

Lemma 7.1 *Let $i,k \in [m]$ with $i \neq k$ and $j,l \in [n]$ with $j \neq l$. Suppose that $p_{ij},p_{kl} \in N$ and $p_{il},p_{kj} \in P$. Then if $\mathcal{M}(N,P)$ is non-empty it is unbounded.*

Proof Let e_{ij} denote the $m \times n$ matrix with a one in the ij position and zeros elsewhere. Suppose that $p \in \mathcal{M}(N,P)$. Then $p + a(e_{il} + e_{kj} - e_{ij} - e_{kl}) \in \mathcal{M}(N,P)$ for all $a > 0$ since adding $a(e_{il} + e_{kj} - e_{ij} - e_{kl})$ does not change the sign of any entry of p nor does it change any of the margins p_{i+} of p_{+j}. Thus $\mathcal{M}(N,P)$ contains matrices with arbitrarily large entries and is unbounded. $\qquad\square$

Let $N' = N \cap \{p_{ij} \,:\, i \in [m], j \in [n]\}$ and $P' = P \cap \{p_{ij} \,:\, i \in [m], j \in [n]\}$. A partition $\lambda = (\lambda_1, \ldots, \lambda_m)$ is a non-increasing sequence of non-negative integers. The length of λ is m (we allow zeros in the partition).

Lemma 7.2 *Suppose that $\mathcal{M}(N, P)$ is non-empty and bounded. There exists a permutation σ of the rows and columns of p and a partition λ such that*

$$\sigma(N') = \{p_{ij} \,:\, j \le \lambda_i\}.$$

The same is true for P' and for every rectangular submatrix of p.

Proof After permuting rows we may assume that the number of elements in row i, λ_i, is a non-increasing sequence. Permuting the columns we may suppose that the only elements of N' in the first row of p are $p_{11}, \ldots, p_{1\lambda_1}$. Permuting columns further, we may assume that the elements in the second row are of the form $p_{21}, \ldots, p_{2\lambda_2}$ with $\lambda_2 \le \lambda_1$. There could not be any element of the form $p_{2j} \in N'$ with $j > \lambda_1$ because otherwise there would be more entries in row two than row one or N' would contain $p_{1\lambda_1}, p_{2j}$ and P' would contain $p_{1j}, p_{2\lambda_1}$ which violates Lemma 7.1. Repeating the argument for each row shows that $\mathcal{M}(N, P)$ can be put into partition form. $\qquad\square$

Lemma 7.3 *Suppose that $\mathcal{M}(N, P)$ is non-empty and bounded. Then $p_{i+}, p_{+j} \in P$ for all i and j.*

Proof Suppose that $\mathcal{M}(N, P)$ is non-empty and N contains, say, p_{+1}. We will show $\mathcal{M}(N, P)$ is unbounded. To do this, it suffices to show that there exist points on the boundary of $\mathcal{M}(N, P)$ with coordinates of arbitrarily large absolute values. Furthermore, we will assume that $\mathcal{M}(N, P)$ is bounded (so that we can make liberal use of Lemmas 7.2 and 7.1) and derive a contradiction. The boundary of $\mathcal{M}(N, P)$ is described by allowing the strict inequalities to become weak inequalities. There are four cases to consider.

Case 1. Suppose that there is no i such that $p_{i+} \in N$. After permuting columns and rows we may suppose that $p_{+j} \in N$ if and only if $j \in [k]$. If $\mathcal{M}(N, P)$ is to be non-empty, we must have $k < m$.

After permuting row and columns in such a way that the set of the first k columns is mapped to itself, we may suppose that the set of variables in N belonging to the submatrix $p[1, m; 1, k]$ is in partition form, according to Lemma 7.2. If $\mathcal{M}(N, P)$ is to be non-empty, it must be the case that $p_{1j} \in N$ for all $j \in [k]$ since the first row is the longest row of the tableau. As $p_{i+} \in P$, there must exist $p_{1l} \in P$ with $l > k$. Then consider the matrix p' with $p'_{11} = -a$, $p_{1j} = a + 1$ and $p_{ij} = 0$ for all other i, j. This matrix satisfies all requirements to belong to the boundary of $\mathcal{M}(N, P)$. Letting a tend to infinity shows that $\mathcal{M}(N, P)$ is unbounded, a contradiction.

For the remaining three cases, we assume that there exists some i and j such that $p_{i+}, p_{+j} \in N$. After permuting rows and columns we may suppose there is $k < m$ and $l < n$ such that $p_{i+} \in N$ if and only if $i \in [k]$ and $p_{+j} \in N$ if and only if $j \in [l]$.

Case 2. Suppose that there is a $p_{ij} \in N$ with $i \in [k]$ and $j \in [l]$ and a $p_{i'j'} \in P$ with $i' \in [k+1,m]$ and $j' \in [l+1,n]$. Then the matrix p' with $p_{ij} = -a$, $p_{i'j'} = a+1$ and all other entries equal satisfies the requirements to belong to the boundary of $\mathcal{M}(N,P)$. Letting a tend to infinity shows that $\mathcal{M}(N,P)$ is unbounded, a contradiction.

Case 3. Suppose that $p_{ij} \in P$ for all $i \in [k]$ and $j \in [l]$. Since $\mathcal{M}(N,P)$ is non-empty, and $p_{i+} \in N$ for all $i \in [k]$, we can find, for each $i \in [k]$, a $j \in [l+1,n]$ such that $p_{ij} \in N$. As $\mathcal{M}(N,P)$ is bounded, this implies that we can permute rows and columns of the matrix p, so that $p[1,k;l+1,n]$ is mapped into itself and so that this submatrix, intersected with N is of tableau form. With these assumptions, we must have $p_{il+1} \in N$ for all $i \in [k]$. Since $p_{+,l+1} \in P$, there must exist $p_{i'l+1} \in P$ with $i' \in [k+1,m]$. Now consider the matrix p' with $p'_{1l+1} = -a$, $p'_{i'l+1} = a+1$ and all other entries equal to zero. This matrix satisfies all requirements for belonging to the boundary of $\mathcal{M}(N,P)$ but as a tends to infinity shows that $\mathcal{M}(N,P)$ is unbounded.

Case 4. Suppose that $p_{ij} \in N$ for all $i \in [k+1,m]$ and $j \in [l+1,n]$. This is equivalent to saying that for all $p_{ij} \in P$, p_{i+} and p_{+j} are not simultaneously in P. If we permute rows and columns of p so that P is in tableau form, this condition is equivalent to saying that there is a $p_{i'j'} \in P$ such that $p_{i'+1j'+1} \notin P$ and none of p_{i+} nor p_{+j} are in P for $i \leq i'$ and $j \leq j'$. (Note that one of i' or j' might be zero, which will work fine in the following argument.) Then for any matrix $p \in \mathcal{M}(N,P)$ we have $0 < \sum_{i=1}^{i'} p_{i+} + \sum_{j=1}^{j'} p_{+j} = 2\sum_{i=1}^{i'} \sum_{j=1}^{j'} p_{ij} + \sum_{i=i'+1}^{m} \sum_{j=1}^{j'} p_{ij} + \sum_{i=1}^{i'} \sum_{j=j'+1}^{n} p_{ij}$. The expression at the end of this equation involves the sum, with positive coefficients, of all $p_{ij} \in P$. Since the p_{ij} in the sum with $p_{ij} \in N$ all occur with coefficient 1, and since $p_{++} = 1$, we deduce that this sum must be strictly greater than 1. Thus $\mathcal{M}(N,P)$ must be empty. $\quad\square$

Lemma 7.4 *Let λ be a partition of length m such that $\lambda_i \leq n-1$ for all i, and $\lambda_m = 0$. Let $N(\lambda) = \{p_{ij} : j \leq \lambda_i\}$ and $P(\lambda) = S \setminus N(\lambda)$. Then $\mathcal{M}(N(\lambda),P(\lambda))$ is non-empty and bounded.*

Proof To show that $\mathcal{M}(N(\lambda),P(\lambda))$ is non-empty amounts to showing that there is a table p with non-zero entries that satisfies all the constraints $p_{ij} < 0$ if $p_{ij} \in N(\lambda)$, $p_{ij} > 0$ if $p_{ij} \in P(\lambda)$ and $p_{++} = 1$. To this end, let $\epsilon > 0$ be a small real number. Define the matrix $p(\epsilon)$ by the following rules:

$$
p(\epsilon)_{ij} = \begin{cases} -\epsilon & \text{if } p_{ij} \in N(\lambda) \\ \epsilon & \text{if } p_{ij} \in P(\lambda),\ i < m, j < n \\ m\epsilon & \text{if } i = m, j < n \\ n\epsilon & \text{if } i < m, j = n \\ 1 - (3mn - 2m - 2n + 1 - 2\sum_k \lambda_k)\epsilon & \text{if } i = m, j = n \end{cases}
$$

By construction, $p(\epsilon) \in \mathcal{M}(N,P)$.

Now we show that $\mathcal{M}(N(\lambda),P(\lambda))$ is bounded. For each $k \in [m-1]$ with $\lambda_k > 0$ we have $0 \leq \sum_{i=1}^{k} p_{i+} + \sum_{j=1}^{\lambda_k} p_{+j} = 2\sum_{i=1}^{k} \sum_{j=1}^{\lambda_k} p_{ij} + \sum_{i=k+1}^{m} \sum_{j=1}^{\lambda_k} p_{ij} +$

$\sum_{i=1}^{k} \sum_{j=\lambda_k+1}^{n} p_{ij}$ which implies that

$$-\left(\sum_{i=1}^{k}\sum_{j=1}^{\lambda_k} p_{ij}\right) \leq \sum_{i=1}^{k}\sum_{j=1}^{\lambda_k} p_{ij} + \sum_{i=k+1}^{m}\sum_{j=1}^{\lambda_k} p_{ij} + \sum_{i=1}^{k}\sum_{j=\lambda_k+1}^{n} p_{ij} \leq \sum_{i=1}^{m}\sum_{j=1}^{n} p_{ij} = 1.$$

Since $p_{ij} \in N(\lambda)$ whenever $i \in [k]$ and $j \in [\lambda_k]$, we deduce that

$$-1 \leq \sum_{i=1}^{k}\sum_{j=1}^{\lambda_k} p_{ij} \leq 0$$

and thus $-1 \leq p_{ij} \leq 0$. Since every $p_{ij} \in N(\lambda)$ belongs to such a sum for some k, we see that p_{ij} is bounded for all $p_{ij} \in N(\lambda)$. This implies that p_{ij} is bounded for all $p_{ij} \in P(\lambda)$ as well, since, $p_{++} = 1$. Thus, $\mathcal{M}(N(\lambda), P(\lambda))$ is bounded. $\qquad\square$

To finish the proof, we use a result from the Master's thesis of Chad Brewbaker (Brewbaker 2005), that counts a family of 0/1 matrices that are closely related to the set N, P that have $\mathcal{M}(N, P)$ bounded.

Theorem 7.4 *The number of 0/1 $m \times n$ matrices A such that no 2×2 submatrix of A is either $\begin{pmatrix} 1 & 0 \\ 0 & 1 \end{pmatrix}$ or $\begin{pmatrix} 0 & 1 \\ 1 & 0 \end{pmatrix}$ is the negative index poly-Bernoulli number $B(m, n)$.*

The 0/1 matrices in the theorem are known as *lonesum matrices* because they are the 0/1 matrices that are uniquely specified by their row and column sums. We are now ready to prove Theorem 7.3.

Proof According to Lemmas 7.1, 7.3 and 7.4, we must count sets $N \subset \{p_{ij} : i \in [m], j \in [n]\}$ with certain properties. Interpreting N as a lonesum 0/1 matrix where M where $M_{ij} = 1$ if $p_{ij} \in N$, we see that we must count the matrices M that do not have any 2×2 submatrices equal to $\begin{pmatrix} 1 & 0 \\ 0 & 1 \end{pmatrix}$ or $\begin{pmatrix} 0 & 1 \\ 1 & 0 \end{pmatrix}$. Furthermore, the fact that no p_{i+} or p_{+j} belongs to N implies that no row or column of M could be all ones (otherwise, we would have, for example, $p_{ij} < 0$ for all j but $p_{i+} > 0$ which implies that $\mathcal{M}(N, P)$ is empty). Because of the fact that each such set N can be rearranged into a partition, and after switching the zeros and ones, this is the same as the number of 0/1 $m \times n$ matrices which have all row and column sums positive. Thus, the number $M(m, n)$ can be obtained from the negative index poly-Bernoulli numbers $B(m, n)$ by inclusion-exclusion which yields the desired formula (7.2). $\quad\square$

References

Brewbaker, C. (2005). Lonesum $(0, 1)$-matrices and poly-Bernoulli numbers. Master's Thesis, Department of Mathematics, Iowa State University.

Buot, M.-L. G. and Richards, D. St. P. (2006). Counting and locating the solutions of polynomial systems of maximum likelihood equations. I, *Journal of Symbolic Computing* **41**, 234–44.

Buot, M.-L. G., Hoşten S. and Richards, D. St. P. (2007). Counting and locating the solutions of polynomial systems of maximum likelihood equations. II. The Behrens-Fisher problem, *Statistica Sinica* **17**, 1343–54.

Catanese, F., Hoşten, S., Khetan, A. and Sturmfels, B. (2006). The maximum likelihood degree, *American Journal of Mathematics* **128**(3), 671–97.

Dempster A. P., Laird, N. M. and Rubin, D. B. (1977). Maximum likelihood from incomplete data via EM algorithm, *Journal of the Royal Statistical Society B* **39**, 1–38.

Greuel, G.-M., Pfister, G. and Schönemann, H. (2005). SINGULAR *3.0. A Computer Algebra System for Polynomial Computations. Centre for Computer Algebra* (available at `www.singular.uni-kl.de`).

Hoşten, S., Khetan, A. and Sturmfels, B. (2005). Solving the likelihood equations, *Foundations of Computational Mathematics* **5**, 389–407.

Little, R. J. A. and Rubin, D. B. (1983). Incomplete data, *Encyclopedia of the Statistical Sciences* **4**, 46–53.

Little, R. J. A. and Rubin, D. B. (2002). *Statistical Analysis with Missing Data*, Series in Probability and Statistics, (Hoboken, NJ, Wiley Interscience).

Rubin, D. B. (1976). Inference and missing data (with discussion), *Biometrika* **63**(3), 581–92.

8

The generalised shuttle algorithm

Adrian Dobra

Stephen E. Fienberg

Abstract

Bounds for the cell counts in multi-way contingency tables given a set of marginal totals arise in a variety of different statistical contexts including disclosure limitation. We describe the Generalised Shuttle Algorithm for computing integer bounds of multi-way contingency tables induced by arbitrary linear constraints on cell counts. We study the convergence properties of our method by exploiting the theory of discrete graphical models and demonstrate the sharpness of the bounds for some specific settings. We give a procedure for adjusting these bounds to the sharp bounds that can also be employed to enumerate all tables consistent with the given constraints. Our algorithm for computing sharp bounds and enumerating multi-way contingency tables is the first approach that relies exclusively on the unique structure of the categorical data and does not employ any other optimisation techniques such as linear or integer programming. We illustrate how our algorithm can be used to compute exact p-values of goodness-of-fit tests in exact conditional inference.

8.1 Introduction

Many statistical research problems involve working with sets of multi-way contingency tables defined by a set of constraints, e.g., marginal totals or structural zeros. Four inter-related aspects involve: (1) the computation of sharp integer bounds, (2) counting, (3) exhaustive enumeration and (4) sampling. Each of these areas or some combination of them play important roles in solving complex data analysis questions arising in seemingly unrelated fields. The computation of bounds is central to the task of assessing the disclosure risk of small cell counts (e.g., cells with entries of 1 or 2) when releasing marginals from a high-dimensional sparse contingency table – for example, see (Fienberg 1999, Dobra and Fienberg 2000) and (Dobra 2001). Another aspect of disclosure risk assessment involves counting feasible tables consistent with the release, see (Fienberg and Slavkovic 2004, Fienberg and Slavkovic 2005), or by estimating probability distributions on multi-way tables as in (Dobra *et al.* 2003b).

Algebraic and Geometric Methods in Statistics, ed. Paolo Gibilisco, Eva Riccomagno, Maria Piera Rogantin and Henry P. Wynn. Published by Cambridge University Press. © Cambridge University Press 2010.

135

(Guo and Thompson 1992) employ sampling from a set of contingency tables to perform exact tests for Hardy–Weinberg proportions. Markov chain Monte Carlo (MCMC) sampling methods depend on the existence of a Markov basis that connects any two feasible tables through a series of Markov moves. (Diaconis and Sturmfels 1998) were the first to show how to produce such moves through algebraic geometry techniques. (Dobra 2003a)gave formulas for Markov bases in the case of decomposable graphical models, while (Dobra and Sullivant 2004) extend this work to reducible graphical models. Markov bases are local moves that change only a relatively small number of cell counts and can be contrasted with global moves that potentially alter all the counts. (Dobra *et al.* 2006) describe how to produce global moves in a set of contingency tables by sequentially adjusting upper and lower bounds as more cells are fixed at certain values. (Chen *et al.* 2006) present a similar method for finding feasible tables. Their sequential importance sampling approach seems to be more efficient than other MCMC techniques and builds on computational commutative algebra techniques to find bounds and to make random draws from the implied marginal cell distributions. Other work on algebraic geometry related to the theory of discrete graphical models includes (Geiger *et al.* 2006) and (Hoşten and Sturmfels 2007).

(Fréchet 1940) presented a special class of bounds for cumulative distribution functions of a random vector $(D_1, D_2, \ldots, D_m)$ in $\mathbb{R}^m$:

$$F_{1,2,\ldots,m}(x_1, x_2, \ldots, x_m) = \Pr(D_1 \leq x_1, D_2 \leq x_2, \ldots, D_m \leq x_m), \qquad (8.1)$$

which are essentially equivalent to contingency tables when the underlying variables are categorical. For example, suppose we have a two-dimensional table of counts, $\{n_{ij}\}$ adding up to the total $n_{++} = n_\phi$. If we normalise each entry by dividing by n and then create a table of partial sums, by cumulating the proportions from the first row and first column to the present ones, we have a set of values of the form (8.1). Thus, Fréchet bound results for distribution functions correspond to bounds for the cell counts where the values $\{x_i\}$ in (8.1) represent 'cut-points' between categories for the ith categorical variable. (Bonferroni 1936) and (Hoeffding 1940) independently developed related results on bounds. When the fixed set of marginals defines a decomposable independence graph, the Fréchet bounds are calculated by the formulas of (Dobra and Fienberg 2000).

In this chapter we propose the generalized shuttle algorithm (GSA) which we can use to compute sharp integer bounds *and* exhaustively enumerate all feasible tables consistent with a set of constraints. (Dobra *et al.* 2003c)provided a brief account of this work, while (Dobra *et al.* 2006) showed its application to sampling contingency tables. Our procedure is deterministic and exploits the special structure of contingency tables, building on the work of (Buzzigoli and Giusti 1999) who proposed the first version of the shuttle algorithm. Their innovative iterative approach simultaneously calculates bounds for all the cells in the table by sequentially alternating between upper and lower bounds; however, their version of the shuttle algorithm fails to converge to the sharp bounds for most configurations of fixed marginal totals, e.g. (Cox 1999). The explanation for this failure lies in the incomplete description of the dependencies among the cells of a contingency table

used by Buzzigoli and Giusti. (Chen *et al.* 2006) give an excellent discussion about
the relationship between linear programming (LP), integer programming (IP) and
the computation of bounds for contingency tables.

This chapter is organised as follows. In Section 8.2 we give the basic definitions and notations. We present the full description of GSA in Section 8.3. In
Sections 8.4 and 8.5, we describe two particular cases when the shuttle procedure converges to the sharp bounds. In Section 8.6 we present an approach for
adjusting the shuttle bounds to the sharp bounds and also show how to transform this procedure to enumerate multi-way tables. In Section 8.7 we show that
GSA is able to efficiently compute bounds for a sixteen-way sparse contingency
table. In Section 8.8 we give six examples that illustrate how GSA can be used
for computing bounds as well as exact p-values based on the hypergeometric distribution. Complete proofs of our theoretical results together with source code implementing GSA are available on the on-line supplement and for download from
`www.stat.washington.edu/adobra/software/gsa/`

8.2 Terminology and notation

Let $X = (X_1, X_2, \ldots, X_k)$ be a vector of k discrete random variables cross-classified
in a frequency count table $\mathbf{n} = \{n(i)\}_{i \in \mathcal{I}}$, where $\mathcal{I} = \mathcal{I}_1 \times \mathcal{I}_2 \times \cdots \times \mathcal{I}_k$ and X_r
takes the values $\mathcal{I}_r := \{1, 2, \ldots, I_r\}$. Denote $K = \{1, 2, \ldots, k\}$. For $r \in K$, denote
by $\mathcal{P}(\mathcal{I}_r)$ the set of all partitions of $\mathcal{I}_r$, i.e.,

$$\mathcal{P}(\mathcal{I}_r) := \Big\{ \{\mathcal{I}_r^1, \mathcal{I}_r^2, \ldots, \mathcal{I}_r^{l_r}\} : \mathcal{I}_r^l \neq \emptyset \text{ for all } l,$$
$$\cup_{j=1}^{l_r} \mathcal{I}_r^j = \mathcal{I}_r, \mathcal{I}_r^{j_1} \cap \mathcal{I}_r^{j_2} = \emptyset \text{ if } j_1 \neq j_2 \Big\}.$$

Let $\mathcal{RD}$ be the set of marginal tables obtainable by aggregating $\mathbf{n}$ not only across
variables, but also across categories within variables. We can uniquely determine a
table $\mathbf{n}' \in \mathcal{RD}$ from $\mathbf{n}$ by choosing $\mathcal{I}_1' \in \mathcal{P}(\mathcal{I}_1)$, $\mathcal{I}_2' \in \mathcal{P}(\mathcal{I}_2), \ldots, \mathcal{I}_k' \in \mathcal{P}(\mathcal{I}_k)$. We
write

$$\mathbf{n}' = \{n'(J_1, J_2, \ldots, J_k) : (J_1, J_2, \ldots, J_k) \in \mathcal{I}_1' \times \mathcal{I}_2' \times \ldots \times \mathcal{I}_k'\},$$

where the entries of $\mathbf{n}'$ are sums of appropriate entries of $\mathbf{n}$:

$$n'(J_1, J_2, \ldots, J_k) := \sum_{i_1 \in J_1} \sum_{i_2 \in J_2} \cdots \sum_{i_k \in J_k} n_K(i_1, i_2, \ldots, i_k).$$

We associate the table $\mathbf{n}$ with $\mathcal{I}_r' = \{\{1\}, \{2\}, \ldots, \{I_r\}\}$, for $r = 1, \ldots, k$. On the
other hand, choosing $\mathcal{I}_r' = \{\mathcal{I}_r\}$ is equivalent to collapsing across the r-th variable.
The *dimension* of $\mathbf{n}' \in \mathcal{RD}$ is the number of variables cross-classified in $\mathbf{n}'$ that
have more than one category. For $C \subset K$, we obtain the C-marginal $\mathbf{n}_C$ of $\mathbf{n}$ by
taking

$$\mathcal{I}_r' = \begin{cases} \{\{1\}, \{2\}, \ldots, \{I_r\}\}, & \text{if } r \in C, \\ \mathcal{I}_r, & \text{otherwise,} \end{cases}$$

for $r = 1, 2, \ldots, k$. The dimension of $\mathbf{n}_C$ is equal to the number of elements in C.
The grand total of $\mathbf{n}$ has dimension zero, while $\mathbf{n}$ has dimension k.

We introduce the set of tables $\mathcal{RD}(\mathbf{n}')$ containing the tables $\mathbf{n}'' \in \mathcal{RD}$ obtainable from $\mathbf{n}'$ by table redesign such that $\mathbf{n}''$ and $\mathbf{n}'$ have the same dimension. We have $\mathbf{n}' \in \mathcal{RD}(\mathbf{n}')$ and $\mathcal{RD}(n_\emptyset) = \{n_\emptyset\}$, where $n_\emptyset$ is the grand total of $\mathbf{n}$. The set $\mathcal{RD}$ itself results from aggregating every marginal $\mathbf{n}_C$ of $\mathbf{n}$ across categories, such that every variable having at least two categories in $\mathbf{n}_C$ also has at least two categories in the new redesigned table:

$$\mathcal{RD} = \bigcup \{\mathcal{RD}(\mathbf{n}_C) : C \subseteq K\}. \tag{8.2}$$

We write $t_{J_1 J_2 \ldots J_k} = \sum_{i_1 \in J_1} \sum_{i_2 \in J_2} \cdots \sum_{i_k \in J_k} n_K(i_1, i_2, \ldots, i_k)$ and we define

$$\mathbf{T} := \{t_{J_1 J_2 \ldots J_k} : \emptyset \neq J_1 \times J_2 \times \cdots \times J_k \subseteq \mathcal{I}_1 \times \mathcal{I}_2 \times \cdots \times \mathcal{I}_k\}. \tag{8.3}$$

The elements in $\mathbf{T}$ are blocks or 'super-cells' formed by joining table entries in $\mathbf{n}$. These blocks can be viewed as entries in a k-dimensional table that cross-classifies the variables $(Y_j : j = 1, 2, \ldots, k)$, where Y_j takes values $y_j \in \{\mathcal{I}'_j : \emptyset \neq \mathcal{I}'_j \subseteq \mathcal{I}_j\}$. The number of elements in $\mathbf{T}$ is $\#(\mathbf{T}) = \prod_{r=1}^{k} \left(2^{I_r} - 1\right)$.

If the set of cell entries in $\mathbf{n}$ that define a 'super-cell' $t_2 = t_{J_1^2 \ldots J_k^2} \in \mathbf{T}$ includes the set of cells defining another 'super-cell' $t_1 = t_{J_1^1 \ldots J_k^1} \in \mathbf{T}$, then we write $t_1 = t_{J_1^1 \ldots J_k^1} \prec t_2 = t_{J_1^2 \ldots J_k^2}$. We formally define the partial ordering $\prec$ on the cells in $\mathbf{T}$ by

$$t_{J_1^1 J_2^1 \ldots J_k^1} \prec t_{J_1^2 J_2^2 \ldots J_k^2} \Leftrightarrow J_1^1 \subseteq J_1^2, J_2^1 \subseteq J_2^2, \ldots, J_k^1 \subseteq J_k^2.$$

This partial ordering, $(\mathbf{T}, \prec)$, has a maximal element, namely the grand total $n_\emptyset = t_{\mathcal{I}_1 \mathcal{I}_2 \ldots \mathcal{I}_k}$ of the table and several minimal elements – the actual cell counts $n(i) = n(i_1, i_2, \ldots, i_k) = t_{\{i_1\}\{i_2\} \ldots \{i_k\}}$. Thus, we can represent the lattice $(\mathbf{T}, \prec)$ as a hierarchy with the grand total at the top level and the cells counts $n(i)$ at the bottom level. If $t_1 = t_{J_1^1 J_2^1 \ldots J_k^1}$ and $t_2 = t_{J_1^2 J_2^2 \ldots J_k^2}$ are such that $t_1 \prec t_2$ with $J_r^1 = J_r^2$, for $r = 1, \ldots, r_0 - 1, r_0 + 1, \ldots, k$ and $J_{r_0}^1 \neq J_{r_0}^2$, we define the *complement* of the cell t_1 with respect to t_2 to be the cell $t_3 = t_{J_1^3 J_2^3 \ldots J_k^3}$, where

$$J_r^3 = \left\{ \begin{array}{ll} J_r^1, & \text{if } r \neq r_0, \\ J_r^2 \setminus J_r^1, & \text{if } r = r_0, \end{array} \right.$$

for $r = 1, 2, \ldots, k$. We write $t_1 \oplus t_3 = t_2$. The elements in $\mathbf{T}$ are blocks formed by joining table entries in $\mathbf{n}$. The operator $\oplus$ is equivalent to joining two blocks of cells in $\mathbf{T}$ to form a third block where the blocks to be joined have the same categories in $(k-1)$ dimensions and they cannot share any categories in the remaining dimension.

8.3　The generalised shuttle algorithm

The fundamental idea behind the generalised shuttle algorithm (GSA) is that the upper and lower bounds for the cells in $\mathbf{T}$ are interlinked, i.e., bounds for some cells in $\mathbf{T}$ induce bounds for some other cells in $\mathbf{T}$. We can improve (tighten) the bounds for all the cells in which we are interested until we can make no further adjustment.

Although (Buzzigoli and Giusti 1999) introduced this innovative idea, they did not fully exploit the special hierarchical structure of $\mathbf{T}$.

Let $L_0(\mathbf{T}) := \{L_0(t) : t \in \mathbf{T}\}$ and $U_0(\mathbf{T}) := \{U_0(t) : t \in \mathbf{T}\}$ be initial upper and lower bounds. By default we set $L_0(t) = 0$ and $U_0(t) = n_\emptyset$, but we can express almost any type of information about the counts in cells $\mathbf{T}$ using these bounds. For example, a known count c in a cell t with a fixed marginal implies that $L_0(t) = U_0(t) = c$. A cell t that can take only two values 0 or 1 has $L_0(t) = 0$ and $U_0(t) = 1$.

We denote by $S[L_0(\mathbf{T}), U_0(\mathbf{T})]$ the set of integer feasible arrays $V(\mathbf{T}) := \{V(t) : t \in \mathbf{T}\}$ consistent with $L_0(\mathbf{T})$ and $U_0(\mathbf{T})$: (i) $L_0(t) \leq V(t) \leq U_0(t)$, for all $t \in \mathbf{T}$ and (ii) $V(t_1) + V(t_3) = V(t_2)$, for all $(t_1, t_2, t_3) \in \mathcal{Q}(\mathbf{T})$, where

$$\mathcal{Q}(\mathbf{T}) := \{(t_1, t_2, t_3) \in \mathbf{T} \times \mathbf{T} \times \mathbf{T} : t_1 \oplus t_3 = t_2\}.$$

We let $\mathcal{N} \subset \mathbf{T}$ be the set of cells in table $\mathbf{n}$. A feasible table consistent with the constraints imposed (e.g., fixed marginals) is $\{V(t) : t \in \mathcal{N}\}$ where $V(\mathbf{T}) \in S[L_0(\mathbf{T}), U_0(\mathbf{T})]$.

The sharp integer bounds $[L(t), U(t)]$, $t \in \mathbf{T}$, are the solution of the integer optimisation problems:

$$\min \{\pm V(t) : V(\mathbf{T}) \in S[L_0(\mathbf{T}), U_0(\mathbf{T})]\}.$$

We initially set $L(\mathbf{T}) = L_0(\mathbf{T})$ and $U(\mathbf{T}) = U_0(\mathbf{T})$ and sequentially improve these loose bounds by GSA until we get convergence. Consider $\mathbf{T}_0 := \{t \in \mathbf{T} : L(t) = U(t)\}$ to be the cells with the current lower and upper bounds equal. We say that the remaining cells in $\mathbf{T} \setminus \mathbf{T}_0$ are *free*. As the algorithm progresses, we improve the bounds for the cells in $\mathbf{T}$ and add more and more cells to $\mathbf{T}_0$. For each t in $\mathbf{T}_0$, we assign a value $V(t) := L(t) = U(t)$.

We sequentially go through the dependencies $\mathcal{Q}(\mathbf{T})$ and update the upper and lower bounds in the following fashion. Consider a triplet $(t_1, t_2, t_3) \in \mathcal{Q}(\mathbf{T})$. We have $t_1 \prec t_2$ and $t_3 \prec t_2$. We update the upper and lower bounds of t_1, t_2 and t_3 so that the new bounds satisfy the dependency $t_1 \oplus t_3 = t_2$.

If all three cells have fixed values, i.e., $t_1, t_2, t_3 \in \mathbf{T}_0$, we check whether $V(t_1) + V(t_3) = V(t_2)$. If this equality does not hold, we stop GSA because $S[L_0(\mathbf{T}), U_0(\mathbf{T})]$ is empty – there is no integer table consistent with the constraints imposed.

Now assume that $t_1, t_3 \in \mathbf{T}_0$ and $t_2 \notin \mathbf{T}_0$. Then t_2 can take only one value, namely $V(t_1) + V(t_3)$. If $V(t_1) + V(t_3) \notin [L(t_2), U(t_2)]$, we encounter an inconsistency and stop. Otherwise we set $V(t_2) = L(t_2) = U(t_2) := V(t_1) + V(t_3)$ and include t_2 in $\mathbf{T}_0$. Similarly, if $t_1, t_2 \in \mathbf{T}_0$ and $t_3 \notin \mathbf{T}_0$, t_3 can only be equal to $V(t_2) - V(t_1)$. If $V(t_2) - V(t_1) \notin [L(t_3), U(t_3)]$, we again discover an inconsistency. If this is not true, we set $V(t_3) = L(t_3) = U(t_3) := V(t_2) - V(t_1)$ and $\mathbf{T}_0 := \mathbf{T}_0 \cup \{t_3\}$. In the case when $t_2, t_3 \in \mathbf{T}_0$ and $t_1 \notin \mathbf{T}_0$, we proceed in an analogous manner.

Next we examine the situation when at least two of the cells t_1, t_2, t_3 do not have a fixed value. Suppose $t_1 \notin \mathbf{T}_0$. The new bounds for t_1 are

$$U(t_1) := \min\{U(t_1), U(t_2) - L(t_3)\}, \qquad L(t_1) := \max\{L(t_1), L(t_2) - U(t_3)\}.$$

If $t_3 \notin \mathbf{T}_0$, we update $L(t_3)$ and $U(t_3)$ in the same way. Finally, if $t_2 \notin \mathbf{T}_0$, we set

$$U(t_2) := \min\{U(t_2), U(t_1) + U(t_3)\}, \qquad L(t_2) := \max\{L(t_2), L(t_1) + L(t_3)\}.$$

After updating the bounds of some cell $t \in \mathbf{T}$, we check whether the new upper bound equals the new lower bound. If this is true, we set $V(t) := L(t) = U(t)$ and include t in $\mathbf{T}_0$.

We continue iterating through all the dependencies in $\mathcal{Q}(\mathbf{T})$ until the upper bounds no longer decrease, the lower bounds no longer increase and no new cells are added to $\mathbf{T}_0$. Therefore the procedure comes to an end if and only if we detect an inconsistency or if we cannot improve the bounds. One of these two events eventually occurs; hence the algorithm stops after a finite number of steps.

If we do not encounter any inconsistencies, the algorithm converges to bounds $L_s(\mathbf{T})$ and $U_s(\mathbf{T})$ that are not necessarily sharp: $L_s(t) \leq L_0(t) \leq U_0(t) \leq U_s(t)$. These arrays define the same feasible set of tables as the arrays $L_0(\mathbf{T})$ and $U_0(\mathbf{T})$ we started with, i.e., $S[L_s(\mathbf{T}), U_s(\mathbf{T})] = S[L_0(\mathbf{T}), U_0(\mathbf{T})]$, since the dependencies $\mathcal{Q}(\mathbf{T})$ need to be satisfied.

There exist two particular cases when we can easily prove that GSA converges to sharp integer bounds: (i) the case of a dichotomous k-dimensional table with all $(k-1)$-dimensional marginals fixed and (ii) the case when the marginals we fix are the minimal sufficient statistics of a decomposable log-linear model. In both instances explicit formulas for the bounds exist. Employing GSA turns out to be equivalent to calculating the bounds directly as we prove in the next two sections.

8.4 Computing bounds for dichotomous k-way cross classifications given all $(k-1)$-dimensional marginals

Consider a k-way table $\mathbf{n} := \{n(i)\}_{i \in \mathcal{I}}$ with $\mathcal{I}_1 = \mathcal{I}_2 = \ldots = \mathcal{I}_k = \{1, 2\}$. The set $\mathbf{T}$ associated with $\mathbf{n}$ is the set of cells of every marginal of $\mathbf{n}$, while the set $\mathbf{T}_0$ of cells having a fixed value is $\mathbf{T}_0 = \{n_C(i_C) : i_C \in \mathcal{I}_C \text{ for some } C \subset K, C \neq K\}$. The only cells in $\mathbf{T}$ that are not fixed are the cells in $\mathbf{n}$: $\mathbf{T} \setminus \mathbf{T}_0 = \{n(i) : i \in \mathcal{I}\}$.

The $(k-1)$-dimensional marginals of $\mathbf{n}$ are the minimal sufficient statistics of the log-linear model of no (k)-way interaction. (Fienberg 1999) pointed that this log-linear model has only one degree of freedom because $\mathbf{n}$ is dichotomous, hence we can uniquely express the count in any cell $n(i)$, $i \in \mathcal{I}$, as a function of one single fixed cell alone.

Let n^* be the unknown count in the $(1, 1, \ldots, 1)$ cell. In Proposition 8.1 we give an explicit formula for computing the count in an arbitrary cell $n(i^0)$, $i^0 \in \mathcal{I}$, based on n^* and on the set of fixed marginals.

Proposition 8.1 *Let n^* be the count in the $(1, 1, \ldots, 1)$ cell. Consider an index $i^0 = (i_1^0, i_2^0, \ldots, i_k^0) \in \mathcal{I}$. Let $\{q_1, q_2, \ldots, q_l\} \subset K$ such that, for $r \in K$, we have*

$$i_r^0 = \begin{cases} 1, & \text{if } r \in K \setminus \{q_1, q_2, \ldots, q_l\}, \\ 2, & \text{if } r \in \{q_1, q_2, \ldots, q_l\}. \end{cases}$$

Table 8.1 *Prognostic factors for coronary heart disease as measured on Czech autoworkers from (Edwards and Havranek 1985).*

				B	no		yes	
F	E	D	C	A	no	yes	no	yes
neg	< 3	< 140	no		44	40	112	67
			yes		129	145	12	23
		≥ 140	no		35	12	80	33
			yes		109	67	7	9
	≥ 3	< 140	no		23	32	70	66
			yes		50	80	7	13
		≥ 140	no		24	25	73	57
			yes		51	63	7	16
pos	< 3	< 140	no		5	7	21	9
			yes		9	17	1	4
		≥ 140	no		4	3	11	8
			yes		14	17	5	2
	≥ 3	< 140	no		7	3	14	14
			yes		9	16	2	3
		≥ 140	no		4	0	13	11
			yes		5	14	4	4

For $s = 1, 2, \ldots, l$, denote $C_s := K \setminus \{q_s\}$. Then

$$n(i^0) = (-1)^l \cdot n^* - \sum_{s=0}^{l-1} (-1)^{l+s} \cdot n_{C_{(l-s)}}(1, \ldots, 1, i^0_{q_{(l-s)}+1}, \ldots, i^0_k). \qquad (8.4)$$

We obtain the upper and lower bounds induced on the $(1, 1, \ldots, 1)$ cell count in table $\mathbf{n}$ by fixing the set of cells $\mathbf{T}_0$ by imposing the non-negativity constraints in Equation (8.4). More explicitly, $n(i^0) \geq 0$ implies that the sharp lower bound for the count n^* is $L(n^*)$ equal to

$$\max \left\{ \sum_{s=0}^{l-1} (-1)^s \cdot n_{C_{(l-s)}}(1, \ldots, 1, i^0_{q_{(l-s)}+1}, \ldots, i^0_k) : l \text{ even} \right\}, \qquad (8.5)$$

whereas the sharp upper bound $U(n^*)$ for the count n^* is equal to

$$\min \left\{ \sum_{s=0}^{l-1} (-1)^s \cdot n_{C_{(l-s)}}(1, \ldots, 1, i^0_{q_{(l-s)}+1}, \ldots, i^0_k) : l \text{ odd} \right\}. \qquad (8.6)$$

We are now ready to give the main result of this section:

Proposition 8.2 *The generalised shuttle algorithm converges to the bounds in Equations (8.5) and (8.6).*

Table 8.2 *Bounds for entries in Table 8.1 induced by fixing the five-way marginals.*

F	E	D	C	A	B no		yes	
					no	yes	no	yes
neg	< 3	< 140	no		[44,45]	[39,40]	[111,112]	[67,68]
			yes		[128,129]	[145,146]	[12,13]	[22,23]
		≥ 140	no		[34,35]	[12,13]	[80,81]	[32,33]
			yes		[109,110]	[66,67]	[6,7]	[9,10]
	≥ 3	< 140	no		[22,23]	[32,33]	[70,71]	[65,66]
			yes		[50,51]	[79,80]	[6,7]	[13,14]
		≥ 140	no		[24,25]	[24,25]	[72,73]	[57,58]
			yes		[50,51]	[63,64]	[7,8]	[15,16]
pos	< 3	< 140	no		[4,5]	[7,8]	[21,22]	[8,9]
			yes		[9,10]	[16,17]	[0,1]	[4,5]
		≥ 140	no		[4,5]	[2,3]	[10,11]	[8,9]
			yes		[13,14]	[17,18]	[5,6]	[1,2]
	≥ 3	< 140	no		[7,8]	[2,3]	[13,14]	[14,15]
			yes		[8,9]	[16,17]	[2,3]	[2,3]
		≥ 140	no		[3,4]	[0,1]	[13,14]	[10,11]
			yes		[5,6]	[13,14]	[3,4]	[4,5]

8.4.1 Example: Bounds for the Czech autoworkers data

Table 8.1 contains a 2^6 table, originally analysed by (Edwards and Havranek 1985) that cross-classifies binary risk factors denoted by A, B, C, D, E, F for coronary thrombosis from a prospective epidemiological study of 1841 workers in a Czechoslovakian car factory. Here A indicates whether or not the worker 'smokes', B corresponds to 'strenuous mental work', C corresponds to 'strenuous physical work', D corresponds to 'systolic blood pressure', E corresponds to 'ratio of β and α lipoproteins' and F represents 'family anamnesis of coronary heart disease'. We use GSA to calculate the bounds induced by fixing the five-way marginals – see Table 8.2. There are only two tables having this set of marginals. The second feasible table is obtained by adding or subtracting one unit from the corresponding entries in Table 8.1.

8.5 Calculating bounds in the decomposable case

Consider p possibly overlapping marginal tables $\mathbf{n}_{C_1}, \mathbf{n}_{C_2}, \ldots, \mathbf{n}_{C_p}$ such that $C_1 \cup C_2 \cup \ldots \cup C_p = K$. Assume that the index sets defining these marginals induce a decomposable independence graph $\mathcal{G}$ with cliques C_j, $j = 1, 2, \ldots, p$ and separators S_j, $j = 2, \ldots, p$. Each separator set S_j is the intersection of two cliques, i.e. $S_j = C_{j_1} \cap C_{j_2}$. The Fréchet bounds induced by this set of marginals are given by the following result due to (Dobra and Fienberg 2000) and (Dobra 2001).

Theorem 8.1 *Equations (8.7) below are sharp bounds given the marginals* $\mathbf{n}_{C_1}$, ..., $\mathbf{n}_{C_p}$:

$$\min\left\{ n_{C_1}\left(i_{C_1}\right), \ldots, n_{C_p}\left(i_{C_p}\right)\right\}$$

$$\geq n(i) \geq \max\left\{ \sum_{j=1}^{p} n_{C_j}\left(i_{C_j}\right) - \sum_{j=2}^{p} n_{S_j}\left(i_{S_j}\right), 0 \right\}. \quad (8.7)$$

We derive analogous Fréchet bounds for each cell in the set of cells $\mathbf{T} = \mathbf{T}^{(\mathbf{n})}$ associated with table $\mathbf{n}$. First we develop inequalities for the cells contained in the marginals of $\mathbf{n}$: $\{n_D(i_D) : i_D \in \mathcal{I}_D \text{ for some } D \subset K\}$.

Proposition 8.3 *For a subset* $D_0 \subset K$ *and an index* $i_{D_0}^0 \in \mathcal{I}_{D_0}$, *the following inequalities hold:*

$$\min\left\{ n_{C \cap D_0}\left(i_{C \cap D_0}^0\right) \mid C \in \mathcal{C}(\mathcal{G})\right\} \geq n_{D_0}\left(i_{D_0}^0\right)$$

$$\geq \max\left\{ 0, \sum_{C \in \mathcal{C}(\mathcal{G})} n_{C \cap D_0}\left(i_{C \cap D_0}^0\right) - \sum_{S \in \mathcal{S}(\mathcal{G})} n_{S \cap D_0}\left(i_{S \cap D_0}^0\right) \right\}. \quad (8.8)$$

The upper and lower bounds in Equation (8.8) are defined to be the Fréchet bounds for the cell entry $n_{D_0}\left(i_{D_0}^0\right)$ *given* $\mathbf{n}_{C_1}, \mathbf{n}_{C_2}, \ldots, \mathbf{n}_{C_p}$.

For $D_0 = K$, Equation (8.8) becomes Equation (8.7). At this point we know how to write Fréchet bounds for cell entries in an arbitrary table $\mathbf{n}' \in \mathcal{RD}$. If $\mathbf{n}'$ is not a proper marginal of $\mathbf{n}$, i.e., $\mathbf{n}' \notin \{\mathbf{n}_D : D \subset K\}$, from Equation (8.2) we deduce that there exists $D_0 \subset K$ such that $\mathbf{n}' \in \mathcal{RD}(\mathbf{n}_{D_0})$. Since the set of fixed marginals $\mathbf{n}_{C_1 \cap D_0}, \mathbf{n}_{C_2 \cap D_0}, \ldots, \mathbf{n}_{C_p \cap D_0}$ of $\mathbf{n}_{D_0}$ induce a decomposable independence graph $\mathcal{G}(D_0)$, we obtain $\mathbf{n}'$ from $\mathbf{n}_{D_0}$ by sequentially joining categories associated with the variables cross-classified in $\mathbf{n}_{D_0}$. If we apply exactly the same sequence of 'join' operations to every marginal $\mathbf{n}_{C_r \cap D_0}$, $r = 1, 2, \ldots, p$, we end up with p fixed marginals $\mathbf{n}'_{C_1 \cap D_0}, \mathbf{n}'_{C_2 \cap D_0}, \ldots, \mathbf{n}'_{C_p \cap D_0}$ of $\mathbf{n}'$. The independence graph induced by those marginals coincides with $\mathcal{G}(D_0)$. Therefore the Fréchet bounds for a cell entry in $\mathbf{n}'$ are given either by Proposition 8.3 or by Theorem 8.1 if $\mathbf{n}' \in \mathcal{RD}(\mathbf{n})$.

The following lemma tells us that the Fréchet bounds for a cell $n_{D_0}\left(i_{D_0}^0\right)$, $D_0 \subset K$, are sharp if $\mathbf{n}$ has two fixed non-overlapping marginals.

Lemma 8.1 *Let* $\mathcal{G} = (K, E)$ *be a decomposable independence graph induced by the marginals* $\mathbf{n}_{C_1}, \mathbf{n}_{C_2}, \ldots, \mathbf{n}_{C_p}$. *Consider a subset* $D_0 \subset K$ *and let* $v \in K \setminus D_0$ *be a simplicial vertex of* $\mathcal{G}$. *It is known that a simplicial vertex belongs to precisely one clique, say* $v \in C_1$. *Then finding bounds for a cell* $n_{D_0}\left(i_{D_0}^0\right)$, $i_{D_0}^0 \in \mathcal{I}_{D_0}$, *given* $\mathbf{n}_{C_1}, \mathbf{n}_{C_2}, \ldots, \mathbf{n}_{C_p}$ *is equivalent to finding bounds for* $n_{D_0}\left(i_{D_0}^0\right)$ *given* $\mathbf{n}_{C_1 \setminus \{v\}}, \mathbf{n}_{C_2}, \ldots, \mathbf{n}_{C_p}$.

The Fréchet bounds for cells in a marginal $\mathbf{n}_{D_0}$ of $\mathbf{n}$ might not be the best bounds possible.

Lemma 8.2 *Assume there are two fixed marginals* $\mathbf{n}_{C_1}$ *and* $\mathbf{n}_{C_2}$ *such that* $C_1 \cup C_2 = K$, *but* $C_1 \cap C_2 = \emptyset$. *Consider* $D_0 \subset K$. *The Fréchet bounds for* $n_{D_0}(i_{D_0}^0)$ *given* $\mathbf{n}_{C_1}$ *and* $\mathbf{n}_{C_2}$

$$\min\left\{n_{C_1 \cap D_0}(i_{C_1 \cap D_0}^0), n_{C_2 \cap D_0}(i_{C_2 \cap D_0}^0)\right\} \geq n_{D_0}(i_{D_0}^0)$$
$$\geq \max\left\{0, n_{C_1 \cap D_0}(i_{C_1 \cap D_0}^0) + n_{C_2 \cap D_0}(i_{C_2 \cap D_0}^0) - n_\emptyset\right\}$$

are sharp given $\mathbf{n}_{C_1}$ *and* $\mathbf{n}_{C_2}$.

If the two marginals are overlapping, Proposition 8.3 states that the Fréchet bounds for $n_{D_0}(i_{D_0}^0)$ are given by

$$\min\left\{n_{C_1 \cap D_0}(i_{C_1 \cap D_0}^0), n_{C_2 \cap D_0}(i_{C_2 \cap D_0}^0)\right\} \text{ and}$$
$$\max\left\{0, n_{C_1 \cap D_0}(i_{C_1 \cap D_0}^0) + n_{C_2 \cap D_0}(i_{C_2 \cap D_0}^0) - n_{C_1 \cap C_2 \cap D_0}(i_{C_1 \cap C_2 \cap D_0}^0)\right\}.$$

It turns out that the bounds in the two equations above are *not* necessarily sharp bounds for $n_{D_0}(i_{D_0}^0)$ given $\mathbf{n}_{C_1}$ and $\mathbf{n}_{C_2}$.

Lemma 8.3 *Let the two fixed marginals* $\mathbf{n}_{C_1}$ *and* $\mathbf{n}_{C_2}$ *be such that* $C_1 \cup C_2 = K$. *Consider* $D_0 \subset K$ *and denote* $D_1 := (C_1 \setminus C_2) \cap D_0$, $D_2 := (C_2 \setminus C_1) \cap D_0$ *and* $D_{12} := (C_1 \cap C_2) \cap D_0$. *Moreover, let* $C_{12} := (C_1 \cap C_2) \setminus D_0$. *Then an upper bound for* $n_{D_0}(i_{D_0}^0)$ *given* $\mathbf{n}_{C_1}$ *and* $\mathbf{n}_{C_2}$ *is:*

$$\sum_{i_{C_{12}}^1 \in \mathcal{I}_{C_{12}}} \min\left\{n_{(C_1 \cap D_0) \cup C_{12}}\left(i_{C_1 \cap D_0}^0, i_{C_{12}}^1\right), n_{(C_2 \cap D_0) \cup C_{12}}\left(i_{C_2 \cap D_0}^0, i_{C_{12}}^1\right)\right\},$$

while a lower bound is

$$\sum_{i_{C_{12}}^1 \in \mathcal{I}_{C_{12}}} \max\left\{0, n_{(C_1 \cap D_0) \cup C_{12}}\left(i_{C_1 \cap D_0}^0, i_{C_{12}}^1\right)\right.$$

$$\left. + n_{(C_2 \cap D_0) \cup C_{12}}\left(i_{C_2 \cap D_0}^0, i_{C_{12}}^1\right) - n_{D_{12}}\left(i_{D_{12}}^0\right)\right\}.$$

The following result characterises the behaviour of GSA in the decomposable case.

Proposition 8.4 *Let* $\mathbf{n}$ *be a k-dimensional table and consider the set of cells* $\mathbf{T} = \mathbf{T}^{(\mathbf{n})}$ *associated with* $\mathbf{n}$ *defined in Equation (8.3). The marginals* $\mathbf{n}_{C_1}, \mathbf{n}_{C_2}, \ldots, \mathbf{n}_{C_p}$ *induce a decomposable independence graph* $\mathcal{G} = (K, E)$ *with* $\mathcal{C}(\mathcal{G}) = \{C_1, C_2, \ldots, C_p\}$ *and* $\mathcal{S}(\mathcal{G}) = \{S_2, \ldots, S_p\}$. *The set of fixed cells* $\mathbf{T}_0 \subset \mathbf{T}^{(\mathbf{n})}$ *is given by the cell entries contained in the tables*

$$\bigcup_{r=1}^p \bigcup_{\{C : C \subseteq C_r\}} \mathcal{RD}(\mathbf{n}_C).$$

For every cell $t \in \mathbf{T}$, *let* $\mathbf{n}_1^{(t)}, \mathbf{n}_2^{(t)}, \ldots, \mathbf{n}_{k_t}^{(t)}$ *be the tables in* $\mathcal{RD}$ *such that* t *is a cell entry in* $\mathbf{n}_r^{(t)}$, $r = 1, 2, \ldots, k_t$. *Then, GSA converges to an upper bound* $U_s(t)$ *and*

Table 8.3 *Bounds for entries in Table 8.1 induced by fixing the marginals [BF],
[ABCE] and [ADE].*

F	E	D	C	A	B no no	yes	yes no	yes
neg	< 3	< 140	no		[0,88]	[0,62]	[0,224]	[0,117]
			yes		[0,261]	[0,246]	[0,25]	[0,38]
		≥ 140	no		[0,88]	[0,62]	[0,224]	[0,117]
			yes		[0,261]	[0,151]	[0,25]	[0,38]
	≥ 3	< 140	no		[0,58]	[0,60]	[0,170]	[0,148]
			yes		[0,115]	[0,173]	[0,20]	[0,36]
		≥ 140	no		[0,58]	[0,60]	[0,170]	[0,148]
			yes		[0,115]	[0,173]	[0,20]	[0,36]
pos	< 3	< 140	no		[0,88]	[0,62]	[0,126]	[0,117]
			yes		[0,134]	[0,134]	[0,25]	[0,38]
		≥ 140	no		[0,88]	[0,62]	[0,126]	[0,117]
			yes		[0,134]	[0,134]	[0,25]	[0,38]
	≥ 3	< 140	no		[0,58]	[0,60]	[0,126]	[0,126]
			yes		[0,115]	[0,134]	[0,20]	[0,36]
		≥ 140	no		[0,58]	[0,60]	[0,126]	[0,126]
			yes		[0,115]	[0,134]	[0,20]	[0,36]

to a lower bound $L_s(t)$ such that

$$\max\{L^r(t) : r = 1, 2, \ldots, k_t\} \leq L_s(t),$$
$$U_s(t) \leq \min\{U^r(t) : r = 1, 2, \ldots, k_t\}, \tag{8.9}$$

where $U^r(t)$ and $L^r(t)$ are the Fréchet bounds of the cell t in table $\mathbf{n}_r^{(t)}$.

Any cell $t_0 \in \mathbf{T}$ can be found in one, two or possibly more tables in $\mathcal{RD}$. It is
sufficient to prove that GSA converges to the Fréchet bounds for t_0 in every table
$\mathbf{n}'$ such that t_0 is a cell of $\mathbf{n}'$. The shuttle procedure updates the bounds for t_0 once
a better upper or lower bound is identified, so Equation (8.9) is true if and only if
the algorithm reaches the Fréchet bounds in every cell of every table in $\mathcal{RD}$. A cell
$n(i^0)$, $i^0 \in \mathcal{I}$, might appear in several tables in $\mathcal{RD}$, but Proposition 8.4 implies
that GSA converges to the Fréchet bounds in Equation (8.7) of $n(i^0)$, and since
from Theorem 8.1 we learn that these bounds are sharp, we deduce that the shuttle
procedure reaches the sharp bounds for $n(i^0)$.

8.5.1 Example: Bounds for the Czech autoworkers data

We return to the 2^6 contingency table given in Table 8.1. (Whittaker 1990, page
263) suggests that an appropriate model for this data is given by the marginals
[BF], [ABCE] and [ADE]. This represents a decomposable log-linear model whose
independence graph has separators [B] and [AE]. The corresponding Fréchet bounds

from Equation (8.7) become:

$$\min\left\{n_{BF}\left(i_{BF}\right), n_{ABCE}\left(i_{ABCE}\right), n_{ADE}\left(i_{ADE}\right)\right\} \geq n(i) \geq$$
$$\max\left\{n_{BF}\left(i_{BF}\right) + n_{ABCE}\left(i_{ABCE}\right) + n_{ADE}\left(i_{ADE}\right) - n_{B}\left(i_{B}\right) - n_{AE}\left(i_{AE}\right), 0\right\}.$$

The bounds computed by GSA are shown in Table 8.3.

8.6 Computing sharp bounds

When the fixed set of marginals defines a decomposable independence graph, GSA converges to the corresponding Fréchet bounds for all the cell entries in the table $\mathbf{n}$. When $\mathbf{n}$ is dichotomous and all the lower-dimensional marginals are fixed, we were also able to explicitly determine the tightest bounds for the cell entries $\mathbf{n}$ and prove that GSA reaches these bounds. Even in these two particular instances GSA is guaranteed to find sharp bounds only for the cells $\mathcal{N}$ in table $\mathbf{n}$. In this section we present a method that sequentially adjusts the bounds $L_s(\mathbf{T})$ and $U_s(\mathbf{T})$ obtained from GSA until they become sharp.

The integer value $U(t_1)$ is a sharp upper bound for a cell $t_1 \in \mathbf{T}$ if and only if there exists an integer array $V(\mathbf{T}) \in S[L_s(\mathbf{T}), U_s(\mathbf{T})]$ with a count of $U(t_1)$ in cell t_1 (i.e., $V(t_1) = U(t_1)$) and if there does not exist another integer array $V'(\mathbf{T}) \in S[L_s(\mathbf{T}), U_s(\mathbf{T})]$ having a count in cell t_1 strictly bigger than $U(t_1)$ (i.e., $V'(t_1) > U(t_1)$). The sharp lower bound $L(t_1)$ can be defined in a similar way.

We know that $L_s(t_1) \leq L(t_1) \leq U(t_1) \leq U_s(t_1)$. This means that the first candidate value for $U(t_1)$ is $U_s(t_1)$. If there is no integer array $V(\mathbf{T}) \in S[L_s(\mathbf{T}), U_s(\mathbf{T})]$ with $V(t_1) = U_s(t_1)$, we sequentially try $U_s(t_1) - 1, U_s(t_1) - 2, \ldots, L_s(t_1)$ and stop when a feasible array with the corresponding count in cell t_1 is determined. The candidate values for the sharp lower bound $L(t_1)$ are $L_s(t_1) + 1, L_s(t_1) + 2, \ldots, U_s(t_1)$ in this particular order. After fixing the count $V(t_1)$ to an integer value between $L_s(t_1)$ and $U_s(t_1)$, we employ GSA to update the upper and lower bounds for all the cells in $\mathbf{T}$. Denote by $L_s^1(\mathbf{T})$ and $U_s^1(\mathbf{T})$ the new bounds identified by GSA. These bounds are tighter than $L_s(\mathbf{T})$ and $U_s(\mathbf{T})$, thus the set of integer arrays $S^1 = S[L_s^1(\mathbf{T}), U_s^1(\mathbf{T})]$ is included in $S[L_s(\mathbf{T}), U_s(\mathbf{T})]$. We reduced the problem of determining sharp bounds for the cell t_1 to the problem of checking whether S^1 is empty. We need to repeat these steps for every cell t_1 for which we want to obtain sharp bounds.

We describe an algorithm for exhaustively enumerating all the integer arrays in $S[L(\mathbf{T}), U(\mathbf{T})]$. Here $L(\mathbf{T})$ and $U(\mathbf{T})$ are arrays of lower and upper bounds for the cells $\mathbf{T}$. We associate with every cell $t = t_{J_1 J_2 \ldots J_k} \in \mathbf{T}$ an index; see (Knuth 1973)

$$IND(t) := \sum_{l=1}^{k} 2^{\sum_{s=l+1}^{k} I_s} \left(\sum_{j_l \in J_l} 2^{j_l - 1} - 1 \right) + 1 \in \{1, 2, \ldots, N\}.$$

We order the cells in $\mathbf{T}$ as a linear list $t_1, t_2, \ldots, t_N$, with $N = 2^{I_1 + I_2 + \ldots + I_k}$. With this ordering, we sequentially attempt to fix every cell at integer values between its current upper and lower bounds and use GSA to update the bounds for the remaining cells. We successfully determined a feasible array when we assigned a value to every cell and GSA did not identify any inconsistencies among these values.

PROCEDURE SharpBounds(k,$L^k(\mathbf{T})$,$U^k(\mathbf{T})$)

(1) IF $k = N + 1$ THEN save the newly identified array
$V(\mathbf{T}) \in S[L(\mathbf{T}), U(\mathbf{T})]$.

(2) FOR every integer $c \in [L^k(t_k), L^k(t_k) + 1, \ldots, U^k(t_k)]$ DO

   (2A) SET $V(t_k)$ to value c.

   (2B) SET $L^{k+1}(t_k) = U^{k+1}(t_k) = c$, $L^{k+1}(t_i) = L^k(t_i)$,
      $U^{k+1}(t_i) = U^k(t_i)$ for $i = 1, \ldots, k-1, k+1, \ldots, N$.

   (2C) Run GSA to update the bounds $L^{k+1}(\mathbf{T})$ and $U^{k+1}(\mathbf{T})$.

   (2D) IF GSA did not identify any inconsistencies THEN
      CALL SharpBounds(k+1,$L^{k+1}(\mathbf{T})$,$U^{k+1}(\mathbf{T})$).

PROCEDURE ENDS

The initial call is SharpBounds(1,$L(\mathbf{T})$,$U(\mathbf{T})$). Note that the updated bounds from step (2C) satisfy

$$L^k(t_i) \leq L^{k+1}(t_i) \leq U^{k+1}(t_i) \leq U^k(t_i),$$

provided that GSA did not report inconsistencies. This sequential improvement of the bounds avoids an exhaustive enumeration of all the combinations of possible values of the cells $\mathbf{T}$ that would lead to a very low computational efficiency of the algorithm.

When computing sharp bounds for a cell t_1, we can stop the SharpBounds procedure after we identified the first table in S^1 or learn that no such table exists.

8.7 Large contingency tables

We demonstrate the scalability of the GSA by computing sharp bounds for the non-zero entries of a 2^{16} contingency table extracted from the 'analytic' data file for National Long-Term Care Survey created by the Center of Demographic Studies at Duke University. Each dimension corresponds to a measure of disability defined by an activity of daily leaving and the table contains information cross-classifying individuals aged 65 and above. The 16 dimensions of this contingency table correspond to 6 activities of daily living (ADLs) and 10 instrumental activities of daily living (IADLs). Specifically, the ADLs are (1) eating, (2) getting in/out of bed, (3) getting around inside, (4) dressing, (5) bathing, (6) getting to the bathroom or using a toilet. The IADLs are (7) doing heavy house work, (8) doing light house work, (9) doing laundry, (10) cooking, (11) grocery shopping, (12) getting about outside, (13) travelling, (14) managing money, (15) taking medicine, (16) telephoning. For each ADL/IADL measure, subjects were classified as being either disabled (level 1) or healthy (level 0) on that measure. For a detailed description of this extract see (Erosheva *et al.* 2007). (Dobra *et al.* 2003a) and Chapter 2 in this volume also consider analyses of these data.

We applied GSA to compute sharp upper and lower bounds for the entries in this table corresponding to a number of different sets of fixed marginals. Here we describe one complex calculation for the set involving three fixed 15-way marginals obtained by collapsing the 16-way table across the variables (14) managing money, (15) taking medicine and (16) telephoning. Of the $2^{16} = 65\,536$ cells, $62\,384$ contain zero entries. Since the target table is so sparse, fixing three marginals of dimension 15 leads to the exact determination (i.e., equal upper and lower bounds) of most of the cell entries. To be more exact, only 128 cells have the upper bounds strictly bigger than the lower bounds! The difference between the upper and lower bounds is equal to 1 for 96 cells, 2 for 16 cells, 6 for 8 cells and 10 for 8 cells.

We take a closer look at the bounds associated with 'small' counts of 1 or 2. There are 1729 cells containing a count of 1. Of these, 1698 cells have the upper bounds equal to the lower bounds. The difference between the bounds is 1 for 28 of the remaining counts of 1, is 2 for two other cells and is equal to 6 for only one entry. As for the 499 cells with a count of 2, the difference between the bounds is zero for 485 cells, is 1 for 10 cells and is 2 for 4 other cells.

GSA converged in approximately 20 iterations to the sharp bounds and it took less than six hours to complete on a single-processor machine at the Department of Statistics, Carnegie Mellon University. We re-checked these bounds by determining the feasible integer tables for which they are attained on the Terascale Computing System at the Pittsburgh Supercomputing Center. We used a parallel implementation of GSA that independently adjust the bounds for various cells and the computations took almost one hour to complete on 56 processors.

8.8 Other examples

In the examples that follow we employ GSA not only to produce sharp bounds, but also to compute exact p-values for conditional inference with the hypergeometric distribution, see (Dobra *et al.* 2006):

$$p(\mathbf{n}) = \left[\prod_{i \in \mathcal{I}} n(i)!\right]^{-1} / \sum_{\mathbf{n}' \in \mathcal{T}} \left[\prod_{i \in \mathcal{I}} n'(i)!\right]^{-1}. \tag{8.10}$$

where $\mathcal{T}$ represents the set of contingency tables consistent with a given set of constraints (e.g., upper and lower bounds for cell entries). The corresponding p-value of the exact test is, see (Guo and Thompson 1992):

$$\sum_{\{\mathbf{n}' \in \mathcal{T}: p(\mathbf{n}') \leq p(\mathbf{n})\}} p(\mathbf{n}'), \tag{8.11}$$

where $\mathbf{n}$ is the observed table. (Sundberg 1975) shows that the normalising constant in Equation (8.10) can be directly evaluated if $\mathcal{T}$ is determined by a decomposable set of marginals, but otherwise it can be computed only if $\mathcal{T}$ can be exhaustively enumerated. GSA can accomplish this task for almost any type of constraints and evaluate $p(\mathbf{n})$ as well as the p-value in Equation (8.11) exactly. We compare our inferences with the results obtained by (Chen *et al.* 2006) who proposed a sequential

Table 8.4 *A sparse 4-way dichotomous table (left panel) from (Sullivant 2005). The right panel gives the MLEs induced by the six 2-way marginals.*

		C	No		Yes		No		Yes	
A	B	D	No	Yes	No	Yes	No	Yes	No	Yes
No	No		0	1	1	0	1.06	0.36	0.36	0.21
	Yes		1	0	0	0	0.36	0.21	0.21	0.21
Yes	No		1	0	0	0	0.36	0.21	0.21	0.21
	Yes		0	0	0	1	0.21	0.21	0.21	0.36

importance sampling method (SIS, henceforth) for approximating exact p-values by randomly sampling from $\mathcal{T}$ and $p(\mathbf{n})$.

Example 8.1 (Vlach 1986) considers the following three matrices:

$$A = \begin{pmatrix} 1 & 1 & 1 \\ 1 & 1 & 1 \\ 1 & 1 & 1 \\ 1 & 1 & 1 \end{pmatrix}, \quad B = \begin{pmatrix} 1 & 0 & 1 \\ 1 & 0 & 1 \\ 0 & 1 & 1 \\ 0 & 1 & 1 \\ 1 & 1 & 0 \\ 1 & 1 & 0 \end{pmatrix}, \quad C = \begin{pmatrix} 0 & 1 & 0 & 1 \\ 1 & 0 & 1 & 0 \\ 1 & 0 & 0 & 1 \\ 0 & 1 & 1 & 0 \\ 1 & 1 & 0 & 0 \\ 0 & 0 & 1 & 1 \end{pmatrix}.$$

Matrices A, B and C appear to be the two-way marginals of a $6 \times 4 \times 3$ contingency table and their one-way marginals coincide; however, there does not exist a $6 \times 4 \times 3$ *integer* table having this set of two-way margins and GSA stopped without producing any bounds due to the inconsistencies it identified.

Example 8.2 (Sullivant 2005) presented a $2 \times 2 \times 2 \times 2$ table with a grand total of 5, reproduced in Table 8.4. This is the only integer table consistent with the six 2-way marginals and GSA correctly identifies it. Fitting the no-3-way interaction model implied by fixing the 2-way margins in R using `loglin` yields the MLEs in the right panel of Table 8.4, but the program reports $d.f. = 5$. The correct number of degrees of freedom is zero since there is only one integer table with these constraints. Testing the significance of the no-3-way interaction model with reference to a χ^2 distribution on 5 degrees of freedom would be erroneous. The lower integer bounds equal the upper integer bounds for all 16 cells. Note the large gaps (up to 1.67) between the integer bounds and the real bounds (see Table 8.5) calculated with the simplex algorithm.

Example 8.3 (Dobra *et al.* 2006) used GSA to determine that there are 810 tables consistent with the set of fixed marginals [ACDEF], [ABDEF], [ABCDE], [BCDF], [ABCF], [BCEF] of Table 8.1. GSA calculates the p-value for the exact goodness-of-fit test in Equation (8.11) to be 0.235. The estimated p-value computed using SIS in (Chen *et al.* 2006) is 0.27, while the estimated number of tables is 840. The `loglin` function in R gives a p-value of 0.21 on 4 degrees of freedom.

Table 8.5 *LP bounds fixing the six 2-way marginals of Table 8.4.*

		C	No		Yes	
A	B	D	No	Yes	No	Yes
No	No		[0, 1.67]	[0, 1]	[0, 1]	[0, 0.67]
	Yes		[0, 1]	[0, 0.67]	[0, 0.67]	[0, 0.67]
Yes	No		[0, 1]	[0, 0.67]	[0, 0.67]	[0, 0.67]
	Yes		[0, 0.67]	[0, 0.67]	[0, 0.67]	[0, 1]

Table 8.6 *The upper panel gives the 4-way abortion option data from (Haberman 1978). The lower panel gives the sharp integer bounds induced by the four 3-way marginals of this table.*

Race	Sex	Opinion	18−25	26−35	36−45	46−55	56−65	66+
White	Male	Yes	96	138	117	75	72	83
		No	44	64	56	48	49	60
		Und	1	2	6	5	6	8
	Female	Yes	140	171	152	101	102	111
		No	43	65	58	51	58	67
		Und	1	4	9	9	10	16
Nonwhite	Male	Yes	24	18	16	12	6	4
		No	5	7	7	6	8	10
		Und	2	1	3	4	3	4
	Female	Yes	21	25	20	17	14	13
		No	4	6	5	5	5	5
		Und	1	2	1	1	1	1
White	Male	Yes	[90, 101]	[130, 146]	[107, 123]	[65, 81]	[61, 78]	[70, 87]
		No	[40, 49]	[58, 71]	[51, 63]	[43, 54]	[44, 57]	[55, 70]
		Und	[0, 2]	[0, 3]	[5, 9]	[4, 9]	[5, 9]	[7, 12]
	Female	Yes	[135, 146]	[163, 179]	[146, 162]	[95, 111]	[96, 113]	[107, 124]
		No	[38, 47]	[58, 71]	[51, 63]	[45, 56]	[50, 63]	[57, 72]
		Und	[0, 2]	[3, 6]	[6, 10]	[5, 10]	[7, 11]	[12, 17]
Nonwhite	Male	Yes	[19, 30]	[10, 26]	[10, 26]	[6, 22]	[0, 17]	[0, 17]
		No	[0, 9]	[0, 13]	[0, 12]	[0, 11]	[0, 13]	[0, 15]
		Und	[1, 3]	[0, 3]	[0, 4]	[0, 5]	[0, 4]	[0, 5]
	Female	Yes	[15, 26]	[17, 33]	[10, 26]	[7, 23]	[3, 10]	[0, 17]
		No	[0, 9]	[0, 13]	[0, 12]	[0, 11]	[0, 13]	[0, 15]
		Und	[0, 2]	[0, 3]	[0, 4]	[0, 5]	[0, 4]	[0, 5]

(Dobra *et al.* 2006) also considered the model determined by fixing the 15 4-way margins. GSA reported a number of 705 884 feasible tables with a corresponding exact p-value in Equation (8.11) equal to 0.432. Fitting the same model with `loglin` yields an approximate p-value of 0.438 by reference to a χ^2 distribution of 7.95 on 8 degrees of freedom.

Example 8.4 Table 8.6 contains a $2 \times 2 \times 3 \times 6$ table from an NORC survey from the 1970s, see (Haberman 1978, p. 291), that cross-classifies race (white, nonwhite), sex (male, female), attitude towards abortion (yes, no, undecided) and age

Table 8.7 *Results of clinical trial for the effectiveness of an analgesic drug from (Koch et al. 1983)*

C	S	R T	1	2	3
1	1	1	3	20	5
1	1	2	11	14	8
1	2	1	3	14	12
1	2	2	6	13	5
2	1	1	12	12	0
2	1	2	11	10	0
2	2	1	3	9	4
2	2	2	6	9	3

(18–25, 26–35, 36–45, 46–55, 56–65, 66+ years). (Christensen 1997, p. 111) considered the log-linear model corresponding to the four 3-way marginals. The `loglin` function in R yields an approximate p-value of 0.807 based on a χ^2 distribution on 6.09 with 10 degrees of freedom. GSA identified 83 087 976 tables consistent with the 3-way marginals and returned an exact p-value for the goodness-of-fit test in Equation (8.11) equal to 0.815. (Chen *et al.* 2006) report that SIS estimated that the number of feasible tables is 9.1×10^7 and that the exact p-value based on the hypergeometric distribution is approximately 0.85. In the bottom panel of Table 8.6 we give the upper and lower bounds computed by GSA. The release of the four 3-way marginals might be problematic from a disclosure limitation perspective due to the tight bounds for some of the small counts of 1 and 2.

Example 8.5 Table 8.7 from (Koch *et al.* 1983) summarises the results of a clinical trial on the effectiveness (R – poor, moderate or excellent) of an analgesic drug (T – 1,2) for patients in two statuses (S) and two centres (C), with a grand total of 193. While most of the counts are relatively large, the table contains two counts of zero that lead to a zero entry in the [CSR] marginal.

(Fienberg and Slavkovic 2004, Fienberg and Slavkovic 2005) discuss several log-linear models associated with this contingency table to illustrate disclosure limitation techniques. The upper and lower bounds presented in their 2004 paper are the same bounds identified by GSA, so we chose not to reproduce them here. The zero entry in the [CSR] marginal leads to the non-existence of MLEs in any log-linear model with a generator [CSR]. This implies that the degrees of freedom for any log-linear model that includes [CSR] as a minimal sufficient statistic needs to be reduced by one – this corresponds to fitting a log-linear model to the incomplete table that does not include the two counts of zero adding up to the zero entry in the [CSR] marginal. For additional details and theoretical considerations, see (Fienberg 1980) and (Fienberg and Rinaldo 2007).

How does the exact goodness-of-fit test in Equation (8.11) perform in this special situation? For the model [CST][CSR], GSA identifies 79 320 780 feasible tables and gives an exact p-value of 0.073. By comparison, the `loglin` function in R yields an approximate p-value of 0.06 based on 7 degrees of freedom. For the model [CST][CSR][TR], GSA finds 155 745 feasible tables with a corresponding p-value of 0.0499, while the `loglin` function gives a p-value of 0.039 based on 5 degrees of freedom. For the model [CST][CSR][CTR], GSA finds 1274 feasible tables with a p-value of 0.152, while the `loglin` function reports a p-value of 0.127 based on 3 degrees of freedom. Finally, for [CST][CSR][SRT] with an exact p-value of 0.093 based on 1022 feasible tables, `loglin` finds an approximate p-value of 0.073 based on 3 degrees of freedom. The discrepancy between the exact and approximate p-values tends to become more significant in degenerate cases when the MLEs do not exist. The model [CST][CSR][TR] seems to fit the data well indicating that there is evidence of a direct relationship between the treatment and response in this clinical trial.

Example 8.6 (Dobra *et al.* 2008) analyse a sparse dichotomous 6-way table from (Edwards 1992) which cross-classifies the parental alleles of six loci along a chromosome strand of a barley powder mildew fungus. The variables are labelled A, B, C, D, E and F and have categories 1 or 2, see Table 8.8. GSA finds a relatively small number 36 453 of tables consistent with the 2-way marginals with an exact p-value of the goodness-of-fit test based on the hypergeometric distribution equal to 0.652. The MLEs for this log-linear model do not exist because of a zero entry in the [AB] marginal; however, the MLEs for the log-linear model [ABCD][CDE][ABCEF] do exist. In this instance, GSA finds 30 tables consistent with the marginals [ABCD], [CDE] and [ABCEF] with an exact p-value of 1.

8.9 Conclusions

We have described the generalised shuttle algorithm that exploits the hierarchical structure of categorical data to compute sharp bounds and enumerate sets of multi-way tables. The constraints defining these sets can appear the form of fixed marginals, upper and lower bounds on blocks of cells or structural zeros. In the most general setting one can restrict the search scope to tables having certain combinations of counts in various cell configurations. GSA produces sharp bounds not only for cells in the multi-way table analysed, but also for any cells that belong to tables obtained through collapsing categories or variables. We showed through several examples that GSA performs very well and leads to valuable results.

We also illustrated that GSA can compute bounds for high-dimensional contingency tables. We are not aware how such computations can be performed through LP or IP methods. No matter how efficient LP/IP might be in solving one optimisation problem, calculating bounds for a 16-dimensional table would involve solving $2 \times 2^{16} = 131,072$ separate optimisation problems and this represents a huge computational undertaking. Instead, GSA computes bounds very close to the sharp bounds in one quick step, then adjusts these bounds to the sharp bounds only for the cells whose value is not uniquely determined by the marginal constraints.

While it is possible to increase the computational efficiency of GSA by adjusting the bounds in parallel or by choosing candidate values for the cell counts starting

Table 8.8 *A sparse genetics 2^6 table from (Edwards 1992). The upper panel gives the cell counts, while the lower panel shows to sharp bounds induced by fixing the two-way marginals.*

| | | | D | 1 | | | | 2 | | | |
| | | | E | 1 | | 2 | | 1 | | 2 | |
A	B	C	F	1	2	1	2	1	2	1	2
1	1	1		0	0	0	0	3	0	1	0
		2		0	1	0	0	0	1	0	0
	2	1		1	0	1	0	7	1	4	0
		2		0	0	0	2	1	3	0	11
2	1	1		16	1	4	0	1	0	0	0
		2		1	4	1	4	0	0	0	1
	2	1		0	0	0	0	0	0	0	0
		2		0	0	0	0	0	0	0	0
1	1	1		[0, 1]	[0, 1]	[0, 1]	[0, 1]	[0, 4]	[0, 2]	[0, 1]	[0, 1]
		2		[0, 1]	[0, 1]	[0, 1]	[0, 1]	[0, 2]	[0, 2]	[0, 1]	[0, 1]
	2	1		[0, 3]	[0, 2]	[0, 3]	[0, 2]	[0, 13]	[0, 2]	[0, 10]	[0, 2]
		2		[0, 1]	[0, 3]	[0, 2]	[0, 4]	[0, 2]	[0, 9]	[0, 2]	[2, 16]
2	1	1		[9, 22]	[0, 2]	[0, 9]	[0, 2]	[0, 2]	[0, 1]	[0, 2]	[0, 1]
		2		[0, 2]	[0, 10]	[0, 3]	[0, 10]	[0, 1]	[0, 2]	[0, 1]	[0, 2]
	2	1		[0, 0]	[0, 0]	[0, 0]	[0, 0]	[0, 0]	[0, 0]	[0, 0]	[0, 0]
		2		[0, 0]	[0, 0]	[0, 0]	[0, 0]	[0, 0]	[0, 0]	[0, 0]	[0, 0]

from the middle of the current feasibility intervals, see (Dobra 2001), we do not make any particular claims about its computational efficiency. The current implementation of the algorithm can be slow for a larger number of dimensions and categories and might need a lot of computer memory. On the other hand, GSA can easily be used as an off-the-shelf method for analysing contingency tables since it is extremely flexible and does not require any additional input (e.g., Markov bases, LP bounds, etc.) or intricate calibration heuristics. GSA is an excellent benchmark for judging the validity and performance of other related methods, e.g., SIS of (Chen *et al.* 2006) that have the potential to properly scale to high-dimensional data.

Acknowledgements

We thank Alessandro Rinaldo for his valuable comments. The preparation of this chapter was supported in part by NSF grants EIA9876619 and IIS0131884 to the National Institute of Statistical Sciences, and Army contract DAAD19-02-1-3-0389, NIH Grant No. R01 AG023141-01, and NSF Grant DMS-0631589 to Carnegie Mellon University.

References

Bonferroni, C. E. (1936). Teoria statistica delle classi e calcolo delle probabilità. In *Publicazioni del R. Instituto Superiore di Scienze Economiche e Commerciali di Firenze*, **8**, 1–62.

Buzzigoli, L. and Giusti, A. (1999). An algorithm to calculate the lower and upper bounds of the elements of an array given its marginals. In *Proc SDP'98*, Eurostat, Luxemburg, 131–47.

Chen, Y., Dinwoodie, I. H., and Sullivant, S. (2006). Sequential Importance Sampling for Multiway Tables, *Annals of Statistics* **34**, 523–45.

Christensen, R. (1997). *Log-linear Models and Logistic Regression*, Springer Series in Statistics 2nd edn (New York, Springer-Verlag).

Cox, L. H. (1999). Some remarks on research directions in statistical data protection, In *Proc. SDP'98*, Eurostat, Luxembourg, 163–76.

Diaconis, P. and Sturmfels, B. (1998). Algebraic algorithms for sampling from conditional distributions, *Annals of Statistics* **26**, 363–97.

Dobra, A. (2001). Statistical tools for disclosure limitation in multi-way contingency tables. PhD thesis, Department of Statistics, Carnegie Mellon University.

Dobra, A. (2003). Markov bases for decomposable graphical models, *Bernoulli* **9**(6), 1–16.

Dobra, A. and Fienberg, S. E. (2000). Bounds for cell entries in contingency tables given marginal totals and decomposable graphs. In *Proc. of the National Academy of Sciences* **97**, 11885–92.

Dobra, A., Erosheva, E. A. and Fienberg, S. E. (2003a). Disclosure limitation methods based on bounds for large contingency tables with application to disability data. In *Proc. of the New Frontiers of Statistical Data Mining*, Bozdogan, H. ed. (New York, CRC Press), 93–116.

Dobra, A., Fienberg, S. E., Rinaldo, A., Slavkovic, A. B. and Zhou, Y. (2008). Algebraic statistics and contingency table problems: estimations and disclosure limitation. In *Emerging Applications of Algebraic Geometry*, Putinar, M. and Sullivant, S. eds. (New York, Springer-Verlag).

Dobra, A., Fienberg, S. E. and Trottini, M. (2003b). Assessing the risk of disclosure of confidential categorical data. In *Bayesian Statistics 7* Bernardo, J., Bayarri, M., Berger, J. O., Dawid, A. P., Heckerman, D., Smith, A. F. M. and West, M. eds. (New York, Oxford University Press), 125–44.

Dobra, A., Karr, A. and Sanil, A. (2003c). Preserving confidentiality of high-dimensional tabulated data: statistical and computational issues, *Statistics and Computing* **13**, 363–70.

Dobra, A. and Sullivant, S. (2004). A divide-and-conquer Algorithm for generating Markov bases of multi-way tables, *Computational Statistics* **19**, 347–66.

Dobra, A., Tebaldi, C. and West, M. (2006). Data augmentation in multi-way contingency tables with fixed marginal totals, *Journal of Statistical Planning and Inference* **136**, 355–72.

Edwards, D. E. (1992). Linkage analysis using log-linear models, *Computational Statistics and Data Analysis* **10**, 281–90.

Edwards, D. E. and Havranek, T. (1985). A fast procedure for model search in multidimensional contingency Tables, *Biometrika* **72**, 339–51.

Erosheva, E. A., Fienberg, S. E. and Joutard, C. (2007). Describing disability through individual-level mixture models for multivariate binary data, *Annals of Applied Statistics* **1**(2) 502–37.

Fienberg, S. E. (1980). *The Analysis of Cross-Classified Categorical Data* 2nd edn (Cambridge, MA, MIT Press). Reprinted (2007) (New York, Springer-Verlag).

Fienberg, S. E. (1999). Fréchet and Bonferroni bounds for multi-way tables of counts with applications to disclosure limitation. In *Proc. SDP'98*, Eurostat, Luxembourg 115–29.

Fienberg, S. E. and Rinaldo, A. (2007). Three centuries of categorical data analysis: log-linear models and maximum likelihood estimation, *Journal of Statistical Planning and Inference* **137**, 3430–45.

Fienberg, S. E. and Slavkovic, A. B. (2004). Making the release of confidential data from multi-way tables count, *Chance* **17**, 5–10.

Fienberg, S. E. and Slavkovic, A. B. (2005). Preserving the confidentiality of categorical databases when releasing information for association rules, *Data Mining and Knowledge Discovery,* **11**, 155–80.

Fréchet, M. (1940). *Les Probabilitiés, Associées a un Système d'Événments Compatibles et Dépendants* (Paris, Hermann & Cie).

Geiger, D., Meek, C. and Sturmfels, B. (2006). On the toric algebra of graphical models, *Annals of Statistics* **34**, 1463–92.

Guo, S. W. and Thompson, E. A. (1992). Performing the exact test of Hardy-Weinberg proportion for multiple alleles, *Biometrics* **48**, 361–72.

Haberman, S. J. (1978). *Analysis of Qualitative Data* (New York, Academic Press).

Hoeffding, W. (1940). Scale-invariant correlation theory. In *Schriften des Mathematischen Instituts und des Instituts für Angewandte Mathematik der Universität Berlin* **5**(3), 181–233.

Hoşten, S. and Sturmfels, B. (2007). Computing the integer programming gap, *Combinatorica* **3** 367–82.

Knuth, D. (1973). *The Art of Computer Programming*, vol. 3 (Upper Saddle River, NJ, Addison-Wesley).

Koch, G., Amara, J., Atkinson, S. and Stanish, W. (1983). Overview of Categorical Analysis Methods, *SAS-SUGI* **8**, 785–95.

Sullivant, S. (2005). Small contingency tables with large gaps, *SIAM Journal of Discrete Mathematics* **18**, 787–93.

Sundberg, R. (1975). Some results about decomposable (or Markov-type) models for multidimensional contingency tables: distribution of marginals and partitioning of tests, *Scandinavian Journal of Statistics* **2**, 71–9.

Vlach, M. (1986). Conditions for the existence of solutions of the three-dimensional planar transportation problem, *Discrete Applied Mathematics* **13**, 61–78.

Whittaker, J. (1990). *Graphical Models in Applied Mathematical Multivariate Statistics* (Chichester, John Wiley & Sons).

Part II

Designed experiments

9

Generalised design: interpolation and statistical modelling over varieties

Hugo Maruri-Aguilar

Henry P. Wynn

Abstract

In the classical formulation an experimental design is a set of sites at each of which an observation is taken on a response Y. The algebraic method treats the design as giving an 'ideal of points' from which potential monomial bases for a polynomial regression can be derived. If the Gröbner basis method is used then the monomial basis depends on the monomial term ordering. The full basis has the same number of terms as the number of design points and gives an exact interpolator for the Y-values over the design points. Here, the notation of design point is generalised to a variety. Observation means, in theory, that one observes the value of the response on the variety. A design is a union of varieties and the assumption is, then, that on each variety we observe the response. The task is to construct an interpolator for the function between the varieties. Motivation is provided by transect sampling in a number of fields. Much of the algebraic theory extends to the general case, but special issues arise including the consistency of interpolation at the intersection of the varieties and the consequences of taking a design of points restricted to the varieties.

9.1 Introduction

Experimental design is defined simply as the choice of sites, or observation points, at which to observe a response, or output. A set of such points is the experimental design. Terminology varies according to the field. Thus, sites may be called 'treatment combinations', 'input configurations', 'runs', 'data points' and so on. For example in interpolation theory 'observation point' is common. Whatever the terminology or field we can nearly always code up the notion of an observation point as a single point in k dimensions which represents a single combination of levels of k independent variables.

The purpose of this chapter is to extend the notation of an observation point to a whole algebraic variety. An experimental design is then a union of such varieties. An observation would be the acquired knowledge of the restriction of the response to the variety. This is an idealisation, but one with considerable utility. It may be,

Algebraic and Geometric Methods in Statistics, ed. Paolo Gibilisco, Eva Riccomagno, Maria Piera Rogantin and Henry P. Wynn. Published by Cambridge University Press. © Cambridge University Press 2010.

159

for example, that one models the restriction of the response to each variety by a separate polynomial.

An important example of sampling via a variety is *transect sampling*. This is a method used in the estimation of species abundance in ecology and geophysics. A key text is (Buckland *et al.* 1993) and the methods are developed further in (Mack and Quang 1998). There one collects information about the distance of objects from the transects and tries to estimate the average density of the objects in the region of interest, namely to say something about a feature connected with the whole region. A useful idea is that of 'reconstruction'; one tries to reconstruct a function given the value on the transects. This reconstruction we interpret here as 'interpolation', or perhaps we should say 'generalised' interpolation. Other examples are tomography, computer vision and imaging.

Our task is to extend the algebraic methods used for observation points to this generalised type of experimental design and interpolation. Within this, the main issue is to create monomial bases to interpolate between the varieties on which we observe. At one level this is a straightforward extension, but there are a number of special constructions and issues the discussion of which should provide an initial guide to the area.

(i) The most natural generalisation is to the case where the varieties are hyperplanes, and therefore we shall be interested in hyperplane arrangements. This covers the case of lines in two dimensions, the traditional transects mentioned above.

(ii) There are consistency issues when the varieties intersect: the observation on the varieties must agree on the intersection.

(iii) Since observing a whole function on a variety may be unrealistic one can consider traditional point designs restricted to the varieties. That is, we may use standard polynomial interpolation on the varieties and then combine the results to interpolate between varieties, but having in mind the consistency issue just mentioned.

(iv) It is also natural to use power series expansions on each variety: is it possible to extend the algebraic interpolation methods to power series? We are here only able to touch on the answer.

We now recall some basic ideas. Interpolation is the construction of a function $f(x)$ that coincides with observed data at n given observation points. That is, for a finite set of distinct points $\mathcal{D} = \{d_1, \ldots, d_n\}$, $d_1, \ldots, d_n \in \mathbb{R}^k$ and observation values $y_1, \ldots, y_n \in \mathbb{R}$, we build a function such that $f(d_i) = y_i$, $i = 1, \ldots, n$. We set our paper within design of experiments theory where the design is a set of points $\mathcal{D}$, n is the design (sample) size and k is the number of factors. Approaches to interpolation range from statistically oriented techniques such as kriging, see (Stein 1999), to more algebraic techniques involving polynomials, splines or operator theory, see (Phillips 2003) and (Sakhnovich 1997).

(Pistone and Wynn 1996) build polynomial interpolators using an isomorphism between the following real vector spaces: the set of real-valued polynomial functions defined over the design, $\phi : \mathcal{D} \longrightarrow \mathbb{R}$, and the quotient ring $\mathbb{R}[x_1, \ldots, x_k]/I(\mathcal{D})$. To construct the quotient ring they first consider the design $\mathcal{D}$ as the set of solutions to a system of polynomial equations. Then this design corresponds to the *design ideal* $I(\mathcal{D})$, that is the set of all polynomials in $\mathbb{R}[x_1,\ldots,x_k]$ that vanish over the points in $\mathcal{D}$. The polynomial interpolator has n terms and is constructed using a basis for $\mathbb{R}[x_1,\ldots,x_k]/I(\mathcal{D})$ called *standard monomials*.

This algebraic method of constructing polynomial interpolators can be applied to, essentially, any finite set of points, see for example (Holliday *et al.* 1999) and (Pistone *et al.* 2009). In fractional factorial designs it has lead to the use of indicator functions, see (Fontana *et al.* 1997, Pistone and Rogantin 2008). Another example arises when the design is a mixture, i.e. the coordinate values of each point in $\mathcal{D}$ add up to one. In such a case the equation $\sum_{i=1}^{k} x_i = 1$ is incorporated into the design ideal, namely the polynomial $\sum_{i=1}^{k} x_i - 1 \in I(\mathcal{D})$, see (Giglio *et al.* 2001). More recently, (Maruri-Aguilar *et al.* 2007) used projective algebraic geometry and considered the projective coordinates of the mixture points. Their technique allows the identification of the support for a homogeneous polynomial model.

If, instead of a set of points, we consider the design as an affine variety, then the algebraic techniques discussed are still valid. As a motivating example, consider the circle in two dimensions with radius two and center at the origin. Take the *radical ideal* generated by the circle as its design ideal, i.e. the ideal generated by $x_1^2+x_2^2-4$. The set of standard monomials is infinite in this case. For a monomial order with initial order $x_2 \prec x_1$, the set of standard monomials is $\{x_2^j, x_1 x_2^j : j \in \mathbb{Z}_{\geq 0}\}$, and can be used to interpolate over the circle. However, a number of questions arise: What is the interpretation of observation on such a variety? What method of statistical analysis should be used?

In this chapter, then, we are concerned with extending interpolation to when the design no longer comprises a finite set of points, but is defined as the union of a finite number of affine varieties, see Definition 9.1. Only real affine varieties (without repetition) and the radical ideals generated by them are considered. Real affine varieties can be linked to complex varieties, see (Whitney 1957) for an early discussion on properties of real varieties. In Section 9.2.2 we study the case when the design $\mathcal{V}$ comprises the union of $(k-1)$-dimensional hyperplanes. In Section 9.2.3 we present the case, when every affine variety is an intersection of hyperplanes. The following is a motivating example of such linear varieties.

Example 9.1 Consider a general bivariate Normal distribution $(X_1, X_2)^{\top} \sim N\left((\mu_1,\mu_2)^{\top}, \Sigma\right)$ with

$$\Sigma = \begin{pmatrix} \sigma_1^2 & \sigma_1\sigma_2\rho \\ \sigma_1\sigma_2\rho & \sigma_2^2 \end{pmatrix},$$

where σ_1,σ_2 are real positive numbers, and $\rho \in [-1,1] \subset \mathbb{R}$. Now when Σ is fixed, $\log p(x_1,x_2)$ is a quadratic form in μ_1,μ_2, where $p(x_1,x_2)$ is the normal bivariate density function. Imagine that, instead of observing at a design point, we are able to observe $\log p(x_1,x_2)$ over a set of lines $\mathcal{V}_i$, $i = 1,\ldots,n$. That is, the design $\mathcal{V}$ is a union of lines (transects), and suppose we have perfect transect sampling on every line on the design. This means that we know the value of $\log p(x_1,x_2)$ on every line.

The question is: how do we reconstruct the entire distribution? Are there any conditions on the transect location?

We do not attempt to resolve these issues here. Rather we present the ideas as a guide to experimentation on varieties in the following sense. If $I(\mathcal{V})$ is the design ideal, then the quotient ring $\mathbb{R}[x_1, \ldots, x_k]/I(\mathcal{V})$ is no longer of finite dimension, but we can still obtain a basis for it and use it to construct statistical models for data observed on $\mathcal{V}$.

Even though we can create a theory of interpolation by specifying, or 'observing' polynomial functions on a fixed variety $\mathcal{V}$, we may wish to observe a point set design $\mathcal{D}$ which is a subset of $\mathcal{V}$. In Section 9.3 we present this alternative, that is, to subsample a set of points $\mathcal{D}$ from a general design $\mathcal{V}$.

If, instead, a polynomial function is given at every point on the algebraic variety, it is often possible to obtain a general interpolator which in turn coincides with the individual given functions. In Section 9.4 we give a simple technique for building an interpolator over a design and in Section 9.5 we survey the interpolation algorithm due to (Becker and Weispfenning 1991). A related approach is to obtain a reduced expression for an analytic function defined over a design, which is discussed in Section 9.6. In Section 9.7 we discuss further extensions.

9.2 Definitions

In this section we restrict to only the essential concepts for the development of the theory, referring the reader to Chapter 1 and references therein; we also refer the reader to the monograph in algebraic statistics by (Pistone *et al.* 2001).

An affine algebraic set is the solution in $\mathbb{R}^k$ of a finite set of polynomials. The affine algebraic set of a polynomial ideal J is $Z(J)$. The set of polynomials which vanish on a set of points W in $\mathbb{R}^k$ is the polynomial ideal $I(W)$, which is radical. Over an algebraically closed field, such as $\mathbb{C}$, the ideal $I(Z(J))$ coincides with the radical ideal $\sqrt{J}$. However, when working on $\mathbb{R}$, which is not algebraically closed, the above does not necessarily hold.

Example 9.2 Take $J = \langle x^3 - 1 \rangle \subset \mathbb{R}[x]$, i.e. the ideal generated by $x^3 - 1$. Therefore $Z(J) = \{1\}$ and $I(Z(J)) = \langle x - 1 \rangle$. However J is a radical ideal and yet $I(Z(J)) \neq J$.

Recall that for $W \subset \mathbb{R}^k$, the set $Z(I(W))$ is the closure of W with respect to the Zariski topology on $\mathbb{R}^k$. There is a one to one correspondence between closed algebraic sets in $\mathbb{R}^k$ and radical ideals in $\mathbb{R}[x_1, \ldots, x_k]$ such that $I(Z(J)) = J$.

Example 9.3 Consider $I = \langle x^2 \rangle \subset \mathbb{R}[x]$. Clearly I is not a radical ideal. However, its affine algebraic set is $Z(I) = \{0\}$, which is irreducible.

A real affine variety $\mathcal{V}$ is the affine algebraic set associated to a prime ideal. Remind that an algebraic variety $\mathcal{V}$ is irreducible, whenever $\mathcal{V}$ is written as the union of two affine varieties $\mathcal{V}_1$ and $\mathcal{V}_2$ then either $\mathcal{V} = \mathcal{V}_1$ or $\mathcal{V} = \mathcal{V}_2$.

Definition 9.1 A design variety $\mathcal{V}$ is affine variety in $\mathbb{R}^k$ which is the union of irreducible varieties, i.e. for $\mathcal{V}_1, \ldots, \mathcal{V}_n$ irreducible varieties, $\mathcal{V} = \bigcup_{i=1}^{n} \mathcal{V}_i$.

We next review quotient rings and normal forms computable with the variety ideal $I(\mathcal{V})$. Two polynomials $f, g \in \mathbb{R}[x_1, \ldots, x_k]$ are congruent modulo $I(\mathcal{V})$ if $f - g \in I(\mathcal{V})$. The quotient ring $\mathbb{R}[x_1, \ldots, x_k]/I(\mathcal{V})$ is the set of equivalence classes for congruence modulo $I(\mathcal{V})$. The ideal of leading terms of $I(\mathcal{V})$ is the monomial ideal generated by the leading terms of polynomials in $I(\mathcal{V})$, which is written as $\langle \mathrm{LT}(I(\mathcal{V})) \rangle = \langle \mathrm{LT}(f) : f \in I(\mathcal{V}) \rangle$.

Two isomorphisms are considered. As real vector space the quotient ring $\mathbb{R}[x_1, \ldots, x_k]/\langle \mathrm{LT}(I(\mathcal{V})) \rangle$ is isomorphic to $\mathbb{R}[x_1, \ldots, x_k]/I(\mathcal{V})$. Secondly, the quotient ring $\mathbb{R}[x_1, \ldots, x_k]/I(\mathcal{V})$ is isomorphic (as real vector space) to $\mathbb{R}[\mathcal{V}]$, the set of polynomial functions defined on $\mathcal{V}$.

For a fixed monomial ordering $\prec$, let G be a Gröbner basis for $I(\mathcal{V})$ and let $L_\prec(I(\mathcal{V}))$ be the set of all monomials that cannot be divided by the leading terms of the Gröbner basis G, that is

$$L_\prec(\mathrm{I}(\mathcal{V})) := \{x^\alpha \in T^k : x^\alpha \text{ is not divisible by } \mathrm{LT}_\prec(g), g \in G\}$$

where T^k is the set of all monomials in $x_1, \ldots, x_k$. This set of monomials is known as the set of standard monomials, and when there is no ambiguity, we refer to it simply as $L(\mathcal{V})$. We reformulate in the setting of interest of this chapter the following proposition (Cox *et al.* 2007, Section 5§3, Proposition 4).

Proposition 9.1 *Let $I(\mathcal{V}) \subset \mathbb{R}[x_1, \ldots, x_k]$ be a radical ideal. Then $\mathbb{R}[x_1, \ldots, x_k]/\langle LT(I(\mathcal{V})) \rangle$ is isomorphic as a $\mathbb{R}$-vector space to the polynomials which are real linear combinations of monomials in $L(\mathcal{V})$.*

In other words, the monomials in $L(\mathcal{V})$ are linearly independent modulo $\langle \mathrm{LT}(I(\mathcal{V})) \rangle$. By the two isomorphisms above, monomials in $L(\mathcal{V})$ form a basis for $\mathbb{R}[x_1, \ldots, x_k]/I(\mathcal{V})$ and for polynomial functions on $\mathcal{V}$. The division of a polynomial f by the elements of a Gröbner basis for $I(\mathcal{V})$ leads to a remainder r which is a linear combinations of monomials in $L(\mathcal{V})$, which is called the normal form of f.

Theorem 9.1 *(Cox et al. 2007, Section 2§3, Theorem 3) Let $I(\mathcal{V})$ be the ideal of a design variety $\mathcal{V}$; let $\prec$ be a fixed monomial order on $\mathbb{R}[x_1, \ldots, x_k]$ and let $G = \{g_1, \ldots, g_m\}$ be a Gröbner basis for $I(\mathcal{V})$ with respect to $\prec$. Then every polynomial $f \in \mathbb{R}[x_1, \ldots, x_k]$ can be expressed as $f = \sum_{i=1}^{m} g_i h_i + r$, where $h_1, \ldots h_m \in \mathbb{R}[x]$ and r is a linear combination of monomials in $L(\mathcal{V})$.*

We have that $f - r \in I(\mathcal{V})$ and, in the spirit of this chapter, we say that the normal form r *interpolates* f on $\mathcal{V}$. That is, f and r coincide over $\mathcal{V}$. We may write $r = \mathrm{NF}_\prec(f, \mathcal{V})$ to denote the normal form of f with respect to the ideal $I(\mathcal{V})$ and the monomial ordering $\prec$.

9.2.1 Designs of points

The most elementary experimental point design has a single point $d_1 = (d_{11}, \ldots, d_{1k}) \in \mathbb{R}^k$, whose ideal is $I(d_1) = \langle x_1 - d_{11}, \ldots, x_k - d_{1k} \rangle$. An experimental design in statistics is the set of distinct points $\mathcal{D} = \{d_1, \ldots, d_n\}$, whose corresponding ideal is $I(\mathcal{D}) = \bigcap_{i=1}^{n} I(d_i)$.

Example 9.4 For $\mathcal{D} = \{(0,0), (1,0), (1,1), (2,1)\} \subset \mathbb{R}^2$, the set $G = \{x_1^3 - 3x_1^2 + 2x_1, x_1^2 - 2x_1 x_2 - x_1 + 2x_2, x_2^2 - x_2\}$ is a Gröbner basis for $I(\mathcal{D})$. If we set a monomial order for which $x_2 \prec x_1$ then the leading terms of G are x_1^3, x_2^2 and x_1^2 and thus $L(\mathcal{D}) = \{1, x_1, x_2, x_1 x_2\}$. Any real-valued polynomial function defined over $\mathcal{D}$ can be expressed as a linear combination of monomials in $L(\mathcal{D})$.

That is, for any function $f : \mathcal{D} \longrightarrow \mathbb{R}$, there is a unique polynomial $r(x_1, x_2) = c_0 + c_1 x_1 + c_2 x_2 + c_{12} x_1 x_2$ where the constants c_0, c_1, c_2, c_{12} are real numbers whose coefficients can be determined by solving the linear system of equations $r(d_i) = f(d_i)$ for $d_i \in \mathcal{D}$. In particular if we observe real values y_i at $d_i \in \mathcal{D}$, in statistical terms, r is a saturated model. For example, if we observe the data $2, 1, 3, -1$ at the points in $\mathcal{D}$ then $r = 2 - x_1 + 5x_2 - 3x_1 x_2$ is the saturated model for the data.

9.2.2 Designs of hyperplane arrangements

Let $H(a, c)$ be the $((k-1)$-dimensional) affine hyperplane directed by a non-zero vector $a \in \mathbb{R}^k$ and with intercept $c \in \mathbb{R}$, i.e.

$$H(a, c) = \left\{ x = (x_1, \ldots, x_k) \in \mathbb{R}^k : l_a(x) - c = 0 \right\}$$

with $l_a(x) := \sum_{i=1}^{n} a_i x_i$. Now for a set of vectors $a_1, \ldots, a_n \in \mathbb{R}^k$, and real scalars $c_1, \ldots, c_n$, the hyperplane arrangement $\mathcal{A}$ is the union of the affine hyperplanes $H(a_i, c_i)$, that is $\mathcal{A} = \bigcup_{i=1}^{n} H(a_i, c_i)$.

We restrict the hyperplane arrangement to consist of distinct hyperplanes, i.e. no repetitions. The polynomial $Q_{\mathcal{A}}(x) := \prod_{i=1}^{n} (l_{a_i}(x) - c_i)$ is called the defining polynomial of $\mathcal{A}$. Combinatorial properties of hyperplane arrangements have been studied extensively in the mathematical literature, see (Grünbaum 2003, Chapter 18).

Clearly $\mathcal{A}$ is a variety as in Definition 9.1, $I(\mathcal{A})$ is a radical ideal and it is generated by $Q_{\mathcal{A}}(x)$. Furthermore for any monomial ordering $\prec$, $\{Q_{\mathcal{A}}(x)\}$ is a Gröbner basis for $I(\mathcal{A})$.

Example 9.5 Let a_i be the i-th unit vector and $c_i = 0$ for $i = 1, \ldots, k$, then $Q_{\mathcal{A}}(x) = x_1 \cdots x_k$ and $\mathcal{A}$ comprises the k coordinate hyperplanes.

Example 9.6 The *braid arrangement* plays an important role in combinatorial studies of arrangements. It has defining polynomial $Q_{\mathcal{A}}(x) = \prod (x_i - x_j - 1)$, where the product is carried on $i, j : 1 \le i < j \le k$, see (Stanley 1996).

In the arrangement generated by the k coordinate hyperplanes of Example 9.5 and for any monomial order, the set of standard monomials comprises all monomials which miss at least one indeterminate, and this set does not depend on the term ordering used. For other hyperplane arrangements, the leading term of $Q_{\mathcal{A}}(x)$ may depend on the actual monomial order used. We have the following elementary result, which we state without proof.

Lemma 9.1 *Let* $\mathcal{A} = \bigcup_{i=1}^{n} H(a_i, c_i)$. *Then for any monomial ordering, the total degree of* $LT_{\prec}(Q_{\mathcal{A}}(x))$ *is* n.

Lemma 9.1 implies that the set of standard monomials for $\mathcal{A}$ contains all monomials up to a total degree $n - 1$. This result can be used in conjunction with the methodology of Section 9.3: an arrangement of n hyperplanes has the potential to identify a full model of total degree $n - 1$.

9.2.3 Generalised linear designs (GLDs)

The design variety in Section 9.2.2 can be generalised to include unions of intersections of distinct hyperplanes. Namely, $\mathcal{V} = \bigcup_{i=1}^{n} \mathcal{V}_i$ where $\mathcal{V}_i = \bigcap_{j=1}^{n_i} H(a_j^i, c_j^i)$ where a_j^i are non-zero vectors in $\mathbb{R}^k$ and $c_j^i \in \mathbb{R}$ for $j = 1, \ldots, n_i$ $i = 1, \ldots, n$ and n and $n_1, \ldots, n_n$ are positive integers. Consequently, the design ideal is the intersection of sums of ideals

$$I(\mathcal{V}) = \bigcap_{i=1}^{n} \sum_{j=1}^{n_i} I(H(a_j^i, c_j^i)).$$

Example 9.7 Let $\mathcal{V} \subset \mathbb{R}^3$ be constructed by the union of the following eleven affine sets: $\mathcal{V}_1, \ldots, \mathcal{V}_8$ are the eight hyperplanes $\pm x_1 \pm x_2 \pm x_3 - 1 = 0$, and $\mathcal{V}_9, \mathcal{V}_{10}, \mathcal{V}_{11}$ are the three lines in direction of the every coordinate axis. The varieties $\mathcal{V}_1, \ldots, \mathcal{V}_8$ form a hyperplane arrangement $\mathcal{A}'$. The variety $\mathcal{V}_9$ is the axis x_1 and thus is the intersection of the hyperplanes $x_2 = 0$ and $x_3 = 0$, i.e $I(\mathcal{V}_9) = \langle x_2, x_3 \rangle$. Similarly $I(\mathcal{V}_{10}) = \langle x_1, x_3 \rangle$ and $I(\mathcal{V}_{11}) = \langle x_1, x_2 \rangle$. The design is $\mathcal{V} = \mathcal{A}' \cup \mathcal{V}_9 \cup \mathcal{V}_{10} \cup \mathcal{V}_{11}$ and the design ideal is $I(\mathcal{V}) = I(\mathcal{A}') \cap I(\mathcal{V}_9) \cap I(\mathcal{V}_{10}) \cap I(\mathcal{V}_{11})$. For the lexicographic monomial ordering in which $x_3 \prec x_2 \prec x_1$, the Gröbner basis of $I(\mathcal{V})$ has three polynomials whose leading terms have total degree ten and are $x_1^9 x_2, x_1^9 x_3, x_1^8 x_2 x_3$ and thus

$$L(\mathcal{V}) = \left\{ 1, x_1, x_1^2, x_1^3, x_1^4, x_1^5, x_1^6, x_1^7 \right\} \otimes \left\{ x_2^i x_3^j : (i, j) \in \mathbb{Z}_{\geq 0}^2 \right\}$$

$$\bigcup \left\{ x_1^{j+9} : j \in \mathbb{Z}_{\geq 0} \right\} \bigcup \left\{ x_1^8 x_2^{j+1} : j \in \mathbb{Z}_{\geq 0} \right\} \bigcup \left\{ x_1^8 x_3^{j+1} : j \in \mathbb{Z}_{\geq 0} \right\} \bigcup \left\{ x_1^8 \right\},$$

where $\otimes$ denotes the Kronecker product of sets. That is, the set of exponents of monomials in $L(\mathcal{V})$ comprises the union of eight shifted copies of $\mathbb{Z}_{\geq 0}^2$, three shifted copies of $\mathbb{Z}_{\geq 0}$ and a finite set of monomials. This finite union of disjoint sets is an example of the Stanley decomposition of an $L(\mathcal{V})$, see (Stanley 1978) and (Sturmfels and White 1991).

9.3 Subsampling from a variety: 'fill-up'

Varieties give a taxonomy which informs experimentation. Indeed, suppose that, for fixed $\mathcal{V}$, we take a finite sample of design points $\mathcal{D}$ from $\mathcal{V}$, i.e. $\mathcal{D} \subset \mathcal{V}$. We have the following inclusion between the quotient rings as real vector spaces

$$\mathbb{R}[x_1, \ldots, x_k]/\langle LT(I(\mathcal{D})) \rangle \subset \mathbb{R}[x_1, \ldots, x_k]/\langle LT(I(\mathcal{V})) \rangle.$$

That is, the basis for the quotient ring $\mathbb{R}[x_1, \ldots, x_k]/I(\mathcal{V})$ provides an indication of the capability of models we can fit over $\mathcal{D}$ by setting the design $\mathcal{D}$ to lie on the affine variety $\mathcal{V}$. In particular, the sets of standard monomials for interpolating over $\mathcal{D}$ and over $\mathcal{V}$ satisfy $L_\prec(\mathcal{D}) \subset L_\prec(\mathcal{V})$. A question of interest is: given any finite subset $L' \subset L_\prec(\mathcal{V})$, can we find a set of points $\mathcal{D} \subset \mathcal{V}$ so that $L' \subseteq L_\prec(\mathcal{D})$?

An interesting case is the circle. Can we 'achieve' a given L' from some finite design of points on the circle? The authors are able, in fact, to answer affirmatively with a sufficiently large equally spaced design around the circle, and a little help from discrete Fourier analysis. For instance set $LT(x_1^2 + x_2^2 - 1) = x_2^2$ and thus $L = \{1, x_2\} \otimes \{x_1^j : j \in \mathbb{Z}_{\geq 0}\}$ and let $L' \subset L$ be the finite sub-basis. For $i = 0, \ldots, n-1$ let $(x_i, y_i) = (\cos(2\pi i/n), \sin(2\pi i/n))$. For n sufficiently large, the design matrix $X = [x_i^u y_i^v]_{(u,v) \in L', i=0,\ldots,n-1}$ has full rank $|L'|$. Indeed we can explicitly compute the non-zero determinant of $X^T X$ using Fourier formulas.

The general case is stated as a conjecture.

Conjecture 9.1 *Let $\mathcal{V}$ be a design variety with set of standard monomials $L_\prec(\mathcal{V})$. Then, for any model with finite support on $L' \subset L_\prec(\mathcal{V})$, there is a finite design with points on the real part of $\mathcal{V}$ such that the model is identifiable.*

This conjecture can be proven when the design $\mathcal{V}$ is in the class of generalised linear designs (GLD) of Section 9.2.3. We believe that the construction may be of some use in the important inverse problem: finding a design which allows identification of a given model.

Proof Let $\mathcal{V} = \bigcup_{i=1}^{n} \mathcal{V}_i$ be a GLD, where the irreducible components are the $\mathcal{V}_i = \bigcap_{j=1}^{n_i} H(a_j^i, c_j^i)$. Take a finite set of monomials $L' \subset L(\mathcal{V})$ and consider a polynomial in this basis:

$$p(x) = \sum_{\alpha \in L'} \theta_\alpha x^\alpha,$$

i.e. $p(x)$ is a polynomial with monomials in L' and real coefficients. Select a $\mathcal{V}_i$ and consider the values of $p(x)$ on this variety. Suppose $\dim(\mathcal{V}_i) = k_i$, then by a linear coordinatisation of the variety we can reduce the design problem on the variety to the identification of a model of a particular order on $\mathbb{R}^{k_i}$. But using the 'design of points' theory and because L' is finite, with a sufficiently large design $\mathcal{D}_i \subset \mathcal{V}_i$ we can carry out this identification and therefore can completely determine the value of $p(x)$ on the variety $\mathcal{V}_i$. Carrying out such a construction for each variety gives the design $\mathcal{D} = \bigcup_{i=1}^{n} \mathcal{D}_i$. Then the values of $p(x)$ are then completely known on each variety and the normal form over $\mathcal{V}$ recaptures $p(x)$, which completes the proof. A shorthand version is: fix a polynomial model on each $\mathcal{V}_i$ and the normal form

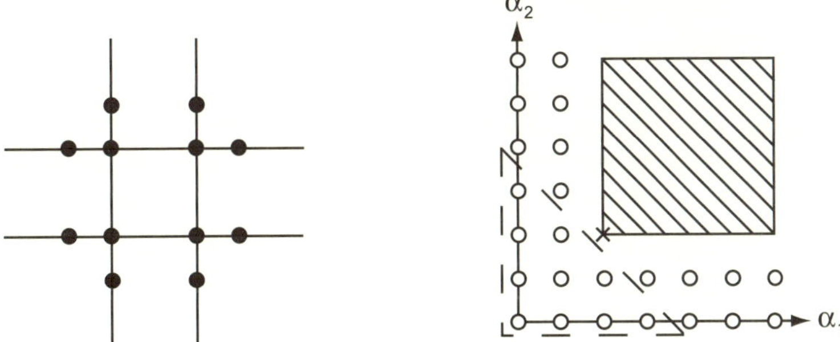

Fig. 9.1 GLDs $\mathcal{V}$ and $\mathcal{D}$ of Example 9.8 (left) and exponents $\alpha = (\alpha_1, \alpha_2)$ for monomials in $L(\mathcal{V})$ (right). The symbol $\times$ corresponds to the leading term $x_1^2 x_2^2$, while the shaded area contains monomials not in $L(\mathcal{V})$.

(remainder) is fixed. The normal form of $p(x)$ with respect to $I(\mathcal{D})$ must agree with the normal forms of $p(x)$ with respect to $I(\mathcal{D}_i)$, for all i, otherwise a contradiction can be shown. This is enough to shown that $p(x)$ can be reconstructed on $\mathcal{V}$ from $\mathcal{D}$. $\qquad\qquad\square$

This points to sequential algorithms in which we 'fix' the values on $\mathcal{V}_1$, reduce the dimension of the model as a result, fix the reduced model on $\mathcal{V}_2$ and so on. Further research is needed to turn such algorithms into a characterisation of designs satisfying Conjecture 9.1 and minimal sample size for the existence of such designs. The following example shows heuristically how such an algorithm might work.

Example 9.8 Take $k = 2$ and the design $\mathcal{V}$ to be the GLD of four lines $x_1 = \pm 1, x_2 = \pm 1$. A Gröbner basis for $I(\mathcal{V})$ is $\left\{ (x_1^2 - 1)(x_2^2 - 1) \right\}$ with leading term $x_1^2 x_2^2$ and

$$L(\mathcal{V}) = \{x_2^2, x_1 x_2^2\} \otimes \{x_2^j : j \in \mathbb{Z}_{\geq 0}\} \bigcup \{x_1^2, x_1^2 x_2\} \otimes \{x_1^j : j \in \mathbb{Z}_{\geq 0}\}$$

$$\bigcup \{1, x_1, x_2, x_1 x_2\}.$$

Take the model with all terms of degree three or less, which has ten terms, see the dashed triangle on the right hand in Figure 9.1. On $x_1 = 1$ the model is cubic in x_2 so that four distinct points are enough to fix it. Thus any design with four distinct points on each line is enough. The design $\mathcal{D} = \{(\pm 1, \pm 1), (\pm 1, \pm 2), (\pm 2, \pm 1)\}$ in Figure 9.1 satisfies our needs.

9.4 Interpolation over varieties

Let $\mathcal{V} = \bigcup_{i=1}^{n} \mathcal{V}_i$ with $\mathcal{V}_i$ irreducible real affine variety and assume that the $\mathcal{V}_i$'s do not intersect i.e. $\mathcal{V}_i \cap \mathcal{V}_j = \emptyset$ for $1 \leq i < j \leq n$. Then the polynomial ideal driving an interpolation on $\mathcal{V}$ can be constructed as the intersection of the n polynomial ideals, each one driving interpolation on a separate $\mathcal{V}_i$. We discuss this approach with an example.

Let $z_1, \ldots, z_4$ be real values observed at design points $(\pm 1, \pm 1) \in \mathbb{R}^2$. Suppose we are able to observe a function over the variety defined by a circle with radius $\sqrt{3}$ and centre at the origin. For simplicity, suppose that we observe the zero function on the circle. We want a polynomial function that interpolates both the values z_i over the factorial points and takes the value zero over the circle. Note that the design $\mathcal{V}$ is the union of five varieties: one for each point, plus the circle. Start by constructing an ideal $I_i \subset \mathbb{R}[x_1, x_2, y]$ for every point d_i, e.g. $I_1 = \langle y - z_1, x_1 - 1, x_2 - 1 \rangle$. A similar approach for the circle gives: $I_C = \langle y, x_1^2 + x_2^2 - 3 \rangle$. Then intersect all the ideals $I^* = I_1 \cap \cdots \cap I_4 \cap I_C$. The ideal I^* contains all the restrictions imposed by all the varieties as well as the restrictions imposed by the observed functions. Then, for a monomial order $x^\alpha \prec y^\beta$, the desired interpolator is $\mathrm{NF}(y, I^*) \in \mathbb{R}[x_1, \ldots, x_k]$. In our current example we have $\mathrm{NF}(y, I^*) = g(x_1, x_2)(x_1^2 + x_2^2 - 3)/4$, where

$$
\begin{aligned}
g(x_1, x_2) = &- (z_1 + z_2 + z_3 + z_4) + (z_2 + z_4 - z_1 - z_3)x_1 \\
&+ (z_3 + z_4 - z_1 - z_2)x_2 + (z_2 + z_3 - z_1 - z_4)x_1 x_2
\end{aligned}
$$

is the interpolator for the four points, adjusted with a negative sign to compensate for the inclusion of $x_1^2 + x_2^2 - 3$. This is the standard formula appearing in books on design of experiments.

The monomial ordering used above is called a blocked ordering; for an application of such type of orders in algebraic statistics see (Pistone *et al.* 2000). This method works well in a number of cases for which the varieties do not intersect, and when the functions defined on each variety are polynomial functions. If the varieties that compose the design intersect, then the methodology needs to ensure compatibility between the observed functions at the intersections. For example, consider again observing the zero function over the circle with radius $\sqrt{3}$; and the function $f(x_1, x_2) = 1$ over the line $x_1 + x_2 - 1 = 0$. The observed functions are not compatible at the two intersection points between the circle and the line, which is reflected in the fact that $\mathrm{NF}(y, I^*) = y \notin \mathbb{R}[x_1, x_2]$.

9.5 Becker–Weispfenning interpolation

(Becker and Weispfenning 1991) define a technique for interpolation on varieties. It develops a polynomial interpolator for a set of pre-specified polynomial functions defined on a set of varieties in $\mathbb{R}^k$.

For a design variety $\mathcal{V} = \bigcup_{i=1}^{n} \mathcal{V}_i$ with $\mathcal{V}_i$ irreducible, the ideal of $\mathcal{V}_i$ is generated in parametric form and a pre-specified polynomial function is determined for each variety. For every variety $\mathcal{V}_i$, let $g_{i1}, \ldots, g_{ik} \in \mathbb{R}[z_1, \ldots, z_m]$ be the set of parametric generators for the ideal $I(\mathcal{V}_i)$ so that $I(\mathcal{V}_i) = \langle x_1 - g_{i1}, \ldots, x_k - g_{ik} \rangle \subset \mathbb{R}[x_1, \ldots, x_k, z_1, \ldots, z_m]$. Also, for every variety $\mathcal{V}_i$, a polynomial function $f_i(z) \in \mathbb{R}[z_1, \ldots, z_m]$ is pre-specified. Now for indeterminates $w_1, \ldots, w_n$, let I^* be the ideal generated by the set of polynomials

$$
\bigcup_{i=1}^{n} \{ w_i (x_1 - g_{i1}), \ldots, w_i (x_k - g_{ik}) \} \bigcup \left\{ \sum_{i=1}^{n} w_i - 1 \right\}. \tag{9.1}
$$

We have $I^* \subset \mathbb{R}[x_1, \ldots, x_k, w_1, \ldots, w_n, z_1, \ldots, z_m]$. The technique of introducing dummy variables w_i is familiar from the specification of point ideals: when any $w_i \neq 0$ we must have $x_j - g_{ij} = 0$ for $j = 1, \ldots, k$, that is, we automatically select the i-th variety ideal. The statement $\sum_{i=1}^{n} w_i - 1 = 0$ prevents all the w_i being zero at the same time. If several w_i are non-zero, the corresponding intersection of $\mathcal{V}_i$ is active. Consistency of the parametrisation is, as Becker and Weispfenning (1991) point out, a necessary, but not sufficient, condition for the method to work.

Let $\prec$ be a block monomial order for which $x^\alpha \prec w^\beta z^\gamma$. Set $f^* = \sum_{i=1}^{m} w_i f_i(z)$ and let $f' = \mathrm{NF}(f^*, I^*)$. The interpolation problem has a solution if the normal form of f^* depends only on x, that is if $f' \in \mathbb{R}[x_1, \ldots, x_k]$. Although the solution does not always exist, an advantage of the approach is the freedom to parametrise each variety separately from a functional point of view, but using a common parameter z.

Example 9.9 (Becker and Weispfenning 1991, Example 3.1) We consider interpolation over $\mathcal{V} = \mathcal{V}_1 \cup \mathcal{V}_2 \cup \mathcal{V}_3 \subset \mathbb{R}^2$ The first variety is the parabola $x_2 = x_1^2 + 1$, defined through the parameter z by $g_{11} = z, g_{12} = z^2 + 1$.

The second and third varieties are the axes x_1 and x_2 and therefore $g_{21} = z$, $g_{22} = 0$ and $g_{31} = 0, g_{32} = z$. The prescribed functions over the varieties are $f_1 = z^2, f_2 = 1$ and $f_3 = z + 1$. The ideal I^* is constructed using the set in Equation (9.1) and we set $f^* = w_1 f_1 + w_2 f_2 + w_3 f_3$. For a block lexicographic monomial order $\prec$ in which $x^\alpha \prec w^\beta z^\gamma$, we compute the normal form of f^* with respect to I^* and obtain $f' = x_2 + 1$.

A variation of the technique of this section leads to an extension of Hermite interpolation, i.e. when derivative values are known over every variety $\mathcal{V}_i$, and a polynomial interpolator is sought. The intuition behind this approach is simple: a multivariate Taylor polynomial is constructed for every variety $\mathcal{V}_i$ using value and derivative information and the algebra is used to obtain the polynomial interpolator. If the varieties $\mathcal{V}_i$ intersect then the Taylor polynomials need to be compatible at intersections, see details in (Becker and Weispfenning 1991).

Example 9.10 Consider interpolating the values $3/5$, 1, 3 and derivative values $9/25, 1, 9$ at design points $-2/3$, 0, $2/3$, respectively. The design points are the varieties $\mathcal{V}_1, \mathcal{V}_2, \mathcal{V}_3$, and the Taylor polynomials for each variety are $3/5 + 9/25(x + 2/3), 1 + x$ and $3 + 9(x - 2/3)$, respectively. The general interpolator is $1 + x + 9/25(x^2 + x^3) + 81/25(x^4 + x^5)$ which at the design points coincides with the given values and derivatives.

9.6 Reduction of power series by ideals

Let us revisit the basic theory. Here $x = (x_1, \ldots, x_k)$. A polynomial $f \in \mathbb{R}[x]$ can be reduced by the ideal $I(\mathcal{V}) \subset \mathbb{R}[x]$ to an equivalent polynomial f' such that $f = f'$ on the affine variety $\mathcal{V}$. By Theorem 9.1, the reduced expression is $f' = \mathrm{NF}(f, \mathcal{V})$ and clearly $f - f' \in I(\mathcal{V})$.

Example 9.11 Consider the hyperplane arrangement $\mathcal{V}$ given by the lines $x_1 = x_2$ and $x_1 = -x_2$. We have $I(\mathcal{V}) = \langle x_1^2 - x_2^2 \rangle$. Now for $i = 1, 2, \ldots$, consider the polynomial $f_i = (x_1 + x_2)^i$. For a monomial ordering in which $x_2 \prec x_1$, we have that $\mathrm{NF}(f_i, \mathcal{V}) = 2^{i-1}(x_1 + x_2)x_2^{i-1}$, for instance $\mathrm{NF}((x_1 + x_2)^5, \mathcal{V}) = 16(x_1 + x_2)x_2^4 = 16x_1 x_2^4 + 16x_2^5$.

A convergent series of the form $f(x) = \sum_{i=0}^{\infty} \alpha_i x^{\alpha_i}$, can be written on the variety $\mathcal{V}$ as

$$\mathrm{NF}(f, \mathcal{V}) = \sum_{i=0}^{\infty} \alpha_i \mathrm{NF}(x^{\alpha_i}, \mathcal{V}). \tag{9.2}$$

See (Apel *et al.* 1996) for a discussion of conditions for the validity of Equation (9.2).

We may also take the normal form of convergent power series with respect to the ideal of an affine variety in $\mathbb{C}$. For example by substituting $x^3 = 1$ in the expansion for e^x we obtain

$$\mathrm{NF}(e^x, \langle x^3 - 1 \rangle) = 1 + \frac{1}{3!} + \frac{1}{6!} + \frac{1}{9!} + \cdots + x \left(1 + \frac{1}{4!} + \frac{1}{7!} + \frac{1}{10!} + \cdots \right)$$
$$+ x^2 \left(\frac{1}{2!} + \frac{1}{5!} + \frac{1}{8!} + \cdots \right)$$
$$= \frac{1}{3}e + \frac{2}{3}e^{-\frac{1}{2}} \cos\left(\frac{\sqrt{3}}{2} \right) + x \left(\frac{1}{3}e - \frac{1}{3}e^{-\frac{1}{2}} \cos\left(\frac{\sqrt{3}}{2} \right) + \frac{1}{3}e^{\frac{1}{2}} \sin\left(\frac{\sqrt{3}}{2} \right) \right)$$
$$+ x^2 \left(\frac{1}{3}e - \frac{1}{3}e^{-\frac{1}{2}} \cos\left(\frac{\sqrt{3}}{2} \right) - \frac{1}{3}e^{\frac{1}{2}} \sin\left(\frac{\sqrt{3}}{2} \right) \right).$$

The relation $\mathrm{NF}(e^x, \langle x^3 - 1 \rangle) = e^x$ holds at the roots d_1, d_2, d_3 of $x^3 - 1 = 0$, with d_1 the only real root. Note that the above series is not the same as the Taylor expansion at, say, 0.

Example 9.12 Consider the ideal $I = \langle x_1^3 + x_2^3 - 3x_1 x_2 \rangle$. The variety $\mathcal{V}$ that corresponds to I is the Descartes' *folium*. For a monomial ordering in which $x_2 \prec x_1$, the leading term of the ideal is x_1^3. Now consider the function $f(x) = \sin(x_1 + x_2)$, whose Taylor expansion is

$$f(x) = (x_1 + x_2) - \frac{1}{3!}(x_1 + x_2)^3 + \frac{1}{5!}(x_1 + x_2)^5 + \cdots \tag{9.3}$$

The coefficients for every term of Equation (9.3) which is divisible by x_1^3 is absorbed into the coefficient of some of the monomials in $L(\mathcal{V})$. For the second term in the summation we have the following remainder

$$\mathrm{NF}\left(-\frac{(x_1 + x_2)^3}{3!}, \mathcal{V} \right) = -\frac{1}{2}\left(x_1^2 x_2 + x_1 x_2^2 + x_1 x_2 \right).$$

Note that different terms of the Taylor series may have normal forms with common terms. For instance the normal form for the third term in the summation is

$$\mathrm{NF}\left(\frac{(x_1 + x_2)^5}{5!}, \mathcal{V} \right) = \frac{3}{40}x_1^2 x_2^3 - \frac{3}{40}x_2^5 + \frac{1}{8}x_1^2 x_2^2 + \frac{1}{4}x_1 x_2^3 - \frac{1}{40}x_2^4 + \frac{3}{40}x_1 x_2^2.$$

The sum of the normal forms for first ten terms of Equation (9.3) is

$$\tilde{f}(x) = x_2 + x_1 - \frac{1}{2}x_1x_2 - \frac{17}{40}x_1x_2^2 - \frac{1}{2}x_1^2x_2 - \frac{1}{40}x_2^4 + \frac{137}{560}x_1x_2^3$$
$$+ \frac{1}{8}x_1^2x_2^2 - \frac{41}{560}x_2^5 - \frac{167}{4480}x_1x_2^4 + \frac{1}{16}x_1^2x_2^3 + \frac{167}{13440}x_2^6$$
$$- \frac{4843}{492800}x_1x_2^5 - \frac{17}{896}x_1^2x_2^4 + \frac{2201}{492800}x_2^7 + \frac{197343}{25625600}x_1x_2^6$$
$$+ \frac{89}{44800}x_1^2x_2^5 - \frac{65783}{76876800}x_2^8 - \frac{4628269}{5381376000}x_1x_2^7 + \frac{1999}{5913600}x_1^2x_2^6$$
$$+ \frac{118301}{1793792000}x_2^9 - \frac{305525333}{1463734272000}x_1x_2^8 - \frac{308387}{1076275200}x_1^2x_2^7 + \cdots$$

The equality $\tilde{f}(x) = \sin(x_1 + x_2)$ is achieved over $\mathcal{V}$ by summing the normal forms for all terms in Equation (9.3): $\tilde{f}(x)$ interpolates $\sin(x_1 + x_2)$ over $\mathcal{V}$.

9.7 Discussion and further work

In this chapter we consider the extension of the theory of interpolation over points to interpolation over varieties with applications in mind to design of experiments in statistics. We associate to the design variety a radical ideal and the quotient ring induced by this variety ideal is a useful source of terms which can be used to form the basis for a (regression) model. In particular, knowledge of the quotient ring for the whole variety can be a useful guide to models which can be identified with a set of points selected from the variety.

If the design variety is not a GLD, the technique still can be applied. As an example consider the structure $\mathcal{V}$ consisting of a circle with a cross, see Figure 9.2. For any monomial ordering, the polynomial $g = x_1x_2(x_1^2 + x_2^2 - 2) = x_1^3x_2 + x_1x_2^3 - 2x_1x_2$ is a Gröbner basis for $I(\mathcal{V})$. Now, for a monomial order in which $x_2 \prec x_1$, we have $\mathrm{LT}_{\prec}(g) = x_1^3x_2$ and $L(\mathcal{D}) = \{x_2, x_1x_2, x_1^2x_2\} \otimes \{x_2^j : j \in \mathbb{Z}_{\geq 0}\} \bigcup \{x_1^{3+j} : j \in \mathbb{Z}_{\geq 0}\} \bigcup \{1, x_1, x_1^2\}$ see Figure 9.2. If we are interested in $L' = \{1, x_1, x_2, x_1^2, x_1x_2, x_2^2\}$ then a good subset of $\mathcal{V}$ which estimates L' is $\mathcal{D} = \{(\pm 1, \pm 1)\} \cup \{(0, \pm\sqrt{2}), (\pm\sqrt{2}, 0)\} \cup \{(0,0)\}$. This is the classic central composite design of response surface methodology.

We have not discussed the issue of statistical variation in interpolation, that is, when observations come with error. In the case of selecting points from $\mathcal{V}$ of Section 9.3, standard models can be used, but when an observation is a whole function as in Sections 9.4 and 9.5, a full statistical theory awaits development. It is likely that such a theory would involve random functions, that is stochastic processes on each variety $\mathcal{V}_i$.

Finally, we note that elsewhere in this volume there is emphasis on probability models defined on discrete sets. Typically the set may be a product set which allows independence and conditional independence statements. A simple approach but with deep consequences is to consider not interpolation of data (y-values) in a variety, but $\log p$ where p is a probability. It is a challenge, therefore, to consider $\log p$ models on varieties, that is, distributions on varieties. One may count occurrences rather than observe real continuous y-values. With counts we may be able to

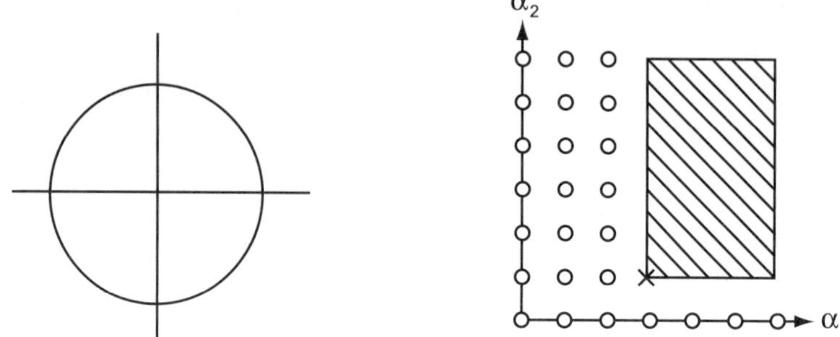

Fig. 9.2 Variety for the ideal $\langle x_1 x_2 (x_1^2 + x_2^2 - 2) \rangle$ (left) and exponents $\alpha = (\alpha_1, \alpha_2)$ for monomials in $L(\mathcal{V})$ (right). The symbol $\times$ in the right diagram corresponds to the leading term $x_1^3 x_2$, while the shaded area contains monomials not in $L(\mathcal{V})$.

reconstruct a distribution on the transect as in Example 9.1. Again the issue would be to reconstruct the full distribution both on and off the transect. This points to a theory of exponential families anchored by prescribing the value on varieties. We trust that the development of such a theory would be in the spirit of this volume and the very valuable work of its dedicatee.

Acknowledgements

The authors acknowledge the EPSRC grant EP/D048893/1, considerable help from referees and an early conversation with Professor V. Weispfenning.

References

Apel, J., Stückrad, J., Tworzewski, P. and Winiarski, T. (1996). Reduction of everywhere convergent power series with respect to Gröbner bases, *J. Pure Appl. Algebra* **110**(2), 113–29.

Becker, T. and Weispfenning, V. (1991). The chinese remainder problem, multivariate interpolation, and Gröbner bases. In *Proc. ISSAC '91* (Bonn, Germany), 64–9.

Buckland, S. T., Anderson, D. R., Burnham, K. P. and Laake, J. L. (1993). *Distance Sampling* (London, Chapman & Hall).

Cox, D., Little, J. and O'Shea, D. (2007). *Ideals, Varieties, and Algorithms* 3rd edn (New York, Springer-Verlag).

Fontana, R., Pistone, G. and Rogantin, M. P. (1997). Algebraic analysis and generation of two level designs, *Statistica Applicata* **9**(1), 15–29.

Giglio, B., Wynn, H. P. and Riccomagno, E. (2001). Gröbner basis methods in mixture experiments and generalisations. In *Optimum design 2000 (Cardiff)* (Dordrecht, Kluwer), 33–44.

Grünbaum, B. (2003). *Convex Polytopes* 2nd edn (New York, Springer-Verlag).

Holliday, T., Pistone, G., Riccomagno, E. and Wynn, H. P. (1999). The application of computational algebraic geometry to the analysis of designed experiments: a case study, *Computational Statistics.* **14**(2), 213–31.

Mack, Y. P. and Quang, P. X. (1998). Kernel methods in line and point transect sampling, *Biometrics* **54**(2), 606–19.

Maruri-Aguilar, H., Notari, R. and Riccomagno, E. (2007). On the description and identifiability analysis of mixture designs, *Statistica Sinica* **17**(4), 1417–40.

Phillips, G. M. (2003). *Interpolation and Approximation by Polynomials* (New York, Springer-Verlag).

Pistone, G., Riccomagno, E. and Rogantin, M.-P. (2009). Methods in algebraic statistics for the design of experiments. In *Search for Optimality in Design and Statistics*, Pronzato, L. and Zhigljavsky, A. A. eds. (Berlin, Springer-Verlag), 97–132.

Pistone, G., Riccomagno, E. and Wynn, H. P. (2000). Gröbner basis methods for structuring and analysing complex industrial experiments, *International Journal of Reliability, Quality and Safety Engineering* **7**(4), 285–300.

Pistone, G., Riccomagno, E. and Wynn, H. P. (2001). *Algebraic Statistics* (Boca Raton, Chapman & Hall/CRC).

Pistone, G. and Rogantin, M. (2008). Indicator function and complex coding for mixed fractional factorial designs, *Journal of Statistical Planning and Inference* **138**(3), 787–802.

Pistone, G. and Wynn, H. P. (1996). Generalised confounding with Gröbner bases, *Biometrika* **83**(3), 653–66.

Sakhnovich, L. A. (1997). *Interpolation theory and its applications* (Dordrecht, Kluwer).

Stanley, R. P. (1978). Hilbert functions of graded algebras, *Advances in Mathematics* **28**(1), 57–83.

Stanley, R. P. (1996). Hyperplane arrangements, interval orders, and trees. In *Proceedings of the National Academy of Sciences of the United States of America* **93**(6), 2620–5.

Stein, M. L. (1999). *Interpolation of Spatial Data* (New York, Springer-Verlag).

Sturmfels, B. and White, N. (1991). Computing combinatorial decompositions of rings, *Combinatorica* **11**(3), 275–93.

Whitney, H. (1957). Elementary structure of real algebraic varieties, *Annals of Mathematics (2)* **66**, 545–56.

10

Design of experiments and biochemical network inference

Reinhard Laubenbacher

Brandilyn Stigler

Abstract

Design of experiments is a branch of statistics that aims to identify efficient procedures for planning experiments in order to optimise knowledge discovery. Network inference is a sub-field of systems biology devoted to the identification of biochemical networks from experimental data. Common to both areas of research is their focus on the maximisation of information gathered from experimentation. The goal of this chapter is to establish a connection between these two areas coming from the common use of polynomial models and techniques from computational algebra.

10.1 Introduction

Originally introduced in (Pistone, Riccomagno and Wynn 2001), the field of *algebraic statistics* focuses on the application of techniques from computational algebra and algebraic geometry to problems in statistics. One initial focus of the field was the design of experiments, beginning with (Pistone and Wynn 1996, Riccomagno 1997). An early exposition of a basic mathematical relationship between problems in the design of experiments and computational commutative algebra appeared in (Robbiano 1998). The basic strategy of (Robbiano 1998) and other works is to construct an algebraic model, in the form of a polynomial function with rational coefficients, of a fractional factorial design. The variables of the polynomial function correspond to the factors of the design. One can then use algorithmic techniques from computational commutative algebra to answer a variety of questions, for instance about the classification of all polynomial models that are identified by a fractional design.

If $\mathbf{p}_1, \ldots, \mathbf{p}_r$ are the points of a fractional design with n levels, then the key algebraic object to be considered is the *ideal of points* I that contains all polynomials with rational coefficients that vanish on all $\mathbf{p}_i$.[1] The form of the polynomials in different generating sets of this ideal is of special interest. In particular, we are interested in so-called interpolating polynomials which have a unique representation,

---

[1] For a review of basic concepts from commutative algebra, refer to Chapter 1.

Algebraic and Geometric Methods in Statistics, ed. Paolo Gibilisco, Eva Riccomagno, Maria Piera Rogantin and Henry P. Wynn. Published by Cambridge University Press. © Cambridge University Press 2010.

175

given an explicit choice of generating set. An interpolating polynomial $f(x_1, \ldots, x_n)$ has the property that if $b_1, \ldots, b_r$ is a response to the design given by the $\mathbf{p}_i$, then $f(\mathbf{p}_i) = b_i$.

Strikingly similar constructions have been used recently to solve an entirely different set of problems related to the inference of intracellular biochemical networks, such as gene regulatory networks, from experimental observations. Relatively recent technological breakthroughs in molecular biology have made possible the simultaneous measurement of many different biochemical species in cell extracts. For instance, using DNA microarrays one can measure the concentration of mRNA molecules, which provide information about the activity levels of the corresponding genes at the time the cell extract was prepared. Such network-level measurements provide the opportunity to construct large-scale models of molecular systems, including gene regulatory networks.

Here, an experimental observation consists of the measurement of n different quantities at r successive time points, resulting in a time course of n-dimensional real-valued vectors $\mathbf{p}_1, \ldots, \mathbf{p}_r$. The number r of experimental observations is typically very small compared to the number n of quantities measured, due in part to the considerable expense of making measurements. In recent years there has been tremendous research activity devoted to the development of mathematical and statistical tools to infer the entire network structure from a limited set of experimental measurements.

Inferring networks from data is a central problem in computational systems biology, and several approaches have been developed using a variety of approaches. Models range from statistical models such as Bayesian networks to dynamic models such as Markov chains and systems of differential equations. Another modelling framework is that of finite dynamical systems such as Boolean networks. A method proposed in (Laubenbacher and Stigler 2004) uses such data to construct a multistate discrete dynamical system

$$f = (f_1, \ldots, f_n) : k^n \longrightarrow k^n$$

over a finite field k such that the coordinate functions f_i are polynomials in variables $x_1, \ldots, x_n$ corresponding to the n biochemical compounds measured. The system f has to fit the given time course data set, that is, $f(\mathbf{p}_i) = \mathbf{p}_{i+1}$ for $i = 1, \ldots, r-1$. The goal is to infer a 'best' or most likely model f from a given data set which specifies a fraction of the possible state transitions of f. An advantage to working in a finite field is that all functions $k^n \to k$ are represented by polynomials. An important, and unanswered, question is to design biological experiments in an optimal way in order to infer a likely model with high probability. One complicating factor is that biochemical networks tend to be highly non-linear.

In this chapter, we describe the two approaches and point out the similarities between the two classes of problems, the techniques used to solve them, and the types of questions asked.

10.2 Design of experiments

In this section we provide a description of the computational algebra approach to experimental design given in (Robbiano 1998, Pistone *et al.* 2001). Let $\mathcal{D}$ be the full factorial design with n factors. We make the additional simplifying assumptions that each factor has the same number p of levels, resulting in p^n points for $\mathcal{D}$. A *model* for the design is a function

$$f : \mathcal{D} \longrightarrow \mathbb{Q},$$

that is, f maps each point of $\mathcal{D}$ to a measurement. Instead of using the field $\mathbb{Q}$ for measurements, one may choose other fields such as $\mathbb{C}$ or a finite field. From here on we will denote the field by k. It is well-known that any function from a finite number of points in k^n to k can be represented by a polynomial, so we may assume that f is a polynomial in variables $x_1, \ldots, x_n$ with coefficients in k.

Definition 10.1 A subset $\mathcal{F} = \{\mathbf{p}_1, \ldots, \mathbf{p}_r\} \subset \mathcal{D}$ is called a *fraction* of $\mathcal{D}$.

We list three important problems in the design of experiments:

(i) Identify a model for the full design $\mathcal{D}$ from a suitably chosen fraction $\mathcal{F}$.
(ii) Given information about features of the model, such as a list of the *monomials* (power products) appearing in it, design a fraction $\mathcal{F}$ which identifies a model for $\mathcal{D}$ with these features.
(iii) Given a fraction $\mathcal{F}$, which models can be identified by it?

These problems can be formulated in the language of computational algebra making them amenable to solution by techniques from this field. The fraction $\mathcal{F}$ is encoded by an algebraic object $I(\mathcal{F})$, an ideal in the polynomial ring $k[x_1, \ldots, x_n]$. This ideal contains all those polynomial functions $g \in k[x_1, \ldots, x_n]$ such that $g(\mathbf{p}_i) = 0$ for all $i = 1, \ldots, r$. It is called the *ideal of points* of the $\mathbf{p}_i$ and contains all polynomials *confounded by* the points in $\mathcal{F}$. Here we assume that the points are distinct. We will see that one can draw conclusions about $\mathcal{F}$ from its ideal of confounding polynomials. In particular, since any two polynomial models on $\mathcal{F}$ that differ by a confounding polynomial are identical on $\mathcal{F}$, it is advantageous to choose models from the quotient ring $R = k[x_1, \ldots, x_n]/I(\mathcal{F})$ rather than from the polynomial ring itself.

It can be shown that the ring R is isomorphic to the vector space k^s, and we need to study possible vector space bases for R consisting of monomials. This can be done using Gröbner bases of the ideal $I(\mathcal{F})$. For each choice of a term order for $k[x_1, \ldots, x_n]$, that is, a special type of total ordering of all monomials, we obtain a canonical generating set $G = \{g_1, \ldots, g_s\}$ for $I(\mathcal{F})$. We obtain a canonical k-basis for the vector space $R \cong k^s$ by choosing all monomials which are not divisible by the leading monomial of any of the g_i. We can then view each polynomial in R as a k-linear combination of the monomials in the basis.

To be precise, let $\{T_1, \ldots, T_t\}$ be the set of all monomials in the variables $x_1, \ldots, x_n$ which are not divisible by the leading monomial of any g_i. Then each element $f \in R$ can be expressed uniquely as a k-linear combination

$$f = \sum_{j=1}^{t} a_j T_j,$$

with $a_j \in k$. Suppose now that we are given a fractional design $\mathcal{F} = \{\mathbf{p}_1, \ldots, \mathbf{p}_r\}$ and an experimental treatment resulting in values $f(\mathbf{p}_i) = b_i$ for $i = 1, \ldots, r$. If we now evaluate the generic polynomial f at the points $\mathbf{p}_i$, we obtain a system of linear equations

$$a_1 T_1(\mathbf{p}_1) + \ldots + a_t T_t(\mathbf{p}_1) = b_1,$$

$$\vdots$$

$$a_1 T_1(\mathbf{p}_r) + \ldots + a_t T_t(\mathbf{p}_r) = b_r.$$

We can view these equations as a system of linear equations in the variables a_j with the coefficients $T_j(\mathbf{p}_i)$. We now obtain the main criterion for the unique identifiability of a model f from the fraction $\mathcal{F}$.

Theorem 10.1 *(Robbiano 1998, Theorem 4.12) Let $\mathcal{X} = \{\mathbf{p}_1, \ldots, \mathbf{p}_r\}$ be a set of distinct points in k^n, and let f be a linear model with monomial support $\mathcal{S} = \{T_1, \ldots, T_t\}$, that is, $f = \sum_i a_i T_i$. Let $X(\mathcal{S}, \mathcal{X})$ be the $(r \times t)$-matrix whose (i, j)-entry is $T_j(\mathbf{p}_i)$. Then the model f is uniquely identifiable by $\mathcal{X}$ if and only if $X(\mathcal{S}, \mathcal{X})$ has full rank.*

In this section we have given a brief outline of a mathematical framework within which one can use tools from computational algebra to address the three experimental design problems listed above. In the next section we will describe a similar set of problems and a similar approach to their solution in the context of biochemical network modelling.

10.3 Biochemical network inference

Molecular biology has seen tremendous advances in recent years due to technological breakthroughs that allow the generation of unprecedented amounts and types of data. For instance, it is now possible to simultaneously measure the activity level of all genes in a cell extract using DNA microarrays. This capability makes it possible to construct large-scale mathematical models of gene regulatory and other types of cellular networks, and the construction of such models is one of the central foci of computational systems biology. The availability of obtaining experimental measurements for large numbers of entities that are presumed to be interconnected in a network drives the need for the development of network inference algorithms. We will focus on the mathematical aspects of this problem for the rest of the section. More biological background can be found in (Laubenbacher and Stigler 2004).

We consider a dynamic network with n variables $x_1, \ldots, x_n$. These could represent products of n genes in a cell extract from a particular organism, say yeast. It is known that cellular metabolism and other functions are regulated by the interaction of genes that activate or suppress other genes and form a complex network. Suppose we are given a collection of pairs of simultaneous measurements of these variables:

$$(\mathbf{p}_1, \mathbf{q}_1), \ldots, (\mathbf{p}_r, \mathbf{q}_r),$$

with $\mathbf{p}_i, \mathbf{q}_i$ points in $\mathbf{R}^n$. For gene networks, each of these measurements could be obtained from a DNA microarray. Each pair $(\mathbf{p}_i, \mathbf{q}_i)$ is to be interpreted as follows. The variables in the network are initialised at $\mathbf{p}_i$ and subsequently the network transitions to $\mathbf{q}_i$. This might be done through a perturbation such as an experimental treatment, and $\mathbf{p}_i$ represents the network state immediately after the perturbation and $\mathbf{q}_i$ represents the network state after the network has responded to the perturbation. Sometimes the measurement pairs are consecutive points in a measured time course. In this case the pairs above consist of consecutive time points. Typically the number n of variables is orders of magnitude larger than the number r of measurements, in contrast to engineering applications where the reverse is true (or where r is on the order of n). For instance the network may contain hundreds or thousands of genes, from which only 10 or 20 experimental measurements are collected.

Example 10.1 Consider the following time course for a biochemical network of three genes, labelled x_1, x_2, and x_3.

x_1	x_2	x_3
1.91	3.30	1.98
1.50	1.42	1.99
1.42	1.31	0.03
0.83	1.96	1.01
0.97	2.08	1.01

Each gene's expression levels were measured at five consecutive time points and each entry represents a measurement. While the data are given in tabular form, we could have also represented the data as the pairs of network states

$$((1.91, 3.30, 1.98), (1.50, 1.42, 1.99))$$
$$((1.50, 1.42, 1.99), (1.42, 1.31, 0.03))$$
$$((1.42, 1.31, 0.03), (0.83, 1.96, 1.01))$$
$$((0.83, 1.96, 1.01), (0.97, 2.08, 1.01)).$$

Network inference problem: given input–output measurements $\{(\mathbf{p}_i, \mathbf{q}_i)\}$, infer a model of the network that produced the data.

One can consider a variety of different model types. First it is of interest to infer the directed graph of causal connections in the network, possibly with signed edges indicating qualitative features of the interactions. Dynamic model types include systems of differential equations, Boolean networks, Bayesian networks, or statistical models, to name a few. In light of the fact that DNA microarray data contain significant amounts of noise and many necessary parameters for models are unknown at this time, it suggests itself to consider a finite number of possible states of the variables x_i rather than treating them as real-valued. This is done by Bayesian network inference methods, for instance. The issue of data discretisation is a very subtle one. On the one hand, discrete data conform more to actual data usage by experimentalists who tend to interpret, e.g., DNA microarray data in terms of fold changes of regulation of genes compared to control. On the other hand, a lot of information is lost in the process of discretising data and the end result typically depends strongly on the method used. In the extreme case, one obtains only two states corresponding to a binary ON/OFF view of gene regulation. In our case, a strong advantage of using discrete data is that it allows us to compute algorithmically the whole space of admissible models for a given data set, as described below. Nonetheless, the result typically depends on the discretisation method and much work remains to be done in understanding the effect of different discretisation methods. Once the variables take on values in a finite set k of states, it is natural to consider discrete dynamical systems

$$F : k^n \longrightarrow k^n.$$

As mentioned, the dynamics is generated by repeated iteration of the mapping F. In order to have mathematical tools available for model construction and analysis, one can make the assumption that k is actually a finite field rather than simply a set. In practice this is easily accomplished, since the only ingredient required is the choice of a finite state set that has cardinality a power of a prime number. With these additional assumptions our models are *polynomial dynamical systems*

$$F = (f_1, \ldots, f_n) : k^n \longrightarrow k^n,$$

with $f_\ell \in k[x_1, \ldots, x_n]$ for $\ell = 1, \ldots, n$. (As remarked above, any function from a finite set of points into a field can be represented as a polynomial function.) The ℓ-th polynomial function f_ℓ describes the transition rule for gene x_ℓ and hence f_ℓ is called the *transition function* for x_ℓ.

Returning to the network inference problem, we can now rephrase it as: *Given the state transitions* $\{(\mathbf{p}_i, \mathbf{q}_i)\}$, *find a polynomial dynamical system (or polynomial model)* F *such that* $F(\mathbf{p}_i) = \mathbf{q}_i$.

This problem can be solved one node at a time, that is, one transition function at a time. This 'local' approach to inference then begins with a collection $\{\mathbf{p}_i\}$ of points, and we are looking for transition functions $f_\ell \in k[x_1, \ldots, x_n]$ that satisfy the condition that $f_\ell(\mathbf{p}_i) = b_i$, where b_i is the ℓ-th entry in $\mathbf{q}_i$.

Example 10.2 Let

$$(\mathbf{p}_1, \mathbf{q}_1) = ((2,2,2),(1,0,2)), \quad (\mathbf{p}_2, \mathbf{q}_2) = ((1,0,2),(1,0,0)),$$
$$(\mathbf{p}_3, \mathbf{q}_3) = ((1,0,0),(0,1,1)), \quad (\mathbf{p}_4, \mathbf{q}_4) = ((0,1,1),(0,1,1)).$$

be the discretisation of the data in Example 10.1 into the three-element field $k = \mathbb{F}_3$ by discretising each coordinate separately, according to the method described in

(Dimitrova *et al.* 2007). Then the goal is to find a polynomial model $F : k^3 \longrightarrow k^3$ such that $F(\mathbf{p}_i) = \mathbf{q}_i$ for $i = 1, \ldots, 4$. Since any such F can be written as $F = (f_1, f_2, f_3)$, we can instead consider the problem of finding transition functions $f_\ell : k^3 \longrightarrow k$ such that $f_\ell(\mathbf{p}_i) = \mathbf{q}_{i\ell}$, for all $1 \le \ell \le 3$ and $1 \le i \le 4$.

The similarity to the experimental design problem in the previous section should now be obvious. Factors correspond to variables x_i representing genes, levels correspond to the elements of the field k representing gene states, the points $\mathbf{p}_i$ of the factorial design correspond to experimental measurements, and the b_i in both cases are the same. As mentioned earlier, the available experimental observations are typically much fewer than the totality of possible system states. Thus, the objective in both cases is the same: Find good polynomial models for the full design from an experimental treatment of a fractional design.

The approach to a solution is quite similar as well. Suppose we are given two transition functions f and g that both agree on the given experimental data, that is, $f(\mathbf{p}_i) = b_i = g(\mathbf{p}_i)$ for all i. Then $(f - g)(\mathbf{p}_i) = 0$, so that any two transition functions differ by a polynomial function that vanishes on all given observations, that is, by a polynomial in the ideal of points $I(\mathbf{p}_1, \ldots, \mathbf{p}_r)$, which we called $I(\mathcal{F})$ in the previous section. If f is a particular transition function that fits the data for some x_ℓ, then the space of all feasible models for x_ℓ is

$$f + I(\mathbf{p}_1, \ldots, \mathbf{p}_r).$$

The problem then is to choose a model from this space. In design of experiments, the single-variable monomials represent the *main effects* and the other monomials represent *interactions*. In the biochemical network case the situation is similar. Single-variable monomials in a model for a gene regulatory network represent the regulation of one gene by another, whereas the other monomials represent the synergistic regulation of one gene by a collection of other genes, for example through the formation of a protein complex. In general, very little theoretical information is available about the absence or presence of any given monomial in the model. One possible choice is to pick the normal form of f with respect to a particular Gröbner basis for the ideal $I(\mathbf{p}_1, \ldots, \mathbf{p}_r)$. However, this normal form depends on the particular choice of Gröbner basis. Other approaches are explored in (Dimitrova *et al.* 2008), in particular an 'averaging' process over several different choices of Gröbner basis.

Example 10.3 In our running example, consider the following polynomials:

$$f_1(x_1, x_2, x_3) = 2x_2 x_3 + 2x_2 + 2x_3,$$
$$f_2(x_1, x_2, x_3) = 2x_3^3 + x_2^2 + x_2 + 2x_3 + 1,$$
$$f_3(x_1, x_2, x_3) = 2x_3^2 + 2x_1 + 2.$$

Each f_ℓ interpolates the discretised data for x_ℓ (see Example 10.2). The ideal of the input points $\mathbf{p}_1, \ldots, \mathbf{p}_4$ is

$$I = \langle x_1 + x_2 + 2, x_2 x_3 + 2x_3^2 + 2x_1 + x_2, x_2^2 + 2x_3^2 + x_2 + 2x_3 \rangle.$$

Then the model space for each x_ℓ is given by $f_\ell + I$. The Gröbner basis G for I w.r.t. the graded reverse lexicographical term order $\succ$ with $x_1 \succ x_2 \succ x_3$ is

$$G = \{x_1 + x_2 + 2, x_2 x_3 + 2x_3^2 + x_2 + 2x_3, x_2^2 + 2x_3^2 + x_2 + 2x_3, x_3^3 + 2x_3\}.$$

To choose a model for each x_ℓ, we compute the normal form $\bar{f}_\ell$ of f_ℓ with respect to $\succ$, resulting in the polynomial dynamical system $F = (\bar{f}_1, \bar{f}_2, \bar{f}_3) : (\mathbb{F}_3)^3 \longrightarrow (\mathbb{F}_3)^3$ with $\bar{f}_1(x_1, x_2, x_3) = 2x_3^2 + x_3$, $\bar{f}_2(x_1, x_2, x_3) = x_3^2 + 2x_3 + 1$, $\bar{f}_3(x_1, x_2, x_3) = 2x_3^2 + x_2 + 1$.

Given a polynomial model $F = (f_1, \ldots, f_n)$ for a network, one can predict the connectivity structure of the nodes by analysing the relationship between the variables and the transition functions. For example, the transition function for x_1 given above is in terms of x_3, but not the other variables. The interpretation is that regulation of the gene represented by x_1 is dependent only on x_3. The dynamic behaviour of the network can be simulated by evaluating F on all possible network states, that is, on all of k^n.

Definition 10.2 Let $F = (f_1, \ldots, f_n) : k^n \longrightarrow k^n$ be a polynomial dynamical system. The *wiring diagram* of F is the directed graph (V, E) with $V = \{x_1, \ldots, x_n\}$ and $E = \{(x_i, x_j) : x_i$ is a variable of $f_j\}$. The *state space* of F is the directed graph (V, E) with $V = k^n$ and $E = \{(\mathbf{a}, F(\mathbf{a}) : \mathbf{a} \in k^n\}$.

Viewing the structure and dynamics of a network via the wiring diagram and state space, respectively, allows one to uncover features of the network, including feedback loops and limit cycles, respectively; for example, see (Laubenbacher and Stigler 2004).

Example 10.4 The polynomial model F in Example 10.3 gives rise to the inferred wiring diagram and state space of the 3-gene network, as displayed in Figure 10.1. The network is predicted to have a feedback loop between x_2 and x_3, and the expression of x_3 is controlled via autoregulation. Furthermore, the network has two possible limit cycles: the fixed point at $(0,1,1)$ and the 3-cycle on $(0,1,0)$, $(0,1,2)$ and $(1,0,1)$. The fixed point is considered to be an equilibrium state of the network, and the 3-cycle represents an oscillation.

While the above polynomial dynamical system may be a reasonable model for the 3-gene network, it is not unique. Recall from Theorem 10.1 that the number of monomials in the basis for $k[x_1, x_2, x_3]/I(\mathbf{p}_1, \ldots, \mathbf{p}_4)$ is the number of data points (four, in this case). Since any transition function can be written as a k-linear combination of the basis monomials, then for a fixed term order there are $|k|^m = 3^4$ possible transition functions where m is the number of data points. In fact there are $(|k|^m)^n = 3^{12}$ possible polynomial models, given a term order. As there are five term orders which produce distinct polynomial models,[2] there are $((|k|^m)^n)^5 = 3^{60}$ possible models for a 3-variable system on three states and four data points.

---

[2] We computed the marked Gröbner bases of the ideal $I(\mathbf{p}_1, \ldots, \mathbf{p}_4)$ via the Gröbner fan and then computed the normal forms of the interpolating polynomials in Example 10.3 with respect to each of these Gröbner bases to obtain the five distinct polynomial models.

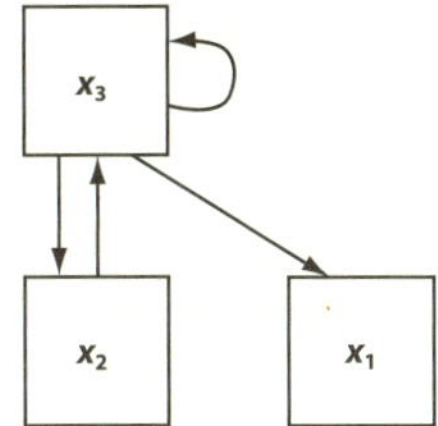

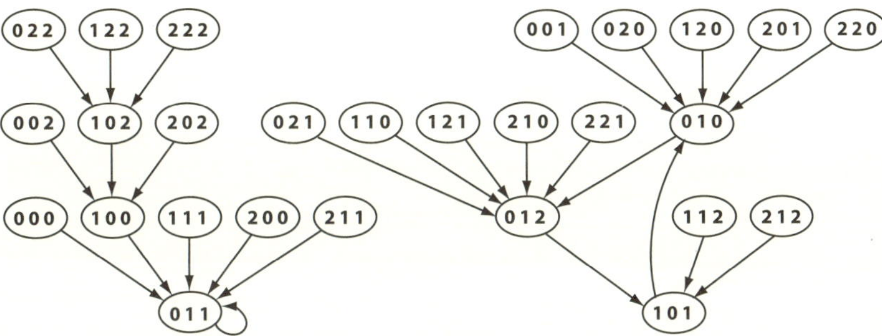

Fig. 10.1 Wiring diagram (top) and state space (bottom) for the polynomial model F in Example 10.3.

An important problem in this context that is common to both design of experiments and biochemical network inference is the construction of good fractional designs that narrow down the model space as much as possible. The challenge in network inference is that experimental observations tend to be very costly, severely limiting the number of points one can collect. Furthermore, many points are impossible to generate biologically or experimentally, which provides an additional constraint on the choice of fractional design.

10.4 Polynomial dynamical systems

It is worth mentioning that polynomial dynamical systems over finite fields (not to be confused with dynamical systems given by differential equations in polynomial form) have been studied in several different contexts. For instance, they have been used to provide state space models for systems for the purpose of developing controllers (Marchand and LeBorgne 1998, Le Borgne 1998) in a variety of contexts, including biological systems (Jarrah *et al.* 2004). Another use for polynomial dynamical systems is as a theoretical framework for agent-based computer simulations (Laubenbacher *et al.* 2009). Note that this class of models includes cellular automata and Boolean networks (choosing the field with two elements as state set), so that general polynomial systems are a natural generalisation. In this context, an important additional feature is the update order of the variables involved.

The dynamical systems in this chapter have been updated in parallel, in the following sense. If $f = (f_1, \ldots, f_n)$ is a polynomial dynamical system and $\mathbf{a} \in k^n$ is a state, then $f(\mathbf{a}) = (f_1(\mathbf{a}), \ldots, f_n(\mathbf{a}))$. By abuse of notation, we can consider each of the f_i as a function on k^n which only changes the ith coordinate. If we now specify a total order of $1, \ldots, n$, represented as a permutation $\sigma \in S_n$, then we can form the dynamical system

$$f_\sigma = f_{\sigma(n)} \circ f_{\sigma(n-1)} \circ \cdots \circ f_{\sigma(1)},$$

which, in general, will be different from f. Thus, f_σ is obtained through *sequential* update of the coordinate functions. Sequential update of variables plays an important role in computer science, e.g., in the context of distributed computation. See (Laubenbacher *et al.* 2009) for details.

Many processes that can be represented as dynamical systems are intrinsically stochastic, and polynomial dynamical systems can be adapted to account for this stochasticity. In the context of biochemical network models, a sequential update order arises naturally through the stochastic nature of biochemical processes within a cell that affects the order in which processes finish. This feature can be incorporated into polynomial dynamical system models through the use of random sequential update. That is, at each update step a sequential update order is chosen at random. It was shown in (Chaves *et al.* 2005) in the context of Boolean networks that such models reflect the biology more accurately than parallel update models. In (Shmulevich *et al.* 2002) a stochastic framework for gene regulatory networks was proposed which introduces stochasticity into Boolean networks by choosing at each update step a random coordinate function for each variable, chosen from a probability space of update functions. Stochastic versions of polynomial dynamical systems have yet to be studied in detail and many interesting problems arise that combine probability theory, combinatorics, and dynamical systems theory, providing a rich source of cross-fertilization between these fields.

10.5 Discussion

This chapter focuses on polynomial models in two fields, design of experiments and inference of biochemical networks. We have shown that the problem of inferring a biochemical network from a collection of experimental observations is a problem in the design of experiments. In particular, the question of an optimal experimental design for the identification of a good model is of considerable importance in the life sciences. When focusing on gene regulatory networks, it has been mentioned that conducting experiments is still very costly, so that the size of a fractional design is typically quite small compared to the number of factors to be considered. Another constraint on experimental design is the fact that there are many limits to an experimental design imposed by the biology, in particular the limited ways in which a biological network can be perturbed in meaningful ways. Much research remains to be done in this direction.

An important technical issue we discussed is the dependence of model choices on the term order used. In particular, the term order choice affects the wiring diagram of the model which represents all the causal interaction among the model variables. Since there is generally no natural way to choose a term order this dependence cannot be avoided. We have discussed available modifications that do not depend on the term order, at the expense of only producing a wiring diagram rather a dynamic model. This issue remains a focus of ongoing research.

As one example, an important way to collect network observations is as a time course of measurements, typically at unevenly spaced time intervals. The network is perturbed in some way, reacts to the perturbation, and then settles down into a steady state. The time scale involved could be on the scale of minutes or days. Computational experiments suggest that, from the point of view of network inference, it is more useful to collect several shorter time courses for different perturbations than to collect one highly resolved time course. A theoretical justification for these observations would aid in the design of time courses that optimise information content of the data versus the number of data points.

Acknowledgements

Laubenbacher was partially supported by NSF Grant DMS-0511441 and NIH Grant R01 GM068947-01. Stigler was supported by the NSF under Agreement No. 0112050.

References

Le Borgne, M. (1998). Partial order control of discrete event systems modeled as polynomial dynamical systems. In *IEEE International conference on control applications*, Trieste, Italy, 770–5.

Chaves, M., Albert, V. and Sontag, E. (2005). Robustness and fragility of Boolean models for genetic regulatory networks, *Journal of Theoretical Biology* **235**, 431–49.

Dimitrova, E. S., Jarrah, A. S., Laubenbacher, R. and Stigler, B. (2008). A Gröbner fan-based method for biochemical network modeling. In *Proceedings of ISSAC'2007* (New York, ACM Press), 122–6.

Dimitrova, E., Vera-Licona, P., McGee, J. and Laubenbacher, R. (2007). Comparison of data discretization methods for inference of biochemical networks (submitted).

Jarrah, A., Vastani, H., Duca, K. and Laubenbacher, R. (2004). An optimal control problem for *in vitro* virus competition. In *43rd IEEE Conference on Decision and Control* (Nassau, Bahamas) 579–84.

Laubenbacher, R., Jarrah, A. S., Mortveitm, H. and Ravi, S. (2009). A mathematical formalism for agent-based modeling. In *Encyclopedia of Complexity and Systems Science*, Meyers, R. ed. (Springer-Verlag).

Laubenbacher, R. and Stigler, B. (2004). A computational algebra approach to the reverse engineering of gene regulatory networks, *Journal of Theoretical Biology* **229**, 523–37.

Marchand, H. and LeBorgne, M. (1998). On the optimal control of polynomial dynamical systems over $\mathbb{Z}/p\mathbb{Z}$. In *Fourth Workshop on Discrete Event Systems, IEEE*, Cagliari, Italy, 385–90.

Pistone, G., Riccomagno, E. and Wynn, H. P. (2001). *Algebraic Statistics* (Boca Raton Chapman & Hall/CRC).

Pistone, G. and Wynn, H. P. (1996). Generalised confounding with Gröbner bases, *Biometrika* **83**, 653–66.

Riccomagno, E. (1997). Algebraic geometry in experimental design and related fields. PhD thesis, Department of Statistics, University of Warwick.

Robbiano, L. (1998). Gröbner bases and statistics. In *Gröbner Bases and Applications* Buchberger, B. and Winkler, F. eds. (Cambridge, Cambridge University Press) 179–204.

Shmulevich, I., Dougherty, E. R., Kim, S. and Zhang, W. (2002). Probabilistic boolean networks: A rule-based uncertainty model for gene regulatory networks, *Bioinformatics* **18**, 261–74.

11

Replicated measurements and algebraic statistics

Roberto Notari

Eva Riccomagno

Abstract

A basic application of algebraic statistics to design and analysis of experiments considers a design as a zero-dimensional variety and identifies it with the ideal of the variety. Then, a subset of a standard basis of the design ideal is used as support for identifiable regression models. Estimation of the model parameters is performed by standard least square techniques. We consider this identifiability problem in the case where more than one measurement is taken at a design point.

11.1 Introduction

The application of algebraic geometry to design and analysis of experiments started with (Pistone and Wynn 1996). There a design $\mathcal{D}$, giving settings for experiments, is seen as a finite set of distinct points in $\mathbb{R}^k$. This is interpreted as the zero set of a system of polynomial equations, which in turn are seen as the generator set of a polynomial ideal (see Chapter 1). The design $\mathcal{D}$ is uniquely identified with this ideal called the design ideal and indicated with $\mathrm{Ideal}(\mathcal{D})$. Operations over designs find a correspondence in operations over ideals, e.g. union of designs corresponds to intersection of ideals; problems of confounding are formulated in algebraic terms and computer algebra software is an aid in finding their solutions; and a large class of linear regression models identifiable by $\mathcal{D}$ is given by vector space bases of a ring, called the quotient ring modulo $\mathrm{Ideal}(\mathcal{D})$ and indicated as $R/\mathrm{Ideal}(\mathcal{D})$. This was the beginning of a successful stream of research which, together with the application of algebraic geometry to contingency table analysis covered in the first part of this volume, went under the heading of Algebraic Statistics (Pistone *et al.* 2001). For a recent review of the foundations of algebraic statistics see (Riccomagno 2008).

In this chapter we consider the problem of determining saturated, linear, regression models identifiable by a design when at each point of the design more than one observation can be taken. In particular we have to look for analogues of $\mathrm{Ideal}(\mathcal{D})$ and $R/\mathrm{Ideal}(\mathcal{D})$. As we are after saturated regression models, from which to obtain a sub-model, this is essentially an interpolation problem. We try to keep the presentation and the proofs as elementary as we can and give only those proofs

Algebraic and Geometric Methods in Statistics, ed. Paolo Gibilisco, Eva Riccomagno, Maria Piera Rogantin and Henry P. Wynn. Published by Cambridge University Press. © Cambridge University Press 2010.

that we deem essential, the other being collected in the on-line supplement. The algebraic construction we provide can be used in different statistical situations, for example when at distinct sample points $\omega_1, \omega_2 \in \Omega$, where Ω is a suitable sample space, the same design point d has been used but the outputs $Y(d(\omega_1))$ and $Y(d(\omega_2))$ can be different (in statistics this is referred to as replication); in a multivariate response situation when at $\omega \in \Omega$ a single design point d is used but more than one output is observed (multi-response models); when a set of sample points $\omega_i \in \Omega$, $i = 1, \ldots, n$, are such that the corresponding design points $d(\omega_i)$ are unknown and identified with the single point d (error-in-variables models and random effect models).

Two papers in the algebraic statistics literature consider replicated points. The technology of indicator functions (see Chapter 12) is employed in (Pistone and Rogantin 2008) where the counting functions of a fraction $\mathcal{D}$, subset of a large design $\mathcal{F}$, is defined as $R : \mathcal{F} \longrightarrow \mathbb{Z}_{\geq 0}$ such that $R(d) = 0$ if $d \in \mathcal{F} \backslash \mathcal{D}$ and otherwise is equal to the number of replicates of d. Information on the geometrical/statistical properties of $\mathcal{D} \subset \mathcal{F}$ are embedded into the coefficients of the indicator function and of the counting function. (Cohen *et al.* 2001) instead consider an extra factor to count the number of replicates, then the results are projected onto the original factor space. Here as well we add a factor but it plays a different role and we refer to Section 11.5 for a comparison.

A helpful picture for our approach is that of a cloud of distinct points lying around a centre point. Each point in the cloud moves towards the centre point along the line connecting them. This movement is described by an extra factor t. A main technique in this chapter is, then, to study the design and interpolation problems as $t \to 0$ by using techniques of linear algebra and Gröbner basis theory. For a related algebraic theory see (Abbott *et al.* 2005).

Specifically, we look for an algebraic method to deal with the error-in-variable case where the points $d_i \in \mathbb{R}^k$ are unknown but close to the point d and $y_i = y(d(\omega_i))$, $i = 1, \ldots, n$, are known values. The other statistical situations indicated above follow straightforwardly. We proceed in two steps: (a) determine a representation of $\mathcal{D}$ which takes into account replicates; (b) determine conditions on the above representation that ensure the good behaviour of the interpolating polynomial.

11.1.1 Outline of the chapter

We develop our results for a multiple point at the origin and then extend them to more multiple points, not necessarily located at the origin. In this outline we consider a single point replicated at the origin 0. Let $\{d_1, \ldots, d_n\} \subset \mathbb{R}^k$ be distinct points close to 0, with $d_i = (a_{1i}, \ldots, a_{ki})$, $i = 1, \ldots, n$, and whose coordinates might be unknown. Let $q_1, \ldots, q_r \subset \mathbb{R}^k$ be other distinct points. For each d_i consider the straight line between d_i and 0. Consider the following construction in which the extra factor t plays an important role:

(i) define $d_i(t) = (ta_{1i}, \ldots, ta_{ki})$ for $t \in \mathbb{R}$;
(ii) consider $\mathcal{D}_t = \{d_1(t), \ldots, d_n(t), q_1, \ldots, q_r\}$, which for each t is a set of distinct points;

(iii) consider the family of polynomial ideals in $\mathbb{R}[x_1, \ldots, x_k, t] = S$

$$\text{Ideal}(\mathcal{D}_t) = \bigcap_{i=1}^{n} \langle x_1 - ta_{1i}, \ldots, x_k - ta_{ki} \rangle \bigcap \text{Ideal}(\{q_1, \ldots, q_r\}).$$

Note $d_i(1) = d_i$ and $d_i(0) = 0$ for all $i = 1, \ldots, n$. We will later observe that

(i) $\text{Ideal}(\mathcal{D}_t)$ defines a flat family and for all $t_0 \in \mathbb{R} \setminus \{0\}$ $n+r$ distinct points are zeros of all polynomials in $\text{Ideal}(\mathcal{D}_t)$; namely $\dim S/\langle \text{Ideal}(\mathcal{D}_t), t - t_0 \rangle = 0$ and $\deg S/\langle \text{Ideal}(\mathcal{D}_t), t - t_0 \rangle = n + r \, (= \dim_{\mathbb{R}} S/\langle \text{Ideal}(\mathcal{D}_t), t - t_0 \rangle)$. We are interested in $t_0 = 0$;

(ii) for almost all $t_0 \in \mathbb{R}$ including $t_0 = 0$ there exists a monomial ideal $I \subset \mathbb{R}[x_1, \ldots, x_k]$ (not depending on t) such that $\text{LT}_{\prec}(\text{Ideal}(\mathcal{D}_t), t - t_0) = \langle t, I \rangle$;

(iii) I can be computed using a local term-ordering for which $x^\alpha t^a \succ x^\beta t^b$ if $a < b$ or $a = b$ and $x^\alpha \succ x^\beta$.

For the definitions of dim and deg see Chapter 1. In particular the following one-to-one maps $S/\langle \text{Ideal}(\mathcal{D}_t), t - t_0 \rangle \sim_{\mathbb{R}} R/I \sim_{\mathbb{R}} \text{Span}(x^\alpha : x^\alpha \notin \text{LT}(I))$ do not depend on t_0. The set $\{x^\alpha : x^\alpha \notin \text{LT}(I)\}$ is called a standard basis. Hence I is a partial analogue of $\text{Ideal}(\mathcal{D})$ of the case of distinct points. Nevertheless, as I is a monomial ideal we have lost information on the aliasing/confounding structure of the design.

Example 11.1 Consider $\mathcal{D}$ formed by the points $q_1 = (1, 2)$, $q_2 = (2, 2)$ and $(0, 0)$ counted twice. The procedure above yields $\{1, x, y, xy\}$. The design/model matrix X below is not full rank and in particular $X^\top X$ is not invertible

$$
\begin{array}{c}
\\
(1,2) \\
(2,2) \\
(0,0) \\
(0,0)
\end{array}
\begin{array}{cccc}
1 & x & y & xy \\
\left(\begin{array}{cccc}
1 & 1 & 2 & 2 \\
1 & 2 & 2 & 4 \\
1 & 0 & 0 & 0 \\
1 & 0 & 0 & 0
\end{array} \right) & & &
\end{array} = X.
$$

Typically the extra degrees of freedom are used in the estimation of model variance.

Example 11.2 (Example 11.1 cont.) The family of matrices X_t, $t \in \mathbb{R}$, below is obtained by evaluating $\{1, x, y, xy\}$ at $\mathcal{D}_t$, $t \in \mathbb{R}$,

$$
X_t = \begin{pmatrix}
1 & 1 & 2 & 2 \\
1 & 2 & 2 & 4 \\
1 & -t & t & -t^2 \\
1 & -2t & t & -2t^2
\end{pmatrix}
$$

which is full rank for $t \in \mathbb{R}$ except a set of zero Lebesgue measure.

Next, assume the value observed at q_i is y_i, for $i = 1, 2$, and y_3 and y_4 are observed at 0. We need to chose 'responses' at the 'moving' points $\mathcal{D}_t$ to determine a vector $Y_t = [y_1, y_2, y_3(t), y_4(t)]^\top$, and consider the linear system $Y_t = X_t \theta$ with symbolic solutions (Cramer rule)

$$\theta_i(t) = \frac{\det(X_{t,i})}{\det(X_t)} = (X_t^{-1} Y_t)_i.$$

We require that $y_3(t), y_4(t)$ are defined so that

(i) $\lim_{t \mapsto 0} \theta_i(t)$ exists finite for $i = 1, \ldots, 4 = n + r$;

(ii) $y_3(1) = y_3$, $y_4(1) = y_4$ and $y_3(0) = y_4(0) = a$. In statistical practice often a is the mean value of the measured responses at $(0,0)$;

(iii) y_i are polynomials of as small as possible degree.

Example 11.3 (Example 11.1 cont.) Observe that

$$\theta_1 = -y_4(t) - 2y_3(t)$$
$$\theta_2 = (y_3(t) - y_4(t))/t$$
$$\theta_3 = (-2y_3(t) + 7 + y_4(t))/2$$
$$\theta_4 = -1 - (y_3(t) - y_4(t))/2t.$$

The order of infinitesimal in $t = 0$ of $\det(X_t) = -t(t-2)^2$ is 1 and we have $Y_t(x,y) = \theta_1(t) + \theta_2(t)x + \theta_3(t)y + \theta_4(t)xy$ whose limit as t goes to zero is $\bar{Y}(x,y) = a + 0.3x - (a-7)/2y + \frac{0.7}{2}xy$.

11.2 Points with multiplicities

We recall the basic notions of algebraic geometry and the definition of point with multiplicity extending the Appendix in Chapter 1, to which we refer for technical terminology. We follow the literature used in the Appendix and also refer to (Cox *et al.* 2007, Cox *et al.* 2008, Hartshorne 1977, Kreuzer and Robbiano 2000, Kreuzer and Robbiano 2005), and do not give further references for the cited results.

Let $\mathcal{K}$ be a field. The affine space of dimension k over $\mathcal{K}$ is defined as the set $\mathbb{A}_{\mathcal{K}}^k = \{(a_1, \ldots, a_k) : a_i \in \mathcal{K}, \text{ for } i = 1, \ldots, k\}$. When no confusion arises, we denote it simply as $\mathbb{A}^k$. In most applications in statistics, $\mathcal{K}$ is the field $\mathbb{Q}$ of rational numbers, the field $\mathbb{R}$ of real numbers or $\mathbb{C}$ of complex numbers (e.g. Chapter 12). At first we assume that $\mathcal{K}$ is an algebraically closed field, e.g. $\mathbb{C}$. This assumption is used to switch from an ideal to the associated locus of zeros. In our application we relax this assumption as we start from the zeros locus, namely the design, and then consider an associated ideal.

We need to have clear the correspondence between algebraic subsets of $\mathbb{A}^k$ and ideals in the polynomial ring $R = \mathcal{K}[x_1, \ldots, x_k]$.

Definition 11.1 Let $f \in R$ be a polynomial and $\mathcal{D}$ *a subset of* $\mathbb{A}^k$.

(i) The zero locus of f is the set $V(f) = \{P \in \mathbb{A}^k : f(P) = 0\}$.

(ii) If $I \subseteq R$ is an ideal, then we define $V(I) = \{P \in \mathbb{A}^k : f(P) = 0 \; \forall f \in I\}$.

(iii) $\mathcal{D}$ is an algebraic set if there exists an ideal $I \subseteq R$ such that $\mathcal{D} = V(I)$.

(iv) If $\mathcal{D}$ is an algebraic set, then we define

$$I(\mathcal{D}) = \{f \in R : f(P) = 0 \text{ for all } P \in \mathcal{D}\}.$$

A topology on $\mathbb{A}^k$, called the Zariski topology, is defined by choosing the algebraic subsets as the closed subsets. It can be shown that $I(\mathcal{D})$ is an ideal and that an algebraic set can be defined by different ideals. For example, both $I = \langle x^2, y \rangle$ and $J = \langle x, y^2 \rangle$ define $\mathcal{D} = \{(0,0)\} \subset \mathbb{A}^2$, but $I(\mathcal{D}) = \langle x, y \rangle$.

A basic result in algebraic geometry is the following Hilbert's Nullstellensatz theorem.

Theorem 11.1 (Hilbert's Nullstellensatz theorem) *Let $\mathcal{K}$ be an algebraically closed field, $I \subseteq R$ an ideal, and let $f \in R$ be a polynomial which vanishes at all points of $V(I)$. Then, $f \in \sqrt{I}$.*

For the definition of $\sqrt{I}$ see the Appendix. An immediate consequence of Theorem 11.1 is that there is a one-to-one inclusion-reversing correspondence between algebraic sets in $\mathbb{A}^k$ and radical ideals in R when $\mathcal{K}$ is algebraically closed. We consider mainly finite subsets of points, which are algebraic sets.

Examples 11.4, 11.5 and 11.6 below illustrate circumstances where it is restrictive to consider radical ideals only.

Example 11.4 (from design theory) In the application of algebraic geometry to design of experiments in (Pistone *et al.* 2001), mainly designs are identified with radical ideals and information is lost on whether more than one observation is taken at the same design point.

Example 11.5 (from geometry) Consider the intersection of the parabola $y - x^2 = 0$ with the tangent line $y = 0$ at the origin. The intersection is associated to the ideal $I = \langle y, y - x^2 \rangle = \langle y, x^2 \rangle$ that is not radical. The ideal $\langle x, y \rangle = \sqrt{I}$ gives the coordinates of the intersection point, but does not describe the geometric situation.

Example 11.6 (from interpolation theory) We want to determine the set of all polynomials in two variables that vanish at the origin together with their first derivatives. If $f \in \mathcal{K}[x, y]$ is such a polynomial then $f(0,0) = f_x(0,0) = f_y(0,0) = 0$ where f_x (resp. f_y) is the partial derivative with respect to x (resp. y). Hence, $f = x^2 f_1 + xy f_2 + y^2 f_3$, with $f_i \in R, i = 1, 2, 3$. Then, the set we want is the ideal $I = \langle x^2, xy, y^2 \rangle$, which is not radical, indeed $\sqrt{I} = \langle x, y \rangle$.

It is a main point of this chapter that the right tool from algebraic geometry to clarify these issues is the primary decomposition of an ideal I such that the quotient ring, R/I, is zero dimensional and has the correct degree. In this way we are able to consider ideals which are not necessarily radical but whose associated algebraic set remains finite. The approach is summarised in Theorem 11.2 below.

Theorem 11.2 *Let $\mathcal{K}$ be algebraically closed. The ring R/I has dimension 0 if, and only if, $V(I) \subset \mathbb{A}^k$ is a finite set of points. Moreover, the following statements are equivalent*

(i) *R/I has dimension 0;*

(ii) *if $I = J_1 \cap \cdots \cap J_n$ is a primary decomposition of I then there exist $P_1, \ldots, P_n$ distinct points in $\mathbb{A}^k$ such that $\sqrt{J_i} = I(P_i)$ for each $i = 1, \ldots, n$, and $V(I) = \{P_1, \ldots, P_n\}$.*

Definition 11.2 If R/I has dimension 0, then we call $V(I) = \{P_1, \ldots, P_n\}$ the *support* of R/I.

An important observation used in (Pistone and Wynn 1996) is that if I is a radical ideal then $\deg(R/I)$ is equal to the number of points in $V(I)$.

Example 11.7 (Example 11.5 cont.) Naively, the intersection of a parabola with the tangent line at a point is a point of multiplicity 2. Indeed, the ideal $I = \langle y, x^2 \rangle$ which describes this intersection is an $\langle x, y \rangle$-primary ideal. The quotient ring R/I has dimension 0 because $V(I) = \{(0,0)\}$ and a basis of R/I as $\mathcal{K}$-vector space is $\{1, x\}$ and so $\deg(R/I) = 2$ which is equal to the multiplicity of the intersection point.

Definition 11.3 The ideal $I \subset R$ defines a point P with multiplicity r if I is $I(P)$-primary, and $\deg(R/I) = r$.

From a design viewpoint, Definition 11.3 means that the support of I is the single point P at which r measurements are taken and that the vector space R/I has dimension r; that is, an associated saturated linear regression model includes r linearly independent terms. In Section 11.3.1 we consider the extension to more than one point. It is worthwhile to note here the use of the words dimension and degree for R/I: the dimension of R/I is zero because I is the ideal of a single point and the degree of R/I is the dimension of R/I as a vector space and is equal to the multiplicity of the point.

Unfortunately, there exist many different ideals that define the same point with the same multiplicity. For example, $I = \langle x, y^3 \rangle$ and $J = \langle x^2, xy, y^2 \rangle$ are $\langle x, y \rangle$-primary ideals, and $\deg(R/I) = \deg(R/J) = 3$. Proposition 11.1 below shows that the Hilbert function of a 0-dimensional ring R/I gives information about the position of the points in the support of R/I. We recall from Chapter 1 the definition of the Hilbert function and refer to it for further details. Here its role is to give the dimension as a vector space of the quotient rings R/I. We use it in Section 11.3 below where, by specialising to the so-called 'flat families', we go some way towards resolving, or at least understanding, the lack of uniqueness just referred to above.

Definition 11.4 Let $I \subset R$ be an ideal. The Hilbert function of R/I is the numerical function $h_{R/I} : \mathbb{Z} \to \mathbb{Z}$ defined as $h_{R/I}(j) = \dim_{\mathcal{K}}(R/I)_{\leq j}$ where $(R/I)_{\leq j}$ is the subset of cosets that contain a polynomial of degree less than or equal to j, and $\dim_{\mathcal{K}}$ is the dimension as a $\mathcal{K}$-vector space.

Proposition 11.1 *Let $\mathcal{D}$ be a finite set of n distinct points in $\mathbb{A}^k$. Then,*

(i) *$\mathcal{D}$ is contained in a line if, and only if, $h_{R/I(\mathcal{D})}(j) = j + 1$ for $0 \le j \le \deg(R/I(\mathcal{D})) - 1$, where h is the Hilbert function.*

(ii) *$\mathcal{D}$ is contained in a smooth irreducible conic if, and only if, $h_{R/I(\mathcal{D})}(j) = 1 + 2j$ for $0 \le j \le m$ where m is the integer part of $(\deg(R/I(\mathcal{D})) - 1)/2$.*

11.3 Flat families

The algebraic background for 'moving points around' relies on the definition of a flat family of 0-dimensional affine schemes over $\mathbb{A}^1$. We discuss, in full details, some special situations and give some examples.

Definition 11.5 The ideal $J \subset S = \mathcal{K}[x_1, \ldots, x_k, t]$ defines a flat family of 0-dimensional rings if

(i) $S/\langle J, t - t_0 \rangle$ is a ring of dimension 0 for every $t_0 \in \mathcal{K}$, and

(ii) $\deg(S/\langle J, t - t_0 \rangle)$ does not depend on t_0.

Example 11.8 The ideal $J = \langle xy, ty^2 + x - t, y^3 - y \rangle \subset \mathcal{K}[x, y, t]$ satisfies Definition 11.5. Indeed, if $t_0 \ne 0$ then, $\langle J, t - t_0 \rangle = \langle x, y - 1, t - t_0 \rangle \cap \langle x, y + 1, t - t_0 \rangle \cap \langle x - t_0, y, t - t_0 \rangle$ and so $\mathcal{K}[x, y, t]/\langle J, t - t_0 \rangle$ has dimension 0 and degree 3. If $t_0 = 0$, then $\langle J, t \rangle = \langle t, x, y^3 - y \rangle$ and $\mathcal{K}[x, y, t]/\langle t, x, y^3 - y \rangle$ has dimension 0 and degree 3. These can be computed with, e.g., the CoCoA commands `Dim` and `Multiplicity`, respectively, see (CoCoA Team 2007).

We can think of it as follows. Let $\mathcal{D} = \{A, B, C\} \subset \mathbb{A}^2$ where $A = (0, 1), B = (0, -1), C = (1, 0)$. We want to move C along the x-axis to the origin $O = (0, 0)$. Hence, we consider the point $C_t = (t, 0)$ and the set $\mathcal{D}_t = \{A, B, C_t\}$. For $t = 1$, we have $\mathcal{D}_1 = \mathcal{D}$, for $t = 0$, we have $\mathcal{D}_0 = \{A, B, O\}$. The ideal $I(\mathcal{D}_t)$ is equal to $I(\mathcal{D}_t) = \langle x, y - 1 \rangle \cap \langle x, y + 1 \rangle \cap \langle x - t, y \rangle \subset \mathcal{K}[x, y, t]$. It defines a flat family because, for any $t \in \mathbb{A}^1$, we have three distinct points. For example for $t_0 = 7$ a Gröbner basis of $\langle I(\mathcal{D}_t), t - t_0 \rangle$ is $\{t - 7, xy, x^2 - 7x, -7y^2 - x + 7\}$ and for $t_0 = 0$ it is $\{t, y^3 - y, -x\}$.

Example 11.9 We can also obtain points with multiplicity, if we move two or more points in such a way that they collapse together. For example, the ideal $J = \langle x, y^2 - ty \rangle$ describes a flat family. For $t \ne 0$ it represents two different points on the y-axis of coordinates $(0, 0)$ and $(0, t)$. For $t = 0$ it represents the origin with multiplicity two.

In a flat family, almost all the choices of $t_0 \in \mathbb{A}^1$ give geometrical objects with the same properties, in particular the same Hilbert function, while for a finite number of values of t, we get different properties. In Example 11.8, the Hilbert function of $\mathcal{D}_t$, for $t \ne 0$, is $1, 3, 3, 3, \ldots$ while the Hilbert function of $\mathcal{D}_0$ is $1, 2, 3, 3 \ldots$. We call those t for which $\mathcal{D}_t$ has different properties, the *special fibers* of the flat family, while those that have the same properties are called *general fibers* of the family.

Usually, the computation of the ideal that defines the special fiber of a flat family is very difficult. We consider a special case in which it is possible to make the computation very easily. We recall two things.

First, a polynomial $F \in \mathcal{K}[x_1, \ldots, x_k]$ is called homogeneous of degree d if it is a finite sum of monomials, each of total degree d, equivalently if $F(zx_1, \ldots, zx_k) = z^d F(x_1, \ldots, x_k)$. Any polynomial $f \in \mathcal{K}[x_1, \ldots, x_k]$ is a finite sum of homogeneous pieces, namely $f = f_0 + \cdots + f_s$ where f_j is homogeneous of degree j and $s = \deg(f)$. We call f_s the leading form $\mathrm{LF}(f)$ of f.

Next, consider an ideal $J \subset S = \mathcal{K}[x_1, \ldots, x_k, t]$ such that for $t_0 \neq 0$ $S/\langle J, t - t_0 \rangle$ has dimension 0 and degree d and for $t_0 = 0$, $S/\langle J, t \rangle$ has not dimension 0 or has not degree d. Then, J does not define a flat family. However, the ideal $J' = \{f \in S : t^a f \in J \text{ for } a \in \mathbb{Z}_{\geq 0}\}$ defines a flat family.

Example 11.10 Let $J = \langle xy, ty^2 + x - t \rangle \subset \mathcal{K}[t, x, y]$. For $t \neq 0$ we have $J = \langle x, y + 1 \rangle \cap \langle x, y - 1 \rangle \cap \langle x - t, y \rangle$ and so $S/\langle J, t - t_0 \rangle$ has dimension 0 and degree 3. For $t = 0$, $\langle J, t \rangle = \langle t, x \rangle$ and $S/\langle J, t - t_0 \rangle$ has dimension different from 0. Hence, J does not define a flat family. Instead the ideal $J' = \langle xy, ty^2 + x - t, y^3 - y \rangle$ defines a flat family as shown in Example 11.8.

In Theorem 11.3, whose proof can be found in the on-line supplement, we collapse n points. It is the specialisation to ideal of points of (Kreuzer and Robbiano 2005, Proposition 4.3.10) to which we refer for a general theory based on Macaulay bases.

Theorem 11.3 *Consider n distinct points $P_1, \ldots, P_n \in \mathbb{A}^k$ with P_i of coordinates $(a_{i1}, \ldots, a_{ik})$, and let $\mathcal{D} = \{P_1, \ldots, P_n\}$. Then $J = \bigcap_{i=1}^{n} \langle x_1 - ta_{ai1}, \ldots, x_k - ta_{ik} \rangle \subset S = \mathcal{K}[x_1, \ldots, x_k, t]$ is a flat family. Its special fiber is the origin with multiplicity n and it is defined by the ideal $I_0 = \{F \in R : F \text{ is homogeneous and there exists } f \in I(\mathcal{D}) \text{ such that } F = \mathrm{LF}(f)\}$. Moreover, the Hilbert function does not depend on t.*

11.3.1 More than one replicated point and some fixed points

In order to generalise the construction behind Theorem 11.3 to the case in which some points are collapsed, and some others remain fixed, we proceed in steps.

Theorem 11.4 *Let $X = \{P_1, \ldots, P_r\}, Y = \{Q_1, \ldots, Q_s\}$ be sets of points in $\mathbb{A}^k$, and assume that $Z = X \cup Y$ has $n = r + s$ distinct points. If P_i has coordinates $(a_{i1}, \ldots, a_{ik})$ then the family*

$$J = \bigcap_{i=1}^{r} \langle x_1 - ta_{i1}, \ldots, x_k - ta_{ik} \rangle \cap I(Q_1) \cap \cdots \cap I(Q_s)$$

is flat, with fibers of dimension 0 and degree $r + s$.

To simplify notation we write J instead of $\mathrm{Ideal}(\mathcal{D})$. The proof relies on Theorem 11.3 and can be found in the on-line supplement. In the setting of Theorem 11.4, the fiber over $t = 0$ can have a different Hilbert function from the general fiber of the family (see Example 11.11 below). Moreover, even if the Hilbert function of the fiber over $t = 0$ is equal to the Hilbert function of the general fiber, it may happen that the initial ideal changes, as Example 11.12 shows.

Example 11.11 Consider $P_1 = (-1, 1)$, $P_2 = (-2, 1)$ and $Q_1 = (1, 0)$, $Q_2 = (2, 0)$ be in $\mathbb{A}^2$. The flat family

$$J = \langle x + t, y - t \rangle \cap \langle x + 2t, y - t \rangle \cap \langle x - 1, y \rangle \cap \langle x - 2, y \rangle$$

describes the collapsing of P_1 and P_2 in the origin $(0, 0)$ along straight lines, while Q_1 and Q_2 remain fixed. The Hilbert function of the general fiber is $H(0) = 1, H(1) = 3, H(j) = 4$ for $j \geq 2$, while the Hilbert function of the fiber over $t = 0$ is $H'(j) = j + 1$ for $j = 0, 1, 2, 3$, and $H'(j) = 4$ for $j \geq 3$. The Hilbert function of an ideal can be computed in CoCoA with the command `HilbertFn`.

Example 11.12 Consider $\mathcal{D} = \{P_1, P_2, Q_1, Q_2\}$ with $P_1 = (-1, 1)$, $P_2 = (-1, -1)$, $Q_1 = (1, 0)$ and $Q_2 = (2, 0)$. Then, the associated flat family J is generated by $xy + yt, y^3 - yt^2, x^3 + x^2 t - 3x^2 - 3xt + 2x + 2t, x^2 t^2 - y^2 t^2 - 3y^2 t - 3xt^2 - 2y^2 + 2t^2$. The Hilbert function of the general fiber is equal to $H(0) = 1, H(1) = 3, H(j) = 4$, for $j \geq 2$, and it is equal to the Hilbert function of the fiber over $t = 0$. The initial ideal of a general fiber is $\langle t, x^2, xy, y^3 \rangle$, while the initial ideal of the fiber over $t = 0$ is $\langle t, xy, y^2, x^3 \rangle$. The computation where performed with respect to the term-order degrevlex with $x > y > t$, see (Cox *et al.* 2008).

Theorem 11.5, which essentially proves that the operation of intersection commutes with taking the fiber over $t = 0$, is useful to perform computations over an intersection of ideals rather than over the quotient ring.

Theorem 11.5 *In the hypotheses of Theorem 11.4, set* $J_1 = \bigcap_{i=1}^{r} \langle x_1 - t a_{i1}, \ldots, x_n - t a_{in} \rangle$ *and* $J_2 = I(Q_1) \cap \cdots \cap I(Q_s)$. *If* $Q_j \neq O$ *for every* $j = 1, \ldots, s$, *then*

$$\langle J, t \rangle / \langle t \rangle = \mathrm{LF}(J_1) \cap J_2.$$

Proof We can identify $\mathcal{K}[x_1, \ldots, x_k, t]$ with the coordinate ring of $\mathbb{A}^{k+1}$. In this larger affine space, J_2 is the ideal of the union of the lines through the points $(Q_j, 0)$ and parallel to the t-axis, while J_1 is the ideal of the union of the lines through the origin and the points $(P_i, 1)$. When we intersect with the hyperplane $t = 0$, we obtain the same ideal both if we consider the union of those $r + s$ lines and if we cut first the r lines and the s lines separately, and then we take their union. Hence, in the hyperplane $t = 0$, the ideals are equal to each other. $\qquad\square$

To complete this section, we analyse the case when some points are collapsed to a first limit point, some others to a second limit point, and so on. Theorem 11.6 relies on the hypothesis, met by the statistical set-up of this chapter, that to start with all the points considered are distinct, although some might be unknown, and then they collapse to form a smaller number of multiple points.

Theorem 11.6 *Let $A_1, \ldots, A_n \in \mathbb{A}^k$ be distinct points and, for $i = 1, \ldots, n$, let $X_i = \{P_{i1}, \ldots, P_{ir_i}\}$ be a set of r_i distinct points. Assume that $Y = X_1 \cup \cdots \cup X_n$ is a set of $r_1 + \cdots + r_n = r$ distinct points. Consider the scheme obtained as X_1 collapses to A_1 keeping fixed the remaining points, X_2 collapses to A_2 keeping fixed the multiple point at A_1 and the remaining points, and so on until X_n collapses to A_n keeping fixed the multiple points at $A_1, \ldots, A_{n-1}$. Then its special fiber is defined by the ideal*

$$J_1 \cap \cdots \cap J_n$$

where J_i is $I(A_i)$–primary, has degree r_i, and it is computed as a leading form ideal.

Proof The argument of the proof of Theorem 11.5 works as well in this more general situation, and so the claim follows. The computation of the leading form ideal $\mathrm{LF}(J_i)$ relies on Theorem 11.3 after a change of coordinates to move A_i to the origin. $\qquad\square$

We end the section with an example that shows how to develop the computation, without explicitly using the new variable t.

Example 11.13 Let $A_1 = (0,0)$, $A_2 = (1,1)$, $A_3 = (-1,1)$, $A_4 = (-1,-1)$, and $A_5 = (1,-1)$ be the limit points, and let

$$\begin{aligned}
X_1 &= \{(0,0), (1,0), (0,1), (-1,0), (0,-1)\}, \quad X_2 = \{(2,1), (1,2)\}, \\
X_3 &= \{(-2,1), (-1,2)\}, \quad X_4 = \{(-2,-1), (-1,-2)\}, \\
X_5 &= \{(1,-2), (2,-1)\}.
\end{aligned}$$

We want to compute the limit ideal when collapsing X_i to $A_i, i = 1, \ldots, 5$, assuming that the collapsing process is independent from one point to the others.

First compute $I(X_1) = \langle xy, x^3 - x, y^3 - y \rangle$ e.g. with `IdealOfPoints` in CoCoA (CoCoATeam 2007) and consider $J_1 = \mathrm{LF}(I(X_1)) = \langle xy, x^3, y^3 \rangle$ with $I(A_1)$–primary of degree 5.

Before computing the ideal J_2, change coordinates and move A_2 to the origin, by setting $x = X + 1, y = Y + 1$. Then, A_2 and X_2 become $(0,0)$ and $\{(1,0),(0,1)\}$, respectively, and $I(X_2) = \langle X + Y - 1, Y^2 - Y \rangle$, giving $\mathrm{LF}(I(X_2)) = \langle X + Y, Y^2 \rangle$. In the old coordinate system, this becomes $J_2 = \langle x + y - 2, y^2 - 2y + 1 \rangle$. To compute J_3, set $x = X - 1, y = Y + 1$, and obtain $I(X_3) = \langle X - Y + 1, Y^2 - Y \rangle$ and thus $J_3 = \langle x - y + 2, y^2 - 2y + 1 \rangle$. Analogously compute $J_4 = \langle x + y + 2, y^2 + 2y + 1 \rangle$ and $I(X_5) = \langle X - Y - 1, Y^2 + Y \rangle$ and finally $J_5 = \langle x - y - 2, y^2 + 2y + 1 \rangle$.

The limit ideal is then $J = J_1 \cap \cdots \cap J_5$ which is generated by

$$x^3 y + xy^3 - 2xy, \; x^4 + 4x^3 y - 2x^2 y^2 + 4xy^3 + y^4 - 8xy,$$
$$2y^5 + x^2 y - 3y^3, \; 2xy^4 + x^3 - 3xy^2, \; 2x^2 y^3 - x^2 y - y^3.$$

The computation were performed using degrevlex with $y > x$.

11.4 Interpolation over points with multiplicity

Consider the set-up of Theorem 11.6. The classical multivariate interpolation problem consists in determining a polynomial $F(x_1, \ldots, x_k)$ such that $F(P_{ij}) = \alpha_{ij}$ for given $\alpha_{ij} \in \mathcal{K}$, $i = 1, \ldots, n$ and $j = 1, \ldots, r_i$. This problem has a unique solution if the monomials in $F(x_1, \ldots, x_k)$ are a $\mathcal{K}$-vector space basis of $R/I(Z)$. Now, we consider the case when X_i collapses to A_i, $i = 1, \ldots, n$. We need to find polynomials $\alpha_{ij}(t) \in \mathcal{K}[t]$ such that $\alpha_{ij}(1) = \alpha_{ij}$ and, if $F(x_1, \ldots, x_k, t)$ interpolates $\alpha_{ij}(t)$ over $(X_i)_t$ then its limit for $t \to 0$ exists and is a polynomial, where $(X_i)_t$ is the set obtained by moving the points in X_i to A_i along the straight line between P_{ij} and A_i for all $P_{ij} \in X_i$.

In Proposition 11.2 we consider the case of only one limit point. Specifically, we start with a cloud of distinct points P_i, $i = 1, \ldots, r$, in $\mathbb{R}^k$, the observed values α_i, $i = 1, \ldots, r$, and a monomial basis, $M_1, \ldots, M_r$, of $R/I(P_1, \ldots, P_r)$. Note that linear regression models based on subsets of $M_1, \ldots, M_r$ are identifiable by $X = \{P_1, \ldots, P_r\}$. Next, we consider r univariate polynomials $\alpha_i(t)$ in the extra factor t such that $\alpha_i(1) = \alpha_i$ for all i; for example, in an error-in-variable set-up we assume that the observations are taken at the unknown points P_i. In Proposition 11.2 we show that there exists a unique polynomial F which is a linear combination of the M_i's and whose coefficients are polynomials in t. In some sense F is a saturated interpolating polynomial which follows the cloud of points while it shrinks toward a centre point and at each t it interpolates the $\alpha_i(t)$'s, which can then be seen as a family of dummy data as t varies.

It is important that t is present only in the coefficient of F and that the construction of F does not depend on the choice of the $\alpha_i(t)$ polynomials. The limit of F as t goes to zero gives the interpolation over the replications at the centre point and is the saturated linear regression model associated to r-replicates at a single point with observed values α_i, $i = 1, \ldots, r$. In Theorem 11.8, our main result, we will provide an easy construction of this limit that does not depend on t.

Proposition 11.2 *Let $X = \{P_1, \ldots, P_r\} \subset \mathbb{A}^k$ be a set of distinct points, and let $\alpha_1, \ldots, \alpha_r$ be in $\mathcal{K}$. Let $M_1, \ldots, M_r$ be a monomial basis of the $\mathcal{K}$-vector space $R/I(X)$, and assume that the total degree of M_j is equal to m_j, and that the monomials are labelled in such a way that $0 = m_1 < m_2 \leq \cdots \leq m_r$. Moreover, let $\alpha_i(t) \in \mathcal{K}[t]$ be a polynomial such that $\alpha_i(1) = \alpha_i$, for $i = 1, \ldots, r$. Then, there exists a unique interpolating polynomial $F(x_1, \ldots, x_n) = c_1 M_1 + \cdots + c_r M_r$ with $c_i \in \mathcal{K}[t]_t$, localization of $\mathcal{K}[t]$ at the polynomial t, such that $F(t_0 P_i) = \alpha_i(t_0)$ for $i = 1, \ldots, r$, and for each $t_0 \neq 0$, where $t_0 P_i$ has coordinates $(t_0 a_{i1}, \ldots, t_0 a_{ik})$.*

Proof For a definition of localisation see Chapter 1 and for a proof in a more general set-up see (Kreuzer and Robbiano 2005, Th. 4.3.22). As in Theorem 11.3, with respect to an elimination order, the initial ideal of $\langle J, t - t_0 \rangle$ is equal to $\mathrm{LT}(\langle t, \mathrm{LF}(I(X)) \rangle)$ and so $M_1, \ldots, M_r$ is a monomial basis of $S/\langle J, t - t_0 \rangle$ for every $t_0 \in \mathcal{K}$. Moreover, for $t_0 \neq 0$, the points $t_0 P_1, \ldots, t_0 P_r$ impose independent conditions on $M_1, \ldots, M_r$, that is to say, the matrix

$$A(t) = \begin{pmatrix} M_1(tP_1) & M_2(tP_1) & \ldots & M_r(tP_1) \\ M_1(tP_2) & M_2(tP_2) & \ldots & M_r(tP_2) \\ \vdots & & & \\ M_1(tP_r) & M_2(tP_r) & \ldots & M_r(tP_r) \end{pmatrix}$$

has rank r for every $t \neq 0$. In fact, from the equality $M_j(tP_i) = t^{m_j} M_j(P_i)$, we obtain that $\det(A(t)) = t^m \det(A(1))$ where $m = m_1 + \cdots + m_r$, and the claim follows because $\det(A(1)) \neq 0$.

The interpolating polynomial F can be computed by solving the linear system $A(t)c = \alpha$ where $c = (c_1, \ldots, c_r)^\top$ and $\alpha = (\alpha_1(t), \ldots, \alpha_r(t))^\top$. By using Cramer's rule, we obtain that the only solution is $c = A(t)^{-1}\alpha$ and so $c_i \in \mathcal{K}[t]_t$ because the entries of $A(t)^{-1}$ are in $\mathcal{K}[t]_t$. $\qquad\square$

The natural interpolating polynomial at the origin is the limit of F. That is to say, we would like to compute the limit $\lim_{t \to 0} c_i$ and obtain an element in $\mathcal{K}$. This is equivalent to requiring that $c_i \in \mathcal{K}[t]$. We need to recall a useful lemma from linear algebra.

Lemma 11.1 *Let $\underline{v}_1, \ldots, \underline{v}_r$ be linearly independent vectors in a $\mathcal{K}$-vector space V, and let V_j be the sub-vector space spanned by $\underline{v}_1, \ldots, \underline{v}_{j-1}, \underline{v}_{j+1}, \ldots, \underline{v}_r$, for $j = 1, \ldots, r$. Then,*

$$\bigcap_{h \geq i} V_h = \mathrm{Span}\langle \underline{v}_1, \ldots, \underline{v}_{i-1} \rangle.$$

Theorem 11.7 *In the hypotheses of Proposition 11.2, let*

$$\alpha = \alpha^0 + t\alpha^1 + \cdots + t^b \alpha^b$$

where $\alpha^h = (\alpha_{1h}, \ldots, \alpha_{rh})^\top$ for some $\alpha_{ij} \in \mathcal{K}$. Then, $c_1, \ldots, c_r \in \mathcal{K}[t]$ if, and only if, $\alpha^j \in \mathrm{Span}\langle A_i : m_i \leq j \rangle$, where A_i is the i-th column of $A(1)$.

Proof We can write the coefficients c_i, $i = 1, \ldots, r$, explicitly. Let $D_i(t)$ be the matrix obtained from $A(t)$ by substituting its i-th column with the column α. We have $\det(D_i(t)) = \sum_{h=1}^{b} t^{m+h-m_i} \det(D_{ih})$ where D_{ih} is the matrix we obtain from $A(1)$ by substituting its i-th column with α^h. Now, c_i is a polynomial if, and only if, $\det(D_{ih}) = 0$ for $h < m_i$, that is to say, $\alpha^h \in \mathrm{Span}\langle A_1, \ldots, A_{i-1}, A_{i+1}, \ldots, A_r \rangle$ for $h < m_i$. Using Lemma 11.1 we conclude the proof. $\qquad\square$

As $A_1 = (1, \ldots, 1)^\top$ and $m_2 \geq 1$, there exists $a \in \mathcal{K}$ such that $\alpha^0 = aA_1$, that is to say, $\alpha_i(0) = a$ for every $i = 1, \ldots, r$. Furthermore, $\alpha_{i0} + \cdots + \alpha_{ib} = \alpha_i$ for every $i = 1, \ldots, r$, and thus it depends on the values which are chosen initially for the interpolation problem. Hence, in general, we can choose $b = m_r$.

Definition 11.6 In the hypotheses of Theorem 11.7, let $F_0(x_1, \ldots, x_n)$ be the limit polynomial of $F(x_1, \ldots, x_n) = c_1 M_1 + \cdots + c_r M_r$ as $t \to 0$.

Theorem 11.8 *In the hypotheses and notation of Theorem 11.7, for every $i = 1, \ldots, r$ it holds*

$$c_i(0) = \frac{\det(D_{i,m_i})}{\det(A(1))}.$$

Note that it is possible to choose $\alpha^h \in \mathrm{Span}\langle A_i : m_i = h \rangle$ because of the way $c_i(0)$ is computed. In fact, the columns of $A(1)$ corresponding to monomials of degree strictly smaller than h cancel the contribution they give to α^h.

11.4.1 Interpolator over multiple points

Now, we analyse the case with various points each of which is obtained by the independent collapsing of a cloud of points. For the proof of Theorem 11.9 see the on-line supplement.

Theorem 11.9 *In the set-up of Theorem 11.6, let J_i be the $I(A_i)$-primary ideal of degree r_i obtained by collapsing X_i to A_i and let $J = J_1 \cap \cdots \cap J_n$. Let $F_i \in R/J_i$ be the limit interpolating polynomial computed in Theorem 11.7. Then there exists a unique polynomial $F \in R/J$ such that $F \bmod J_i = F_i$.*

Example 11.14 (Example 11.13 cont.) The values to be interpolated are given in Display (11.1)

$$\begin{array}{c|c|c|c|c} X_1 & X_2 & X_3 & X_4 & X_5 \\ 1,2,-1,1,0 & -2,-1 & 2,3 & -3,-1 & 1,0 \end{array} \tag{11.1}$$

By Theorem 11.8 we compute the limit interpolating polynomial, when collapsing X_1 to $(0,0)$. The monomial basis of R/J_1 is $\{1, x, y, x^2, y^2\}$ and the matrix $A(1)$ is

$$A(1) = \begin{pmatrix} 1 & 0 & 0 & 0 & 0 \\ 1 & 1 & 0 & 1 & 0 \\ 1 & 0 & 1 & 0 & 1 \\ 1 & -1 & 0 & 1 & 0 \\ 1 & 0 & -1 & 0 & 1 \end{pmatrix}.$$

By Theorem 11.7, we change the values according to the polynomials

$$\alpha(t) = d_1 \begin{pmatrix} 1 \\ 1 \\ 1 \\ 1 \\ 1 \end{pmatrix} + t \begin{pmatrix} 0 \\ b \\ c \\ -b \\ -c \end{pmatrix} + t^2 \begin{pmatrix} 1 - d_1 \\ 2 - b - d_1 \\ -1 - c - d_1 \\ 1 + b - d_1 \\ c - d_1 \end{pmatrix}$$

and the limit polynomial is

$$F_1 = d_1 + bx + cy + \frac{1}{2}x^2 - \frac{3}{2}y^2 \in \frac{R}{J_1}.$$

Analogously, to compute the limit polynomials $F_2, \ldots, F_5$, we change coordinate system, using the one in which the limit point is the origin. By Theorems 11.7

and 11.8, we obtain in the original coordinate system, $F_2 = d_2 - 1 + y, F_3 = d_3 - 1 + y, F_4 = d_4 + 2 + 2y, F_5 = d_5 - 1 - y$.

A monomial basis of R/J is $\{1, x, y, x^2, xy, y^2, x^3, x^2y, xy^2, y^3, x^2y^2, xy^3, y^4\}$, and thus the polynomial H, as described in the proof of Theorem 11.9, must be of the form

$$H = a_1 xy + a_2 x^3 + a_3 x^2 y + a_4 xy^2 + a_5 y^3 + a_6 x^2 y^2 + a_7 xy^3 + a_8 y^4.$$

By imposing that the normal form of $F_1 + H - F_i$ in R/J_i is zero, for $i = 2, \ldots, 5$, we obtain a linear system in the a_i's, whose only solution gives

$$H = \frac{2d_2 - 2d_3 + 2d_4 - 2d_5 + 3}{8} xy + \frac{d_2 - d_3 - d_4 + d_5 - 8b - 3}{16} x^3$$
$$+ \frac{3d_2 + 3d_3 - 3d_4 - 3d_5 - 8c - 3}{16} x^2 y + \frac{3d_2 - 3d_3 - 3d_4 + 3d_5 - 8b + 3}{16} xy^2$$
$$+ \frac{d_2 + d_3 - d_4 - d_5 - 8c + 3}{16} y^3 + \frac{-16d_1 + 4d_2 + 4d_3 + 4d_4 + 4d_5 - 1}{16} x^2 y^2$$
$$- \frac{3}{8} xy^3 + \frac{17}{16} y^4$$

and so the interpolating polynomial we are looking for is $F_1 + H$, where in practice the d_i are the mean of the observed values over X_i, $i = 1, \ldots, 5$.

11.5 Projection to the support

To conclude, we consider the set $Y = \{A_1, \ldots, A_n\}$ and compare the rings R/J and $R/I(Y)$, where J is the ideal that describes the union of the multiple points over $A_1, \ldots, A_n$. In few words, we will show that projecting the interpolating polynomial obtained in Section 11.4 and computing the interpolating polynomial over $A_1, \ldots, A_n$ directly yield the same set of identifiable monomials. This supports standard practice.

Proposition 11.3 *The inclusion $J \subset I(Y)$ induces a surjective map*

$$\psi : \frac{R}{J} \to \frac{R}{I(Y)}$$

defined as $\psi(G) = G \mod I(Y)$.

Proof The ideal J has the following primary decomposition: $J = J_1 \cap \cdots \cap J_n$ where J_i is $I(A_i)$–primary. Hence, $J_i \subset I(A_i)$ and so $J \subset I(A_1) \cap \cdots \cap I(A_n) = I(Y)$. The second part of the statement is then easy to check. $\qquad\square$

Theorem 11.10 *Let $F_i \in R/J_i$ be the limit interpolating polynomial for $i = 1, \ldots, n$, and let $F \in R/J$ be the limit polynomial interpolating the values d_i over A_i for $i = 1, \ldots, n$. Let $F_i(A_i) \in \mathcal{K}$ and let $G \in R/I(Y)$ be the interpolating polynomial such that $G(A_i) = F_i(A_i)$, for $i = 1, \ldots, m$. Then, $\psi(F) = G$.*

The interpolated values d_i will be some average of the α_{ij} observed at the replicated point A_i.

Proof Now, R/J (resp. $R/I(Y)$) is isomorphic to $R/J_1 \oplus \cdots \oplus R/J_n$ (resp. $R/I(A_1) \oplus \cdots \oplus R/I(A_n)$). The map ψ acts on $(F_1, \ldots, F_n) \in R/J_1 \oplus \cdots \oplus R/J_n$ as $\psi(F_1, \ldots, F_n) = (F_1(A_1), \ldots F_n(A_n))$ and so the claim follows. $\square$

Example 11.15 (Example 11.14 cont.) The set of limit points is $Y = \{(0,0), (1,1), (-1,1), (-1,-1), (1,-1)\}$ and its ideal is $I(Y) = \langle x^2 - y^2, xy^2 - x, y^3 - y \rangle$. The normal form of $F_1 + H$ modulo $I(Y)$ is

$$G = d_1 + \frac{d_2 - d_3 - d_4 + d_5}{4}x + \frac{d_2 + d_3 - d_4 - d_5}{4}y$$
$$+ \frac{d_2 - d_3 + d_4 - d_5}{4}xy + \frac{-4d_1 + d_2 + d_3 + d_4 + d_5}{4}y^2.$$

An easy calculation confirms the statement of Theorem 11.10. In fact $G(0,0) = d_1, G(1,1) = d_2, G(-1,1) = d_3, G(-1,-1) = d_4, G(1,-1) = d_5$, and so G interpolates the values $d_1, \ldots; d_5$ over Y.

11.6 Further comments

There are a series of hypotheses underpinning this work which could be relaxed for more generality. Foremost is the fact that the points in a cloud are moved towards the common point along straight lines. In a first approximation, we can assume that this occurs. Rather than fully developing a theory, this chapter aims to provide a novel framework for thinking about design and modelling issues in the presence of replications. In particular, it wants to outline some geometric aspects which so far have been obscured in favour of computational algebraic arguments.

This research project is still at its onset and there are many issues that have to be thought through. In particular a satisfactory description of the aliasing structure of a design with replicated points is missing, unless one only considers aliasing and confounding on the un-replicated design. Next, to derive a sub-model from the saturated model/interpolating polynomial one could use standard least squares techniques or techniques to record information about the derivatives. More relevant for a statistical analysis, could be to devise ways to partition the X_t matrices and use a part in the estimation of the regression parameters, θ, and a part in the estimation of the variance parameter, σ^2, driving the regression model under the standard Gauss–Markov distributional assumptions. It might be that in developing our research we shall have to make complete use of the theory of Hilbert schemes and Macauley bases. Here we have preferred arguments of linear algebra.

Finally, we hinted at a connection with derivation. Theorem 11.3 shows that the ideal of a multiple point obtained in the way we consider is homogeneous. A zero-dimensional ring R/J with J homogeneous has always a description via derivatives. The construction is known but we have not investigated if there is a relation between the two representations of the ideal J. When we consider finitely many multiple points in Theorem 11.6, we can obtain, at least in principle, a differential description of the total ideal by considering all the differential descriptions at the various points, each one being local.

Example 11.16 (Example 11.13 cont.) Consider X_1 collapsing on A_1. The ideal of the multiple point is $J = \langle xy, x^3, y^3 \rangle$ and it is a homogeneous ideal. As a homogeneous ideal, the Hilbert function of R/J is $H(0) = 1, H(1) = 2, H(2) = 2, H(j) = 0$ for $j \neq 0, 1, 2$, and it is equal to the first difference of the Hilbert function as a non-homogeneous ideal. Let $p = \frac{\partial}{\partial x}$ and $q = \frac{\partial}{\partial y}$. We want to find homogeneous polynomials in p, q to be interpreted as differential equations that, evaluated at the origin $(0, 0)$, are satisfied by all and only the polynomials in J. We have to find as many differential equations of degree j as the value of $H(j)$, for every $j \in \mathbb{Z}$. In degree 0, we have only one relation that is $f(0, 0) = 0$, and this is always the case. In degree 1 we have to find two independent equations: of course, they are $p(f)(0, 0) = q(f)(0, 0) = 0$, i.e. $f_x(0, 0) = f_y(0, 0) = 0$. In degree 2 we need two more independent equations. A general differential equation of second order has the form $ap^2 + bpq + cq^2$ for some $a, b, c \in \mathcal{K}$. We want xy to satisfy it, and so $0 = (ap^2 + bpq + cq^2)(xy) = ap^2(xy) + bpq(xy) + cq^2(xy)$. But $p^2(xy) = \frac{\partial^2}{\partial x^2}(xy) = 0$, and analogously the other terms, and so $b = 0$. Hence, the two equations we are looking for are $p^2(f)(0, 0) = q^2(f)(0, 0) = 0$, i.e. $f_{xx}(0, 0) = f_{yy}(0, 0) = 0$. Finally, we see that J contains all the polynomials that verify the following equations:

$$f(0, 0) = f_x(0, 0) = f_y(0, 0) = f_{xx}(0, 0) = f_{yy}(0, 0) = 0$$

which is the description of J via derivatives.

References

Abbott, J., Kreuzer, J. M. and Robbiano, L. (2005). Computing zero-dimensional schemes. *Journal of Symbolic Computation* **39**(1), 31–49.

CoCoATeam (2007). *CoCoA, a system for doing Computations in Commutative Algebra*, 4.7 edn (available at http://cocoa.dima.unige.it).

Cohen, A. M. and Di Bucchianico, A. and Riccomagno, E. (2001). Replications with Gröbner bases. In *m*ODa 6 Atkinson, A.C., Hackl, P. and Müller, W.G. eds. (Puchberg/Schneeberg) 37–44.

Cox, D., Little, J. and O'Shea, D. (2007). *Ideals, Varieties, and Algorithms* 3rd edn (New York, Springer-Verlag).

Cox, D., Little, J. and O'Shea, D. (2008). *Using Algebraic Geometry* 2nd edn (New York, Springer-Verlag).

Hartshorne, R. (1977). *Algebraic Geometry*, GTM 52 (New York, Springer-Verlag).

Kreuzer, M. and Robbiano, L. (2000). *Computational Commutative Algebra. 1* (Berlin, Springer-Verlag).

Kreuzer, M. and Robbiano, L. (2005). *Computational Commutative Algebra. 2* (Berlin, Springer-Verlag).

Pistone, G., Riccomagno, E. and Wynn, H. P. (2001). *Algebraic Statistics* (Boca Raton, Chapman & Hall/CRC).

Pistone, G. and Rogantin, M. P. (2008). Indicator function and complex coding for mixed fractional factorial designs, *Journal of Statistical Planning and Inference* **138**, 787–802.

Pistone, G. and Wynn, H. P. (1996). Generalised confounding with Gröbner bases, *Biometrika* **83**(3), 653–66.

Riccomagno, E. (2008). A short history of algebraic statistics, *Metrika* **69**, 397–418.

12

Indicator function and sudoku designs

Roberto Fontana

Maria Piera Rogantin

Abstract

In this chapter algebraic statistics methods are used for design of experiments generation. In particular the class of Gerechte designs, that includes the game of sudoku, has been studied.

The first part provides a review of the algebraic theory of indicator functions of fractional factorial designs. Then, a system of polynomial equations whose solutions are the coefficients of the indicator functions of all the sudoku fractions is given for the general $p^2 \times p^2$ case (p integer). The subclass of symmetric sudoku is also studied. The 4×4 case has been solved using CoCoA. In the second part the concept of move between sudoku has been investigated. The polynomial form of some types of moves between sudoku grids has been constructed.

Finally, the key points of a future research on the link between sudoku, contingency tables and Markov basis are summarised.

12.1 Introduction

Sudoku is currently a very popular game. Every day many newspapers all over the world propose such puzzles to their readers. From wikipedia we read:

Sudoku is a logic-based number placement puzzle. The objective is to fill a 9×9 grid so that each column, each row, and each of the nine 3×3 boxes (also called blocks or regions) contains the digits from 1 to 9, only one time each (that is, exclusively). The puzzle setter provides a partially completed grid. (`http://en.wikipedia.org/wiki/Sudoku`)

This description refers to the standard game but also 4×4, 6×6, 12×12 and 16×16 grids are played.

Sudoku can be considered as a special design of experiment and in particular a special Latin square in the class of *gerechte designs*, introduced in 1956 by W.U. Behrens. A recent paper (Bailey *et al.* 2008) gives an overview of relations among sudoku and gerechte designs, and provides computational techniques for finding and classifying them, using tools from group theory.

Algebraic and Geometric Methods in Statistics, ed. Paolo Gibilisco, Eva Riccomagno, Maria Piera Rogantin and Henry P. Wynn. Published by Cambridge University Press. © Cambridge University Press 2010.

The aim of this chapter is twofold: mainly, to use the sudoku game to illustrate the power of the indicator function method for experimental designs, and then to make a link to the Diaconis–Sturmfels algorithm for contingency tables, thus connecting design of experiment and contingency table analysis. On the link between contingency tables and designs see also Chapter 13 in this volume and (Aoki and Takemura 2006).

In Section 12.2 we review the algebraic theory of indicator function; for simplicity we consider single replicate fractions. The rules of the game are translated into conditions on the coefficients of the indicator function of a sudoku in Section 12.3 and we characterise all the possible $p^2 \times p^2$ sudoku as solutions of a system of polynomial equations. In Section 12.4 we analyse the moves between different sudoku. Examples for the 4×4 and 9×9 cases are given throughout. An on-line supplement provides some proofs, all the 4×4 sudoku grids and the algorithms used for their generation, implemented in CoCoA (CoCoATeam 2007). We conclude this introduction with a review of the literature on the indicator function for experimental designs.

The polynomial indicator function for two-level fractional factorial designs was introduced in (Fontana *et al.* 1997) and (Fontana *et al.* 2000). Independently, (Tang and Deng 1999) introduced quantities related to coefficients of the indicator function, called J-characteristics in (Tang 2001). Generalisation to two-level designs with replications is due to (Ye 2003) and extension to three-level factors, using orthogonal polynomials with an integer coding of levels, is in (Cheng and Ye 2004). In (Pistone and Rogantin 2008) a full generalisation to mixed (or asymmetrical) designs with replicates was given, coding the levels with the m-th roots of unity.

With this complex coding, the coefficients of the indicator function are related to many properties of the fraction in a simple way. In particular orthogonality among the factors and interactions, projectivity, aberration and regularity can be deduced from the values of the coefficients of the indicator function.

Further results for two-level designs with replicates are in (Li *et al.* 2003, Balakrishnan and Yang 2006b) and (Balakrishnan and Yang 2006a), where some general properties of foldover designs are obtained form the pattern of the terms of the indicator function. (Kotsireas *et al.* 2004) give an algorithm to check the equivalence between Hadamard matrices.

Elsewhere in this volume a design of experiment is represented using Gröbner bases. The two representations show different characteristics of a design and they are compared in (Notari *et al.* 2007), where algorithms to switch between them are provided, see also (Pistone *et al.* 2009).

12.2 Notation and background

12.2.1 Full factorial design

We adopt and summarise below the notation in (Pistone and Rogantin 2008). If not otherwise stated the proof of the reported results can be found in (Pistone and Rogantin 2007) and (Pistone and Rogantin 2008).

– $\mathcal{D}_j$: *factor* with m_j levels coded with the m_j-th roots of unity:

$$\mathcal{D}_j = \{\omega_0,\ldots,\omega_{m_j-1}\} \qquad \omega_h = \exp\left(i\frac{2\pi}{m_j}\,h\right)\ h = 0,\ldots,m_j-1;$$

– $\mathcal{D}$: *full factorial design* in *complex coding*, $\mathcal{D} = \mathcal{D}_1 \times \cdots \mathcal{D}_j \cdots \times \mathcal{D}_k$;
– $|\mathcal{D}|$: cardinality of $\mathcal{D}$;
– L: *full factorial design* in *integer coding*, $L = \mathbb{Z}_{m_1} \times \cdots \times \mathbb{Z}_{m_j} \cdots \times \mathbb{Z}_{m_k}$;
– α: element of L, $\alpha = (\alpha_1,\ldots,\alpha_k)$, $\alpha_j = 0,\ldots,m_j-1, j = 1,\ldots,k$;
– $[\alpha - \beta]$: component-wise difference the k-tuple

$$\left([\alpha_1 - \beta_1]_{m_1},\ldots,[\alpha_j - \beta_j]_{m_j},\ldots,[\alpha_k - \beta_k]_{m_k}\right),$$

where the computation of the j-th element is in the ring $\mathbb{Z}_{m_j}$;
– X_j: j-th component function, which maps a point to its i-th component: X_j : $\mathcal{D} \ni (\zeta_1,\ldots,\zeta_k) \longmapsto \zeta_j \in \mathcal{D}_j$; the function X_j is called a *simple term* or, by abuse of terminology, a *factor*;
– X^α: *interaction term* $X_1^{\alpha_1} \cdots X_k^{\alpha_k}$, i.e. the function

$$X^\alpha : \quad \mathcal{D} \ni (\zeta_1,\ldots,\zeta_k) \mapsto \zeta_1^{\alpha_1} \cdots \zeta_k^{\alpha_k}.$$

Notice that L is both the full factorial design with integer coding and *the exponent set of all the simple factors and interaction terms* and α is both a treatment combination in the integer coding and a multi-exponent of an interaction term. The full factorial design in complex coding is identified as the zero-set in $\mathbb{C}^k$ of the system of polynomial equations

$$X_j^{m_j} - 1 = 0 \quad \text{for} \quad j = 1,\ldots,k. \tag{12.1}$$

Definition 12.1

(i) A *response* f on a design $\mathcal{D}$ is a $\mathbb{C}$-valued polynomial function defined on $\mathcal{D}$.

(ii) The *mean value* on $\mathcal{D}$ of a response f, denoted by $\mathrm{E}_{\mathcal{D}}[f]$, is:

$$\mathrm{E}_{\mathcal{D}}[f] = \frac{1}{|\mathcal{D}|}\sum_{\zeta \in \mathcal{D}} f(\zeta).$$

(iii) A response f is *centred* on $\mathcal{D}$ if $\mathrm{E}_{\mathcal{D}}[f] = 0$. Two responses f and g are *orthogonal* on $\mathcal{D}$ if $\mathrm{E}_{\mathcal{D}}[f\,\bar{g}] = 0$, where $\bar{g}$ is the complex conjugate of g.

Notice that the set of all the responses is a complex Hilbert space with the Hermitian product: $f \cdot g = \mathrm{E}_{\mathcal{D}}[f\,\bar{g}]$. Moreover, (i) $X^\alpha \overline{X^\beta} = X^{[\alpha-\beta]}$; (ii) $\mathrm{E}_{\mathcal{D}}[X^0] = 1$, and $\mathrm{E}_{\mathcal{D}}[X^\alpha] = 0$.

The set of functions $\{X^\alpha,\ \alpha \in L\}$ is an orthonormal basis of the responses on $\mathcal{D}$. In fact $|L| = |\mathcal{D}|$ and, from (i) and (ii) above, we have

$$\mathrm{E}_{\mathcal{D}}[X^\alpha \overline{X^\beta}] = \mathrm{E}_{\mathcal{D}}[X^{[\alpha-\beta]}] = \begin{cases} 1 & \text{if } \alpha = \beta \\ 0 & \text{if } \alpha \neq \beta. \end{cases}$$

Each response f can be written as a unique $\mathbb{C}$-linear combination of constant, simple and interaction terms, by repeated applications of the re-writing rules derived from Equations (12.1). Such a polynomial is called the *normal form* of f on $\mathcal{D}$. In this chapter we intend that all the computation are performed, and all results presented, in normal form.

Example 12.1 If $\mathcal{D}$ is the 2^3 full factorial design, then the monomial responses are $1, X_1, X_2, X_3, X_1 X_2, X_1 X_3, X_2 X_3, X_1 X_2 X_3$ and L is

$$\{(0,0,0),(1,0,0),(0,1,0),(0,0,1),(1,1,0),(1,0,1),(0,1,1),(1,1,1)\}.$$

12.2.2 Fractions of a full factorial design

A fraction $\mathcal{F}$ is a subset of the design, $\mathcal{F} \subseteq \mathcal{D}$, and can be obtained as the solution set of a system of polynomial equations formed by Equations (12.1) and other equations, called *generating equations*. Definition 12.1 specialises to $\mathcal{F} \subseteq \mathcal{D}$. Note that with the complex coding the vector orthogonality of X^α and X^β is equivalent to their combinatorial orthogonality, namely all the level combinations appear equally often in $X^\alpha X^\beta$.

Definition 12.2 The *indicator function* F of a fraction $\mathcal{F}$ is a response defined on $\mathcal{D}$ such that $F(\zeta) = 1$ if $\zeta \in \mathcal{F}$ and $F(\zeta) = 0$ if $\zeta \in \mathcal{D} \setminus \mathcal{F}$.

Denote by b_α the coefficients of the representation of F on $\mathcal{D}$ using the monomial basis $\{X^\alpha,\ \alpha \in L\}$:

$$F(\zeta) = \sum_{\alpha \in L} b_\alpha X^\alpha(\zeta) \qquad \zeta \in \mathcal{D} \quad b_\alpha \in \mathbb{C}.$$

The equation $F - 1 = 0$ is a generating equation of the fraction $\mathcal{F}$. As the indicator function is real valued, we have $\overline{b_\alpha} = b_{[-\alpha]}$.

Proposition 12.1 *The following facts hold*

   (i) $b_\alpha = \frac{1}{|\mathcal{D}|} \sum_{\zeta \in \mathcal{F}} \overline{X^\alpha(\zeta)}$; *in particular*, b_0 *is the ratio between the number of points of the fraction and that of the design;*

   (ii) $b_\alpha = \sum_{\beta \in L} b_\beta\, b_{[\alpha - \beta]}$;

   (iii) X^α *is centred on* $\mathcal{F}$, *i.e.* $\mathrm{E}_\mathcal{F}[X^\alpha]$, *if, and only if,* $b_\alpha = b_{[-\alpha]} = 0$

   (iv) X^α *and* X^β *are orthogonal on* $\mathcal{F}$, *i.e.* $\mathrm{E}_\mathcal{F}[X^\alpha\, \overline{X^\beta}]$, *if, and only if,* $b_{[\alpha - \beta]} = 0$.

Example 12.2 Consider the fraction $\mathcal{F} = \{(-1,-1,1),(-1,1,-1)\}$ of the design in Example 12.1. All monomial responses on $\mathcal{F}$ and their values on the points are

ζ	1	X_1	X_2	X_3	$X_1 X_2$	$X_1 X_3$	$X_2 X_3$	$X_1 X_2 X_3$
$(-1,-1,1)$	1	-1	-1	1	1	-1	-1	1
$(-1,1,-1)$	1	-1	1	-1	-1	1	-1	1

By Item (i) of Proposition 12.1, compute $b_{(0,1,0)} = b_{(0,0,1)} = b_{(1,1,0)} = b_{(1,0,1)} = 0$, $b_{(0,0,0)} = b_{(1,1,1)} = 2/4$ and $b_{(1,0,0)} = b_{(0,1,1)} = -2/4$. Hence, the indicator function is

$$F = \frac{1}{2}\left(1 - X_1 - X_2 X_3 + X_1 X_2 X_3\right).$$

As $b_{(0,1,0)} = 0 = b_{(0,0,1)}$, then X_1 and X_3 are centred; as $b_{(1,1,0)} = 0 = b_{(1,0,1)}$, then X_1 is orthogonal to both X_2 and X_3.

12.2.3 Projectivity and orthogonal arrays

Definition 12.3 A fraction $\mathcal{F}$ *factorially projects* onto the I-factors, $I \subset \{1,\dots,k\}$, if the projection is a full factorial design where each point appears equally often. A fraction $\mathcal{F}$ is a *mixed orthogonal array* of strength t if it factorially projects onto any I-factors with $|I| = t$.

Strength t means that, for any choice of t columns of the matrix design, all possible combinations of symbols appear equally often.

Proposition 12.2 (Projectivity)

(i) *A fraction factorially projects onto the I-factors if, and only if, the coefficients of the indicator function involving only the I-factors are zero.*

(ii) *If there exists a subset J of $\{1,\dots,k\}$ such that the J-factors appear in all the non null elements of the indicator function, the fraction factorially projects onto the I-factors, where I is the complementary set of J, $I = J^c$.*

(iii) *A fraction is an orthogonal array of strength t if, and only if, all the coefficients of the indicator function up to order t are zero.*

Example 12.3 (Orthogonal array) The fraction of a 2^6 full factorial design

$$\mathcal{F}_O = \{(-1,-1,-1,-1,-1,1),(-1,-1,-1,1,1,1),(-1,-1,1,-1,-1,-1),(-1,-1,1,1,1,-1),$$
$$(-1,1,-1,-1,-1,-1),(-1,1,-1,1,1,-1),(-1,1,1,-1,1,1),(-1,1,1,1,-1,1),$$
$$(1,-1,-1,-1,1,1),(1,-1,-1,1,-1,1),(1,-1,1,-1,1,-1),(1,-1,1,1,-1,-1),$$
$$(1,1,-1,-1,1,-1),(1,1,-1,1,-1,-1),(1,1,1,-1,-1,1),(1,1,1,1,1,1)\}$$

is an orthogonal array of strength 2; in fact, its indicator function

$$F = \frac{1}{4} + \frac{1}{4}X_2 X_3 X_6 - \frac{1}{8}X_1 X_4 X_5 + \frac{1}{8}X_1 X_4 X_5 X_6 + \frac{1}{8}X_1 X_3 X_4 X_5 + \frac{1}{8}X_1 X_2 X_4 X_5$$
$$+ \frac{1}{8}X_1 X_3 X_4 X_5 X_6 + \frac{1}{8}X_1 X_2 X_4 X_5 X_6 + \frac{1}{8}X_1 X_2 X_3 X_4 X_5 - \frac{1}{8}X_1 X_2 X_3 X_4 X_5 X_6$$

contains only terms of order greater than 2 and the constant term.

12.2.4 Regular fractions

Let m be the least common multiple of $\{m_1,\dots,m_k\} \subset \mathbb{Z}_{>0}$ and $\mathcal{D}_m$ the set of the m-th roots of unity. Let $\mathcal{L}$ be a subset of L, containing $(0,\dots,0)$ and let $l = |\mathcal{L}|$. Let e be a map from $\mathcal{L}$ to $\mathcal{D}_m$, $e : \mathcal{L} \to \mathcal{D}_m$.

Definition 12.4 A fraction $\mathcal{F}$ is *regular* if

(i) $\mathcal{L}$ is a sub-group of L,

(ii) e is a group homomorphism, $e([\alpha + \beta]) = e(\alpha)\, e(\beta)$ for each $\alpha, \beta \in \mathcal{L}$,

(iii) the equations $X^\alpha = e(\alpha)$, with $\alpha \in \mathcal{L}$ are a set of generating equations.

In the literature the terms X^α appearing in Item (iii) are called defining words; so we call $X^\alpha = e(\alpha)$, $\alpha \in \mathcal{L}$, *defining* equations of $\mathcal{F}$. If $\mathcal{H}$ is a minimal generator of the group $\mathcal{L}$, then equations $X^\alpha = e(\alpha)$, $\alpha \in \mathcal{H} \subset \mathcal{L}$, are called a minimal set of *generating* equations.

Proposition 12.3 compares different definitions of regular fractions.

Proposition 12.3 (Regularity) *The following statements are equivalent*

(i) *$\mathcal{F}$ is regular according to Definition 12.4.*

(ii) *The indicator function of the fraction has the form*

$$F(\zeta) = \frac{1}{l} \sum_{\alpha \in \mathcal{L}} \overline{e(\alpha)}\, X^\alpha(\zeta) \qquad \zeta \in \mathcal{D}$$

where $\mathcal{L}$ is a given subset of L and $e : \mathcal{L} \to \mathcal{D}_m$ is a given mapping.

(iii) *For each $\alpha, \beta \in L$, the parametric functions represented on $\mathcal{F}$ by the terms X^α and X^β are either orthogonal or totally confounded.*

(iv) *$\mathcal{F}$ is either a subgroup or a lateral of a subgroup of the multiplicative group $\mathcal{D}$.*

Example 12.4 (Regular fraction) The fraction of a 3^4 full factorial design

$$\begin{aligned}
\mathcal{F}_R = \{&(1,1,1,1), (1,\omega_1,\omega_1,\omega_1), (1,\omega_2,\omega_2,\omega_2), (\omega_1,1,\omega_1,\omega_2), (\omega_1,\omega_1,\omega_2,1),\\
&(\omega_1,\omega_2,1,\omega_1), (\omega_2,1,\omega_2,\omega_1), (\omega_2,\omega_1,1,\omega_2), (\omega_2,\omega_2,\omega_1,1)\}
\end{aligned}$$

is regular; in fact, its indicator function is

$$\begin{aligned}
F = \frac{1}{9} \Big(&1 + X_2 X_3 X_4 + X_2^2 X_3^2 X_4^2 + X_1 X_2 X_3^2 + X_1^2 X_2^2 X_3 \\
&+ X_1 X_2^2 X_4 + X_1^2 X_2 X_4^2 + X_1 X_3 X_4^2 + X_1^2 X_3^2 X_4 \Big).
\end{aligned}$$

Furthermore, $\mathcal{H} = \{(1,1,2,0), (1,2,0,1)\}$, $e(1,1,2,0) = e(1,2,0,1) = \omega_0 = 1$, and $\mathcal{L}$ is $\{(0,0,0,0), (0,1,1,1), (0,2,2,2), (1,1,2,0), (2,2,1,0), (1,2,0,1), (2,1,0,2), (1,0,1,2), (2,0,2,1)\}$. From the values of the coefficients of F, we deduce that the fraction has nine points, because $b_{0,0,0,0} = |\mathcal{F}|/3^4$; each factor is orthogonal to the constant term, as the coefficients of the terms of order 1 are 0; any two factors are mutually orthogonal, as the coefficients of the terms of order 2 are 0. The interaction terms appearing in the indicator function are the defining words.

The indicator function of a p-level regular fraction can be written using a set of generating equations. This generalises the two-level case in (Fontana and Pistone 2008).

Corollary 12.1 *The indicator function of a p^{k-r} regular fraction with generating equations $X^{\alpha_1} = e(\alpha_1)$, $\ldots$, $X^{\alpha_r} = e(\alpha_r)$, with $\alpha_1, \ldots, \alpha_r \in \mathcal{H} \subset \mathcal{L}$, and $e(\alpha_i) \in \mathcal{D}_p$, can be written as*

$$F(\zeta) = \frac{1}{p^r} \prod_{j=1}^{r} \left(\sum_{i=0}^{p-1} \left(\overline{e(\alpha_j)} X^{\alpha_j}(\zeta) \right)^i \right) \qquad \zeta \in \mathcal{D}.$$

Proof The indicator function of a fraction $\mathcal{F}_j$ defined by a single equation is $F_j(\zeta) = \frac{1}{p} \sum_{i=0}^{p-1} \left(\overline{e(\alpha_j)} X^{\alpha_j}(\zeta) \right)^i$ and the indicator function of $\mathcal{F} = \cap_{j=1}^r \mathcal{F}_j$ is $F = \prod_{j=1}^r F_j$. $\square$

Proposition 12.4 (Regularity under permutation of levels) *A regular fraction is mapped into another regular fraction by the group of transformations generated by the following level permutations:*

(i) *Cyclical permutations on the factor X_j:*

$$(\zeta_1, \ldots, \zeta_j, \ldots, \zeta_k) \mapsto (\zeta_1, \ldots, \omega_h \zeta_j, \ldots, \zeta_k) \qquad h = 0, \ldots, m_j - 1.$$

(ii) *If m_j is a prime number, permutations on the factor X_j:*

$$(\zeta_1, \ldots, \zeta_j, \ldots, \zeta_k) \mapsto (\zeta_1, \ldots, \omega_h \zeta_j^r, \ldots, \zeta_m)$$

with $h = 0, \ldots, m_j - 1$ and $r = 1, \ldots, m_j - 1$.

Permutations of type (i) *and* (ii) *on all the factors produce, on the transformed fraction, the monomials:*

$$\left(\prod_{j=1}^{k} \omega_{h_j}^{\alpha_j} \right) X^{\alpha} \qquad and \qquad \prod_{j=1}^{k} \omega_{h_j}^{\alpha_j} X_j^{[\alpha_j r_j]}$$

respectively.

Note that all the m-level cyclical permutations are obtained as in (i) and that a sub-group of permutation of order $m_j(m_j - 1)$ is obtained as in (ii). In particular, if $m = 2$ or $m = 3$ all the level permutations are of type 2.

Example 12.5 (Permutation of levels – Example 12.4 cont.) The transformation $(\zeta_1, \zeta_2, \zeta_3, \zeta_4) \mapsto (\zeta_1, \zeta_2, \zeta_3, \omega_1 \zeta_4^2)$ permutes the levels ω_0 and ω_1 of the last factor X_4. The indicator function of the transformed, regular, fraction is:

$$\begin{aligned}
F = {} & \frac{1}{9} \big(1 + \omega_1 X_2 X_3 X_4^2 + \omega_2 X_2^2 X_3^2 X_4 + X_1 X_2 X_3^2 + X_1^2 X_2^2 X_3 \\
& + \omega_1 X_1 X_2^2 X_4^2 + \omega_2 X_1^2 X_2 X_4 + \omega_2 X_1 X_3 X_4 + \omega_1 X_1^2 X_3^2 X_4^2 \big).
\end{aligned}$$

The generating equations of the starting fraction are transformed into the generating equations of the transformed fraction as $X_1 X_2 X_3^2 = 1$ and $X_1 X_2^2 X_4^2 = \omega_2$.

12.3 Sudoku fraction and indicator functions

We consider $p^2 \times p^2$ square sudoku, with $p \in \mathbb{Z}_{\geq 2}$. A sudoku is a particular subset of cardinality $p^2 \times p^2$ of the $p^2 \times p^2 \times p^2$ possible assignments of a digit between 1 and p^2 to the cells of a $p^2 \times p^2$ grid.

We consider a sudoku as a fraction $\mathcal{F}$ of a factorial design $\mathcal{D}$ with four factors R, C, B, S, corresponding to rows, columns, boxes and symbols, with p^2 levels each. The three 'position' factors are dependent; in fact a row and a column identify a box, but the polynomial relation between B and R, C is fairly complicated.

As well known, when the number of factor levels is not a prime, a factor can be split into pseudo-factors. This is not necessary for applying the theory in Section 12.2. But for sudoku designs it has the advantage of specifying the box factor in a simple way. If the row factor R levels splits into R_1 and R_2 pseudo-factors with p levels each, and analogously the column factor C splits into C_1 and C_2, then the box factor B corresponds to R_1 and C_1. Pseudo-factors for symbols are introduced for symmetry of representation. Hence,

$$\mathcal{D} = R_1 \times R_2 \times C_1 \times C_2 \times S_1 \times S_2$$

where each factor is coded with the p-th roots of unity. The factor R_1 identifies the 'band' and C_1 the 'stack'; R_2 and C_2 identify rows within a band and columns within a stack respectively, see (Bailey *et al.* 2008).

A row r of the sudoku grid is coded by the levels of the pseudo-factors R_1 and R_2 $(\omega_{r_1}, \omega_{r_2})$ with $r_i \in \mathbb{Z}_p$ and $r - 1 = p\, r_1 + r_2$. Similarly, for columns and symbols. For example, the symbol 5 in the first row, first column, and first box for $p = 3$ corresponds to the point $(\omega_0, \omega_0, \omega_0, \omega_0, \omega_1, \omega_1)$. See also Example 24.1 in the on-line supplement.

The game rules translate into:

(i) the fraction has p^4 points: the number of the cells of the grid;

(ii)   (a) all the cells appears exactly once: $R_1 \times R_2 \times C_1 \times C_2$ is a full factorial design;

      (b) each symbol appears exactly once in each row: $R_1 \times R_2 \times S_1 \times S_2$ is a full factorial design,

      (c) each symbol appears exactly once in each column: $C_1 \times C_2 \times S_1 \times S_2$ is a full factorial design,

      (d) each symbol appears exactly once in each box: $R_1 \times C_1 \times S_1 \times S_2$ is a full factorial design.

Proposition 12.5 re-writes the games rules into conditions on the coefficients of the indicator function F of $\mathcal{F}$. We shall indifferently use the equivalent notations

$$X^{\alpha} \quad \text{or} \quad R_1^{\alpha_1} R_2^{\alpha_2} C_1^{\alpha_3} C_2^{\alpha_4} S_1^{\alpha_5} S_2^{\alpha_6} \quad \text{or} \quad X_1^{\alpha_1} X_2^{\alpha_2} X_3^{\alpha_3} X_4^{\alpha_4} X_5^{\alpha_5} X_6^{\alpha_6}.$$

Proposition 12.5 (Sudoku fractions) *A fraction $\mathcal{F}$ corresponds to a sudoku grid if, and only if, the coefficients b_{α} of its indicator function satisfy the following conditions:*

(i) $b_{000000} = 1/p^2$;

	00	01	02	10	11	12	20	21	22
00	3	5	9	**2**	**4**	**8**	1	6	7
01	4	8	1	6	7	3	5	9	2
02	⑦	2	6	⑨	1	5	⑧	3	4
10	8	1	4	**7**	**3**	**6**	9	2	5
11	2	6	7	1	5	9	3	4	8
12	⑤	9	3	④	8	2	⑥	7	1
20	6	7	2	**5**	**9**	**1**	4	8	3
21	9	3	5	8	2	4	7	1	6
22	①	4	8	③	6	7	②	5	9

Fig. 12.1 A symmetric sudoku presented in (Bailey *et al.* 2008).

(ii) *for all* $i_j \in \{0, 1, \ldots, p-1\}$

    (a) $b_{i_1 i_2 i_3 i_4 00} = 0$ *for* $(i_1, i_2, i_3, i_4) \neq (0, 0, 0, 0)$,

    (b) $b_{i_1 i_2 00 i_5 i_6} = 0$ *for* $(i_1, i_2, i_5, i_6) \neq (0, 0, 0, 0)$,

    (c) $b_{00 i_3 i_4 i_5 i_6} = 0$ *for* $(i_3, i_4, i_5, i_6) \neq (0, 0, 0, 0)$,

    (d) $b_{i_1 0 i_3 0 i_5 i_6} = 0$ *for* $(i_1, i_3, i_5, i_6) \neq (0, 0, 0, 0)$.

Proof Items (i) and (ii) follow from Proposition 12.1(i) and 12.2(i), respectively.
$\qquad\square$

Definition 12.5 (Sudoku fraction) A fraction of a p^6 full factorial design is a *sudoku fraction* if its indicator function satisfies the conditions of Proposition 12.5.

From Proposition 12.5, two remarks follow. First, each interaction term of the indicator function of a sudoku contains at least one of the factors corresponding to rows, R_1 or R_2, one to columns, C_1 or C_2, and one corresponding to symbols, S_1 or S_2, but not only R_1 and C_1. Next, conditions (a)–(c) of Proposition 12.5 characterise Latin square designs and (a)–(d) gerechte designs.

12.3.1 Symmetric sudoku fraction

We consider a variant of sudoku, called symmetric sudoku and proposed in (Bailey *et al.* 2008). A *broken row* is the union of p rows occurring in the same position in each box of a stack. A *broken column* is the union of p columns occurring in the same position in each box of a band. A *location* is a set of p cells occurring in a fixed position in all the boxes (for example, the cells on last row and last column of each box). Broken rows correspond to factors R_2 and C_1, broken columns to factors R_1 and C_2, and locations to factors R_2 and C_2. Figure 12.1 reproduces a symmetric sudoku presented in (Bailey *et al.* 2008). The bold face numbers are a broken row and the circled numbers a location.

A symmetric sudoku fraction is a sudoku for which each symbol appears exactly once

(a) in each broken row: $R_2 \times C_1 \times S_1 \times S_2$ is a full factorial design,
(b) in each broken column: $R_1 \times C_2 \times S_1 \times S_2$ is a full factorial design,
(c) in each location: $R_2 \times C_2 \times S_1 \times S_2$ is a full factorial design.

Proposition 12.6 (Symmetric sudoku fractions) *A fraction $\mathcal{F}$ corresponds to a symmetric sudoku grid if, and only if, the coefficients b_α of its indicator function satisfy the conditions of Proposition 12.5 and, for all $i_j \in \{0, 1, \ldots, p-1\}$,*

(a) $b_{0i_2 i_3 0 i_5 i_6} = 0$ *for* $(i_2, i_3, i_5, i_6) \neq (0,0,0,0)$,
(b) $b_{i_1 0 0 i_4 i_5 i_6} = 0$ *for* $(i_1, i_4, i_5, i_6) \neq (0,0,0,0)$,
(c) $b_{0 i_2 0 i_4 i_5 i_6} = 0$ *for* $(i_2, i_4, i_5, i_6) \neq (0,0,0,0)$.

Definition 12.6 (Symmetric sudoku fraction) A sudoku fraction is a *symmetric sudoku fraction* if its indicator function satisfies the previous conditions (besides those of Proposition 12.5).

From Proposition 12.6 it follows that each interaction term of the indicator function of a symmetric sudoku contains at least three of the factors corresponding to rows and columns, R_1, R_2, C_1 and C_2, and one corresponding to symbols, S_1 or S_2, but not only R_1 and C_1.

Example 12.6 The indicator function of the symmetric sudoku of Figure 12.1, computed using Item (i) of Proposition 12.1, is $F = 1/81 + F_s + \overline{F_s}$ with

$$
\begin{aligned}
F_s = {} & \frac{1}{81} R_1 R_2^2 C_2^2 S_1 - \frac{1}{27} \left(R_1 C_1^2 C_2 S_1 S_2^2 + R_1 C_1^2 C_2 S_1^2 S_2^2 + R_1 R_2 C_1 S_2 \right. \\
& + R_1 R_2 C_1 S_1 S_2 + R_2 C_1^2 C_2^2 S_2^2 + R_2 C_1^2 C_2^2 S_1 S_2^2 \left. \right) + \frac{2}{27} \left(R_2 C_1^2 C_2^2 S_1^2 S_2^2 \right. \\
& + R_1 C_1^2 C_2 S_2^2 + R_1 R_2 C_1 S_1^2 S_2 \left. \right) + \frac{\omega_2}{27} \left(R_2 C_1 C_2^2 S_2 + R_1 C_1 C_2 S_2 \right. \\
& + R_1^2 R_2^2 C_1 S_2^1 + R_2 C_1 C_2^2 S_1 S_2 + R_1^2 R_2^2 C_1 S_1 S_2 + R_1 C_1 C_2 S_1 S_2 \\
& + R_2 C_1 C_2^2 S_1^2 S_2 + R_1 C_1 C_2 S_1^2 S_2 + R_1^2 R_2^2 C_1 S_1^2 S_2 \left. \right)
\end{aligned}
$$

and $\overline{F_s}$ is the conjugate polynomial of F_s. The conditions on the coefficients of Propositions 12.5 and 12.6 are satisfied.

12.3.2 Generating and solving sudoku

The previous algebraic framework allows us both to characterise all the possible $p^2 \times p^2$ sudoku and to solve a partially filled grid.

Proposition 12.7 *Let L_G and L_{SG} be the subsets of L whose multi-exponents α correspond to null b_α of Propositions 12.5 and 12.6, respectively. The solutions of the following system of polynomial equations*

$$
\begin{cases}
b_\alpha = \sum_{\beta \in L} b_\beta \, b_{[\alpha - \beta]} & \text{with } \alpha \in L \\
b_\alpha = 0 & \text{with } \alpha \in M \subset L
\end{cases}
$$

are the coefficients of the indicator functions of all sudoku fractions if $M = L_G$ and of all symmetric sudoku fractions if $M = L_{SG}$.

Proof The equations $b_\alpha = \sum_{\beta \in L} b_\beta\, b_{[\alpha - \beta]}$, with $\alpha \in L$, characterise the coefficients of an indicator function, by Item (ii) of Proposition 12.1. The equations $b_\alpha = 0$, with $\alpha \in L_G$ or $\alpha \in L_{SG}$, are the conditions for sudoku fractions and symmetric sudoku fractions, respectively, by Propositions 12.5 and 12.6. $\qquad\square$

In principle, Proposition 12.7 provides the possibility to generate all the sudoku of a given dimension. But, in practice, software able to deal with complex numbers and a high number of indeterminates is not available.

Using CoCoA all the 288 possible 4×4 sudoku have been found, see the on-line supplement. Among them, 96 sudoku correspond to regular fractions and the other 192 to non-regular fractions. There are no 4×4 symmetric sudoku. Removing one or two of the symmetry conditions (a)–(c) of Proposition 12.6 there are 24 sudoku in each case; all of them correspond to regular fractions. The indicator functions of non regular fractions have 10 terms: the constant $(1/4)$, one interaction with coefficient $1/4$, two with coefficients $-1/8$ and six with coefficients $1/8$.

Proposition 12.7 allows us also to know how many and which solutions has a partially filled puzzle. It is enough to add to the system of the coefficients the conditions $F(x_j) = 1$, where x_j are the points of $\mathcal{F}$ already known. For instance, among the 72 sudoku with the symbol 4 in position $(4, 4)$ of the sudoku grid, there are 18 sudoku grids with the symbol 3 in position $(1, 1)$ and, among them, there are 9 sudoku with the symbol 2 in position $(2, 3)$.

12.4 Moves between sudoku fractions

Most sudoku players, probably, know that applying one of the following moves to a sudoku grid generates another sudoku grid

(1) permutation of symbols, bands, rows within a band, stacks, columns within a stack;
(2) transposition between rows and columns;
(3) moves acting on special parts of the sudoku grid.

All these moves, being represented by functions over $\mathcal{D}$, can be written as polynomials. In this section we provide these polynomials, study the composition of moves and analyse their effects on a sudoku regular fraction. Let F be the indicator function of a sudoku fraction. Denote by $\mathcal{M}(F)$ the set of the polynomials corresponding to the previous moves, by $\mathcal{M}_1(F)$, $\mathcal{M}_2(F)$ and $\mathcal{M}_3(F)$ the polynomial moves described in Items (1), (2) and (3), respectively. The above states

$$\mathcal{M}(F) = \mathcal{M}_1(F) \cup \mathcal{M}_2(F) \cup \mathcal{M}_3(F).$$

Definition 12.7 The polynomial $M(F)$ is a *valid move* if the polynomial $F_1 = F + M(F)$ is the indicator function of a sudoku fraction $\mathcal{F}_1$.

Let b_α, b_α^1, m_α be the coefficients of F, F_1 and $M(F)$, respectively. Then

$$M(F) = \sum_\alpha m_\alpha X^\alpha = \sum_\alpha (b_\alpha^1 - b_\alpha) X^\alpha.$$

From Proposition 12.5 we derive the following conditions on the coefficients of the moves.

Corollary 12.2 *The coefficients of a polynomial move satisfy the following conditions:*

   (i) $m_{000000} = 0$ *and*
  (ii) *for all $i_j \in \{0, 1, \ldots, p-1\}$:*

       (a) $m_{i_1 i_2 i_3 i_4 00} = 0$ *for* $(i_1, i_2, i_3, i_4) \neq (0,0,0,0)$,
       (b) $m_{i_1 i_2 00 i_5 i_6} = 0$ *for* $(i_1, i_2, i_5, i_6) \neq (0,0,0,0)$,
       (c) $m_{00 i_3 i_4 i_5 i_6} = 0$ *for* $(i_3, i_4, i_5, i_6) \neq (0,0,0,0)$,
       (d) $m_{i_1 0 i_3 0 i_5 i_6} = 0$ *for* $(i_1, i_3, i_5, i_6) \neq (0,0,0,0)$.

Observe that $M(F)$ takes values $\{-1, 0, 1\}$ over $\mathcal{D}$ depending on which point should be removed, left or added. Moreover, it holds $M(F) = (1 - 2F)(F_1 - F)^2$.

12.4.1 Polynomial form of $\mathcal{M}_1$ and $\mathcal{M}_2$ moves

Any permutation can be decomposed into a finite number of exchanges, the so-called 2-cycles. For $\mathcal{M}_1$ moves, these exchanges involves specific factors: symbols S_1, S_2, bands R_1, rows within a band R_1, R_2, stacks C_1 or columns within a stack C_1, C_2. Denote by s the set of factor indices involved in one such exchange and by g its complementary set. For instance, if we consider the exchange of two symbols, $s = \{5, 6\}$ and $g = \{1, 2, 3, 4\}$. Denote by:

- $\mathcal{D}_g$ and $\mathcal{D}_s$ the corresponding split of the full factorial design: $\mathcal{D} = \mathcal{D}_g \times \mathcal{D}_s$;
- ζ_g a point of $\mathcal{D}_g$ and ζ_u and ζ_v the points of $\mathcal{D}_s$ to be exchanged; for an exchange involving two factors we have $\zeta_u = (\omega_{u_1}, \omega_{u_2})$ and $\zeta_v = (\omega_{v_1}, \omega_{v_2})$, while for an exchange involving a single factor we have $\zeta_u = \omega_u$ and $\zeta_v = \omega_v$;
- L_g and L_s the split of the set of the exponents: $L = L_g \times L_s$;
- α_g and α_s the elements of L_g and L_s;
- X^{α_g} and X^{α_s} the corresponding simple or interaction terms;
- $e_{\alpha_s, uv}$ the complex number

$$e_{\alpha_s, uv} = X^{\alpha_s}(\zeta_u) - X^{\alpha_s}(\zeta_v);$$

for example, in the symbol exchange case $e_{\alpha_s, uv} = \omega_{u_1}^{\alpha_5} \omega_{u_2}^{\alpha_6} - \omega_{v_1}^{\alpha_5} \omega_{v_2}^{\alpha_6}$; in the band exchange case $e_{\alpha_s, uv} = \omega_u^{\alpha_1} - \omega_v^{\alpha_1}$.

In Proposition 12.8 we find the indicator function of the fraction obtained by the exchange of the symbols u and v. We consider only briefly the other exchanges. In Lemma 12.1 we construct the polynomial, $E_{s,uv}$, taking value -1 at the points u to be deleted, 1 at the points v to be added and 0 otherwise and the indicator functions of the cells containing the symbols u and v, respectively.

Lemma 12.1

(i) *The replacement of u by v is represented by the polynomial function $E_{s,uv}$ in the factors S_1 and S_2 defined as*

$$E_{s,uv} = \frac{1}{p^2} \sum_{i=0}^{p-1} \sum_{j=0}^{p-1} \left(\overline{\omega}_{v_1}^i \, \overline{\omega}_{v_2}^j - \overline{\omega}_{u_1}^i \, \overline{\omega}_{u_2}^j \right) S_1^i S_2^j = \frac{1}{p^2} \sum_{\alpha_s \in L_s} \left(- \overline{e_{\alpha_s,uv}} \right) X^{\alpha_s}.$$

(ii) *The indicator function P_u of the points of $\mathcal{D}$ corresponding to the cells containing u is obtained substituting the couple ζ_u in the indicator function F,*
$$P_u(\zeta_g) = F(\zeta_g, \zeta_u).$$

Proof The polynomial $E_{s,uv}$ is obtained by difference between the indicator functions G_u and G_v of the symbols u and v; with

$$G_u = \frac{1}{p^2} \left(\sum_{i=0}^{p-1} (\overline{\omega}_{u_1} S_1)^i \right) \left(\sum_{i=0}^{p-1} (\overline{\omega}_{u_2} S_2)^i \right).$$

$\square$

Note that $E_{s,vu} = -E_{s,uv}$ and the constant term is 0. Denote by $P_{g,uv}(F)$ the polynomial:

$$P_{g,uv}(F) = P_u - P_v = \sum_{\alpha_g \in L_g; \alpha_s \in L_s} b_{(\alpha_g, \alpha_s)} \, e_{\alpha_s,uv} \, X^{\alpha_g}.$$

Proposition 12.8 *The move corresponding to the exchange of the symbol u with the symbol v is*

$$M(F) = E_{s,uv} \, P_{g,uv}(F) = \sum_{\alpha_g \in L_g} \sum_{\beta_s \in L_s} m_{\alpha_g,\beta_s} \, X_g^{\alpha_g} \, X_s^{\beta_s}$$

where $m_{\alpha_g,\beta_s} = \frac{1}{p^2} \left(- \overline{e_{\beta_s,uv}} \right) \sum_{\alpha_s \in L_s} b_{(\alpha_g,\alpha_s)} e_{\alpha_s,uv}$.

Proof A full proof is in the on-line supplement. Here we provide an outline. First, we prove that the polynomial $M(F)$ takes value -1 in the points of $\mathcal{F}$ to be deleted, 1 in the points to be added, and 0 otherwise. Then $F_1 = F + M(F)$ is the indicator function of the fraction where the symbol u and v has been exchanged. Finally, we derive the form of the coefficients of $M(F)$ and we prove that they meet the conditions of Corollary 12.2. $\square$

Analogue results hold for exchanges of bands or stacks or rows within a band or columns within a stack. One needs only to define properly the set of factor indices involved in the exchange and related entities, as at the beginning of this subsection. Observe that exchanges of rows (resp. columns) must be within a band (resp. stack). Example 12.7 shows that an exchange between rows belonging to different bands is not a valid move.

Example 12.7 Consider the following 4×4 sudoku grid

1	2	3	4
3	4	1	2
2	1	4	3
4	3	2	1

The corresponding indicator function is $F = \frac{1}{4}(1 - R_1 C_2 S_2)(1 - R_2 C_1 S_1)$. If we exchange the second row of the grid with the third one, the coefficient m_{101010} of $M(F)$ is $1/4$ and conditions of Corollary 12.2 are not satisfied, see the on-line supplement.

Now we turn to a general permutation, that is a composition of exchanges.

Corollary 12.3 *The composition of exchanges is a valid move.*

Proof First, exchange u and v (from fraction $\mathcal{F}$ to fraction $\mathcal{F}_1$), then exchange l and m (from fraction $\mathcal{F}_1$ to fraction $\mathcal{F}_2$). The indicator function of $\mathcal{F}_2$ is $F_2 = F_1 + M_{lm}(F_1) = F_1 + M_{lm}(F + M_{uv}(F))$ where the sub-indices of M identify the exchange. The coefficients of $M_{lm}(F_1)$ satisfy the conditions of Corollary 12.2, in fact F_1 is a sudoku fraction. $\qquad\square$

It follows that moves in $\mathcal{M}_1$ are valid. Proposition 12.9 shows that also moves in $\mathcal{M}_2$ are valid.

Proposition 12.9 *In a sudoku fraction, the transposition of rows with columns leads to a sudoku fraction.*

Proof Given $F = \sum_\alpha b_\alpha X^\alpha$, the indicator function of the transposed grid, $F^\top = \sum_\alpha b_\alpha^\top X^\alpha$, has the following coefficients $b_{ijklmn}^\top = b_{klijmn}$ that satisfy the requirements of Proposition 12.5. $\qquad\square$

The 'inverse move' both for permutation moves $\mathcal{M}_1$ and transposition move $\mathcal{M}_2$ coincides with the move itself and we can check, for example, for the exchange of symbols u and v, it holds: $M_{kh}(F_1) = E_{s,vu}\, P_{g,vu}(F_1) = -E_{s,uv}\, P_{g,vu}(F_1) = E_{s,uv}\, P_{g,uv}(F) = M_{uv}(F)$. The transposition case is straightforward.

12.4.2 Polynomial form of $\mathcal{M}_3$ moves

We introduce this kind of move with an example.

Example 12.8 The sudoku grid below on the right is obtained by exchanging the symbols 1 and 2 only in the first stack.

<table>
<tr><td>1</td><td>2</td><td>3</td><td>4</td></tr>
<tr><td>3</td><td>4</td><td>1</td><td>2</td></tr>
<tr><td>4</td><td>3</td><td>2</td><td>1</td></tr>
<tr><td>2</td><td>1</td><td>4</td><td>3</td></tr>
</table>

$\Longrightarrow$

<table>
<tr><td>2</td><td>1</td><td>3</td><td>4</td></tr>
<tr><td>3</td><td>4</td><td>1</td><td>2</td></tr>
<tr><td>4</td><td>3</td><td>2</td><td>1</td></tr>
<tr><td>1</td><td>2</td><td>4</td><td>3</td></tr>
</table>

The move works because it involves one stack and two rows of two different bands. Non valid moves on the first stack are, for example, the following:

- exchange of the symbols 1 and 4, because they are in different rows;
- exchange of the row 2 and 4, because they contain different symbols.

We identify the parts of the sudoku grid where the $\mathcal{M}_3$ moves are applied. Fix

- a stack: $C_1 = \omega_t$,
- two columns of this stack $C_2 = \omega_{c_u}$ and $C_2 = \omega_{c_v}$,
- two boxes of this stack: $(R_1, C_1) = (\omega_{b_m}, \omega_t)$ and $(R_1, C_1) = (\omega_{b_n}, \omega_t)$.
- a row in each box: $(R_1, R_2, C_1) = (\omega_{b_m}, \omega_{r_p}, \omega_t)$ and $(R_1, R_2, C_1) = (\omega_{b_n}, \omega_{r_q}, \omega_t)$.

In this way we select two couples of cells, as shown in the following table

R_1	R_2	C_1	C_2	symbol
ω_{b_m}	ω_{r_p}	ω_t	ω_{c_u}	a_1
ω_{b_m}	ω_{r_p}	ω_t	ω_{c_v}	a_2
ω_{b_n}	ω_{r_q}	ω_t	ω_{c_u}	a_3
ω_{b_n}	ω_{r_q}	ω_t	ω_{c_v}	a_4

Clearly, analogue identification holds by fixing a band, and then two rows of this band, etc. Moreover, this kind of exchange can be generalised to more than two symbols, simultaneously.

Proposition 12.10 *The two couples of cells selected above can be exchanged only if they contain exactly two symbols a_1 and a_2 (i.e. $a_4 = a_1$ and $a_3 = a_2$). The coefficients of the move are*

$$m_{i_1 i_2 i_3 i_4 i_5 i_6} = \frac{1}{p^4} \, \overline{\omega}_t^{i_3} \, \left(-\overline{e}_{i_1 i_2, uv}\right) \, n_{i_4 i_5 i_6}$$

where

$$n_{i_4 i_5 i_6} = \sum_{\alpha_s} e_{\alpha_s, uv} \sum_{\alpha_3} \omega_t^{\alpha_3} \sum_{\alpha_4} b_{\alpha_s, \alpha_3, \alpha_4, i_5, i_6} \left(\omega_{c_u}^{[\alpha_4 - i_4]} + \omega_{c_v}^{[\alpha_4 - i_4]}\right).$$

Moreover, it holds $n_{0 i_5 i_6} = 0$ for all $(i_5, i_6) \in \{0, \cdots, p-1\}^2 \setminus \{(0,0)\}$.

	00	01	02	10	11	12	20	21	22
00	5	3	4	6	7	8	9	1	2
01	6	7	2	1	9	5	3	4	8
02	1	9	8	3	4	2	5	6	7
10	8	5	9	7	6	1	4	2	3
11	4	2	6	8	5	3	7	9	1
12	7	1	3	9	2	4	8	5	6
20	9	6	1	5	3	7	2	8	4
21	2	8	7	4	1	9	6	3	5
22	3	4	5	2	8	6	1	7	9

$\Longrightarrow$

	00	01	02	10	11	12	20	21	22
00	5	3	4	6	7	8	9	1	2
01	6	7	2	1	9	5	3	8	4
02	1	9	8	3	4	2	5	6	7
10	8	5	9	7	6	1	4	2	3
11	4	2	6	8	5	3	7	9	1
12	7	1	3	9	2	4	8	5	6
20	9	6	1	5	3	7	2	4	8
21	2	8	7	4	1	9	6	3	5
22	3	4	5	2	8	6	1	7	9

Fig. 12.2 An example for Proposition 12.10.

Example 12.9 Figure 12.2 provides an example where stack: $C_1 = \omega_2$, columns: $(C_1, C_2) = (\omega_2, \omega_1)$ and $(C_1, C_2) = (\omega_2, \omega_2)$, boxes: $(R_1, C_1) = (\omega_0, \omega_2)$ and $(R_1, C_1) = (\omega_2, \omega_2)$, rows: $(R_1, R_2) = (\omega_0, \omega_1)$ and $(R_1, R_2) = (\omega_2, \omega_0)$, symbols: 4 and 8.

Proof For the complete proof see the on-line supplement. Here we provide an outline. The new grid has both the boxes, the rows and the columns involved in the moves that still contain all the symbols repeated exactly once. Let F be the indicator function of the starting sudoku fraction. We define the following indicator functions of specific parts of the grid: S identifying the cells of the stack represented by $C_1 = \omega_t$, K_1 and K_2 identifying the cells of the columns represented by $C_2 = \omega_{c_u}$ and $C_2 = \omega_{c_v}$, K identifying the cells of both the columns represented by $C_2 = \omega_{c_u}$ and $C_2 = \omega_{c_v}$. The polynomial $F \cdot S \cdot K$ is the indicator function of the cells of the specific sudoku grid in the stack and in both the columns identified by S and K respectively.

The coefficients of the polynomial move are obtained as in Proposition 12.8, where the coefficients of the indicator function are those of $F \cdot S \cdot K$. $\square$

Example 12.10 (Example 12.8 cont.) The indicator function of the sudoku fraction is:

$$F = \frac{1}{4} - \frac{1}{4} R_1 C_2 S_2 + \frac{1}{4} R_1 R_2 C_1 S_1 - \frac{1}{4} R_2 C_1 C_2 S_1 S_2. \tag{12.2}$$

Observe that in the 4×4 sudoku grids, there are only two columns and two boxes given a stack, so we can suppose $\omega_{c_u} = -1$ and $\omega_{c_v} = 1$ and $\omega_{b_m} = -1$ and $\omega_{b_n} = 1$. The system of condition equations becomes:

$$(\omega_{r_p} - \omega_{r_q})(2\, b_{0100 i_5 i_6} + 2\, \omega_s b_{0110 i_5 i_6}) - 2(2\, b_{1000 i_5 i_6} + 2\, \omega_s b_{1010 i_5 i_6}) +$$
$$(-\omega_{r_p} - \omega_{r_q})(2\, b_{1100 i_5 i_6} + 2\, \omega_s b_{1110 i_5 i_6}) = 0,$$

for $i_5, i_6 \in \{0, 1\}$. We notice that the coefficients $b_{0100 i_5 i_6}$, $b_{1000 i_5 i_6}$, $b_{1100 i_5 i_6}$ and $b_{1010 i_5 i_6}$ are 0, being a sudoku fraction. Then the condition equations are

$$(\omega_{r_p} - \omega_{r_q}) b_{0110 i_5 i_6} - (\omega_{r_p} + \omega_{r_q}) b_{1110 i_5 i_6} = 0.$$

From Equation (12.2), $b_{0110i_5i_6} = 0$. Hence the system reduces to $\omega_{r_p} + \omega_{r_q} = 0$. This condition corresponds to four valid moves: for each of the two stacks, exchange of the first and the fourth row and exchange of the second and the third row.

We finally observe that in the $p^2 \times p^2$ case a similar move can be generalised to q symbols, $(2 \leq q \leq p)$. In Example 12.9, we can exchange the symbols $5, 3, 4$ of the first row of the first stack with the symbols $3, 4, 5$ of the last row of the same stack.

12.4.3 Composition of moves

We explore what happens when we compose two moves in $\mathcal{M}_1(F) \cup \mathcal{M}_2(F)$, namely the permutations and the transposition. Composition means that first we move from $\mathcal{F}$ to $\mathcal{F}_1$ using a move, let's say $M(F)$, and then we move from $\mathcal{F}_1$ to $\mathcal{F}_2$ using the move $M_1(F_1)$:

$$F_2 = F_1 + M_1(F_1) = F_1 + M_1(F + M(F)).$$

In general, the composition is not commutative. Propositions 12.11 and 12.12 give commutative cases. The proof of Proposition 12.11 is in the on-line supplement.

Proposition 12.11 *Let σ_1, σ_2 be two exchanges in $\mathcal{M}_1(F)$ and write $\sigma_1(F) = F + E_{s_1,u_1v_1} \, P_{g_1,u_1v_1}$ and $\sigma_2(F) = F + E_{s_2,u_2v_2} \, P_{g_2,u_2v_2}$, where E_{s_i,u_iv_i} and P_{g_i,u_iv_i}, $i = 1, 2$, are defined in Lemma 12.1. The composed move $\sigma_1 \circ \sigma_2$ equals to $\sigma_2 \circ \sigma_1$ if one of the two following conditions holds:*

- *$s_1 \cap s_2 = \emptyset$, i.e. the moves act on different factors,*
- *$s_1 = s_2$ and $\{u_1, v_1\} \cap \{u_2, v_2\} = \emptyset$, i.e. the moves act on the same factors and on different bands/rows/stacks/columns/symbols.*

Proposition 12.12 *Let σ_P be in $\mathcal{M}_1(F)$ and σ_T the transposition between rows and columns in $\mathcal{M}_2(F)$ and write*

$$\sigma_P(F) = F + E_{s_1,u_1v_1} \, P_{g_1,u_1v_1} \qquad \sigma_T(F) = F^{\top}.$$

The composed move $\sigma_P \circ \sigma_T$ equals $\sigma_T \circ \sigma_P$ if $s_1 = \{5, 6\}$.

Proof We have:

$$(\sigma_T \circ \sigma_P)(F) = \sigma_T(F + E_{s_1,u_1v_1} \, P_{g_1,u_1v_1}) = F^{\top} + E_{s_1,u_1v_1} (P_{g_1,u_1v_1})^{\top}$$
$$(\sigma_P \circ \sigma_T)(F) = \sigma_P(F^{\top}) = F^{\top} + E_{s_1,u_1v_1} (P_{g_1,u_1v_1})^{\top}.$$

$\square$

The composition between a move in $\mathcal{M}_3(F)$ (acting on a part of the sudoku grid) and another move can lead to a non sudoku fraction. For instance, if we consider the move of Example 12.8 (σ_1) and the move exchanging the first and the second row (σ_2), the move $\sigma_1 \circ \sigma_2$ leads to a non sudoku fraction.

12.4.4 Moves applied to a sudoku regular fraction

It is easy to check that the conditions of Proposition 12.5 are consistent with the existence of sudoku regular fractions. In this section we analyse which moves applied to a sudoku regular fraction preserve regularity.

Proposition 12.13 *Let $\mathcal{F}$ be a $p^2 \times p^2$ sudoku regular fraction.*

(i) *The transposition applied to $\mathcal{F}$ preserves the regularity.*

(ii) *Moves in $\mathcal{M}_1(\mathcal{F})$ applied to $\mathcal{F}$ preserve the regularity in the 4×4 and 9×9 cases.*

Proof Let R_i' and C_i', $i = 1, 2$, be the factors of the new fraction corresponding to factors R_i and C_i, $i = 1, 2$, of the starting fraction, respectively. (i) The transposition corresponds to the monomial transformation: $R_1' = C_1$, $R_2' = C_2$, $C_1' = R_1$, $C_2' = R_2$. (ii) For the 4×4 and 9×9 cases, permutations of bands, stacks, rows within band, columns within stack and symbols preserve the regularity according to Proposition 12.4. $\qquad\square$

The following example shows the indicator function of a 9×9 sudoku regular fraction obtained exchanging two symbols in a sudoku regular fraction.

Example 12.11 Consider the following indicator function of a sudoku regular fraction

$$F = \frac{1}{9}(1 + R_1 C_2 S_2 + R_1^2 C_2^2 S_2^2)(1 + R_2 C_1 S_1 + R_2^2 C_2^2 S_1^2).$$

We exchange the symbol 1, corresponding to the point $\zeta_u = (1, 1)$ of $\mathcal{D}_{56}$, with the symbol 6, corresponding to the point $\zeta_v = (\omega_1, \omega_2)$. From Proposition 12.8 the coefficients of $M(F)$ are

$$m_{i_1 i_2 i_3 i_4 i_5 i_6} = \frac{1}{p^2}(-\overline{e_{i_5 i_6, hk}}) \sum_{\alpha_s \in L_s} b_{(i_1 i_2 i_3 i_4, \alpha_s)} e_{\alpha_s, uv}.$$

The non null coefficients of $M(F)$ are in $\left\{ \frac{1}{27}, \frac{\omega_1}{27}, \frac{\omega_2}{27} \right\}$ and they lead to an indicator function of a regular fraction, by Proposition 12.3.

Proposition 12.14 generalises Example 12.10. For the proof see the on-line supplement.

Proposition 12.14 *Let $\mathcal{F}$ be a 4×4 sudoku regular fraction. A move in $\mathcal{M}_3(F)$ must satisfy the equation system:*

$$(\omega_{r_p} - \omega_{r_q})b_{0110 i_5 i_6} - (\omega_{r_p} + \omega_{r_q})b_{1110 i_5 i_6} = 0 \quad \text{for all } i_5, i_6 \in \{0, 1\}.$$

It leads to a non regular fraction.

We summarise the 4×4 case. Propositions 12.13 and 12.14 show that all the moves in $\mathcal{M}_1(F)$ and $\mathcal{M}_2(F)$ maintain the regularity, while the moves in $\mathcal{M}_3(F)$ do not. Then, by contradiction, applying a move in $\mathcal{M}_1(F) \cup \mathcal{M}_2(F)$ to sudoku non regular fractions, we still obtain a sudoku non regular fraction, because the 'inverse

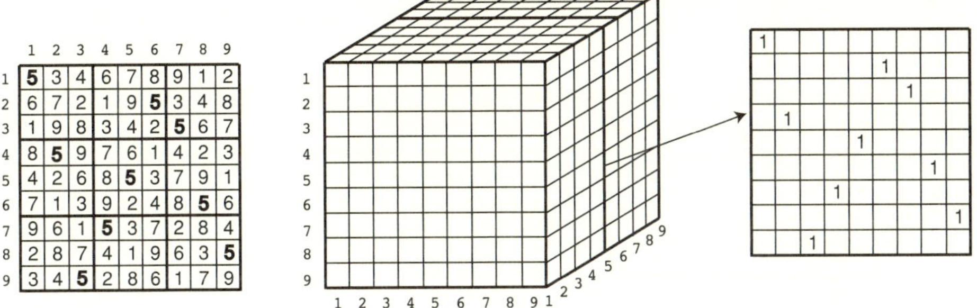

Fig. 12.3 A sudoku grid, the sudoku contingency table and its slice for the symbol 5.

move' is the move itself. It follows, and it is known in the literature, that all sudoku split into two orbits:

- starting from a regular fraction and applying moves in $\mathcal{M}_1(F) \cup \mathcal{M}_2(F)$ we get all the 96 regular fractions;
- starting from a non regular fraction and applying moves in $\mathcal{M}_1(F) \cup \mathcal{M}_2(F)$ we get all the 192 non regular fractions;
- applying moves in $\mathcal{M}_3(F)$ we switch from one orbit to the other.

For the general $p^2 \times p^2$ case $(p > 2)$, at the moment, we can only conjecture that the moves in $\mathcal{M}_1(F) \cup \mathcal{M}_2(F) \cup \mathcal{M}_3(F)$ connect all the sudoku.

12.5 Sudoku and contingency table (joint with Fabio Rapallo)

Sudoku moves can be studied also using Markov basis, a fundamental tool in algebraic statistics. We indicate here the main steps of such development. To translate the problem in terms of counts, a sudoku (filled) grid can be viewed as a $0-1$ three-way contingency table $\mathbf{n}$ with size $p^2 \times p^2 \times p^2$, which we call a sudoku contingency table. The three dimensions correspond to the factors R, C, S in the design framework. The entry $\mathbf{n}_{rcs}$ is 1 if, and only if, the symbol s appears in the r-th row and the c-th column. The link between contingency table and indicator function is strong and specifically it is given by the equality below

$$\mathbf{n}_{rcs} = F(\omega_{r_1}, \omega_{r_2}, \omega_{c_1}, \omega_{c_2}, \omega_{s_1}, \omega_{s_2})$$

with $r = 1 + p\, r_1 + r_2$, $c = 1 + p\, c_1 + c_2$ and $s = 1 + p\, s_1 + s_2$.

Example 12.12 Figure 12.3 illustrates a sudoku grid, the sudoku contingency table and its slice for the symbol 5. For instance, we have $\mathbf{n}_{115} = 1$ and $\mathbf{n}_{125} = 0$. In general, the entry $\mathbf{n}_{rc5}$ is 1 if, and only if, the symbol 5 appears in the r-th row and c-th column, or equivalently, if $F(\omega_{r_1}, \omega_{r_2}, \omega_{c_1}, \omega_{c_2}, \omega_1, \omega_1) = 1$.

The set of all sudoku tables corresponds to a set of contingency tables defined through linear constraints in the entries of the table. For instance, a symbol must

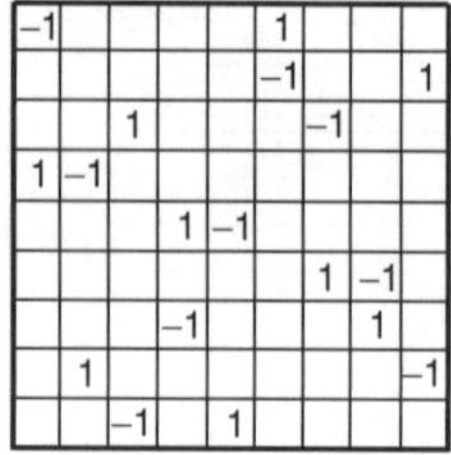

Fig. 12.4 The 5-th slice of the move for the exchange of 5 and 8 in Example 12.13.

appear exactly once in each row and this translates into the marginal constraints $\sum_{cs} \mathbf{n}_{rcs} = 1$, for all r.

The notion of Markov basis as introduced in (Diaconis and Sturmfels 1998) allows us to define a path between any two sudoku tables through tables with non-negative integer entries. This suggests how this approach enables us to generate all the sudoku grids starting from a given grid. The sudoku moves described in Section 12.4 can be translated into a linear combination of moves of a Markov basis. This is because a sudoku move takes from a sudoku fraction to a sudoku fraction, both of which correspond to a sudoku table. These two tables can be joined by a path of Markov moves through sudoku tables. The exact details of this correspondence are under investigation. If $p = 2$ the sudoku moves in $\mathcal{M}_1(F) \cup \mathcal{M}_2(F) \cup \mathcal{M}_3(F)$ span the space of all sudoku and hence there is an exact correspondence between sudoku moves and Markov moves. For $p > 2$ we conjecture an analogue correspondence.

Example 12.13 (Example 12.12 cont.) Figure 12.4 shows the 5-th slice of the move for the exchange between the symbol 5 and 8, a 2-cycle. It is a linear combination of moves of a Markov basis.

The use of the Markov basis method could allow the generation of all the sudoku grids of size $p^2 \times p^2$, but some practical problems arise. The computation of the relevant Markov basis involves symbolic computations in a polynomial ring with p^6 indeterminates and thus it is unfeasible to solve the problem by a straightforward applications of the Diaconis–Sturmfels algorithm, also for $p = 2$. Rather, we need specific algorithms exploiting the sparseness of sudoku contingency tables and the strong symmetries imposed by the linear constraints. This approach has been fruitful in other applications, see e.g. (Aoki and Takemura 2008) for tables with fixed one-way marginal totals.

12.6 Conclusions

In this chapter, after a review of the properties of the indicator function of a fraction, we applied this tool to sudoku. We characterised all the possible sudoku as the solutions of a system of polynomial equations and we solved it for the 4×4 case. We constructed the polynomial form of different kind of moves between sudoku and analysed their properties, showing that for $p = 2$ these moves span the space of all sudoku. Future investigations will concern the connectivity of all sudoku grids via the studied moves, for p larger than 2.

A new approach to sudoku moves in the framework of Markov basis for contingency tables has been outlined. This is a promising research area that could lead to a stronger link between design of experiments and contingency tables and could potentially bring some new results in both fields. A contribution in this direction is Chapter 13 by Takemura and Aoki in this volume. Clearly, computational issues will play an extremely relevant role, in particular for the construction of Markov chains, see (Aoki *et al.* 2008)

Acknowledgement

We wish to thank Professor Giovanni Pistone for his continuous support and helpful hints. We thank also Eva Riccomagno for her useful comments and suggestions.

References

Aoki, S. and Takemura, A. (2006). Markov chain Monte Carlo tests for designed experiments, METR Technical Report, 2006-56 (available at arXiv:math/0611463v1 [math.ST]).

Aoki, S. and Takemura, A. (2008). The largest group of invariance for Markov bases and toric ideals, *Journal of Symbolic Computing* **43**(5), 342–58.

Aoki, S., Takemura, A. and Yoshida, R. (2008). Indispensable monomials of toric ideals and Markov bases, *Journal of Symbolic Computing* **43**(5), 490–509.

Bailey, R. A., Cameron, P. J. and Connelly, R. (2008). Sudoku, Gerechte Designs, Resolutions, Affine Space, Spreads, Reguli, and Hamming Codesread, *American Mathematics Monthly*.

Balakrishnan, N. and Yang, P. (2006a). Classification of three-word indicator functions of two-level factorial designs, *Annals of the Institute of Statistical Mathematics* **58**(3), 595–608.

Balakrishnan, N. and Yang, P. (2006b). Connections between the resolutions of general two-level factorial designs, *Annals Institute of Statistical Mathematics* **58**(3), 609–18.

Cheng, S.-W. and Ye, K. Q. (2004). Geometric isomorphism and minimum aberration for factorial designs with quantitative factors, *Annals of Statistics* **32**(5), 2168–85.

CoCoATeam (2007). *CoCoA, a system for doing Computations in Commutative Algebra*, 4.7 edn (available at `http://cocoa.dima.unige.it`).

Diaconis, P. and Sturmfels, B. (1998). Algebraic algorithms for sampling from conditional distributions, *Annals of Statistics* **26**(1), 363–97.

Fontana, R. and Pistone, G. (2008). 2-level factorial fractions which are the union of non trivial regular design, Dipartimento di Matematica, Politecnico di Torino, Technical Report 3. (available at `arXiv:0710.5838v1`).

Fontana, R., Pistone, G. and Rogantin, M. P. (1997). Algebraic analysis and generation of two-levels designs, *Statistica Applicata* **9**(1), 15–29.

Fontana, R., Pistone, G. and Rogantin, M. P. (2000). Classification of two-level factorial fractions, *Journal of Statistical Planning and Inference* **87**(1), 149–72.

Kotsireas, I. S., Koukouvinos, C. and Rogantin, M. P. (2004). Inequivalent Hadamard matrices via indicator functions, *International Journal of Applied Mathematics* **16**(3), 355–63.

Li, W., Lin, D. K. J. and Ye, K. Q. (2003). Optimal foldover plans for two-level nonregular designs, *Technometrics* **45**(4), 347–51.

Notari, R., Riccomagno, E. and Rogantin, M. P. (2007). Two polynomial representations of experimental design, *Journal of Statistical Theory and Practice* **1**(3-4), 329–46.

Pistone, G., Riccomagno, E. and Rogantin, M. P. (2009). Methods in algebraic statistics for the design of experiments. In *Search for Optimality in Design and Statistics: Algebraic and Dynamical System Methods*, Pronzato, L. and Zhigljavsky, A. eds. (Berlin, Springer-Verlag) 97–132.

Pistone, G. and Rogantin, M. P. (2007). Comparison of different definitions of regular fraction, Dipartimento di Matematica del Politecnico di Torino, Technical report.

Pistone, G. and Rogantin, M. P. (2008). Indicator function and complex coding for mixed fractional factorial designs, *Journal of Statistical Planning and Inference* **138**(3), 787–802.

Tang, B. (2001). Theory of J-characteristics for fractional factorial designs and projection justification of minimum G_2-aberration, *Biometrika* **88**(2), 401–7.

Tang, B. and Deng, L. Y. (1999). Minimum G_2-aberration for nonregular fractional factorial designs, *Annals of Statistics* **27**(6), 1914–26.

Ye, K. Q. (2003). Indicator function and its application in two-level factorial designs, *Annals of Statistics* **31**(3), 984–94.

13
Markov basis for design of experiments with three-level factors

Satoshi Aoki

Akimichi Takemura

Abstract

We consider Markov bases arising from regular fractional factorial designs with three-level factors. They are used in a Markov chain Monte Carlo procedure to estimate p-values for various conditional tests. For designed experiments with a single observation for each run, we formulate a generalised linear model and consider a sample space with the same values of that sufficient statistic for the parameters under the null model as for the observed data. Each model is characterised by a covariate matrix, which is constructed from the main and the interaction effects. We investigate fractional factorial designs with 3^{p-q} runs and underline a correspondence with models for 3^{p-q} contingency tables.

13.1 Introduction

In the past decade, a new application of computational algebraic techniques to statistics has been developed rapidly. On one hand, (Diaconis and Sturmfels 1998) introduced the notion of *Markov basis* and presented a procedure for sampling from discrete conditional distributions by constructing a connected, aperiodic and reversible Markov chain on a given sample space. Since then, many works have been published on the topic of the Markov basis by both algebraists and statisticians. Contributions of the present authors on Markov bases can be found in (Aoki *et al.* 2008, Aoki and Takemura 2003, Aoki and Takemura 2005, Aoki and Takemura 2006, Aoki and Takemura 2008a, Aoki and Takemura 2008b, Aoki *et al.* 2008, Hara *et al.* 2009, Takemura and Aoki 2004) and (Takemura and Aoki 2005). On the other hand, series of works by Pistone and his collaborators, e.g. (Pistone and Wynn 1996, Robbiano and Rogantin 1998, Pistone *et al.* 2001, Galetto *et al.* 2003) and (Pistone and Rogantin 2008b), successfully applied the theory of Gröbner bases to designed experiments. In these works, a design is represented as the variety defined by a set of polynomial equations.

It is of interest to investigate statistical problems which are related to both designed experiments and Markov bases. In (Aoki and Takemura 2006) we initiated the study of conditional tests for main effects and interaction effects when count

Algebraic and Geometric Methods in Statistics, ed. Paolo Gibilisco, Eva Riccomagno, Maria Piera Rogantin and Henry P. Wynn. Published by Cambridge University Press. © Cambridge University Press 2010.

data are observed from a designed experiment. We investigated Markov bases arising from fractional factorial designs with two-level factors. In this chapter, extending those results, we consider Markov bases for fractional factorial designs with three-level factors. Motivated by comments by a referee, we also start to discuss relations between the Markov basis approach and the Gröbner basis approach to designed experiments. In considering alias relations for regular fractional factorial designs, we mainly use a classical notation, as explained in standard textbooks on designed experiments such as (Wu and Hamada 2000). We think that the classical notation is more familiar to practitioners of experimental designs and our proposed method is useful for practical applications. However, mathematically the aliasing relations can be more elegantly expressed in the framework of algebraic statistics by Pistone *et al.* We make this connection clear in Section 13.2.

We relate models for regular fractional factorial designs to models for contingency tables. In the literature most Markov basis models for contingency tables are hierarchical. But when we map models for fractional factorial designs to models for contingency tables, the resulting models are not necessarily hierarchical. Therefore Markov bases for the case of fractional factorial designs often have different features than Markov bases for hierarchical models. In particular here we find interesting degree three moves and indispensable fibers with three elements. These are of interest also from the algebraic viewpoint.

In Section 13.2, we illustrate the problem and describe the testing procedure for evaluating p-values of the main and the interaction effects for controllable factors in designed experiments. Similarly to the preceding works on Markov basis for contingency tables, our approach is to construct a connected Markov chain for an appropriate conditional sample space. We explain how to define this sample space corresponding to various null hypotheses. In Section 13.3, we consider the relation between models for contingency tables and models for designed experiments for fractional factorial designs with three-level factors. Then we state properties of Markov bases for designs which are practically important. In Section 13.4, we give some discussion.

13.2 Markov chain Monte Carlo tests for designed experiments

We consider the Markov chain Monte Carlo procedure for conditional tests for main and interaction effects of controllable factors for discrete observations derived from various designed experiments. Our arguments are based on the theory of generalised linear models (McCullagh and Nelder 1989).

13.2.1 Conditional tests for discrete observations

Suppose that the observations are counts of some events and one observation is obtained for each run of a regular designed experiment, defined by some *aliasing relation*. (In Section 13.4 we also consider observations which are the ratio of counts.) Table 13.1 gives a 1/8 fraction of a two-level full factorial design defined

Table 13.1 *Design and number of defects y for the wave-solder experiment.*

			Factor					y
Run	A	B	C	D	E	F	G	
1	0	0	0	0	0	0	0	69
2	0	0	0	1	1	1	1	31
3	0	0	1	0	0	1	1	55
4	0	0	1	1	1	0	0	149
5	0	1	0	0	1	0	1	46
6	0	1	0	1	0	1	0	43
7	0	1	1	0	1	1	0	118
8	0	1	1	1	0	0	1	30
9	1	0	0	0	1	1	0	43
10	1	0	0	1	0	0	1	45
11	1	0	1	0	1	0	1	71
12	1	0	1	1	0	1	0	380
13	1	1	0	0	0	1	1	37
14	1	1	0	1	1	0	0	36
15	1	1	1	0	0	0	0	212
16	1	1	1	1	1	1	1	52

by the aliasing relations

$$\mathrm{ABDE} = \mathrm{ACDF} = \mathrm{BCDG} = I.$$

This data set was considered in (Aoki and Takemura 2006, Condra 1993, Hamada and Nelder 1997). The observation y is the number of defects found in a wave-soldering process in attaching components to an electronic circuit card and the seven factors are: (A) prebake condition, (B) flux density, (C) conveyor speed, (D) preheat condition, (E) cooling time, (F) ultrasonic solder agitator and (G) solder temperature. The aim of the experiment is to decide which levels for each factor are desirable to reduce solder defects.

The standard approach to two-levels designs is to code the levels with ± 1, use the multiplicative notations and often exploit group theory (Wu and Hamada 2000). A main observation in algebraic statistics is that the aliasing relations are more elegantly expressed as a set of polynomials defining an ideal in a polynomial ring (see Section 1.3 and Section 4.6 of (Pistone *et al.* 2001)). Consider $A, B, \ldots, G$ as indeterminates and let $\mathbb{C}[A, B, \ldots, G]$ be the ring of polynomials in $A, B, \ldots, G$ with complex coefficients. Then the ideal

$$\langle A^2 - 1, B^2 - 1, \ldots, G^2 - 1, \mathrm{ABDE} - 1, \mathrm{ACDF} - 1, \mathrm{BCDG} - 1 \rangle \qquad (13.1)$$

determines the aliasing relations. For this design, two interaction effects are aliased with each other if and only if the difference of the corresponding monomials belongs to the ideal (13.1). Given a particular term order, the set of standard monomials corresponds to a particular saturated model, which can be estimated from the experiment.

Table 13.2 *Design and observations for a 3^{4-2} fractional factorial design.*

| | | Factor | | | y |
Run	A	B	C	D	
1	0	0	0	0	y_1
2	0	1	1	2	y_2
3	0	2	2	1	y_3
4	1	0	1	1	y_4
5	1	1	2	0	y_5
6	1	2	0	2	y_6
7	2	0	2	2	y_7
8	2	1	0	1	y_8
9	2	2	1	0	y_9

Table 13.2 shows a 3^{4-2} fractional factorial design with levels in $\{0, 1, 2\}$. Note that it is derived from the aliasing relations, $C = AB$, $D = AB^2$. We give a more detailed explanation of these aliasing relations in Section 13.2.2.

For count data, it is natural to consider the Poisson model (McCullagh and Nelder 1989). Write the observations as $\mathbf{y} = (y_1, \ldots, y_k)'$, where k is the number of runs. The observations are realisations from k random variables Y_i which are mutually independently distributed with the mean parameter $\mu_i = E[Y_i], i = 1, \ldots, k$. We express the mean parameter μ_i as

$$g(\mu_i) = \beta_0 + \beta_1 x_{i1} + \cdots + \beta_{\nu-1} x_{i\nu-1},$$

where $g(\cdot)$ is the link function and $x_{i1}, \ldots, x_{i\nu-1}$ are the $\nu - 1$ covariates. The sufficient statistic is written as $\sum_{i=1}^{k} x_{ij} y_i, j = 1, \ldots, \nu - 1$. For later use, we write the ν-dimensional parameter β and the covariate matrix X as

$$\beta = (\beta_0, \beta_1, \ldots, \beta_{\nu-1})' \tag{13.2}$$

and

$$X = \begin{pmatrix} 1 & x_{11} & \cdots & x_{1\nu-1} \\ \vdots & \vdots & \ldots & \vdots \\ 1 & x_{k1} & \cdots & x_{k\nu-1} \end{pmatrix} = \begin{pmatrix} \mathbf{1}_k & \mathbf{x}_1 & \cdots & \mathbf{x}_{\nu-1} \end{pmatrix}, \tag{13.3}$$

where $\mathbf{1}_k = (1, \ldots, 1)'$ is the k-dimensional column vector consisting of 1's. Using the canonical link function, which is $g(\mu_i) = \log(\mu_i)$ for the Poisson distribution, $X'\mathbf{y} = (\mathbf{1}_k'\mathbf{y}, \mathbf{x}_1'\mathbf{y}, \ldots, \mathbf{x}_{\nu-1}'\mathbf{y})$ is the sufficient statistic for β.

To define a conditional test, we specify the *null model* and the *alternative model* in terms of the parameter vector β. To avoid confusion, we express the free parameters under the null model as the ν-dimensional parameter (13.2) in this chapter. Alternative hypotheses are usually expressed in terms of additional parameters. For example, in various goodness-of-fit tests with an alternative saturated model with k parameters, we write

$$H_0 : (\beta_\nu, \ldots, \beta_{k-1}) = (0, \ldots, 0),$$
$$H_1 : (\beta_\nu, \ldots, \beta_{k-1}) \neq (0, \ldots, 0).$$

Depending on the hypotheses, we also specify an appropriate test statistic $T(\mathbf{y})$. The likelihood ratio statistics or the Pearson goodness-of-fit statistics are frequently used. Once we specify the null model and the test statistic, our purpose is to calculate the p-value. Here the Markov chain Monte Carlo procedure is a valuable tool, especially when the traditional large-sample approximation is inadequate and the exact calculation of the p-value is unfeasible. To perform the Markov chain Monte Carlo procedure, the key idea is to calculate a Markov basis over the sample space

$$\mathcal{F}(X'\mathbf{y}^o) = \{\mathbf{y} \mid X'\mathbf{y} = X'\mathbf{y}^o, \; y_i \text{ is a non-negative integer, } i = 1, \ldots, k\}, \quad (13.4)$$

where $\mathbf{y}^o$ is the observed count vector. Once a Markov basis is calculated, we can construct a connected, aperiodic and reversible Markov chain over the space in (13.4). By the Metropolis–Hastings procedure, the chain can be modified so that the stationary distribution is the conditional distribution under the null model, written as

$$f(\mathbf{y} \mid X'\mathbf{y} = X'\mathbf{y}^o) = C(X'\mathbf{y}^o) \prod_{i=1}^{k} \frac{1}{y_i!},$$

where $C(X'\mathbf{y}^o)$ is the normalising constant defined as

$$C(X'\mathbf{y}^o)^{-1} = \sum_{\mathbf{y} \in \mathcal{F}(X'\mathbf{y}^o)} \left(\prod_{i=1}^{k} \frac{1}{y_i!} \right).$$

For the definition of Markov basis see (Diaconis and Sturmfels 1998) and for computational details of Markov chains see (Ripley 1987). In applications, it is most convenient to rely on algebraic computational software such as 4ti2 (4ti2 Team 2006) to derive a Markov basis.

13.2.2 How to define the covariate matrix

In (13.3) the matrix X is constructed from the design matrix to reflect the presence of the main and the interaction effects.

For two-level factors, each main effect and interaction effect can be represented as one column of X because each of them has one degree of freedom. For the design of Table 13.1, the main effect model of the seven factors, A, B, C, D, E, F, G can be represented as the 16×8 covariate matrix by defining $\mathbf{x}_j \in \{0, 1\}^{16}$ in (13.3) as the levels for the j-th factor given in Table 13.1. Note that, for each column $\mathbf{x}_j$ of X, $\mathbf{x}_j'\mathbf{y}$ is a sufficient statistic for the parameter β_j. We regard β_j as a *contrast* $\alpha_{j1} - \alpha_{j2}$ of the main effect parameters, where α_{j1}, α_{j2} are the main effect parameters of the jth factor. In the following, we use the word 'contrast' to indicate a column of the matrix X in this sense. If we intend to include, for example, the interaction effect of A × B, the column

$$(1, 1, 1, 1, 0, 0, 0, 0, 0, 0, 0, 0, 1, 1, 1, 1)'$$

is added to X, which represents the contrast of A × B. It is calculated as $a + b$ mod (2), where a and b represent the levels of the factors A and B. It should be noted that the Markov basis for testing the null hypothesis depends on the model, namely the choice of various interaction effects included in X.

In this chapter, we consider the case of three-level designs. We do not assume ordering relations among three levels. First we consider 3^p full factorial designs. It is a special case of a multi-way layout, hence we can use the notions of ANOVA model. Each main effect has two degrees of freedom since each factor has three levels. Similarly, an interaction of order h, $h = 1, \ldots, p$, has $(3 - 1)^h$ degrees of freedom. We write the levels of the factors A, B, C, $\ldots$ as $a, b, c \ldots \in \{0, 1, 2\}$ hereafter. For example the A × B interaction effect is decomposed into two components denoted AB and AB^2, each of them with two degrees of freedom, where AB represents the contrasts satisfying

$$a + b(\mathrm{mod}\ 3) \quad \text{and} \quad 2a + 2b(\mathrm{mod}\ 3)$$

and AB^2 represents the contrasts satisfying

$$a + 2b(\mathrm{mod}\ 3) \quad \text{and} \quad 2a + b(\mathrm{mod}\ 3).$$

We follow the standard convention in (Wu and Hamada 2000) and we set *the coefficient for the first non-zero factor 1*. Similarly, n-factor interaction effects, which have 2^n degrees of freedom, can be decomposed to 2^{n-1} components with two degrees of freedom.

The covariate matrix X for the full factorial designs is constructed splitting each 3-level factor into two 2-level factors, as in the ANOVA decomposition. The corresponding model can be written as

$$\log E\left[Y_{ijk}\right] = (\mu + \rho_3 + \psi_3 + \omega_3) + (\rho_i - \rho_3) + (\psi_j - \psi_3) + (\omega_k - \omega_3),$$

where ρ_i, ψ_j and ω_k $(i, j, k = 1, 2, 3)$ are the effects of the factors A, B and C, respectively. Other parametrisations are possible, see p. 59 of (Wu and Hamada 2000). The first column represents the total mean effect, the second and the third columns represent the contrasts of the main effect of A and so on. We see, for example, the sufficient statistics $\mathbf{x}_1'\mathbf{y}, \mathbf{x}_2'\mathbf{y}$ for β_1, β_2 are written as $y_{1..}, y_{2..}$, respectively. When we consider also the interaction A × B, the four columns are added to X, where each pair of columns represents the contrasts of AB and AB^2, respectively, as explained before. The covariate matrix X for the saturated model has 27 columns, i.e., one column for the total mean effect, 6 columns for the contrasts of the main effects, $2^h \times \binom{3}{h}$ columns for the contrasts of interaction effects of order h.

Now we consider regular fractional factorial designs. In the 3^{4-2} fractional factorial design in Table 13.2 of Section 13.2.1, the model of the main effects for all factors, A, B, C, D, is nothing but the saturated model. For models with interaction effects, we need to consider designs with at least 27 runs. For example, a 3^{4-1} fractional factorial design of resolution IV is defined by the aliasing relation D = ABC which means that the level d of the factor D is determined by the relation $d = a + b + c$ (mod 3), equivalently written as $a + b + c + 2d = 0$ (mod 3). Therefore this aliasing relation is also written, using the *multiplicative notation*,

Table 13.3 *Aliasing structure for the design in Table 13.2.*

$$
\begin{aligned}
&I = ABCD^2 \\
&A = BCD^2 = AB^2C^2D \qquad B = ACD^2 = AB^2CD^2 \\
&C = ABD^2 = ABC^2D^2 \qquad D = ABC = ABCD \\
&AB = CD^2 = ABC^2D \qquad AB^2 = AC^2D = BC^2D \\
&AC = BD^2 = AB^2CD \qquad AC^2 = AB^2D = BC^2D^2 \\
&AD = AB^2C^2 = BCD \qquad AD^2 = BC = AB^2C^2D^2 \\
&BC^2 = AB^2D^2 = AC^2D^2 \qquad BD = AB^2C = ACD \\
&CD = ABC^2 = ABD
\end{aligned}
$$

as $ABCD^2 = I$. By the similar *modulus 3 calculus*, we can derive all the aliasing relations as follows. Note that, following (Wu and Hamada 2000), we treat a term and its square as the same and use the notational convention that the coefficient for the first non-zero factor is 1. The full table would have had first row $I = ABCD^2 = A^2B^2C^2D$. The equivalence can be explained as follows. For BCD^2, the three groups satisfying

$$
b + c + 2d = 2(2b + 2c + d) = 0, 1, 2 \ (\mathrm{mod}\ 3)
$$

can be equivalently defined by

$$
2b + 2c + d = 0, 1, 2 \ (\mathrm{mod}\ 3)
$$

by relabelling groups. From Table 13.3, we can clarify the models where all the effects are estimable. For example, the model of the main effects for the factors A, B, C, D and the interaction effects $A \times B$ are estimable, since the two components of $A \times B$, AB and AB^2 are not confounded to any main effect. Among the model of the main effects and two two-factor interaction effects, the model with $A \times B$ and $A \times C$ is estimable, while the model with $A \times B$ and $C \times D$ is not estimable since the components AB and CD^2 are confounded. In (Wu and Hamada 2000), main effects or components of two-factor interaction effects are called *clear* if they are not confounded to any other main effects or components of two-factor interaction effects. Moreover, a two-factor interaction effect, say $A \times B$ is called *clear* if both of its components, AB and AB^2, are clear. Therefore Table 13.3 implies that each of the main effect and the components, $AB^2, AC^2, AD, BC^2, BD, CD$ are clear, while there is no clear two-factor interaction effect.

It is not easy to derive structures of Markov bases from the aliasing relations in Table 13.3 directly. Note that the Markov bases ensure the connectivity, preserving the condition that each entry is positive, and in general have more complicated structure than the lattice bases which could be read from the unused rows of the alias table.

Aliasing relations can be more elegantly described in the framework of (Pistone *et al.* 2001). We consider the polynomial ring $\mathbb{C}[A, B, C, D]$ in indeterminates

A, B, C, D and the polynomials defining the full factorial design:

$$\mathrm{A}^3 - 1, \ \mathrm{B}^3 - 1, \ \mathrm{C}^3 - 1, \ \mathrm{D}^3 - 1. \tag{13.5}$$

Note that the roots of $x^3 = 1$ are $1, \omega, \omega^2$, where $\omega = \cos(2\pi/3) + i\sin(2\pi/3)$ is the principal cube root of the unity. Therefore (13.5) corresponds to labelling the three levels of the factors $\mathrm{A}, \ldots, \mathrm{D}$ as $1, \omega$ or ω^2. An important note here is that, when we consider polynomials in $\mathbb{C}[\mathrm{A}, \mathrm{B}, \mathrm{C}, \mathrm{D}]$, we cannot treat two monomials as the same even if they designate the same contrast by relabelling indices (and hence we cannot use the notational convention of (Wu and Hamada 2000)). The ideal

$$\langle \mathrm{A}^3 - 1, \mathrm{B}^3 - 1, \mathrm{C}^3 - 1, \mathrm{D}^3 - 1, \mathrm{D} - \mathrm{ABC} \rangle \tag{13.6}$$

determines the aliasing relations on the fraction, i.e., two interaction effects are aliased if and only if the difference of the corresponding monomials belongs to (13.6). For example, A and $\mathrm{B}^2\mathrm{C}^2\mathrm{D}$ are aliased since

$$\mathrm{A} - \mathrm{B}^2\mathrm{C}^2\mathrm{D} = (\mathrm{B}^2\mathrm{C}^2\mathrm{D} - \mathrm{A})(\mathrm{A}^3 - 1) - \mathrm{A}^4\mathrm{C}^3(\mathrm{B}^3 - 1) - \mathrm{A}^4(\mathrm{C}^3 - 1)$$
$$- \mathrm{A}^3\mathrm{B}^2\mathrm{C}^2(\mathrm{D} - \mathrm{ABC}) \in \langle \mathrm{A}^3 - 1, \mathrm{B}^3 - 1, \mathrm{C}^3 - 1, \mathrm{D}^3 - 1, \mathrm{D} - \mathrm{ABC} \rangle \ .$$

In Example 29 of (Pistone *et al.* 2001), the three levels are coded as $\{-1, 0, 1\}$ and the polynomials $\mathrm{A}^3 - \mathrm{A}, \ldots, \mathrm{D}^3 - \mathrm{D}$ are used for determining the design ideal. The complex coding allows us to better understand properties of fractional factorial designs. See also (Pistone and Rogantin 2008a).

13.3 Correspondence to the models for contingency tables

In this section, we investigate the relation between regular fractional factorial designs with 3^{p-q} runs and contingency tables. Given a model on a regular fractional factorial design, described by a covariate matrix X, and an observation vector $\mathbf{y}$, we want to find Markov bases connecting all the possible observations producing the same minimal sufficient statistic $X'\mathbf{y}$, which is called a *fiber* (Diaconis and Sturmfels 1998), to perform various tests for the coefficients of the model. Moreover, we want to analyse the structure of the Markov bases. Since Markov bases have been mainly considered in the context of contingency tables, it is convenient to characterise the relations from the viewpoint of hierarchical models of contingency tables. The 2^{p-q} fractional factorial design has been considered in (Aoki and Takemura 2006). In this chapter, we show that many interesting indispensable fibers with three elements appear from the three-level designs.

13.3.1 Models for the full factorial designs

First we consider 3^p full factorial design and prepare a fundamental fact. We index observations as $\mathbf{y} = (y_{i_1 \cdots i_p})$, where i_j corresponds to the level of the j-th factor, instead of $\mathbf{y} = (y_1, \ldots, y_{3^p})'$, to investigate the correspondence to the 3^p contingency table. We consider the fractional design of Table 13.2. The projection of the fraction

Table 13.4 *Contrasts for each factor and observations.*

Run	A	B	AB	AB^2	y
1	0	0	0	0	y_{11}
2	0	1	1	2	y_{12}
3	0	2	2	1	y_{13}
4	1	0	1	1	y_{21}
5	1	1	2	0	y_{22}
6	1	2	0	2	y_{23}
7	2	0	2	2	y_{31}
8	2	1	0	1	y_{32}
9	2	2	1	0	y_{33}

onto the first two factors is the 3^2 full factorial design. The contrasts for each factor and the observation are written as in Table 13.4. In this case, we see that, under the saturated model, the sufficient statistic for the parameter of the total mean is expressed as $y_{..}$ and, under given $y_{..}$, the sufficient statistic for the parameter of the main effects of the factors A and B are expressed as $y_{i.}$ and $y_{.j}$, respectively. Moreover, as the defining relations of the fraction are C = AB and D = AB^2, the saturated model is obtained by adding the contrasts for AB and AB^2 to the full factorial design formed by the first two factors. Note that this relation, i.e., that a higher marginal table is uniquely determined from the sufficient statistics for the lower contrasts, also holds for higher-dimensional contingency tables, which we summarise in the following. We write the controllable factors as $A_1, A_2, A_3, \ldots$ instead of A, B, C $\ldots$ here. We also use the notation of D-marginal in the p-dimensional contingency tables for $D \subset \{1, \ldots, p\}$ here. For example, $\{1\}$-marginal, $\{2\}$-marginal, $\{3\}$-marginal of $\mathbf{y} = (y_{ijk})$ are the one-dimensional tables $\{y_{i..}\}$, $\{y_{.j.}\}$, $\{y_{..k}\}$, respectively, and $\{1,2\}$-marginal, $\{1,3\}$-marginal, $\{2,3\}$-marginal of $\mathbf{y} = (y_{ijk})$ are the two-dimensional tables $\{y_{ij.}\}$, $\{y_{i.k}\}$, $\{y_{.jk}\}$, respectively. See (Dobra 2003) for the formal definition.

Proposition 13.1 *For 3^p full factorial design, write observations as* $\mathbf{y} = (y_{i_1 \cdots i_p})$, *where i_j corresponds to the level of the j-th factor. Then the necessary and the sufficient condition that the $\{i_1, \ldots, i_n\}$-marginal n-dimensional table $(n \leq p)$ is uniquely determined from $X'\mathbf{y}$ is that the covariate matrix X includes the contrasts for all the components of m-factor interaction effects $A_{j_1} \times A_{j_2} \times \cdots \times A_{j_m}$ for all* $\{j_1, \ldots, j_m\} \subset \{i_1, \ldots, i_n\}, m \leq n$.

Proof The saturated model for the 3^n full factorial design is expressed as the contrast for the total mean, $2 \times n$ contrasts for the main effects, $2^m \times \binom{n}{m}$ contrasts for the m-factor interaction effects for $m = 2, \ldots, n$, since they are linearly independent and $\sum_{m=0}^{n} 2^m \binom{n}{m} = (1+2)^n = 3^n$. $\qquad\square$

13.3.2 Models for the regular fractional factorial designs

Proposition 13.1 states that hierarchical models for the controllable factors in the 3^p full factorial design corresponds to the hierarchical models for the 3^p contingency table completely. On the other hand, hierarchical models for the controllable factors in the 3^{p-q} fractional factorial design do not correspond to the hierarchical models for the 3^p contingency table in general. This is because X contains only part of the contrasts of interaction elements in the case of fractional factorial designs.

As a simplest example, we first consider a design with nine runs with the three controllable factors A, B, C, and defined by $C = AB$. The design is represented in Table 13.2 by ignoring the factor D. The covariate matrix for the main effects model of A, B, C is defined as

$$
X' = \begin{bmatrix}
1 & 1 & 1 & 1 & 1 & 1 & 1 & 1 & 1 \\
1 & 1 & 1 & 0 & 0 & 0 & 0 & 0 & 0 \\
0 & 0 & 0 & 1 & 1 & 1 & 0 & 0 & 0 \\
1 & 0 & 0 & 1 & 0 & 0 & 1 & 0 & 0 \\
0 & 1 & 0 & 0 & 1 & 0 & 0 & 1 & 0 \\
1 & 0 & 0 & 0 & 0 & 1 & 0 & 1 & 0 \\
0 & 1 & 0 & 1 & 0 & 0 & 0 & 0 & 1
\end{bmatrix}.
$$

To investigate the structure of the fiber, write the observation as a frequency of the 3×3 contingency table, $y_{11}, \ldots, y_{33}$. Then the fiber is the set of tables with the same row sums $\{y_{i\cdot}\}$, column sums $\{y_{\cdot j}\}$ and the contrast displayed as

$$
\begin{bmatrix}
0 & 1 & 2 \\
1 & 2 & 0 \\
2 & 0 & 1
\end{bmatrix}.
$$

Note that the three groups defined by

$$
a + b = 0, 1, 2 \ (\mathrm{mod}\ 3)
$$

are displayed as $y_{a+1, b+1}$. To construct a minimal Markov basis, we see that the moves to connect the following three-elements fiber are sufficient

$$
\left\{
\begin{bmatrix} 1 & 0 & 0 \\ 0 & 1 & 0 \\ 0 & 0 & 1 \end{bmatrix},
\begin{bmatrix} 0 & 1 & 0 \\ 0 & 0 & 1 \\ 1 & 0 & 0 \end{bmatrix},
\begin{bmatrix} 0 & 0 & 1 \\ 1 & 0 & 0 \\ 0 & 1 & 0 \end{bmatrix}
\right\}.
$$

Therefore any two moves from the set

$$
\left\{
\begin{bmatrix} +1 & -1 & 0 \\ 0 & +1 & -1 \\ -1 & 0 & +1 \end{bmatrix},
\begin{bmatrix} +1 & 0 & -1 \\ -1 & +1 & 0 \\ 0 & -1 & +1 \end{bmatrix},
\begin{bmatrix} 0 & +1 & -1 \\ -1 & 0 & +1 \\ +1 & -1 & 0 \end{bmatrix}
\right\}
$$

is a minimal Markov basis. In the following, to save the space, we use a binomial representation. For example, the above three moves are

$$
y_{11} y_{22} y_{33} - y_{12} y_{23} y_{31}, \quad y_{11} y_{22} y_{33} - y_{13} y_{21} y_{32}, \quad y_{12} y_{23} y_{31} - y_{13} y_{21} y_{32}.
$$

In this chapter, we consider three types of regular fractional factorial designs with 27 runs, which are important for practical applications. We investigate the relations between various models for the fractional factorial designs and the $3 \times 3 \times 3$ contingency table. Markov bases for the $3 \times 3 \times 3$ contingency tables have been investigated by many researchers, especially for the no three-factor interaction model by (Aoki and Takemura 2003). In the following, we investigate Markov bases for some models, especially we are concerned about their *minimality, unique minimality* and *indispensability* of their elements. These concepts are presented in (Takemura and Aoki 2004, Aoki *et al.* 2008). In this chapter, we define that a Markov basis is minimal if no proper subset of it is a Markov basis. A minimal Markov basis is unique if there is only one minimal Markov basis except for sign changes of their elements. An element of a Markov basis is represented as a binomial. We call it a move following our previous papers. A move $\mathbf{z}$ is indispensable if $\mathbf{z}$ or $-\mathbf{z}$ belongs to every Markov basis.

3_{IV}^{4-1} **fractional factorial design defined from** $\mathrm{D = ABC}$ In the case of four controllable factors for design with 27 runs, we have a resolution IV design, for instance, by setting $\mathrm{D = ABC}$. As seen in Section 13.2.2, all main effects are clear, whereas all two-factor interactions are not clear in this design.

For the main effect model in this design, the sufficient statistic is written as $\{y_{i\cdot\cdot}\}$, $\{y_{\cdot j\cdot}\}$, $\{y_{\cdot\cdot k}\}$ and for the contrasts of ABC,

$$y_{111} + y_{123} + y_{132} + y_{213} + y_{222} + y_{231} + y_{312} + y_{321} + y_{333},$$
$$y_{112} + y_{121} + y_{133} + y_{211} + y_{223} + y_{232} + y_{313} + y_{322} + y_{331},$$
$$y_{113} + y_{122} + y_{131} + y_{212} + y_{221} + y_{233} + y_{311} + y_{323} + y_{332}.$$

By calculation by 4ti2, we see that the minimal Markov basis for this model consists of 54 degree 2 moves and 24 degree 3 moves. All the elements of the same degrees are on the same orbit, see (Aoki and Takemura 2008a, Aoki and Takemura 2008b). The elements of degree 2 connect three-elements fibers such as

$$\{y_{112}y_{221}, \ y_{121}y_{212}, \ y_{122}y_{211}\} \tag{13.7}$$

into a tree, and the elements of degree 3 connect three-elements fibers such as

$$\{y_{111}y_{122}y_{133}, \ y_{112}y_{123}y_{131}, \ y_{113}y_{121}y_{132}\} \tag{13.8}$$

into a tree. For the fiber (13.7), for example, two moves such as

$$y_{121}y_{212} - y_{112}y_{221}, \ y_{122}y_{211} - y_{112}y_{221}$$

are needed for a Markov basis. See (Takemura and Aoki 2004) for detail on the structure of a minimal Markov basis.

Considering the aliasing relations given in Table 13.3, we can consider models with interaction effects. We see by running 4ti2 that the structures of the minimal Markov bases for each model are given as follows.

- For the model of the main effects and the interaction effect $\mathrm{A \times B}$, 27 indispensable moves of degree 2 such as $y_{113}y_{321} - y_{111}y_{323}$ and 54 dispensable moves of degree

3 constitute a minimal Markov basis. The degree 3 elements are on two orbits, one connects 9 three-elements fibers such as (13.8) and the other connects 18 three-elements fibers such as $\{y_{111}y_{133}y_{212},\ y_{112}y_{131}y_{213},\ y_{113}y_{132}y_{211}\}$.

- For the model of the main effects and the interaction effects $A \times B, A \times C$, 6 dispensable moves of degree 3, 81 indispensable moves of degree 4 such as

$$y_{112}y_{121}y_{213}y_{221} - y_{111}y_{122}y_{211}y_{223}$$

  and 171 indispensable moves of degree 6, 63 moves such as

$$y_{112}y_{121}y_{133}y_{213}y_{222}y_{231} - y_{111}y_{123}y_{132}y_{211}y_{223}y_{232}$$

  and 108 moves such as

$$y_{112}y_{121}y_{213}y_{231}y_{311}y_{323} - y_{111}y_{122}y_{211}y_{233}y_{313}y_{321}$$

  constitute a minimal Markov basis. The degree 3 elements connect three-elements fibers such as (13.8).

- For the model of the main effects and the interaction effects $A \times B, A \times C, B \times C$, 27 indispensable moves of degree 6 such as

$$y_{113}y_{121}y_{132}y_{211}y_{222}y_{233} - y_{111}y_{122}y_{133}y_{213}y_{221}y_{232}$$

  and 27 indispensable moves of degree 8 such as

$$y_{111}^{2}y_{122}y_{133}y_{212}y_{221}y_{313}y_{331} - y_{112}y_{113}y_{121}y_{131}y_{211}y_{222}y_{311}y_{333}$$

  constitute a unique minimal Markov basis.

- For the model of the main effect and the interaction effects $A \times B, A \times C, A \times D$, 6 dispensable moves of degree 3 constitute a minimal Markov basis, which connect three-elements fibers such as (13.8).

Two 3_{III}^{5-2} fractional factorial designs   Similarly, for the case of five controllable factors for designs with 27 runs, we consider two 3_{III}^{5-2} fractional factorial designs from Table 5A.2 of (Wu and Hamada 2000), defined from $D = AB, E = AB^2C$ and $D = AB, E = AB^2$, respectively. For each design, we can consider nine and four distinct hierarchical models (except for the saturated model), respectively, and calculate minimal Markov bases by 4ti2. We see that in the six models of the former design and all the four models of the latter design, a unique minimal Markov basis exists. For details of these results, see (Aoki and Takemura 2007).

13.4 Discussion

In this chapter, we investigate a Markov basis arising from regular fractional factorial designs with three-level factors. As noted in Section 13.1, the notion of a Markov basis is fundamental in the first work in computational algebraic statistics. Moreover, the designed experiment is also one of the areas in statistics where the theory of Gröbner bases found applications. Since we give a different application of the theory of Gröbner bases to the designed experiments, this chapter relates to both the works (Diaconis and Sturmfels 1998) and (Pistone and Wynn 1996).

One of the aims of this work is to propose a method to construct models and test their fitting in the framework of the conditional tests. In most of the classical literatures on designed experiments with non-normal data, exact testing procedures based on the conditional sampling space are not considered. Since the experimental design is used when the cost of obtaining data is relatively high, it is very important to develop techniques for exact testing. Another aim of this work is to give a general method to specify our models to the corresponding models of 3^p contingency tables, to make use of general results for the Markov bases of contingency tables.

Though in Section 13.2, we suppose that the observations are counts, our arguments can also be applied to the case that the observations are ratios of counts. In this case, we consider the logistic link function instead of the logit link, and investigate the relation between 3^{p-q} fractional factorial designs to the 3^{p-q+1} contingency tables. See (Aoki and Takemura 2006) for the two-level case.

One of the interesting observations of this chapter is that many three-elements fibers arise in considering minimal Markov bases. In fact, in the examples considered in Section 13.3.2, all the dispensable moves of minimal Markov bases are needed for connecting three-elements fibers, where each element of the fibers does not share support with other elements of the same fiber. This shows that every positive and negative part of every dispensable move is an indispensable monomial. See the notion of the *indispensable monomial* in (Aoki *et al.* 2008).

It is of great interest to clarify relationships between our approach and the works by Pistone, Riccomagno and Wynn. In (Pistone *et al.* 2001), designs are defined as the set of points (i.e., the affine variety), and the set of polynomials vanishing at these points (i.e., the design ideal) are considered. They calculate a Gröbner basis of the design ideal, which is used to specify the identifiable models and confounding relations. In Section 13.2 we explained that the aliasing relations for fractional factorial designs specified in the classical notation can be more elegantly described in the framework of (Pistone *et al.* 2001). It is important to study whether a closer connection can be established between a design ideal and the Markov basis (toric ideal). It should be noted, however, that a Markov basis depends on the covariate matrix X, which incorporates the statistical model we aim to test, whereas the Gröbner basis depends only on the design points and a given term order.

Finally as suggested by a referee, it may be valuable to consider relations between the arguments of this chapter and designs other than fractional factorial designs, such as the Plackett–Burman designs or balanced incomplete block designs. These topics are left to future work.

References

4ti2 Team (2006). *4ti2 – A software package for algebraic, geometric and combinatorial problems on linear spaces* (available at www.4ti2.de).

Aoki, S., Hibi, T., Ohsugi, H. and Takemura, A. (2008). Markov basis and Gröbner basis of Segre-Veronese configuration for testing independence in group-wise selections, *Annals of the Institute of Statistical Mathematics*, to appear. (available at arXiv:math/0704.1074 [math.ST]).

Aoki, S. and Takemura, A. (2003). Minimal basis for a connected Markov chain over

$3 \times 3 \times K$ contingency tables with fixed two-dimensional marginals, *Australian and New Zealand Journal of Statistics* **45**, 229–49.

Aoki, S. and Takemura, A. (2005). Markov chain Monte Carlo exact tests for incomplete two-way contingency tables, *Journal of Statistical Computation and Simulation* **75**, 787–812.

Aoki, S. and Takemura, A. (2006). Markov chain Monte Carlo tests for designed experiments, METR Technical Report, 2006-56 (available at arXiv:math/0611463v1 [math.ST]).

Aoki, S. and Takemura, A. (2007). Markov basis for design of experiments with three-level factors, METR Technical Report, 2007-54 (available at arXiv:math/0709.4323v2 [stat.ME]).

Aoki, S. and Takemura, A. (2008a). Minimal invariant Markov basis for sampling contingency tables with fixed marginals, *Annals of the Institute of Statistical Mathematics* **60**, 229–56.

Aoki, S. and Takemura, A. (2008b). The largest group of invariance for Markov bases and toric ideals, *Journal of Symbolic Computing* **43**(5), 342–58.

Aoki, S., Takemura, A. and Yoshida, R. (2008). Indispensable monomials of toric ideals and Markov bases, *Journal of Symbolic Computing* **43**(5), 490–509.

Condra, L. W. (1993). *Reliability Improvement with Design of Experiments* (New York, Marcel Dekker).

Diaconis, P., and Sturmfels, B. (1998). Algebraic methods for sampling from conditional distributions, *Annals of Statistics* **26**, 363–97.

Dobra, A. (2003). Markov bases for decomposable graphical models, *Bernoulli* **9**(6), 1–16.

Galetto, F., Pistone, G. and Rogantin, M. P. (2003). Confounding revisited with commutative computational algebra, *Journal of Statistical Planning and Inference* **117**, 345–63.

Hamada, M. and Nelder, J. A. (1997). Generalized linear models for quality-improvement experiments, *Journal of Quality Technology* **29**, 292–304.

Hara, H., Aoki, S. and Takemura, A. (2009). Minimal and minimal invariant Markov bases of decomposable models for contingency tables, *Bernoulli*, to appear. METR Technical Report, 2006-66 (available at arXiv:math/0701429 [math.ST]).

McCullagh, P. and Nelder, J. A. (1989). *Generalized Linear Models* 2nd edn (London, Chapman & Hall).

Pistone, G., Riccomagno, E., and Wynn, H. P. (2001). *Algebraic Statistics* (Boca Raton, Chapman & Hall).

Pistone, G. and Rogantin, M. P. (2008a). Algebraic statistics of codings for fractional factorial designs, *Journal of Statistical Planning and Inference*, **138**, 234–244.

Pistone, G. and Rogantin, M. P. (2008b). Indicator function and complex coding for mixed fractional factorial designs, *Journal of Statistical Planning Inference* **138**(3), 787–802.

Pistone, G. and Wynn, H. P. (1996). Generalised confounding with Gröbner bases, *Biometrika* **83**, 653–66.

Ripley, B. D. (1987). *Stochastic Simulation* (New York, John Wiley & Sons).

Robbiano, L. and Rogantin, M. P. (1998). Full factorial designs and distracted fractions. In *Gröbner Bases and Applications*, Buchberger, B. and Winkler, F. eds. (Cambridge, Cambridge University Press) 473–82.

Takemura, A. and Aoki, S. (2004). Some characterizations of minimal Markov basis for sampling from discrete conditional distributions, *Annals of the Institute of Statistical Mathematics* **56**, 1–17.

Takemura, A. and Aoki, S. (2005). Distance reducing Markov bases for sampling from a discrete sample space, *Bernoulli* **11**, 793–813.

Wu, C. F. J. and Hamada, M. (2000). *Experiments: Planning, Analysis, and Parameter Design Optimization* (New York, John Wiley & Sons).

Part III

Information geometry

14

Introduction to non-parametric estimation

Raymond F. Streater

14.1 Parametric estimation; the Cramér–Rao inequality

Information geometry had its roots in Fisher's theory of estimation. Let $\rho_\eta(x)$, $x \in \mathbb{R}$, be a strictly positive differentiable probability density, depending on a parameter $\eta \in \mathbb{R}$. To stress the analogy between the classical case and quantum case a density is also referred to as a state. The *Fisher information* of ρ_η is defined to be (Fisher 1925)

$$G := \int \rho_\eta(x) \left(\frac{\partial \log \rho_\eta(x)}{\partial \eta} \right)^2 dx.$$

We note that this is the variance of the random variable $Y = \partial \log \rho_\eta / \partial \eta$, which has mean zero. Furthermore, G is associated with the *family* $\mathcal{M} = \{\rho_\eta\}$ of distributions, rather than any one of them. This concept arises in the theory of estimation as follows. Let X be a random variable whose distribution is believed or hoped to be one of those in $\mathcal{M}$. We estimate the value of η by measuring X independently m times, getting the data $x_1, \ldots, x_m$. An *estimator* f is a function of $(x_1, \ldots, x_m)$ that is used for this estimate. So f is a function of m independent copies of X, and so is a random variable. To be useful, the estimator must be a known function of X, not depending of η, which we do not (yet) know. We say that an estimator is *unbiased* if its mean is the desired parameter; it is usual to take f as a function of X and to regard $f(x_i)$, $i = 1, \ldots, m$ as samples of f. Then the condition that f is unbiased becomes

$$\rho_\eta \cdot f := \int \rho_\eta(x) f(x) dx = \eta.$$

A good estimator should also have only a small chance of being far from the correct value, which is its mean if it is unbiased. This chance is measured by the variance. (Fisher 1925) proved that the variance V of an unbiased estimator f obeys the inequality $V \geq G^{-1}$. This is called the Cramér–Rao inequality and its proof is based on the Cauchy–Schwarz inequality. We shall show how this is done.

If we do N independent measurements for the estimator, and average them, we improve the inequality to $V \geq G^{-1}/N$. This inequality expresses that, given the family ρ_η, there is a limit to the reliability with which we can estimate η. Fisher

Algebraic and Geometric Methods in Statistics, ed. Paolo Gibilisco, Eva Riccomagno, Maria Piera Rogantin and Henry P. Wynn. Published by Cambridge University Press. © Cambridge University Press 2010.

241

termed VG^{-1} the *efficiency* of the estimator f. Equality in the Schwarz inequality occurs if and only if the two functions are proportional. In this case, let $-\partial\xi/\partial\eta$ denote the factor of proportionality. Then the optimal estimator occurs when

$$\log\rho_\eta(x) = -\int \partial\xi/\partial\eta(f(x) - \eta)\,d\eta.$$

Doing the integral, and adjusting the integration constant by normalisation, leads to

$$\rho_\eta(x) = Z^{-1}\exp\{-\xi f(x)\}$$

which defines the 'exponential family'.

This can be generalised to any n-parameter manifold $\mathcal{M} = \{\rho_\eta\}$ of distributions, $\eta = (\eta_1,\ldots,\eta_n)$ with $\eta \in \mathbb{R}^n$. Suppose we have unbiased estimators $(X_1,\ldots,X_n)$, with covariance matrix V. Fisher introduced the *information matrix*

$$G^{ij} = \int \rho_\eta(x)\frac{\partial\log\rho_\eta(x)}{\partial\eta_i}\frac{\partial\log\rho_\eta(x)}{\partial\eta_j}\,dx. \tag{14.1}$$

(Rao 1945) remarked that G^{ij} provides a Riemannian metric for $\mathcal{M}$. Cramér and Rao obtained the analogue of the inequality $V \geq G^{-1}$ when $n > 1$. Put $V_{ij} = \rho_\eta\cdot[(X_i - \eta_i)(X_j - \eta_j)]$, the covariance matrix of the estimators $\{X_i\}, i = 1,\ldots,n$, and $Y^i = \partial\rho_\eta/\partial\eta_i$. We say that the estimators are *locally unbiased* if

$$\int \rho_\eta(x)Y^i(x)(X_j(x) - \eta_j)\,dx = \delta_{ij}. \tag{14.2}$$

Then we get the *Cramér–Rao* matrix inequality $V \geq G^{-1}$ as a matrix. For, Equation (14.2) shows that the covariance of X_j with Y^i is δ_{ij}, so the covariance matrix of X_j and Y^i is

$$K := \begin{pmatrix} V & I \\ I & G \end{pmatrix}. \tag{14.3}$$

It follows that the matrix (14.3) is positive semi-definite; let us treat the case when it is definite. Then its inverse exists, and is

$$K^{-1} = \begin{pmatrix} (G - V^{-1})^{-1} & -G^{-1}(V - G^{-1})^{-1} \\ -V^{-1}(G - V^{-1})^{-1} & (V - G^{-1})^{-1} \end{pmatrix}.$$

This is positive semi-definite. Hence, both diagonal $n \times n$ submatrices are positive semi-definite; thus their inverses are too, giving $VG \geq I$. By taking limits, one can then treat the cases where (14.3) is positive semi-definite. Again, one can easily see that the only state that gives equality $VG = I$ is in the exponential family: that $VG = I$ for the exponential family is proved below. That this is the only way that $VG = I$ can be achieved follows from the definiteness of the Schwarz inequality. Thus, the theory of Cramér–Rao justifies the method of maximum entropy of (Jaynes 1957). There, if the experimenter measures the random variables $X_1,\ldots,X_n$, Jaynes postulates that the best estimate for the state is that of the greatest entropy, given the measured values η_i for the means of X_i; we now see that this leads to the exponential family of states, generalising the work of Gibbs from one variable, the energy, to n.

Let us consider the discrete case and call the density p. Indeed, to maximise $S :=$ $-\sum_\omega p(\omega)\log p(\omega)$ subject to the constraints $\sum_\omega p(\omega) = 1$ and $\sum_\omega p(\omega)X_i(\omega) = \eta_i$, $i = 1,\dots,n$ we use the method of Lagrange multipliers λ, ξ^j, and maximise

$$-\sum_\omega p(\omega)\log p(\omega) - \lambda \sum_\omega p(\omega) - \sum_{j=1}^n \xi^j p(\omega)X_j(\omega) \qquad (14.4)$$

subject to no constraints. We then find λ and ξ^j by the conditions

$$\sum_\omega p(\omega) = 1 \text{ and } \sum_\omega p(\omega)X_j(\omega) = \eta_j, j = 1,\dots,n.$$

The expression in Equation (14.4) is a maximum when its derivatives with respect to $p(\omega)$ are all zero; solving the equations obtained, we see that the entropy is a maximum on the *exponential manifold* of probabilities of the form

$$p_\xi(\omega) = Z^{-1}\exp\left\{-\sum_j \xi^j X_j(\omega)\right\}$$

where

$$Z = \sum_\omega \exp\left\{-\sum_j \xi^j X_j(\omega)\right\}.$$

It is easy to show that

$$\eta_j = -\frac{\partial\Psi}{\partial\xi^j} \qquad V_{jk} = -\frac{\partial\eta_j}{\partial\xi^k}, \qquad (14.5)$$

for $j,k = 1,\dots,n$, where $\Psi = \log Z$, and that Ψ is a convex function of ξ^j. The Legendre dual to Ψ is $\Psi - \sum \xi^i \eta_i$ and this is the entropy $S = -p \cdot \log p$. The dual relations are

$$\xi^j = \frac{\partial S}{\partial\eta_j} \qquad G^{jk} = -\frac{\partial\xi^j}{\partial\eta_k}. \qquad (14.6)$$

By the rule for Jacobians, V and G are mutual inverses: $V = G^{-1}$ and we have achieved the Cramér–Rao bound. This gives us estimators of 100% efficiency. Thus Jaynes's methods (maximising entropy subject to maintaining observed means) does give us the best estimate. We can paraphrase Jaynes, and say that in settling for the Gibbs state, Nature is making the best estimate, given the information available, the mean energy. More, in settling for the grand canonical state, Nature is making the best choice, given the mean energy and mean particle number. We do not agree with Jaynes that this is the reason why so many states are at or close to equilibrium. We usually measure much more than the mean energy and density of a state. For example, the energy of the cosmic background radiation, as found in the COBE experiment, is very close to the Planck distribution, the thermal state for a system of free photons. The whole shape of the distribution is Planckian (to a close approximation); it is not just that the mean energy $p \cdot \mathcal{E}$ is the same as predicted by Planck's formula. By measuring, and thus knowing, the moments $p \cdot \mathcal{E}$,

$p \cdot \mathcal{E}^2, \ldots, p \cdot \mathcal{E}^n$, Jaynes would say that the best state is the multiple exponential state

$$p = Z^{-1} \exp \left\{ -\beta_1 \mathcal{E} - \beta_2 \mathcal{E}^2 - \ldots - \beta_n \mathcal{E}^n \right\}.$$

Ingarden (Ingarden 1992) has called these extra parameters, $\beta_2, \ldots, \beta_n$, the generalised inverse temperatures. When Jaynes finds that for the background radiation, all the higher terms $\beta_2, \ldots, \beta_n$ are very nearly zero, he cannot explain why. This is why Jaynes and Ingarden do not solve the problem of statistical mechanics, i.e. why do systems approach equilibrium, by their work. On this also see (Grünwald and Dawid 2004).

14.2 Manifolds modelled by Orlicz spaces

(Pistone and Sempi 1995) have developed a version of information geometry, which does not depend on a choice of the span of a finite number of estimators. Let $(\Omega, \mathcal{B}, \mu)$ be a measure space; thus, Ω is the sample space, and $\mathcal{B}$ is a given σ-algebra defining the measurable sets, the events. The measure μ, used to specify the sets of measure zero, the impossible events, is non-negative, but need not be normalised to 1. The probabilities on Ω, which represent the possible states of the system, are positive, normalised measures ν on Ω that are equivalent to μ. Let $\mathcal{M}$ be the set of all probability measures ν that are equivalent to μ; such a measure is determined by its Radon–Nikodym derivative ρ relative to μ:

$$d\nu = \rho d\mu.$$

Here, the probability density ρ satisfies $\rho(x) > 0$ μ-almost everywhere, and

$$\mathrm{E}_{d\mu}[\rho] := \int_\Omega \rho(x) \mu(dx) = 1.$$

Let ρ_0 be such a density. Pistone and Sempi sought a family of sets $\mathcal{N}$ containing ρ_0, and which obey the axioms of neighbourhoods of the state defined by ρ_0. They then did the same for each point of $\mathcal{N}$, and added these to the set connected to ρ_0, and so on with each new point added, thus constructing a topological space $\mathcal{M}$. They showed that $\mathcal{M}$ has the structure of a Banach manifold. In their construction, the topology on $\mathcal{M}$ is not given by the L^1-distance defined by $d\mu$, or by $\rho_0 d\mu$, but by an Orlicz norm (Rao and Ren 1992), as follows.

Let u be a random variable on $(\Omega, \mathcal{B})$, and consider the class of measures whose density ρ has the form

$$\rho = \rho_0 \exp\{u - \psi_{\rho_0}(u)\}$$

in which ψ, called the free energy, is finite for all states of a one-parameter exponential family:

$$\psi_{\rho_0}(\lambda u) := \log \mathrm{E}_{\rho_0 d\mu}[e^{-\lambda u}] < \infty \text{ for all } \lambda \in [-\epsilon, \epsilon]. \tag{14.7}$$

Here, $\epsilon > 0$. This implies that all moments of u exist in the probability measure $d\nu = \rho_0 d\mu$ and that the moment-generating function is analytic in a neighbourhood of $\lambda = 0$. The random variables satisfying Equation (14.7) for some $\epsilon > 0$ are said to lie in the *Cramér class*. The (real) span of this class was shown to be a Banach space by (Pistone and Sempi 1995), and so to be complete, when furnished with the norm

$$\|u\|_{L} := \inf \left\{ r > 0 : \mathrm{E}_{d\mu} \left[\rho_0 \left(\cosh \frac{u}{r} - 1 \right) \right] < 1 \right\}. \tag{14.8}$$

The map

$$u \mapsto \exp\{u - \psi_{\rho_0}(u)\}\rho_0 =: e_{\rho_0}(u)$$

maps the unit ball in the Cramér class into the class of probability distributions that are absolutely continuous relative to μ. We can identify ψ as the 'free energy' by writing $\rho_0 = \exp\{-h_0\}$. Then $\rho = \exp\{-h_0 + u - \psi_\rho(u)\}$ and h_0 appears as the 'free Hamiltonian' and $-u$ as the 'perturbing potential', of the 'Gibbs state' $\rho d\mu$.

The function $\Phi(x) = \cosh x - 1$ used in the Definition 14.8 of the norm, is a Young function. That is, Φ is convex, and obeys

(i) $\Phi(x) = \Phi(-x)$ for all x
(ii) $\Phi(0) = 0$
(iii) $\lim_{x \to \infty} \Phi(x) = +\infty$

The epigraph of Φ is the set of points $\{(x, y) : y \geq \Phi(x)\}$. The epigraph is convex, and is closed if and only if Φ is lower semicontinuous. If so, the map $\lambda \mapsto \Phi(\lambda x)$ is continuous on any open set on which it is finite (Krasnoselski and Ruticki 1961, Rao and Ren 1992). Examples of Young functions are

$$\Phi_1(x) := \cosh x - 1$$
$$\Phi_2(x) := e^{|x|} - |x| - 1$$
$$\Phi_3(x) := (1 + |x|) \log(1 + |x|) - |x|$$
$$\Phi^p(x) := |x|^p \qquad \text{defined for } 1 \leq p < \infty.$$

Let Φ be a Young function. Then its Legendre–Fenchel dual,

$$\Phi^*(y) := \sup_x \{xy - \Phi(x)\}$$

is also a Young function. It is lower semicontinuous, being the supremum of linear functions over a convex set. So Φ^{**} is lower semicontinuous; its epigraph is the closure of the epigraph of Φ (which is always the epigraph of a Young function, known as the lower semicontinuous version of Φ). For example, $\Phi_2 = \Phi_3^*$ and $\Phi^p = \Phi^{q*}$ when $p^{-1} + q^{-1} = 1$.

The theory of Orlicz spaces shows that given a Young function Φ, one can define a norm on the Cramér class by

$$\|u\|_{\Phi} := \sup_v \left\{ \int |uv| d\nu : v \in L^{\Phi^*}, \int \Phi^*(v(x)) d\nu \leq 1 \right\},$$

or with the equivalent *gauge norm*, also known as a Luxemburg norm: for some $a > 0$,

$$\|u\|_{L,a} := \inf \left\{ r > 0 : \int \Phi \left(\frac{u(x)}{r} \right) \nu(dx) < a \right\}. \tag{14.9}$$

For a given Φ, all the Luxemburg norms are equivalent, whatever a is chosen. By *the* Luxemburg norm, denoted $\|u\|_L$, we shall mean the case when $a = 1$.

Equivalence. We say that two Young functions Φ and Ψ are equivalent if there exist $0 < c < C < \infty$ and $x_0 > 0$ such that

$$\Phi(cx) \le \Psi(x) \le \Phi(Cx)$$

holds for all $x \ge x_0$. We then write $\Phi \equiv \Psi$; the scale of x is then not relevant. For example, $\Phi_1 \equiv \Phi_2$. Duality is an operation on the equivalence class:

$$\Phi \equiv \Psi \implies \Phi^* \equiv \Psi^*.$$

Equivalent Young functions give equivalent norms.

The Δ_2-class. We say that a Young function Φ satisfies the Δ_2-condition if and only if there exist $\kappa > 0$ and $x_0 > 0$ such that

$$\Phi(2x) \le \kappa\Phi(x) \qquad \text{for all } x \ge x_0.$$

For example, Φ^p and Φ_3 satisfy Δ_2, but Φ_1 and Φ_2 do not.

The Orlicz space and the Orlicz class. Let $(\Omega, \mathcal{B}, \nu)$ be a measurable space obeying some mild conditions, and let Φ be a Young function. The *Orlicz class* defined by $(\Omega, \mathcal{B}, \nu), \Phi$ is the set $\hat{L}^\Phi(\nu)$ of real-valued measurable functions u on Ω obeying

$$\int_\Omega \Phi(u(x))\nu(dx) < \infty.$$

It is a convex space of random variables, and is a vector space if and only if $\Phi \in \Delta_2$. The span of $\hat{L}^\Phi(\nu)$ is called the *Orlicz space*, L^Φ, and can be written as

$$L^\Phi := \{u : \Omega \to \mathbf{R}, \text{ measurable, and}$$

$$\int_\Omega \Phi(\alpha u(x))\nu(dx) < \infty \text{ for some } \alpha \in \mathbf{R}\}.$$

The Orlicz space L^Φ is separable if and only if $\Phi \in \Delta_2$. Thus with the choice Φ_1 of Pistone and Sempi, the space of states near a point ρ_0 becomes a convex subset of a non-separable Banach space.

Analogue of Hölder's inequality. One can prove the inequality

$$\int_\Omega |uv|\nu(dx) \le 2\|u\|_L \|v\|_{L^*},$$

where $\|v\|_{L^*}$ uses Φ^* in Equation (14.9).

Example 14.1 For $\Omega = \mathbb{R}$ and $\Phi(u) = \Phi^p(u) = |u|^p$, the Orlicz class is the Lebesgue space L^p, and the dual Orlicz space is L^q, where $p^{-1} + q^{-1} = 1$. The Orlicz norms are equivalent to the corresponding Hölder norm. We see that the Orlicz classes are the same as the Orlicz spaces, and that these are separable Banach spaces. The space associated to the function $\Phi(u) = \Phi_1(u) = \cosh u - 1$ is the dual of L^{Φ_3}, also known as the space $L \log L$ of distributions having finite differential entropy. The spaces L^{Φ_1} and L^{Φ_3} are known as Zygmund spaces. Thus, Pistone and Sempi have an infinite-dimensional version, giving a topology on observables whose dual is the space of states having finite differential entropy. The same phenomenon arises in our choice of quantum Young function: it gives a topology on the space of (generalised) observables dual to the set of states of finite von Neumann entropy.

The centred Cramér class C_0 is defined as the subset of the Cramér class C at ρ with zero mean in the state ρ; this is a closed subspace. A sufficiently small ball in the quotient Banach space $C_0 := C/\mathbb{R}$ then parametrises a neighbourhood of ρ, and can be identified with the tangent space at ρ; namely, the neighbourhood contains those points σ of $\mathcal{M}$ such that

$$\sigma = Z^{-1} e^{-X} \rho \qquad \text{for some } X \in C$$

where Z is a normalising factor. Pistone and Sempi show that the Luxemburg norm based on any point, say ρ_1 in the neighbourhood $\mathcal{N}$ of ρ_0 is equivalent to the norm given by basing it on ρ_0. Points in the intersection of two neighbourhoods, about ρ_0 and ρ_1 can therefore be given equivalent topologies, as required in the definition of a Banach manifold. Thus, they prove that the set of states in the Cramér class of any point form a Banach manifold. It is not a trivial manifold, however: the set of coordinates of any point lies in the Orlicz *class* of some point; not all points in the Orlicz *space* lie in the manifold. This is developed in Chapter 15 by R. F. Streater.

(Pistone and Sempi 1995) show that the bilinear form

$$G(X, Y) = E_\rho [XY] \tag{14.10}$$

is a Riemannian metric on the tangent space C_0, thus generalising the Fisher–Rao theory. Given n estimators, $X_1, \ldots, X_n$, they show that the Cramér–Rao inequality holds, and the the most efficient of these is 100% efficient, and it lies in the exponential space $\{e^Y\}$, where $Y \in \mathrm{Span}\{X_1, \ldots, X_n\}$; it is the state of greatest differential entropy among the allowed family, thus confirming Jaynes's ideas.

This theory is called *non-parametric* estimation theory, because we do not limit the distributions to those specified by a finite number of parameters, but allow any 'shape' for the density ρ. It is this construction that we take over to the quantum case, except that the spectrum is discrete.

14.3 Efron, Dawid and Amari

A Riemannian metric G, given by Equation (14.10), gives us a notion of parallel transport, namely that given by the Levi-Civita affine connection. Recall that an affine map, U acting on the right, from one vector space $\mathcal{T}_1$ to another $\mathcal{T}_2$, is one that obeys

$$(\lambda X + (1 - \lambda)Y)U = \lambda XU + (1 - \lambda)YU,$$

for all $X, Y \in \mathcal{T}_1$ and all $\lambda \in [0, 1]$. The same definition works on an *affine space*, that is, a convex subset of a vector space. This leads to the concept of an affine connection, which we now give.

Let $\mathcal{M}$ be a manifold and denote by T_ρ the tangent space at $\rho \in \mathcal{M}$. Consider an affine map $U_\gamma(\rho, \sigma) : T_\rho \to T_\sigma$ defined for each pair of points ρ, σ and each (continuous) path γ in the manifold starting at ρ and ending at σ. Let ρ, σ and τ be any three points in $\mathcal{M}$ and γ_1 any path from ρ to σ and γ_2 any path from σ to τ.

Definition 14.1 We say that U is an *affine connection* if $U_\emptyset = Id$ and

$$U_{\gamma_1 \cup \gamma_2} = U_{\gamma_1} \circ U_{\gamma_2}$$

where $\cup$ stands for path composition. Let X be a tangent vector at ρ_1; we call XU_{γ_1} the parallel transport of X to σ along the path γ_1.

We also require U to be smooth in ρ in a neighbourhood of the point ρ. A given metric g defines a special connection (that of Levi-Civita), and its geodesics are lines of minimal length, as measured by the metric.

Estimation theory might be considered geometrically as follows. Our model is that the distribution of a random variable lies on a submanifold $\mathcal{M}_0 \subseteq \mathcal{M}$ of states. The data give us a histogram, which is an empirical distribution. We seek the point on $\mathcal{M}_0$ that is 'closest' to the data. Suppose that the sample space is Ω, with $|\Omega| < \infty$. Let us place all positive distributions, including the experimental one, in a common manifold, $\mathcal{M}$. This manifold will be endowed with the Riemannian structure, G, provided by the Fisher metric. We then draw the geodesic curve through the data point that has shortest distance to the submanifold $\mathcal{M}_0$; where it cuts $\mathcal{M}_0$ is our estimate for the state. This procedure, however, does not always lead to unbiased estimators, if the Levi-Civita connection is used. (Dawid 1975) noticed that the Levi-Civita connection is not the only useful one. First, the ordinary mixtures of densities ρ_1, ρ_2 leads to

$$\rho = \lambda\rho_1 + (1 - \lambda)\rho_2, \qquad 0 < \lambda < 1. \tag{14.11}$$

Done locally, this leads to a connection on the manifold, now called the (-1)-Amari connection: two tangents, one at ρ_1 given by the density $\rho - \rho_1$ and the other at ρ_2 given by $\sigma - \rho_2$, are parallel if the functions on the sample space, $\rho - \rho_1$ and $\sigma - \rho_2$ are proportional as functions of ω. This differs from the parallelism given by the Levi-Civita connection.

There is another obvious convex structure, that obtained from the linear structure of the space of centred random variables, the scores. Take $\rho_0 \in \mathcal{M}$ and write $f_0 = -\log \rho_0$. Consider a perturbation ρ_X of ρ_0, which we write as

$$\rho_X = Z_X^{-1} e^{-f_0 - X}.$$

The random variable X is not uniquely defined by ρ_X, since by adding a constant to X, we can adjust the partition function to give the same ρ_X. Among all these equivalent X we can choose the *score* which has zero expectation in the state ρ_0: $\rho_0.X := \mathrm{E}_{\rho_0}(X) = 0$. The space of (zero-mean) scores was denoted C_0 above. We can define a sort of mixture of two such perturbed states, ρ_X and ρ_Y by

$$`\lambda\rho_X + (1 - \lambda)\rho_Y` := \rho_{\lambda X + (1-\lambda)Y}.$$

This is a convex structure on the space of states, and differs from that given in Equation (14.11). It leads to an affine connection, defined as follows. Let γ be any path from ρ to σ. Let X be a score at ρ. Then the parallel transport of X from ρ to σ (along γ) is

$$U_\gamma^+ X = X - \sigma \cdot X$$

now called the $(+1)$-Amari connection. Clearly, U_γ^+ does not depend on γ, and it maps the score at ρ into the score at σ.

Neither of these two connections, $U^\pm$ is metric relative to the Fisher metric, according to the following definition.

Definition 14.2 Let G be a Riemannian metric on the manifold $\mathcal{M}$. A connection $\gamma \mapsto U_\gamma$ is called a *metric connection* if

$$G_\sigma(XU_\gamma, YU_\gamma) = G_\rho(X,Y)$$

for all tangent vectors X, Y and all paths γ from ρ to σ.

The Levi-Civita connection is a metric connection; the $(\pm)$ Amari connections, while not metric, are dual relative to the Rao–Fisher metric; that is, let γ be a path connecting ρ with σ. Then for all X, Y:

$$G_\sigma(XU_\gamma^+, YU_\gamma^-) = G_\rho(X,Y)$$

where $U^\star$ is the parallel transport for the $(\star)$-connection with $\star \in \{\pm 1, 0\}$ and 0 stands for the Levi-Civita connection. Let $\nabla^\pm$ be the two covariant derivatives obtained from the connections $U^\pm$. Then $\nabla^0 =: \frac{1}{2}(\nabla^+ + \nabla^-)$ is self-dual and therefore metric, as is known. (Amari 1985) shows that $\nabla^\pm$ define flat connections without torsion. Flat means that the transport is independent of the path, and 'no torsion' means that U takes the origin of T_μ to the origin of T_ρ around any loop: it is linear. In that case there are affine coordinates, that is, global coordinates in which the respective convex structure is obtained by simply mixing coordinates linearly. Also, the geodesics of flat connections are straight lines when written in affine coordinates. Amari shows that ∇^0 is not flat, but that the manifold is a sphere in the Hilbert space L^2, and the Levi-Civita parallel transport is vector translation in this space, followed by projection back onto the sphere. The resulting affine connection is not flat, because the sphere is not flat.

In the Orlicz theory, when ν is discrete with countable support, the Orlicz spaces associated with Φ^p are the p-summable sequences $\ell^p, 1 \le p \le \infty$. These form a nested family of Banach spaces, with ℓ^1 the smallest and ℓ^∞ the largest. However, this is not the best way to look at Orlicz spaces. Legendre transforms come into their own in the context of a manifold, as a transform between the tangent space and the cotangent spaces at each point. There is only one manifold, but many co-ordinatisations. For the information manifold of Pistone and Sempi, the points of the manifold are the probability measures ν equivalent to μ, and can be coordinatised by the Radon–Nikodym derivatives $\rho = d\nu/d\mu$. In finite dimensions, the linear structure of $L^1(\Omega, d\mu)$ provides the tangent space with an affine structure, which is called the (-)-affine structure in Amari's notation. (Amari 1985) has suggested that we might also use the coordinates

$$\ell_\alpha(\rho) := \frac{2}{1-\alpha}\rho^{(1-\alpha)/2}, \qquad -1 < \alpha < 1,$$

known as the Amari embeddings of the manifold into L^p, where $p = 2/(1-\alpha)$. Then, since $\rho \in L^1$, we have $u = \rho^{(1-\alpha)/2} \in L^p$. However, in infinite dimension, the space L^1 is too big, as it contains states with infinite entropy. The Amari coordinates do provide us with an interesting family of connections, $\nabla_\alpha := \partial/\partial\ell_\alpha$, which define the Amari affine structures (for finite dimensions). The formal limit $p \to \infty$ is the case $\alpha = 1$. This corresponds to the embedding

$$\ell_1(\rho) := \log\rho.$$

This gives us the connection $(+1)$ of Amari; its geodesics are straight lines in the log coordinates. The relative entropy $S(\rho|\sigma)$ is the 'divergence' of the Fisher metric along the $(+)$-geodesic from ρ to σ. The $(+)$ affine structure corresponds to the linear structure of the random variables u, where $\rho = \rho_0 e^u$, as in the theory of Pistone and Sempi. The Orlicz topology on state space is not equivalent to that of L^1, but gives the Orlicz space corresponding to $L\log L$, as desired.

In estimation theory, the method of maximum entropy for unbiased estimators, described above, makes use of the ∇^+ connection, and gives the same answer as finding the max-entropy state on the manifold of all states; this has the same expectations for the chosen variables as the true state, and so cannot be distinguished from it by the measurements made. The same idea will be given below when the theory is governed by quantum dynamics.

14.4 The finite quantum information manifold

In the classical case (and later in the quantum case too) (Čencov 1982) asked whether the Fisher–Rao metric, Equations (14.1), was unique. Any manifold has a large number of different metrics on it; apart from those that differ just by a constant factor, one can multiply a metric by a positive space-dependent factor. There are many others. Čencov therefore imposed conditions on the metric. He saw the metric (and the Fisher metric in particular) as a measure of the distinguishability of two states. He argued that if this is to be true, then the distance between two states must be *reduced* by any stochastic map; for, a stochastic map must 'muddy the waters', reducing our ability to distinguish states. He therefore considered the class of metrics G that are reduced by any stochastic map on the random variables. Recall that in classical probability

Definition 14.3 A stochastic map is a linear map on the algebra of random variables that preserves positivity and takes 1 to itself.

Čencov was able to prove that the Fisher–Rao metric is unique, among all metrics, being the only one (up to a constant multiple) that is reduced by any stochastic map.

In finite-dimensional quantum mechanics, instead of the algebra of random variables we use the non-commutative algebra of complex matrices M_n. Measures on Ω are replaced by 'states', that is, $n \times n$ density matrices. A density matrix is a positive semi-definite matrix that replaces the density distribution function of probability theory. These are dual concepts: a state ρ determines a real number from any observable, A; the value is interpreted as the expectation of that observable A when the state of the system is ρ. It is given by

$$\rho \cdot A := Tr(\rho A).$$

The positivity of ρ gives that, if A is a positive observable, that is, operator, then $\rho \cdot A \geq 0$ for all states ρ. We limit discussion to the faithful states, which means that if A is a non-zero positive operator, then $\rho \cdot A > 0$. We take the manifold $\mathcal{M}$ to comprise the faithful states; it is a genuine manifold, and not one of the non-commutative manifolds without points that occur in Connes's theory (Connes 1994). In infinite dimensions, we choose a C^*-algebra and are able to add the requirement that the states have finite entropy. The natural morphisms in the quantum case are the completely positive maps that preserve the identity. We call these the quantum stochastic maps.

Definition 14.4 A linear map $T : M_n \to M_n$ is said to be quantum stochastic if

   (i) $TI = I$;
   (ii) $T \otimes I_j$ is positive on $M_n \otimes M_j$ for all integers $j = 1, 2, \ldots$

where $I \in M_n$ is the unit observable and I_j is the $j \times j$ unit matrix.

(Morozova and Čencov 1991) consider that the uniqueness theorem of Čencov did not extend to quantum theory: uniqueness of the metric (up to a multiple) does not follow from the requirement that the distance between any two states is reduced or left the same by every quantum stochastic map T. They do not quite prove this. Čencov passed away before they were able to complete the work. This was carried out by (Petz 1996), who has constructed all metrics on M_n with the Čencov property. As two examples, the GNS (short for Gelfand–Naimark–Segal) and BKM (short for Bogoliubov–Kubo–Mori) metrics are in common use in quantum estimation, and both are decreased, or left the same, by every such T. However, these two metrics are not proportional.

As in the classical case, there are several affine structures on the manifold of density matrices. The first one comes from the mixing of the states, and is called the (-1)-affine structure. Coordinates for a state ρ in a neighbourhood of ρ_0 are provided by $\rho - \rho_0$, a traceless matrix (with trace equal to zero) which can be taken to be small in norm. The whole tangent space at ρ is thus identified with the set of traceless matrices, and this is a vector space with the usual rules for adding matrices. Obviously, the manifold is flat relative to this affine structure.

The $(+1)$-affine structure is constructed as follows. Since a state $\rho_0 \in \mathcal{M}$ is faithful we can write $H_0 := -\log \rho_0$ and any ρ near $\rho_0 \in \mathcal{M}$ as

$$\rho = Z_X^{-1} \exp(-H_0 - X)$$

for some Hermitian matrix X. We see that X is ambiguous up to the addition of a multiple of the identity. We choose to fix X by requiring $\rho \cdot X = 0$, and call X the 'score' of ρ (in analogy to the classical case). Then the tangent space at ρ can be identified with the set of scores. Let us denote this tangent space by $\partial \mathcal{M}_\rho$. The $+1$-linear structure on $\partial \mathcal{M}_\rho$ is given by matrix addition of the scores. If the quantum Hilbert space is of infinite dimension, so that $\dim \mathcal{H} = \infty$, we shall require that X be a small form-perturbation of H_0. We also require that the generalised mean of X be zero. Corresponding to these two affine structures, there are two affine connections, whose covariant derivatives are denoted $\nabla^\pm$.

The affine structures ∇^α corresponding to Amari's family ℓ_α can be studied, but not here, see (Gibilisco and Isola 1999).

As an example of a metric on $\mathcal{M}$, let $\rho \in \mathcal{M}$, and for X, Y in $\partial \mathcal{M}_\rho$ define the GNS metric by

$$G_\rho(X, Y) = \operatorname{Re} \operatorname{Tr}[\rho X Y].$$

We remarked above that this metric is reduced by all completely positive stochastic maps T; that is, it obeys

$$G_{T\rho}(TX, TX)) \leq G_\rho(X, X),$$

in accordance with Čencov's idea. Now G is positive definite since ρ is faithful. This has been adopted by (Helstrom 1976) in the theory of quantum estimation. However, (Nagaoka 1995) has noted that if we take this metric, then the $(\pm)$-affine connections are not dual; the dual to the (-1)-affine connection, relative to this metric, is not flat and has torsion. This might lead one to choose a different metric, with respect to which these two connections are dual. In fact the BKM metric has this property, as well as being a Čencov metric. It is the only Čencov metric, up to a factor, for which this is true (Grasselli and Streater 2001).

14.4.1 Quantum Cramér–Rao inequality

We seek a quantum analogue of the Cramér–Rao inequality. Given a family $\mathcal{M}$ of density operators, parametrised by a real parameter η, we seek an estimator X whose mean we can measure in the true state ρ_η. To be unbiased, we would require $\operatorname{Tr} \rho_\eta X = \eta$, which, as in the classical case implies the weaker condition of being locally unbiased:

$$\operatorname{Tr} \left\{ \rho_\eta \rho_\eta^{-1} \frac{\partial \rho_\eta}{\partial \eta} (X - \eta) \right\} \bigg|_{\eta=0} = 1. \tag{14.12}$$

Here, we have used $\eta \operatorname{Tr} \partial \rho / \partial \eta = 0$. We adopt this weaker condition.

It is tempting to regard $L_r = \rho^{-1} \partial \rho / \partial \eta$ as a quantum analogue of the Fisher information; it has zero mean, and the above equation says that its covariance with $X - \eta$ is equal to 1. However, ρ and its derivative need not commute, so L_r is not Hermitian, and is not popular as a measure of quantum information. Instead we could use any of the Čencov metrics found by (Petz 1996), for example, the BKM metric. Let X and Y have zero mean in the state ρ. Then put

$$g_\rho(X, Y) = \int_0^1 \mathrm{Tr}\left[\rho^\alpha X \rho^{1-\alpha} Y\right] d\alpha.$$

This is a positive definite scalar product on the space of self-adjoint matrices, known as the BKM metric. Each metric leads to a Cramér–Rao inequality. Thus, the map

$$A \mapsto \frac{\partial}{\partial \eta} \mathrm{Tr}\, \rho_\eta A \Big|_{\eta=0}$$

is a linear functional on the space of self-adjoint matrices, and so must be of the form $g_\rho(A, L)$ for some $L = L^*$. This remark together with Equation (14.12) leads to $g_\rho(X, L) = 1$, which from the Schwarz inequality gives the quantum Cramér–Rao inequality (Petz 2002)

$$1 = g_\rho(X, L) \le g_\rho(X, X)^{\frac{1}{2}} g_\rho(L, L)^{\frac{1}{2}}.$$

For estimators for several parameters we get a version of the inequality in matrix form (Petz 2002). In this case, we must allow that the estimators need not commute with other. For, given a large collection of copies of the same density operator, we can find the means of X, from a subset, the mean of Y from another subset, and so on, even when X and Y do not commute.

We can reach the quantum Cramér–Rao bound when the estimator X is proportional to L. This leads to the quantum form of the exponential family. Indeed, for the exponential family,

$$\rho_\eta \cdot X = Z_\eta^{-1} \mathrm{Tr}\left[\exp\{-H - \eta X\} X\right]$$

for which

$$\frac{\partial \rho_\eta \cdot X}{\partial \eta}\Big|_{\eta=0} = Z_0^{-2} \frac{\partial Z}{\partial \eta} \exp\{-H\} \cdot X + Z_0^{-1} \mathrm{Tr} \int_0^1 d\alpha e^{-\alpha H} X e^{-(1-\alpha)H} X$$

$$= g_{\rho_0}(X, X)$$

since $e^{-H} \cdot X = 0$, as X is a score. Thus $X = L$ for the exponential family. Furthermore, by the sharpness of the Cramér–Rao inequality (which is derived from the sharpness of the Schwarz inequality) the only solution that maximises the efficiency leads to a member of the exponential family, and we are led to Jaynes's proposed solution (Jaynes 1957).

The BKM metric g is the second-degree term of the expansion of the Massieu function $\log Z$ (the Umegaki relative entropy) as in Equation (14.5). The entropy is the Legendre transform of the Massieu function,

$$S(X) = \inf\{g_\rho(X, Y) - \log Z_Y\}$$

and the reciprocal relations of Equation (14.6) hold. We have used the relative entropy, $S(\sigma|\rho) := \mathrm{Tr}\, \rho(\log \rho - \log \sigma)$, and Theorem 14.1.

Theorem 14.1 $S(\sigma|\rho) = \log Z_X$.

Proof We have

$$S(\sigma|\rho) = \operatorname{Tr}\rho\left(-H + H + X + \log \operatorname{Tr} e^{-H-X}\right),$$

giving the result, since X is a score, and so obeys $\rho \cdot X = 0$. $\square$

14.5 Perturbations by forms

We now extend the class of perturbations X to *forms* that are small relative to H (Streater 2000). The special case of analytic perturbations is covered in (Grasselli and Streater 2000, Streater 2004).

Let Σ be the set of density operators on $\mathcal{H}$, and let int Σ be its interior, the faithful states. We shall deal only with systems described by $\rho \in \operatorname{int}\Sigma$. The following class of states turns out to be tractable. Let $p \in (0,1)$ and let $\mathcal{C}_p$, denote the set of operators C such that $|C|^p$ is of trace class. This is like the Schatten class, except that we are in the less popular case, $0 < p < 1$, for which $C \mapsto (\operatorname{Tr}[|C|^p])^{1/p}$ is only a quasi-norm. Let

$$\mathcal{C}_< = \bigcup_{0<p<1} \mathcal{C}_p.$$

One can show that the entropy

$$S(\rho) := -\operatorname{Tr}[\rho \log \rho] \tag{14.13}$$

is finite for all states in $\mathcal{C}_<$. We take the *underlying set* of the quantum info manifold to be

$$\mathcal{M} = \mathcal{C}_< \cap \operatorname{int}\Sigma.$$

For example, this set contains the case $\rho = \exp\{-H_0 - \psi_0\}$, where H_0 is the Hamiltonian of the quantum harmonic oscillator, and $\psi_0 = \operatorname{Tr}\exp\{-H_0\}$. The set $\mathcal{M}$ includes most other examples of non-relativistic physics. It contains also the case where H_0 is the Hamiltonian of the free relativistic field, in a box with periodic boundary conditions. More, all these states have finite von Neumann entropy, Equation (14.13). In limiting the theory to faithful states, we are imitating the decision of Pistone and Sempi that the probability measures of the information manifold should be equivalent to the guiding measure μ, rather than, say, merely absolutely continuous. Here, the trace is the quantum analogue of the measure μ. Thus in general, an element ρ of $\mathcal{M}$ has a self-adjoint logarithm, and can be written

$$\rho = \exp(-H)$$

for some self-adjoint H, which is non-negative, since $\operatorname{Tr}\exp(-H) = 1$. Note that the set $\mathcal{M}$ is not complete relative to any quasi-norm.

Our aim is to cover $\mathcal{M}$ with balls with centre at a point $\rho \in \mathcal{M}$, each belonging to a Banach space; we have a Banach manifold when $\mathcal{M}$ is furnished with the topology induced by the norms; for this, the main problem is to ensure that various Banach norms, associated with points in $\mathcal{M}$, are equivalent at points in the overlaps of the balls. This is a main idea in (Pistone and Sempi 1995).

Let $\rho_0 \in \mathcal{M}$ and write $H_0 = -\log \rho_0 + cI$. We choose c so that $H_0 - I$ is positive definite, and we write $R_0 = H_0^{-1}$ for the resolvent at 0. We define a neighbourhood of ρ_0 to be the set of states of the form

$$\rho_V = Z_V^{-1} \exp -(H_0 + V),$$

where V is a sufficiently small H_0-bounded form perturbation of H_0. The necessary and sufficient condition to be Kato-bounded is that

$$\|V\|_0 := \|R_0^{1/2} V R_0^{1/2}\|_\infty < \infty. \tag{14.14}$$

The set of such V makes a Banach space, which we shall identify with the tangent space $\mathcal{T}(0)$ of a manifold, in analogy with the construction of Pistone and Sempi. Instead of the norm given here in Equation (14.14) we shall construct a quantum analogue of the Orlicz norm of (Pistone and Sempi 1995). It remains an open question whether these norms are equivalent. The first result is that $\rho_V \in \mathcal{M}$ for V inside a small ball in $\mathcal{T}(0)$, whichever norm is used.

The expectation value of a form V is defined as

$$\rho \cdot V := \mathrm{Tr}\{\rho^{\frac{1}{2}} V \rho^{\frac{1}{2}}\},$$

which can be shown to by finite for all states $\rho \in \mathcal{M}$. We can then define the $(+1)$-affine connection by transporting the score $V - \rho \cdot V$ at the point ρ to the score $V - \sigma \cdot V$ at σ. This connection is flat; it is also torsion-free, since it patently does not depend on the path between ρ and σ. The (-1)-connection can also be defined in $\mathcal{M}$ since each $\mathcal{C}_p$ is a vector space. However, I do not see a proof that it is continuous in the Orlicz norm. In (Streater 2009) we show that Jaynes method gives the best estimators for n commuting observables in the general case.

14.6 Conclusion

We have shown how the geometric approach to non-parametric estimation needs non-trivial analytic instruments such as the Orlicz spaces, Sections 14.2 and 14.3. This is due to the fact the the L^p topology-geometry cannot be used to model the neighbourhood of an arbitrary state (density) when we are in infinite dimension, if we wish to include only states of finite entropy. As for other parts of information geometry, e.g. Čencov's theorem in Section 14.4, it is natural to seek quantum analogues for the Pistone–Sempi construction of a manifold structure on the space of all densities in an arbitrary measure space. Different approaches to this problem outlined in Section 14.5, will be discussed in detail in Chapter 15.

References

Amari, S.-I. (1985). *Differential-geometrical Methods in Statistics, Lecture Notes in Statistics* 28 (New York, Springer-Verlag).

Čencov, N. N. (1982). *Statistical Decision Rules and Optimal Inference* (Providence, RI, American Mathematical Society). Translation from the Russian edited by Lev J. Leifman.

Connes, A. (1994). *Noncommutative Geometry* (San Diego, CA, Academic Press).

Dawid, A. (1975). Discussion of a paper by Bradley Efron, *Annals of Statistics* **3**, 1231–4.

Fisher, R. A. (1925). The theory of statistical estimation, *Proceedings of the Cambridge Philosophical Society* **22**, 700–25.

Gibilisco, P. and Isola, T. (1999). Connections on statistical manifolds of density operators by geometry of non-commutative L^p-spaces, *Infinite Dimensional Analysis, Quantum Probability and Related Topics* **2**, 169.

Grasselli, M. R., and Streater, R. F. (2000). The quantum information manifold for epsilon-bounded forms, *Reports on Mathematical Physics* **46**, 325–35 (available at arXiv:math-phys/9910031).

Grasselli, M. R. and Streater, R. F. (2001). The uniqueness of the Čencov metric in quantum information theory, *Infinite Dimensional Analysis, Quantum Probability and Related Topics* **4**, 173–82.

Grünwald, P. D. and Dawid, P. (2004). Game theory, maximum entropy, minimum discrepancy and robust Bayesian decision theory, *Annals of Statistics* **32**(4), 1367–1433.

Helstrom, C. W. (1976). *Quantum Detection and Estimation Theory* (New York, Academic Press).

Ingarden, R. S. (1992). Towards mesoscopic thermodynamics: small systems in higher-order states, *Open Systems and Information Dynamics* Vol. 1, 75–102.

Jaynes, E. T. (1957). Information theory and statistical mechanics, I, II, *Physical Review* **106**, 620–30, and **108**, 171–90,

Krasnoselski M. A. and Ruticki, Ya. B. (1961). *Convex Functions and Orlicz Spaces* (P. Noordhoff).

Morozova, E. A. and Čencov N. N. (1991). Markov invariant geometry on state manifolds (in Russian), *Itogi Nauki i Tekhniki* **36**, 69–102.

Nagaoka, H. (1995). Differential geometrical aspects of quantum states estimation and relative entropy. In *Quantum Communications and Measurements* Belavkin, V. P., Hirota, O. and Hudson, R. L. eds. (New York, Plenum).

Petz, D. (1996). Monotone metrics on matrix spaces, *Linear Algebra and Applications* **244**, 81–96.

Petz, D. (2002). Covariance and Fisher information in quantum mechanics, *Journal of Physics A, Mathematical and General* **35**, 929–39.

Pistone, G. and Sempi, C. (1995). An infinite-dimensional geometric structure on the space of all probability measures equivalent to a given one, *Annals of Statistics* **33**, 1543–61.

Rao, C. R. (1945). Information and accuracy attainable in the estimation of statistical parameters, *Bulletin Calcutta Mathematics Society* **37**, 81–91,

Rao, M. M. and Ren, Z. D. (1992). *Theory of Orlicz spaces* (New York, Marcel Decker).

Streater, R. F. (2000). The information manifold for relatively bounded potentials, *Proc. Steklov Institute of Mathematics* **228**, 217–35. arXiv:math-ph/9910035.

Streater, R. F. (2004). Quantum Orlicz spaces in information geometry, *Open Systems and Information Dynamics* **11**, 359–75.

Streater, R. F. (2009). *Statistical Dynamics* (Imperial College, London).

15

The Banach manifold of quantum states

Raymond F. Streater

Abstract

We show that a choice of Young function, for quantum states given by density operators, leads to an Orlicz norm such that the set of states of Cramér class becomes a Banach manifold. A comparison is made with the case studied by Pistone and Sempi, which arises in the theory of non-parametric estimation in classical statistics.

15.1 The work of Pistone and Sempi

The work of (Pistone and Sempi 1995) arises as a generalisation to infinitely many parameters of the theory of the best estimation of parameters of a probability distribution, using the data obtained by sampling. It is also sometimes called 'non-parametric estimation'. In 1995, Pistone and Sempi obtained a notable formalism, making use of an Orlicz space. From the point of view of quantum mechanics, the classical case corresponds to the special case where all observables generate an abelian algebra. The quantum case of a finite-dimensional Hilbert space leads to the theory of quantum information, but does not involve delicate questions of topology; this is because all norms on a space of finite dimension are equivalent. The question arises, whether we can imitate the use of an Orlicz norm in the infinite-dimensional case. We here show that this is possible, by completing the outline made earlier (Streater 2004a). We must start with a brief review of the classical case. We follow (Streater 2004a), with minor corrections.

(Pistone and Sempi 1995) develop a theory of best estimators (of minimum variance) among all locally unbiased estimators, in classical statistical theory. Thus, there is a sample space, $\mathcal{X}$, and a given σ-ring $\mathcal{B}$ of subsets of $\mathcal{X}$, the measurable sets, representing the possible events. On $\mathcal{X}$ is given a positive measure μ, which is used to specify the sets of zero measure, that is, the impossible events. It may not be true that μ is normalised, so it is not a probability. The probabilities on $\mathcal{X}$, which represent the possible states of the system, are positive, normalised measures ν on $\mathcal{X}$ that are equivalent to μ. By the Radon–Nikodym theorem, we may write

$$dv = f d\mu$$

Algebraic and Geometric Methods in Statistics, ed. Paolo Gibilisco, Eva Riccomagno, Maria Piera Rogantin and Henry P. Wynn. Published by Cambridge University Press. © Cambridge University Press 2010.

where $f(x) > 0$ μ-almost everywhere, and $\mathrm{E}_{d\,mu}[f] := \int_{\mathcal{X}} f(x)\mu(dx) = 1$. Let f_0 be such a density. (Pistone and Sempi 1995) seek a family of sets N containing f_0, and which can be taken to define the neighbourhoods of the state defined by f_0. They then did the same for each point of N, and so on, thus constructing a topological space which had the structure of a Banach manifold. Their construction is as follows. Let u be a random variable on $(\mathcal{X}, \mathcal{B})$, and consider the class of measures whose density f has the form

$$f = f_0 \exp\{u - \psi_{f_0}(u)\},$$

in which ψ, called the free energy, is finite for all states of a one-parameter exponential family:

$$\psi_{f_0}(\lambda u) := \log \mathrm{E}_{f_0 d\mu}[e^{-\lambda u}] < \infty \text{ for all } \lambda \in [-\epsilon, \epsilon], \quad \epsilon > 0. \tag{15.1}$$

This implies that all moments of u exist in the probability measure $d\nu = f_0 d\mu$ and the moment-generating function is analytic in a neighbourhood of $\lambda = 0$. The random variables u satisfying (15.1) for some $\epsilon > 0$ are said to lie in the Cramér class. This class was shown (Pistone and Sempi 1995) to be a Banach space, and so to be complete, when furnished with the norm

$$\|u\|_L := \inf \left\{ r > 0 : \mathrm{E}_{d\mu} \left[f_0 \left(\cosh \frac{u}{r} - 1 \right) \right] < 1 \right\}. \tag{15.2}$$

The map

$$u \mapsto \exp\left\{u - \psi_{f_0}(u)\right\} f_0 := f_0(u) \tag{15.3}$$

maps the unit ball in the Cramér class into the class of probability distributions that are absolutely continuous relative to μ. We can identify ψ as the free energy by writing $f_0 = \exp\{-h_0\}$. Then

$$f = \exp\left\{-h_0 - u - \psi_f(u)\right\}$$

and h_0 appears as the 'free Hamiltonian', and u as the perturbing potential, of the Gibbs state $f d\mu$. Random variables u and v that differ by a constant give rise to the same distribution. The map (15.3) becomes bijective if we adjust u so that $\mathrm{E}_{d\mu}[f_0 u] = 0$; that is, u has zero mean in the measure $f_0 d\mu$. Such a u is called a *score* in statistics. The corresponding family of measures, $f_0(\lambda u)d\mu$, is called a one-parameter exponential family. In (Pistone and Sempi 1995), a neighbourhood N of f_0 consists of all distributions in some exponential family, as u runs over the Cramér class at f_0. Similarly, Pistone and Sempi define the neighbourhood of any $f \in N$, and so on; consistency is shown by proving that the norms are equivalent on overlapping neighbourhoods. They thus construct the information manifold $\mathcal{M}$, modelled on the Banach space functions of Cramér class. This Banach space is identified with the tangent space at any $f \in \mathcal{M}$. The manifold $\mathcal{M}$ is furnished with a Riemannian metric, the Fisher metric, which at $f \in \mathcal{M}$ is the second Fréchet differential of $\psi_f(u)$.

Here we construct a quantum analogue of this manifold, following (Streater 2000, Streater 2004a), and complete it by proving that the topology is consistent, in that the norms used are equivalent on overlapping neighbourhoods of any two points. We thus extend the theory of (Gibilisco and Isola 1999) from the cases with Schatten class L_p to a suitable limit $p \to \infty$, in order to include the case analogous to the Zygmund space.

15.2 Quantum Orlicz spaces

15.2.1 The underlying set of the information manifold

The function

$$\Phi(x) = \cosh x - 1$$

used in Definition (15.2) of the Orlicz norm, is a Young function. That is, Φ is convex, and obeys

    (i) $\Phi(x) = \Phi(-x)$,

    (ii) $\Phi(0) = 0$,

    (iii) $\lim_{x \to \infty} \Phi(x) = +\infty$.

The classical theory of Orlicz spaces can use any Young function; see (Krasnoselski and Ruticki 1961, Rao and Ren 1992). It would appear, then, that to define a quantum Orlicz space would require the definition of quantum Young functions. Possibly the first attempt to do this was done in (Kunze 1990). This author takes a classical Young function Φ, and writes the corresponding quantum Young function $\Phi(X)$ as a function of the operator, X, but considers only functions of the form $\Phi(X) = \Phi(|\tilde{X}|)$, where the tilde denotes the 'reordered' value of the modulus. This is well defined for any classical Young function Φ, as we can use the spectral theorem for the self-adjoint operator $|\tilde{X}|$ to define the function. This gives rise to a norm, but it would seem that it fails to take account of the quantum phase between operators, and so might not be the correct quantum version. However, some use of this idea has been used in (Al-Rashid and Zegarlinski 2007).

The author has proposed (Streater 2000) a quantum Young function, which might be the non-commutative version of the classical Young function $\cosh x - 1$. (Jenčová 2003) has proposed a different function, closer to that in (Al-Rashid and Zegarlinski 2007), and has obtained a theory which is worth studying more closely. Let us here present our new version, which uses a different Young function.

Let $\mathcal{H}$ be a separable Hilbert space, with $\mathcal{B}(\mathcal{H})$ denoting the algebra of bounded operators on $\mathcal{H}$, and denote by Σ_+ the set of faithful normal states on $\mathcal{B}(\mathcal{H})$. In (Streater 2000) it was suggested that the quantum information manifold $\mathcal{M}$ in infinite dimensions should consist of $\rho \in \Sigma_+$ with the property that there exists $\beta_0 \in [0, 1)$ such that ρ^β is of trace class for all $\beta > \beta_0$. That is, states in $\mathcal{M}$ lie in the class $\mathcal{C}_\beta$ of Schatten, in the unfashionable case $\beta < 1$; this is a complete metrisable space of compact operators furnished by the quasi-norm

$$\rho \mapsto \|\rho\|_\beta := (\mathrm{Tr}\rho^\beta)^{1/\beta}.$$

In (Streater 2000) we took the underlying set of the quantum information manifold to be

$$\mathcal{M} := \bigcup_{0 < \beta < 1} \mathcal{C}_\beta \cap \Sigma_+.$$

For example, this set contains the case $\rho = \exp\{-H_0 - \psi_0\}$, where H_0 is the Hamiltonian of the quantum harmonic oscillator, and $\psi_0 = \mathrm{Tr}\exp\{-H_0\}$. In this example, we may take $\beta_0 = 0$. The set $\mathcal{M}$ includes most other examples of non-relativistic physics. It contains also the case where H_0 is the Hamiltonian of the free relativistic field, in a box with periodic boundary conditions. More, all these states have finite von Neumann entropy. In limiting the theory to faithful states, we are imitating the decision of Pistone and Sempi that the probability measures of the information manifold should be equivalent to the guiding measure μ, rather than, say, merely absolutely continuous. Here, the trace is the quantum analogue of the measure μ. Thus in general, an element ρ of $\mathcal{M}$ has a self-adjoint logarithm, and can be written

$$\rho = \exp(-H)$$

for some self-adjoint H, which is non-negative, since $\mathrm{Tr}\exp(-H) = 1$. Note that the set $\mathcal{M}$ is not complete relative to any given quasi-norm $\| \ . \ \|_\beta$.

Let us add to the remarks in (Streater 2000, Streater 2004a). First, we may write $\beta H = H - (1 - \beta)H$; then we have that the operator $(1 - \beta)H$ is H-small. Thus the perturbation theory of (Streater 2004b) shows that the free energy $\log \mathrm{Tr}\exp\{\beta H\}$ is indeed analytic in β lying in a neighbourhood of $\beta = 1$. We conclude that the function $\mathrm{Tr}\exp\{-\beta H\}$ is analytic if it is finite in a neighbourhood of $\beta = 1$. Note that in this theory, H is not a given Hamiltonian of some dynamics in the theory; rather, H is a positive self-adjoint operator that determines the state ρ of interest.

15.2.2 The quantum Cramér class

We perturb a given state $\rho \in \mathcal{M}$ by adding a potential X say, to H, in analogy with the classical theory where the potential is u as in (15.1). Suppose that X is a quadratic form on $\mathcal{H}$ such that $\mathrm{Dom}X \subseteq \mathrm{Dom}H^{1/2}$ and there exist positive a, b such that

$$|X(\phi, \phi)| \leq a \left\langle H^{1/2}\phi, H^{1/2}\phi \right\rangle + b\|\phi\|^2 \tag{15.4}$$

for all $\phi \in \mathrm{Dom}H^{1/2}$. Then we say that X is form-bounded relative to H. The infimum of all a satisfying (15.4) for some $b > 0$ is called the H-form bound of X; we shall denote the form bound by $\|X\|_K$, in honour of T. Kato. It is a semi-norm on the linear set of forms bounded relative to H. It is well known that if $\|X\|_K < 1$, then $H + X$ defines a semi-bounded self-adjoint operator. More, if $\|X\|_K$ is small enough, less than $a < 1 - \beta_0$, then by Lemma 4 of (Streater 2000), we have

$$e^b \beta \mathrm{Tr}\left(e^{-(1-a)H\beta}\right) \geq \mathrm{Tr}\left(e^{-(H+X)\beta}\right) \geq e^{-b\beta}\mathrm{Tr}\left(e^{-(1+a)H\beta}\right). \tag{15.5}$$

It follows that $\exp(-\beta(H+X))$ is of trace class for all $\beta > \beta_x := \beta_0/(1-a)$, which is less than 1. Thus $\rho_x := \exp-(H+X+\psi(X)) \in \mathcal{M}$ for all forms X with form-bound less that $1-\beta_0$. Here, $\psi(X) := \mathrm{Tr}[\exp-(H+X)]$.

In (Streater 2000) we defined the Cramér class (for the state $\rho = \exp\{-H\}$) to be the set of all H-form-bounded forms X of small enough semi-norm $\|X\|_K$. In (Streater 2004a) we defined the Cramér class to be the (smaller) one for which we had a proof of the analyticity condition, namely: $\psi(\lambda X)$ is analytic in a neighbourhood of $\lambda = 0$. In the commutative case, in which X commutes with H, both definitions reduce to that of Cramér, so either is a possible definition. We here revert to a definition related to that in (Streater 2000). We note that in (Streater 2000) we defined a norm on the set of H-form-bounded forms X by

$$\|X\|_0 := \|(H+1)^{-1/2}X(H+1)^{-1/2}\|.$$

Here the norm is the operator norm. We showed that the set of H-small forms was complete in this norm. Moreover, the norm of Y about a point ρ_x was shown to be equivalent to the norm of Y about ρ_0, if the point lies in both neighbourhoods. In this way, the set of states $\mathcal{M}$ is furnished with a topology making it into a Banach manifold. This theory is not related in a simple way to the classical theory of Pistone and Sempi, since the norm is not given by a Young function. We can assume that if $\rho = \exp(-H)$, then the lowest point of the spectrum of H is positive; for if it is zero, then the trace of ρ would be greater than 1. Thus H^{-1} is bounded, and we may take the Cramér class (of the state $\rho = \exp(-H)$) to consist of H-small forms X with

$$\|X\|_1 := \|H^{-1/2}XH^{-1/2}\| < 1.$$

For then we see that

$$X(\phi,\phi) = \langle H^{1/2}\phi, H^{-1/2}XH^{-1/2}H^{1/2}\phi\rangle \le \|X\|_1 \langle H^{1/2}\phi, H^{1/2}\phi\rangle.$$

This tells us that $a = \|X\|_1$ and $b = 0$; and see from (15.5) that $\psi(\lambda X)$ is finite if $|\lambda| < 1$ and continuous at $\lambda = 0$, since its value is sandwiched between $\mathrm{Tr}\left(\exp\left((1-\lambda)\beta H\right)\right)$ and $\mathrm{Tr}\left(\exp\left((1+\lambda)\beta H\right)\right)$.

15.3 The Orlicz norm

In (Streater 2000) it is proposed that the quantum analogue of the Young function $\cosh x - 1$ could be

$$\Phi_H(X) := \frac{1}{2}\mathrm{Tr}\left[(\exp\{-H+X\} + \exp\{-H-X\})\right] - 1.$$

In (Streater 2004a) this map is shown to obey the axioms

   (i) $\Phi(X)$ is finite for all forms with sufficiently small Kato bound,

   (ii) $X \mapsto \Phi(X)$ is convex,

   (iii) $\Phi(X) = \Phi(-X)$,

   (iv) $\Phi(0) = 0$ and if $X \ne 0$, $\Phi(X) > 0$, including ∞ as a possible value.

It is shown that the Luxemburg definition

$$\|X\|_H := \inf_r \left\{ r : \Phi_H\left(X/r\right) < a \right\}$$

defines a norm on the space of H-bounded forms, and that all norms obtained by different choices of $a > 0$ are equivalent. It was not proved that the two norms of a form in the neighbourhood of two states are equivalent, and this is main purpose of the present chapter.

Theorem 15.1 Let $\rho := \exp -H \in \mathcal{M}$ and let X be a form which is small relative to H. Then the Luxemburg norms relative to both H and $H + X + \psi(X)$ are equivalent: there exists a constant C such that

$$C^{-1}\|Y\|_H \leq \|Y\|_{H+X} \leq C\|Y\|_H$$

holds for all forms Y that are bounded relative to both H and $H + X$.

Proof It is known that two norms are equivalent if and only if they define the same topology on the vector space. Furthermore, it is enough to prove this at the origin, since the space is a vector space. So it is enough to prove that any convergent net $\{Y_n\} n \in \mathcal{N}$, going to zero relative to one norm, goes to zero relative to the other.

(1) Suppose that $\|Y_n\|_H \to 0$ as $n \to \infty$; then $\|Y_n\|_{H+X} \to 0$ as $n \to \infty$. Suppose not. Then there exists a net Y_n such that $\|Y_n\|_H \to 0$ but $\|Y_n\|_{H+X}$ does not go to zero. Then there exists $\delta > 0$ and a subnet $Y_{n'}$ such that $\|Y_{n'}\|_H \to 0$ but for all n' we have

$$\|Y_{n'}\|_{H+X} \geq \delta.$$

The net

$$Z_{n'} := \frac{Y_{n'}}{\|Y_{n'}\|_{H+X}}$$

still goes to zero in $\| \cdot \|_H$ but has $\|Z_{n'}\|_{H+X} = 1$. Let us drop the prime from n. We have thus proved that there exists a net Z_n such that as $n \to \infty$, $\|Z_n\|_H \to 0$ while $\|Z_n\|_{H+X} = 1$. In terms of the Young function, this gives

$$1 = \inf_r \left\{ r : \Phi_{H+X}\left(\frac{Z_n}{r}\right) < 1 \right\}$$

$$1 = \inf_s \left\{ s : \Phi_H\left(\frac{Z_n}{s\|Z_n\|_H}\right) < 1 \right\}.$$

We may choose our Φ to be lower semi-continuous (Streater 2004a), which is continuous where it is finite. So the inf is achieved at $r = 1$ and $s = 1$, to give the equations

$$1 = \frac{1}{2}\left\{ \mathrm{Tr}\left(\exp -(H + X + \psi(X) + Z_n)\right) \exp -(H + X + \psi(X) - Z_n)\right\} - 1$$

$$1 = \frac{1}{2}\left\{ \mathrm{Tr}\left(\exp -(H + \frac{Z_n}{\|Z_n\|_H})\right) + \exp -(H - Z_n/\|Z_n\|_H)\right\} - 1.$$

Therefore for all n,

$$4 = \mathrm{Tr}\exp\{-(H + X + \psi(X) + Z_n)\} + \mathrm{Tr}\exp\{-(H + X + \psi(X) - Z_n))\}$$

$$4 = \mathrm{Tr}\exp\left\{-\left(H + \frac{Z_n}{\|Z_n\|_H}\right)\right\} + \mathrm{Tr}\exp-\left\{-\left(H - \frac{Z_n}{\|Z_n\|_H}\right)\right\}.$$

We now show that it is not possible for these equations to hold for any sequence $\{Z_n\}$ with $\|Z_n\|_H \to 0$. Indeed, each exponential on the right is less than 4. Then

$$\mathrm{Tr}\exp-(H + X + \psi(X) + Z_n)$$
$$= \mathrm{Tr}\exp-\left\{\|Z_n\|\left(H + \frac{Z_n}{\|Z_n\|_H}\right) - (1 - \|Z_n\|_H)\left(H + \frac{X + \psi(X)}{1 - \|Z_n\|_H}\right)\right\}$$
$$\leq \mathrm{Tr}\left[\exp\left\{-\|Z_n\|_H\left(H + \frac{Z_n}{\|Z_n\|_H}\right)\right\}\exp\left\{-(1 - \|Z_n\|_H)(H + \frac{X + \psi(X)}{1 - \|Z_n\|_H}\}\right]$$

by the Golden–Thompson inequality, and by using the Holder inequality with $p = 1/\|Z_n\|_H$ and $q = 1/(1 - \|Z_n\|_H)$ this is smaller than or equal to the product of

$$\left\{\mathrm{Tr}\exp\left(-H - \frac{Z_n}{\|Z_n\|_H}\right)\right\}^{\|Z_n\|_H}$$

and

$$\epsilon_n := \left\{\mathrm{Tr}\exp\left(-H - \frac{X + \psi(X)}{1 - \|Z_n\|_H}\right)\right\}^{1 - \|Z_n\|_H}.$$

Now, $\mathrm{Tr}\exp\{-H - X - \psi(X)\} = 1$ and so $\epsilon_n \to 1$ as $n \to \infty$.

Since $\mathrm{Tr}\exp\{-H + Z_n/\|Z_n\|_H\} < 4$, we get, by letting $n \to \infty$, that the left-hand side obeys

$$\mathrm{Tr}\exp\{-H - X - \psi(X) - Z_n\} \leq 4^{\|Z_n\|_H}\,\epsilon_n^{1 - \|Z_n\|_H}.$$

Similarly, $\mathrm{Tr}\exp\{-H - X - \psi(X) + Z_n\} \to 1$ as $n \to \infty$, so the sum converges to 2 at most. This contradicts (15.6). Therefore our assumption, (1), that there exists a net Y_n with $\|Y_n\|_H \to 0$, but $\|Y_n\|_{H+X} \not\to 0$, is false, and we have proved that the topology given by $\|Y\|_H$ is stronger than the topology given by $\|Y\|_{H+X}$.

(2) We may replace H by $H + X + \psi(X)$ in the above argument, since by Lemma (4) of (Streater 2000), the state $\exp\{-H + X - \psi(-X)\}$ lies in $\mathcal{M}$. Then the state $\exp - H$ gets replaced by $\exp\{-H - X + \psi(X)\}$ and the same argument then shows that the topology given by $\|Y\|_{H+X}$ is stronger than that given by $\|Y\|_H$.

Hence the topologies are equivalent, by combining (1) and (2), and so the norms are equivalent, proving the theorem. $\square$

More details can be found in (Streater 2009).

References

Al-Rashid, M. H. A. and Zegarlinski, B. (2007). Non-commutative Orlicz Spaces associated with a state. *Studia Mathematica* **180**(3), 199–209.

Gibilisco, P. and Isola, T. (1999). Connections on statistical manifolds of density operators by geometry of non-commutative L^p-spaces. *Infinite Dimensional Analysis, Quantum Probability, and Related Topics*, **2**, 169–78.

Jenčová, A. (2003). Affine connections, duality and divergences for a von Neumann algebra ArXiv/math-ph/0311004.

Krasnoselski, M. A. and Ruticki, Ya. B. (1961). *Convex Functions and Orlicz Spaces*. (Gronigen, P. Noordhoff).

Kunze, W. (1990). Noncommutative Orlicz spaces and generalized Arens algebras. *Math. Nachrichten* **147**, 123–38.

Pistone, G. and Sempi, C. (1995). An infinite-dimensional geometric structure on the space of all the probability measures equivalent to the given one. *Annals of Statistics* **23**, 1543–61.

Rao, M. M. and Ren, Z. D. (1992). *Theory of Orlicz Spaces*. (New York, Marcel Decker).

Streater. R. F. (2000). The information manifold for relatively bounded potentials. *Proc. Steklov Institute of Mathematics* **228**, 205–23.

Streater, R. F. (2004a). Quantum Orlicz spaces in information geometry. *Open Systems and Information Dynamics* **11**, 359–375.

Streater, R. F. (2004b). Duality in quantum information geometry. *Open Systems and Information Dynamics* **11**, 71–77.

Streater, R. F. (2009). *Statistical Dynamics* (Imperial College, London).

16

On quantum information manifolds

Anna Jenčová

16.1 Introduction

The aim of information geometry is to introduce a suitable geometrical structure on families of probability distributions or quantum states. For parametrised statistical models, such structure is based on two fundamental notions: the Fisher information and the exponential family with its dual mixed parametrisation, see for example (Amari 1985, Amari and Nagaoka 2000).

For the non-parametric situation, the solution was given by Pistone and Sempi (Pistone and Sempi 1995, Pistone and Rogantin 1999), who introduced a Banach manifold structure on the set $\mathcal{P}$ of probability distributions, equivalent to a given one. For each $\mu \in \mathcal{P}$, the authors considered the non-parametric exponential family at μ. As it turned out, this provides a C^∞-atlas on $\mathcal{P}$, with the exponential Orlicz spaces $L_\Phi(\mu)$ as the underlying Banach spaces, here Φ is the Young function of the form $\Phi(x) = \cosh(x) - 1$.

The present contribution deals with the case of quantum states: we want to introduce a similar manifold structure on the set of faithful normal states of a von Neumann algebra $\mathcal{M}$. Since there is no suitable definition of a non-commutative Orlicz space with respect to a state φ, it is not clear how to choose the Banach space for the manifold. Of course, there is a natural Banach space structure, inherited from the predual $\mathcal{M}_*$. But, as it was already pointed out in (Streater 2004), this structure is not suitable to define the geometry of states: for example, any neighbourhood of a state φ contains states such that the relative entropy with respect to φ is infinite.

In (Jenčová 2006), we suggest the following construction. We define a Luxemburg norm using a quantum Young function, similar to that in (Streater 2004) but restricted to the space of self-adjoint operators in $\mathcal{M}$. Then we take the completion under this norm. In the classical case, this norm coincides with the norm of Pistone and Sempi, restricted to bounded measurable functions. This is described in Section 16.2. In Section 16.3, we show that an equivalent Banach space can be obtained in a more natural and easier way, using some results of convex analysis. In the following sections, we use the results in (Jenčová 2006) to introduce the manifold, and discuss possible extensions.

Algebraic and Geometric Methods in Statistics, ed. Paolo Gibilisco, Eva Riccomagno, Maria Piera Rogantin and Henry P. Wynn. Published by Cambridge University Press. © Cambridge University Press 2010.

265

Section 16.6 is devoted to channels, that is, completely positive unital maps between the algebras. We show that the structures we introduced are closely related to sufficiency of channels and a new characterisation of sufficiency is given. As it turns out, the new definition of the spaces provides a convenient way to deal with these problems.

16.2 The quantum Orlicz space

We recall the definition and some properties of the quantum exponential Orlicz space, as given in (Jenčová 2006).

16.2.1 Young functions and associated norms

Let V be a real Banach space and let V^* be its dual. We say that a function $\Phi : V \to \mathbb{R} \cup \{\infty\}$ is a *Young function*, if it satisfies:

(i) Φ is convex and lower semicontinuous;
(ii) $\Phi(x) \geq 0$ for all $x \in V$ and $\Phi(0) = 0$,
(iii) $\Phi(x) = \Phi(-x)$ for all $x \in V$,
(iv) if $x \neq 0$, then $\lim_{t \to \infty} \Phi(tx) = \infty$.

Since Φ is convex, its effective domain

$$\mathrm{dom}(\Phi) := \{x \in V, \ \Phi(x) < \infty\}$$

is a convex set. Let us define the sets

$$C_\Phi := \{x \in V, \Phi(x) \leq 1\},$$
$$L_\Phi := \{x \in V, \exists s > 0, \ \text{such that } \Phi(sx) < \infty\}.$$

Then L_Φ is the smallest vector space, containing $\mathrm{dom}(\Phi)$. Moreover, the Minkowski functional of C_Φ,

$$\|x\|_\Phi := \inf\{\rho > 0, x \in \rho C_\Phi\} = \inf\{\rho > 0, \Phi(\rho^{-1}x) \leq 1\}$$

defines a norm in L_Φ.

Let B_Φ be the completion of L_Φ under $\|\cdot\|_\Phi$. If the function Φ is finite valued, $\Phi : V \to \mathbb{R}$, (or, more generally, $0 \in \mathrm{int}\,\mathrm{dom}(\Phi)$), then $L_\phi = V$ and the norm $\|\cdot\|_\Phi$ is continuous with respect to the original norm in V, so that we have the continuous inclusion $V \subseteq B_\Phi$.

Let now $\Phi : V \to \mathbb{R}$ be a Young function and let the function $\Phi^* : V^* \to \mathbb{R} \cup \{\infty\}$ be the conjugate of Φ,

$$\Phi^*(v) = \sup_{x \in V} v(x) - \Phi(x)$$

then Φ^* is a Young function as well. The associated norm satisfies

$$|v(x)| \leq 2\|x\|_\Phi \|v\|_{\Phi^*} \qquad x \in B_\Phi, \ v \in B_{\Phi^*}$$

(the Hölder inequality), so that each $v \in B_{\Phi^*}$ defines a continuous linear functional on B_Φ, in fact, it can be shown that

$$L_{\Phi^*} = B_{\Phi^*} = B_\Phi^* \subseteq V^*$$

in the sense that the norm $\|\cdot\|_{\phi^*}$ is equivalent with the usual norm in B_{Φ}^*. Similarly, we have $L_{\Phi} = V \subseteq B_{\Phi} \subseteq B_{\Phi^*}^*$.

16.2.2 Relative entropy

Let $\mathcal{M}$ be a von Neumann algebra in standard form. Let $\mathcal{M}_*^+$ be the set of normal positive linear functionals and $\mathfrak{S}_*$ be the set of normal states on $\mathcal{M}$. For ω and φ in $\mathcal{M}_*^+$, the relative entropy is defined as

$$
S(\omega, \varphi) = \begin{cases} -\langle \log(\Delta_{\varphi, \xi_\omega})\xi_\omega, \xi_\omega \rangle & \text{if } \operatorname{supp}\omega \leq \operatorname{supp}\varphi \\[2ex] \infty & \text{otherwise} \end{cases}
$$

where ξ_ω is the representing vector of ω in a natural positive cone and $\Delta_{\varphi, \xi_\omega}$ is the relative modular operator. Then S is jointly convex and weakly lower semicontinuous. We will also need the following identity

$$
S(\psi_\lambda, \varphi) + \lambda S(\psi_1, \psi_\lambda) + (1-\lambda)S(\psi_2, \psi_\lambda) = \lambda S(\psi_1, \varphi) + (1-\lambda)S(\psi_2, \varphi) \quad (16.1)
$$

where ψ_1, ψ_2 are normal states and $\psi_\lambda = \lambda\psi_1 + (1-\lambda)\psi_2$, $0 \leq \lambda \leq 1$. This implies that S is strictly convex in the first variable.

Let us denote

$$
\mathcal{P}_\varphi := \{\omega \in \mathcal{M}_*^+, S(\omega, \varphi) < \infty\}
$$
$$
\mathcal{S}_\varphi := \{\omega \in \mathfrak{S}_*, \ S(\omega, \varphi) < \infty\}
$$
$$
K_{\varphi, C} := \{\omega \in \mathfrak{S}_*, \ S(\omega, \varphi) \leq C\}, \qquad C > 0.
$$

Then $\mathcal{P}_\varphi$ is a convex cone dense in $\mathcal{M}_*^+$ and $\mathcal{S}_\varphi$ is a convex set generating $\mathcal{P}_\varphi$. By (16.1), $\mathcal{S}_\varphi$ is a face in $\mathfrak{S}_*$. For any $C > 0$, the set $K_{\varphi, C}$ separates the elements in $\mathcal{M}$ and it is convex and compact in the $\sigma(\mathcal{M}_*, \mathcal{M})$-topology.

16.2.3 The quantum exponential Orlicz space and its dual

Let $\mathcal{M}_s$ be the real Banach subspace of self-adjoint elements in $\mathcal{M}$, then the dual $\mathcal{M}_s^*$ is the subspace of Hermitian (not necessarily normal) functionals in $\mathcal{M}^*$. We define the functional $F_\varphi : \mathcal{M}_s^* \to \mathbb{R} \cup \{\infty\}$ by

$$
F_\varphi(\omega) = \begin{cases} S(\omega, \varphi) & \text{if } \omega \in \mathfrak{S}_* \\ \infty & \text{otherwise.} \end{cases}
$$

Then F_φ is strictly convex and lower semicontinuous; with $\operatorname{dom}(F_\varphi) = \mathcal{S}_\varphi$. Its conjugate

$$
F_\varphi^*(h) = \sup_{\omega \in \mathfrak{S}_*} \omega(h) - S(\omega, \varphi)
$$

is convex and lower semicontinuous; in fact, being finite valued, it is continuous on $\mathcal{M}_s$. We have $F_\varphi^{**} = F_\varphi$ on $\mathcal{M}_s^*$.

We define the function $\Phi_\varphi : \mathcal{M}_s \to \mathbb{R}$ by

$$
\Phi_\varphi(h) = \frac{\exp(F_\varphi^*(h)) + \exp(F_\varphi^*(-h))}{2} - 1.
$$

Then Φ_φ is a Young function. Let us denote $\|h\|_\varphi := \|h\|_{\Phi_\varphi}$ and $B_\varphi := B_{\Phi_\varphi}$, then we call B_φ the *quantum exponential Orlicz space*.

Let $h \in \mathcal{M}_s$, $\|h\|_\varphi \le 1$. Then

$$\cosh(\omega(h)) \le 2e^{S(\omega,\varphi)}.$$

It follows that each $\omega \in \mathcal{S}_\varphi$ defines a continuous linear functional on B_φ. We denote by $B_{\varphi,0}$ the Banach subspace of centred elements in B_φ, that is, $h \in B_\varphi$ with $\varphi(h) = 0$. Then

$$\Phi_{\varphi,0}(h) = \frac{F_\varphi^*(h) + F_\varphi^*(-h)}{2}$$

is a Young function on $\mathcal{M}_{s,0} := \{h \in \mathcal{M}_s, \varphi(h) = 0\}$ and it defines an equivalent norm in $B_{\varphi,0}$.

Remark 16.1 Let $\mathcal{M}$ be commutative, then $\mathcal{M} = L_\infty(X, \Sigma, \mu)$ for some measure space (X, Σ, μ) with σ-finite measure μ. Then φ is a probability measure on Σ, with the density $p := d\varphi/d\mu \in L_1(X, \Sigma, \mu)$. For any Hermitian element $u \in \mathcal{M}$, $F_\varphi^*(u) = \log \int \exp(u) p \, d\mu$, so that

$$\Phi_\varphi(u) = \int \cosh(u) p \, d\mu - 1.$$

It follows that in this case, our space B_φ coincides with the closure $M_\Phi(\varphi)$ of $L_\infty(X, \Sigma, \varphi)$ in $L_\Phi(\varphi)$.

Let us now describe the dual space $B_{\varphi,0}^*$. It was proved that $B_\varphi^* = \mathcal{P}_\varphi - \mathcal{P}_\varphi$ and $B_{\varphi,0}^* = \cup_n n(K_{\varphi,1} - K_{\varphi,1})$. If we denote by $C_{\varphi,0}$ the closed unit ball in $B_{\varphi,0}^*$, then

$$C_{\varphi,0} \subseteq K_{\varphi,1} - K_{\varphi,1} \subseteq 4C_{\varphi,0} \tag{16.2}$$

so that any element in $C_{\varphi,0}$ can be written as a difference of two states in $K_{\varphi,1}$.

For v in $\mathcal{S}_\varphi - \mathcal{S}_\varphi$, let $L_v := \{\omega_1, \omega_2 \in \mathcal{S}_\varphi,\ v = \omega_1 - \omega_2\}$. We define the function $\Psi_{\varphi,0} : \mathcal{M}_{s,0}^* \to \mathbb{R}^+$ by

$$\Psi_{\varphi,0}(v) = \begin{cases} \inf_{L_v} S(\omega_1, \varphi) + S(\omega_2, \varphi) & \text{if } v \in \mathcal{S}_\varphi - \mathcal{S}_\varphi \\ \\ \infty & \text{otherwise.} \end{cases}$$

Then $\Psi_{\varphi,0}$ is a Young function and it was proved that

$$\Phi_{\varphi,0}^*(v) = 1/2\Psi_{\varphi,0}(2v)$$

for $v \in \mathcal{M}_{s,0}^*$. It follows that the norm in $B_{\varphi,0}^*$ is equivalent with $\|\cdot\|_{\Psi_{\varphi,0}}$.

16.3 The spaces $A(K_\varphi)$ and $A(K_\varphi)^{**}$

In this section, we use a well-known representation of compact convex sets, see for example (Asimow and Ellis 1980) for details. We obtain a Banach space, which turns out to be equivalent to $B_{\varphi,0}$.

Let $K \subset \mathfrak{S}_*$ be a convex set, compact in the $\sigma(\mathcal{M}_*, \mathcal{M})$-topology and separating the points in $\mathcal{M}_s$. In particular, let $K_\varphi := K_{\varphi,1}$. Let $A(K)$ be the Banach space of continuous affine functions $f : K \to \mathbb{R}$, with the supremum norm. Then K can be identified with the set of states on $A(K)$, where each element $\omega \in K$ acts on $A(K)$ by evaluation $f \mapsto f(\omega)$. Moreover, the topology of K coincides with the weak*-topology of the state space.

It is clear that any self-adjoint element in $\mathcal{M}$ belongs to $A(K)$, moreover, $\mathcal{M}_s$ is a linear subspace in $A(K)$, separating the points in K and containing all the constant functions. It follows that $\mathcal{M}_s$ is norm-dense in $A(K)$.

The dual space $A(K)^*$ is the set of all elements of the form

$$p : f \mapsto a_1 f(\omega_1) - a_2 f(\omega_2)$$

for some $\omega_1, \omega_2 \in K$, $a_1, a_2 \in \mathbb{R}^+$, so that $A(K)^*$ is a real linear subspace in $\mathcal{M}_*$. The embedding of $A(K)^*$ to $\mathcal{M}_*$ is continuous and the weak*-topology on $A(K)^*$ coincides with $\sigma(\mathcal{M}_*, \mathcal{M})$ on bounded subsets. It is also easy to see that the second dual $A(K)^{**}$ is the set of all bounded affine functionals on K.

Let $L \subseteq K$ be convex and compact. For $f \in A(K)^{**}$, the restriction to L is in $A(L)^{**}$, continuous if $f \in A(K)$ and such that $\|f|_L\|_L \leq \|f\|_K$.

Lemma 16.1 *Let $a, b > 0$, then $A(K_{\varphi,a}) = A(K_{\varphi,b})$ and $A(K_{\varphi,a})^{**} = A(K_{\varphi,b})^{**}$, in the sense that the corresponding norms are equivalent.*

Proof Suppose that $a \geq b$. Since $K_{\varphi,b} \subseteq K_{\varphi,a}$, it follows that $A(K_{\varphi,a}) \subseteq A(K_{\varphi,b})$ and $A(K_{\varphi,a})^{**} \subseteq A(K_{\varphi,b})^{**}$ with $\|f\|_{\varphi,b} \leq \|f\|_{\varphi,a}$ for $f \in A(K_{\varphi,a})^{**}$. On the other hand, let $\omega \in K_{\varphi,a}$, then $\omega_t := t\omega + (1-t)\varphi \in K_{\varphi,b}$ whenever $t \leq b/a$. Then

$$\omega = a/b\,\omega_{b/a} - (a/b - 1)\varphi$$

so that $K_{\varphi,a}$ is contained in the closed ball with radius $(2a - b)/b$ in $A(K_{\varphi,b})^*$. It follows that any $f \in A(K_{\varphi,b})^{**}$ defines a bounded affine functional over $K_{\varphi,a}$, continuous if $f \in A(K_{\varphi,b})$ and

$$\|f\|_{\varphi,a} = \sup_{\omega \in K_{\varphi,a}} |f(\omega)| \leq \|f\|_{\varphi,b}(2a - b)/b.$$

$\square$

We see from the above proof that $\mathcal{S}_\varphi \subset A(K_{\varphi,b})^*$ and each $K_{\varphi,a}$ is weak*-compact in $A(K_{\varphi,b})^*$. It follows that $\mathrm{dom}\,(F_\varphi) \subset A(K_{\varphi,b})^*$ and F_φ is a convex weak*-lower semicontinuous functional on $A(K_{\varphi,b})^*$.

Let us denote by $A_0(K)$ the subspace of elements $f \in A(K)$, such that $f(\varphi) = 0$. Then we have

Theorem 16.1 $A_0(K_\varphi) = B_{\varphi,0}$, *with equivalent norms.*

Proof We have by (16.2) that the norms are equivalent on $\mathcal{M}_s$. The statement follows from the fact that $\mathcal{M}_s$ is dense in both spaces. $\square$

16.4 The perturbed states

As we have seen, $F_\varphi(\omega) = S(\omega, \varphi)$ defines a convex lower semicontinuous functional $A(K_\varphi)^* \to \mathbb{R}$. Let $f \in A(K_\varphi)^{**}$. We denote

$$c_\varphi(f) := \inf_{\omega \in \mathcal{S}_\varphi} f(\omega) + S(\omega, \varphi).$$

Then $-c_\varphi(-f)$ is the conjugate functional $F_\varphi^*(f)$, so that c_φ is concave and upper semicontinuous, with values in $\mathbb{R} \cup \{-\infty\}$.

Suppose that $c_\varphi(f)$ is finite and that there is a state $\psi \in \mathcal{S}_\varphi$, such that

$$c_\varphi(f) = f(\psi) + S(\psi, \varphi).$$

Then this state is unique, this follows from the fact that S is strictly convex in the first variable. Let us denote this state by φ^f. Note that if $f \in \mathcal{M}_s$, then φ^f exists and it is the perturbed state (Ohya and Petz 1993), so that we can see the mapping $f \mapsto \varphi^f$ as an extension of state perturbation.

In (Jenčová 2006), we defined the perturbed state for elements in $B_{\varphi,0}$; we remark that there we used the notation $c_\varphi = F_\varphi^*$ and the state was denoted by $[\varphi^h]$, $h \in B_\varphi$. It was shown that $[\varphi^h]$ is defined for all $h \in B_\varphi$ and that the map

$$B_{\varphi,0} \in h \mapsto [\varphi^h]$$

can be used to define a C^∞-atlas on the set of faithful states on $\mathcal{M}$. By Theorem 16.1, we have the same for $A_0(K_\varphi)$. We will recall the construction below, but before that, we give some results obtained for $f \in A(K_\varphi)^{**}$.

First of all, it is clear that $c_\varphi(f + c) = c_\varphi(f) + c$ for any real c and $\varphi^f = \varphi^{f+c}$ if φ^f is defined. We may therefore suppose that $f \in A_0(K_\varphi)^{**}$.

Lemma 16.2 *Let $f \in A(K_\varphi)^{**}$ be such that φ^f exists. Then for all $\omega \in \mathcal{S}_\varphi$,*

$$S(\omega, \varphi) + f(\omega) \geq S(\omega, \varphi^f) + c_\varphi(f).$$

Equality is attained on the face in $\mathcal{S}_\varphi$, generated by φ^f.

Proof The statement is proved using the identity (16.1), the same way as Lemmas 12.1 and 12.2 in (Ohya and Petz 1993). $\qquad\square$

The previous lemma has several consequences. For example, it follows that $c_\varphi(f) \leq -S(\varphi, \varphi^f) \leq 0$ if $f \in A_0(K_\varphi)^{**}$. Further, $S(\omega, \varphi^f)$ is bounded on K_φ, so that $K_\varphi \subseteq K_{\varphi^f, C}$ for some $C > 0$. It also follows that $\mathcal{S}_\varphi \subseteq \mathcal{S}_{\varphi^f}$. In particular, $S(\varphi, \varphi^f) < \infty$ and since also $S(\varphi^f, \varphi) < \infty$, the states φ and φ^f have the same support.

Lemma 16.3 *Let $\psi = \varphi^f$ for some $f \in A(K_\varphi)^{**}$. Then we have the continuous embeddings $A(K_\psi) \sqsubseteq A(K_\varphi)$ and $A(K_\psi)^{**} \sqsubseteq A(K_\varphi)^{**}$.*

Proof Follows from $K_\varphi \subseteq K_{\psi, C}$ and Lemma 16.1. $\qquad\square$

We will now consider the set of all states φ^f, with some $f \in A(K_\varphi)^{**}$. Let ψ be a normal state, such that $\varphi \in \mathcal{S}_\psi$. We denote

$$f_\psi(\omega) := S(\omega, \psi) - S(\omega, \varphi) - S(\varphi, \psi).$$

By identity (16.1), f_ψ is an affine functional $K_\varphi \to \mathbb{R} \cup \{\infty\}$, such that $f_\psi(\varphi) = 0$.

Theorem 16.2 *Let ψ be a normal state. Then $\psi = \varphi^f$ for some $f \in A(K_\varphi)^{**}$ if and only if $\psi \in \mathcal{S}_\varphi$ and $K_\varphi \subseteq K_{\psi,C}$ for some $C > 0$.*

Proof It is clear that $\psi \in \mathcal{S}_\varphi$ if $\psi = \varphi^f$ and we have seen that also $K_\varphi \subseteq K_{\psi,C}$. Conversely, if $K_\varphi \subseteq K_{\psi,C}$, then $f_\psi \in A_0(K_\varphi)^{**}$ and

$$f_\psi(\omega) + S(\omega, \varphi) = S(\omega, \psi) - S(\varphi, \psi) \geq -S(\varphi, \psi)$$

for all $\omega \in \mathcal{S}_\varphi$. Since equality is attained for $\omega = \psi$, $\psi = \varphi^{f_\psi}$. $\qquad\square$

Note also that, by the above proof, $c_\varphi(f_\psi) = -S(\varphi, \psi)$.

16.4.1 The subdifferential

Let $\psi \in \mathcal{S}_\varphi$. The *subdifferential* at ψ is the set of elements $f \in A_0(K_\varphi)^{**}$, such that $\psi = \varphi^f$. Let us denote the subdifferential by $\partial_\varphi(\psi)$. By Theorem 16.2, the subdifferential at ψ is non-empty if and only if $K_\varphi \subseteq K_{\psi,C}$.

Lemma 16.4 *If $\partial_\varphi(\psi) \neq \emptyset$, then it is a closed convex subset in $A_0(K_\varphi)^{**}$. Moreover, c_φ is affine over $\partial_\varphi(\psi)$.*

Proof Let $f, g \in \partial_\varphi(\psi)$ and let $g_\lambda = \lambda g + (1 - \lambda)f$, $\lambda \in (0, 1)$. Then

$$g_\lambda(\psi) + S(\psi, \varphi) = \lambda c_\varphi(g) + (1 - \lambda)c_\varphi(f).$$

Since c_φ is concave, this implies that $\psi = \varphi^{g_\lambda}$ and that $c_\varphi(g_\lambda) = \lambda c_\varphi(g) + (1 - \lambda)c_\varphi(f)$. Moreover, we can write

$$\partial_\varphi(\psi) = \{g \in A(K_\varphi)^{**}, \ c_\varphi(g) - g(\psi) \geq S(\psi, \varphi)\}$$

and this set is closed, since c_φ is upper semicontinuous. $\qquad\square$

Lemma 16.5 *Let $\psi \in \mathcal{S}_\varphi$, $\partial_\varphi(\psi) \neq \emptyset$ and let $g \in A_0(K_\varphi)^{**}$. Then $g \in \partial_\varphi(\psi)$ if and only if there is some $k \in \mathbb{R}$, such that*

$$g(\omega) - f_\psi(\omega) \geq k, \ \omega \in \mathcal{S}_\varphi \quad \text{and} \quad g(\psi) - f(\psi) = k. \tag{16.3}$$

In this case, $k = c_\varphi(g) - c_\varphi(f_\psi) \leq 0$.

Proof If $g \in \partial_\varphi(\psi)$, then (16.3) follows from Lemma 16.2 and $k \leq 0$ is obtained by putting $\omega = \varphi$. Conversely, suppose that (16.3) is true, then we have for $\omega \in \mathcal{S}_\varphi$

$$g(\omega) + S(\omega, \varphi) = g(\omega) - f_\psi(\omega) + f_\psi(\omega) + S(\omega, \varphi) \geq k + c_\varphi(f_\psi)$$

and $g(\psi) + S(\psi, \omega) = k + c_\varphi(f_\psi)$, this implies that $\psi = \varphi^g$ and $c_\varphi(g) = k + c_\varphi(f_\psi)$.
$\square$

16.4.2 The chain rule

Let $\mathcal{C}_\varphi := \{\psi \in \mathcal{S}_\varphi, \ K_\varphi \subseteq K_{\psi,C}, \ K_\psi \subseteq K_{\varphi,A}$ for some $A, C > 0\}$.

Theorem 16.3 *Let $\psi \in \mathcal{C}_\varphi$. Then*

   (i) $\mathcal{S}_\varphi = \mathcal{S}_\psi$,
   (ii) $A(K_\varphi) = A(K_\psi)$, $A(K_\varphi)^{**} = A(K_\psi)^{**}$, *with equivalent norms,*
   (iii) $\varphi \in \mathcal{C}_\psi$.

Proof Let $\psi \in \mathcal{C}_\varphi$. By Theorem 16.2, $\psi = \varphi^f$, $f \in A(K_\varphi)^{**}$ and also $\varphi = \psi^g$ for some $g \in A(K_\psi)^{**}$. Now we have (i) by Lemma 16.2 and (ii) by Lemma 16.3, (iii) is obvious.
$\square$

We also have the following chain rule.

Theorem 16.4 *Let $\psi \in \mathcal{C}_\varphi$ and let $g \in A(K_\varphi)^{**}$ be such that ψ^g exists. Then*

$$c_\psi(g) = c_\varphi(g + f) - c_\varphi(f), \qquad \psi^g = \varphi^{f+g} \tag{16.4}$$

holds for $f = f_\psi$.

Proof Suppose that ψ^g exists, then

$$g(\omega) + f_\psi(\omega) + S(\omega, \varphi) = g(\omega) + S(\omega, \psi) + c_\varphi(f_\psi) \geq c_\psi(g) + c_\varphi(f_\psi)$$

for all $\omega \in \mathcal{S}_\varphi = \mathcal{S}_\psi$ and equality is attained at $\omega = \psi^g$. This implies $c_\psi(g) = c_\varphi(g + f) - c_\varphi(f)$ and $\psi^g = \varphi^{f+g}$.
$\square$

16.5 The manifold structure

Let $\mathcal{F}$ be the set of faithful normal states on $\mathcal{M}$. Let $\varphi \in \mathcal{F}$. In this section we show that we can use the map $f \mapsto \varphi^f$ to define the manifold structure on $\mathcal{F}$. So far, it is not clear if this map is well-defined or one-to-one on $A_0(K_\varphi)^{**}$. The situation is better if we restrict to $A_0(K_\varphi)$, as Theorem 16.5 shows.

Theorem 16.5 *Let $f \in A(K_\varphi)$. Then*

   (i) φ^f *exists and $\varphi^f \in \mathcal{C}_\varphi$.*
   (ii) *If $g \in A(K_\varphi)$ is such that $\varphi^g = \varphi^f$, then $f - g = \varphi(f - g)$.*
   (iii) *In Lemma 16.2, equality is attained for all $\omega \in \mathcal{S}_\varphi$, in particular,*

$$f - f(\varphi) = f_{\varphi^f}.$$

   (iv) *The chain rule (16.4) holds for all $f, g \in A(K_\varphi)$.*

Proof We may suppose that $f \in A_0(K_\varphi)$. By the results in (Jenčová 2006) and Theorem 16.1, if $f \in A_0(K_\varphi) = B_{\varphi,0}$, then $\psi = \varphi^f$ exists, $f - f(\psi) \in A_0(K_\psi) = B_{\psi,0}$ and $\varphi = \psi^{-f}$. By Theorem 16.2, $\psi \in C_\varphi$ and (i) is proved. (ii),(iii) and (iv) were proved in (Jenčová 2006). $\qquad\square$

Proposition 16.1 is not needed in our construction. It shows that each $\psi \in C_\varphi$ is faithful on $A(K_\varphi)$.

Proposition 16.1 *Let $\psi \in C_\varphi$ and let $g \in A(K_\varphi)$ be positive. Then $g(\psi) = 0$ implies $g = 0$.*

Proof Let g be a positive element in $A(K_\varphi) = A(K_\psi)$, with $g(\psi) = 0$, then by Lemma 16.5, $f_\psi + g \in \partial_\varphi(\psi)$. Since ψ^g exists, we have by the chain rule that $\psi^g = \varphi^{f_\psi + g} = \psi$. Since $g \in A_0(K_\psi)$, $g = 0$. $\qquad\square$

Let us recall that a C^p-atlas on a set X is a family of pairs $\{(U_i, e_i)\}$, such that

(i) $U_i \subset X$ for all i and $\cup U_i = X$;
(ii) for all i, e_i is a bijection of U_i onto an open subset $e_i(U_i)$ in some Banach space B_i, and for all i, j, $e_i(U_i \cap U_j)$ is open in B_i;
(iii) the map $e_j e_i^{-1} : e_i(U_i \cap U_j) \to e_j(U_i \cap U_j)$ is a C^p-isomorphism for all i, j.

Let now $X = \mathcal{F}$. For $\varphi \in \mathcal{F}$, let V_φ be the open unit ball in $A_0(K_\varphi)$ and let $s_\varphi : V_\varphi \to \mathcal{F}$ be the map $f \mapsto \varphi^f$. By Theorem 16.5, s_φ is a bijection onto the set $U_\varphi := s_\varphi(V_\varphi)$. Let e_φ be the map $U_\varphi \ni \psi \mapsto f_\psi \in V_\varphi$. Then we have

Theorem 16.6 *(Jenčová 2006)* $\{(U_\varphi, e_\varphi), \varphi \in \mathcal{F}\}$ *is a C^∞-atlas on $\mathcal{F}$.*

In the commutative case, the space corresponding to $A(K_\varphi)$ is not the exponential Orlicz space L_Φ, but the subspace M_Φ, see Remark 16.1. The corresponding commutative information manifold structure was considered in (Grasselli 2009). It follows from the theory of Orlicz spaces that (under some reasonable conditions on the base measure μ)

$$M_\Phi(\mu)^* = L_{\Phi^*}(\mu), \qquad L_{\Phi^*}(\mu)^* = L_\Phi(\mu).$$

By comparing $A(K_\varphi)$ with these results, it seems that the quantum exponential Orlicz space should be the second dual $A(K_\varphi)^{**}$, rather than $A(K_\varphi)$.

To get the counterpart of the Pistone and Sempi manifold, we would need to extend the map s_φ to the unit ball V_φ^{**} in $A_0(K_\varphi)^{**}$ and show that it is one-to-one. At present, it is not clear how to prove this. At least, we can prove that c_φ is finite on V_φ^{**}.

Lemma 16.6 *Let $f \in A_0(K_\varphi)^{**}$, $\|f\| \le 1$. Then $0 \ge c_\varphi(f) \ge -1$ and the infimum can be taken over K_φ.*

Proof Let $\omega \in \mathcal{S}_\varphi$ be such that $S(\omega, \varphi) > 1$. Since the function $t \mapsto S(\omega_t, \varphi)$ is convex and lower semicontinuous in $(0, 1)$, it is continuous and there is some

$t \in (0,1)$ such that $S(\omega_t, \varphi) = 1$, recall that $\omega_t = t\omega + (1-t)\varphi$. By strict convexity, it follows that $1 = S(\omega_t, \varphi) < tS(\omega, \varphi)$ and $S(\omega, \varphi) > 1/t$. On the other hand, $\omega_t \in K_\varphi$ and therefore $-1 \le f(\omega_t) = tf(\omega)$. It follows that

$$f(\omega) + S(\omega, \varphi) > -1/t + 1/t = 0 = f(\varphi) + S(\varphi, \varphi) \ge c_\varphi(f).$$

From this, $c_\varphi(f) = \inf_{\omega \in K_\varphi} f(\omega) + S(\omega, \varphi) \ge -1$. $\qquad\square$

16.6 Channels and sufficiency

Let $\mathcal{N}$ be another von Neumann algebra. A *channel* from $\mathcal{N}$ to $\mathcal{M}$ is a completely positive, unital map $\alpha : \mathcal{N} \to \mathcal{M}$. We will also require that a channel is normal, then its dual $\alpha^* : \varphi \mapsto \varphi \circ \alpha$ maps normal states on $\mathcal{M}$ to normal states on $\mathcal{N}$.

An important property of such channels is that the relative entropy is monotone under these maps:

$$S(\omega \circ \alpha, \varphi \circ \alpha) \le S(\omega, \varphi), \qquad \omega, \varphi \in \mathfrak{S}_*.$$

This implies that α^* defines a continuous affine map $K_\varphi \to K_{\varphi \circ \alpha}$. If $f_0 \in A(K_{\varphi \circ \alpha})^{**}$, then composition with α^* defines a bounded affine functional over K_φ, which we denote by $\alpha(f_0)$. Then $\alpha(f_0)$ is continuous if $f_0 \in A(K_{\varphi \circ \alpha})$ and

$$\|\alpha(f_0)\| = \sup_{\omega \in K_\varphi} |f_0(\omega \circ \alpha)| \le \sup_{\omega_0 \in K_{\varphi \circ \alpha}} |f_0(\omega_0)| = \|f_0\|$$

so that α is a contraction $A(K_{\varphi \circ \alpha})^{**} \to A(K_\varphi)^{**}$ and $A(K_{\varphi \circ \alpha}) \to A(K_\varphi)$.

Lemma 16.7 *Let $\alpha : \mathcal{N} \to \mathcal{M}$ be a channel and let $g_0 \in A(K_{\varphi \circ \alpha})^{**}$. Then $c_{\varphi \circ \alpha}(g_0) \le c_\varphi(\alpha(g_0))$.*

Proof We compute

$$c_\varphi(\alpha(g_0)) = \inf_{\omega \in \mathcal{S}_\varphi} g_0(\omega \circ \alpha) + S(\omega, \varphi) \ge \inf_{\omega \in \mathcal{S}_\varphi} g_0(\omega \circ \alpha) + S(\omega \circ \alpha, \varphi \circ \alpha) \ge c_{\varphi \circ \alpha}(g_0).$$

$\qquad\square$

Let $\mathcal{S}$ be a set of states in $\mathfrak{S}_*(\mathcal{M})$. We say that the channel $\alpha : \mathcal{N} \to \mathcal{M}$ is *sufficient for* $\mathcal{S}$ if there is a channel $\beta : \mathcal{M} \to \mathcal{N}$, such that

$$\omega \circ \alpha \circ \beta = \omega, \qquad \omega \in \mathcal{S}.$$

This definition of sufficient channels was introduced in (Petz 1986), see also (Jenčová and Petz 2006a), and several characterisations of sufficiency were given. Here we are interested in the following two characterisations. For simplicity, we will assume that the states, as well as the channel, are faithful.

Theorem 16.7 *(Petz 1986) Let $\psi \in \mathcal{S}_\varphi$. The channel α is sufficient for the pair $\{\psi, \varphi\}$ if and only if $S(\psi, \varphi) = S(\psi \circ \alpha, \varphi \circ \alpha)$.*

Theorem 16.8 *(Jenčová and Petz 2006b) Let $\psi = \varphi^f$ for some $f \in \mathcal{M}_s$. Then α is sufficient for $\{\psi, \varphi\}$ if and only if there is some $g_0 \in \mathcal{N}_s$, such that $\psi \circ \alpha = (\varphi \circ \alpha)^{g_0}$ and $f = \alpha(g_0)$.*

In this section we show how Theorem 16.8 can be extended to pairs $\{\psi, \varphi\}$ such that $\partial_\varphi(\psi) \neq \emptyset$.

So let $\psi = \varphi^f$ for some $f \in A(K_\varphi)^{**}$ and suppose that $\alpha : \mathcal{N} \to \mathcal{M}$ is a sufficient channel for the set $\{\psi, \varphi\}$. Let us denote $\varphi_0 := \varphi \circ \alpha$, $\psi_0 := \psi \circ \alpha$. Let $\beta : \mathcal{M} \to \mathcal{N}$ be the channel such that $\varphi_0 \circ \beta = \varphi$, $\psi_0 \circ \beta = \psi$. We will show that $\psi_0 = \varphi_0^{\beta(f_\psi)}$. To see this, note that for $\omega_0 \in \mathcal{S}_{\varphi_0}$,

$$\beta(f_\psi)(\omega_0) = f_\psi(\omega_0 \circ \beta) = S(\omega_0 \circ \beta, \psi) - S(\omega_0 \circ \beta, \varphi) - S(\psi, \varphi).$$

Then

$$\beta(f_\psi)(\omega_0) + S(\omega_0, \varphi_0)$$
$$= S(\omega_0 \circ \beta, \psi) + S(\omega_0, \varphi_0) - S(\omega_0 \circ \beta, \varphi_0 \circ \beta) - S(\psi, \varphi) \geq -S(\psi, \varphi) = c_\varphi(f_\psi)$$

by positivity and monotonicity of the relative entropy, and

$$\beta(f_\psi)(\psi_0) + S(\psi_0, \varphi_0) = -S(\psi, \varphi)$$

so that $c_{\varphi_0}(\beta(f_\psi)) = c_\varphi(f_\psi)$ and $\psi_0 = \varphi_0^{\beta(f_\psi)}$.

On the other hand, this implies by Theorem 16.2 that $f_{\psi_0} \in A(K_{\varphi_0})^{**}$ and we obtain in the same way that $\psi = \varphi^{\alpha(f_{\psi_0})}$ and $c_\varphi(\alpha(f_{\psi_0})) = c_{\varphi_0}(f_{\psi_0})$.

Theorem 16.9 *Let ψ be such that $\partial_\varphi(\psi) \neq \emptyset$ and let $\alpha : \mathcal{N} \to \mathcal{M}$ be a channel. Let $\varphi_0 = \varphi \circ \alpha$, $\psi_0 = \psi \circ \alpha$ The following are equivalent*

    (i) *α is sufficient for the pair $\{\varphi, \psi\}$,*
    (ii) *$f_{\psi_0} \in A(K_{\varphi_0})^{**}$ and $\psi = \varphi^{\alpha(f_{\psi_0})}$,*
    (iii) *$c_{\varphi_0}(f_{\psi_0}) = c_\varphi(f_\psi)$.*

Proof The implication (i) $\to$ (ii) was already proved above. Suppose (ii) holds, then

$$c_\varphi(\alpha(f_{\psi_0})) = \alpha(f_{\psi_0})(\psi) + S(\psi, \varphi) =$$
$$= -S(\psi_0, \varphi_0) - S(\varphi_0, \psi_0) + S(\psi, \varphi).$$

By putting $\omega = \varphi$ in Lemma 16.2, we obtain $c_\varphi(\alpha(f_{\psi_0})) \leq -S(\varphi, \psi)$. Then

$$0 \leq S(\psi, \varphi) - S(\psi_0, \varphi_0) \leq S(\varphi_0, \psi_0) - S(\varphi, \psi) \leq 0.$$

It follows that $c_\varphi(f_\psi) = -S(\varphi, \psi) = -S(\varphi_0, \psi_0) = c_{\varphi_0}(f_{\psi_0})$, hence (iii) holds. The implication (iii) $\to$ (i) follows from Theorem 16.7. $\qquad\square$

In particular, if $\psi = \varphi^f$ for $f \in A(K_\varphi)$, the above theorem can be formulated as follows.

Theorem 16.10 *Let $\psi = \varphi^f$, $f \in A(K_\varphi)$ and $\alpha : \mathcal{N} \to \mathcal{M}$ be a channel. Then α is sufficient for $\{\psi, \varphi\}$ if and only if there is some $g_0 \in A(K_{\varphi_0})$, such that $\psi_0 = \varphi_0^{g_0}$ and $f = \alpha(g_0)$.*

Proof The statement follows from Theorem 16.9 and the fact that if $\psi = \varphi^f$ for $f \in A_0(K_\varphi)$, then we must have $f = f_\psi$, by Theorem 16.5. $\qquad\square$

Acknowledgement

Supported by the Center of Excellence EU-QUTE and SAS-Quantum Technologies.

References

Amari, S. (1985). *Differential-geometrical Methods in Statistics* (New York, Springer-Verlag).

Amari, S. and Nagaoka, H. (2000). *Methods of Information Geometry* (AMS monograph, Oxford University Press).

Asimow, L. and Ellis, A. J. (1980). *Convexity Theory and its Applications in Functional Analysis* (London, Academic Press).

Grasselli, M. R. (2009). Dual connections in nonparametric classical information geometry, *Annals of the Institute of Statistical Mathematics* (to appear) (available at arXiv:math-ph/0104031v1).

Jenčová, A. (2006). A construction of a nonparametric quantum information manifold, *Journal of Functional Analysis* **239**, 1–20.

Jenčová, A. and Petz, D. (2006a). Sufficiency in quantum statistical inference, *Communications in Mathematical Physics* **263**, 259–76.

Jenčová, A. and Petz, D. (2006b). Sufficiency in quantum statistical inference. A survey with examples, *Infinite Dimensional Analysis, Quantum Probability and Related Topics* **9**, 331–51.

Ohya, M. and Petz, D. (1993). *Quantum Entropy and Its Use* (Heidelberg, Springer-Verlag).

Petz, D. (1986). Sufficient subalgebras and the relative entropy of states of a von Neumann algebra, *Communications in Mathematical Physics* **105**, 123–31.

Pistone, G. and Rogantin, M. P. (1999). The exponential statistical manifold, mean parameters, orthogonality and space transformations, *Bernoulli* **5**, 721–60.

Pistone, G. and Sempi, C. (1995). An infinite-dimensional geometric structure on the space of all the probability measures equivalent to the given one, *Annals of Statistics* **23**, 1543–61.

Streater, R. F. (2004). Quantum Orlicz spaces in information geometry, *Open Systems and Information Dynamics* **11**, 359–75.

17
Axiomatic geometries for text documents

Guy Lebanon

Abstract

High-dimensional structured data such as text and images is often poorly understood and misrepresented in statistical modelling. Typical approaches to modelling such data involve, either explicitly or implicitly, arbitrary geometric assumptions. In this chapter, we consider statistical modelling of non-Euclidean data whose geometry is obtained by embedding the data in a statistical manifold. The resulting models perform better than their Euclidean counterparts on real world data and draw an interesting connection between Čencov and Campbell's axiomatic characterisation of the Fisher information and the recently proposed diffusion kernels and square root embedding.

17.1 Introduction

Geometry is ubiquitous in many aspects of statistical modelling. During the last half century a geometrical theory of statistical inference has been constructed by Rao, Efron, Amari, and others. This theory, commonly referred to as information geometry, describes many aspects of statistical modelling through the use of Riemannian geometric notions such as distance, curvature and connections (Amari and Nagaoka 2000). Information geometry has been mostly involved with the geometric interpretations of asymptotic inference. Focusing on the geometry of parametric statistical families $\mathcal{P} = \{p_\theta : \theta \in \Theta\}$, information geometry has had relatively little influence on the geometrical analysis of data. In particular, it has largely ignored the role of the geometry of the data space X in statistical inference and algorithmic data analysis.

On the other hand, the recent growth in computing resources and data availability has lead to widespread analysis and modelling of structured data such as text and images. Such data does not naturally lie in $\mathbb{R}^n$ and the Euclidean distance and its corresponding geometry do not describe it well. In this chapter, we address the issue of modelling structured data using non-Euclidean geometries. In particular, by embedding data $x \in X$ into a statistical manifold, we draw a connection between

Algebraic and Geometric Methods in Statistics, ed. Paolo Gibilisco, Eva Riccomagno, Maria Piera Rogantin and Henry P. Wynn. Published by Cambridge University Press. © Cambridge University Press 2010.

the geometry of the data space X and the information geometric theory of statistical manifolds.

We begin by discussing the role of the geometry of data spaces in statistical modelling and then proceed to discuss the question of how to select an appropriate geometry. We then move on to discuss the geometric characterisations due to Čencov, Campbell and Lebanon and their applications to modelling structured data. While much of this chapter is relevant for a wide variety of data, we focus on the specific case of text data.

17.2 The role of geometry in statistical modelling

Statistical modelling often involves making assumptions concerning the geometry of the data space X. Such assumptions are sometimes made explicitly as in the case of nearest neighbour classifiers and information retrieval in search engines. In these cases the model or learning algorithm makes direct use of a distance function d on X. In other cases, geometric assumptions are implicitly made, and are revealed only after a careful examination. For example, the choice of a parametric statistical family such as the Gaussian family carries clear geometrical assumptions. Other examples include the choice of a smoothing kernel in non-parametric smoothing and the parametric form of logistic regression

$$p(y|x\,;\theta) \propto \exp(-y\langle x,\theta\rangle), \quad x,\theta \in \mathbb{R}^n, y \in \{+1,-1\} \tag{17.1}$$

where careful examination reveals its dependence on the Euclidean margin

$$\langle x,\theta\rangle = \|\theta\|_2\langle x,\hat{\theta}\rangle = \|\theta\|_2\left(\|x\|_2 - d(x,H_\theta)\right) \tag{17.2}$$

where $d(x,H_\theta) = \inf_{y \in H_\theta}\|x - y\|_2$ is the Euclidean distance of x from the flat decision hyperplane orthogonal to the unit vector $\hat{\theta} = \theta/\|\theta\|_2$ (Lebanon and Lafferty 2004).

Before proceeding we pause to informally describe the geometric notions that we will use later on in this chapter. More details concerning basic Riemannian geometry may be found in general introduction to the field such as (Spivak 1975) or the statistically oriented monographs (Kass and Voss 1997, Amari and Nagaoka 2000).

A smooth manifold X is a continuous set of points on which differentiation and other smooth operations can take place. While a smooth manifold X by itself does not carry any geometrical properties, considering it in conjunction with a local inner product g turns the topological structure X into a geometric space (X,g) called a Riemannian manifold.

The local inner product or Riemannian metric g is defined as a smooth symmetric, bilinear and positive definite function $g_x(\cdot,\cdot)$, $g_x : T_xX \times T_xX \to \mathbb{R}$ where T_xX is the tangent space of a manifold X at $x \in X$. Assuming that X is a smooth surface in $\mathbb{R}^N$, the tangent space T_xX intuitively corresponds to the subspace of vectors in $\mathbb{R}^N$ that are centred at x and are tangent to the surface X at $x \in X$. The smoothness requirement refers to smoothness of $g_x(u,v)$ as a function of $x \in X$.

The inner product leads to the notion of lengths of parametrised curves $\alpha : I \to X$

$$l(\alpha) \stackrel{\text{def}}{=} \int_I \sqrt{g_{\alpha(t)}(\dot{\alpha}(t), \dot{\alpha}(t))}\, dt \tag{17.3}$$

where $\dot{\alpha}(t)$ is the tangent vector to the curve α at time t. Using the definition of curve lengths (17.3), the local metric g leads to a distance function d on X defined as the length of the shortest curve connecting the two points

$$d(x, y) \stackrel{\text{def}}{=} \inf_{\alpha : x, y \in \alpha} l(\alpha). \tag{17.4}$$

The simplest example of a Riemannian manifold is of course $(\mathbb{R}^n, \delta)$ where $\delta_x(u, v) = \sum u_i v_i$ is a metric that is constant in $x \in X$. Curve lengths (17.3) in this case become the Euclidean curve lengths from calculus and the distance function $d(x, y)$ in (17.4) becomes the Euclidean or L_2 distance $d(x, y) = \|x - y\|_2 \stackrel{\text{def}}{=} \sqrt{\sum(x_i - y_i)^2}$.

In general, expressions (17.3), (17.4) do not have closed form expressions which may make their calculation slow and impractical, especially when the dimensionality of X is high. There are, however, parametric families of metrics $\mathcal{G} = \{g^\theta : \theta \in \Theta\}$ possessing efficient closed form expressions for (17.3), (17.4). Fortunately, such metric classes are often quite flexible and contain many popular distance functions e.g., (Lebanon 2006).

The local metric g associated with a Riemannian manifold (X, g) provides additional geometric structure beyond the concept of a distance function. This additional structure includes concepts such as curvature, flatness and angles and provides a full geometric characterisation of X.

Once a metric g on X has been identified it can be used in parametric modelling to define the parametric family under consideration. For example, the family

$$p(x\,; \mu, c) = \exp(-c\, d^2(x, \mu) - \log \psi(c, \mu)) \quad \mu \in X, c \in \mathbb{R}_{>0} \tag{17.5}$$

where d is given by (17.4) and $\psi(c, \mu) = \int \exp(-c\, d^2(x, \mu))\, dx$ generalises the Gaussian distribution to arbitrary Riemannian spaces (X, g). Inference on (X, g) using the family $\{p(\cdot\,; \mu, c) : \mu \in X, c \in \mathbb{R}_{>} 0\}$ can then proceed according to standard statistical procedures such as maximum likelihood or Bayesian analysis.

The distribution (17.5) may also be used to define a geometric smoothing kernel for use in non-parametric smoothing (Wand and Jones 1995)

$$\hat{p}(x) = \frac{1}{m} \sum_{i=1}^{m} K_c(x, x_i) = \frac{1}{m} \sum_{i=1}^{m} p(x\,; x_i, c) \quad x_1, \ldots, x_m \in X. \tag{17.6}$$

Distributions such as (17.5) and the estimator (17.6) express an explicit dependence on the data manifold geometry (X, g) which may or may not be Euclidean.

The metric g can also be used in regression or classification where we estimate a conditional model $p(y|x), x \in X$. For example, following the reasoning in (17.1), (17.2) we can define the natural extension of logistic regression to (X, g) as

$$p(y|x\,; \theta, \eta) \propto \exp(\theta\, s(x, \eta)\, d(x, H_\eta)) \quad \theta \in \mathbb{R} \tag{17.7}$$

where H_η is a decision boundary that is a flat submanifold in (X, g) (parametrised by η) and $s(x, \eta) = +1$ or -1 depending on the location of x with respect to the decision boundary H_η (Lebanon and Lafferty 2004). The notation $d(x, H_\eta)$ refers to the geometric generalisation of the margin $d(x, A) \stackrel{\text{def}}{=} \min_{y \in A} d(x, y)$ with $d(x, y)$ defined in (17.4). Note that the metric g is expressed in this case through the distance function d and the definition of flat decision boundaries H_η. Similarly, the geometric analogue of non-probabilistic classifiers such as nearest neighbours or support vector machines (SVM) corresponding to (X, g) may be defined using the distance function (17.4) and the approximated geometric diffusion kernel $K_c(x, y) = \exp(-c\, d^2(x, y))$ (Lafferty and Lebanon 2005).

In many cases, the geometric models defined above and others reduce to well-known statistical procedures when the data space is assumed to have a Euclidean geometry. This emphasises the arbitrariness associated with the standard practice of assuming the data lies in $(X, g) = (\mathbb{R}, \delta)$. The non-Euclidean analogues mentioned above demonstrate the relaxation of this assumption in favour of arbitrary geometries.

In principle, the ideas above are not entirely new. The issue of which parametric family to select or which kernel to use in smoothing have been studied extensively in statistics. Our goal in this chapter is to examine these issues from a geometric perspective. This perspective emphasises the geometric assumptions on X which are often made implicitly and without much consideration. Bringing the geometry to the forefront enables us to discover new distances, parametric families and kernels that are more appropriate for data than their commonly used Euclidean counterparts. The benefit associated with the geometric viewpoint is particularly strong in the case of structured data such as text where it is often difficult to motivate the specific selection of distances, parametric families and kernels.

17.3 Geometry selection

We turn in this section to the problem of obtaining a suitable local metric g for a given data space X. Methods for obtaining the local metric may be roughly classified according to three categories: elicitation from a domain expert, estimation from data and axiomatic characterisation. We briefly describe these methods and then proceed to concentrate on the axiomatic characterisation category in the remainder of this chapter.

17.3.1 Geometry elicitation

The most straightforward way to obtain g is to have the statistician or a domain expert define it explicitly. Unfortunately, a complete specification of the geometry is a difficult task as the inner product function g_x is local and needs to be specified at each point $x \in X$ in a smooth manner. Another source of difficulty is that it is not always easy for non-experts to understand what is the meaning or role of the local inner product g_x and specify it accordingly.

The problem of eliciting a geometry is similar to prior elicitation in subjective Bayesian analysis. In order to successfully use domain knowledge in specifying a geometry, a statistician or a geometer needs to interact with a domain expert.

The two experts work as a team with the statistician posing carefully thought-out questions to the domain experts. The responses made by the domain expert are used to obtain a relatively small class of appropriate geometries for use later on in the modelling process.

The following example makes this process more concrete. For simplicity we assume that $X = \mathbb{R}^n$ making the metric g_x a symmetric bilinear positive definite function $g_x : \mathbb{R}^n \times \mathbb{R}^n \to \mathbb{R}$. Through interaction with the domain expert, the metric search can start with a very simple metric and progressively consider more complicated forms. For example, the search may start with a constant diagonal metric $g_x(u, v) = g(u, v) = \sum_{j=1}^{n} g_j u_j v_j$ whose parameters $g_1, \ldots, g_n$ represent the importance of different dimensions and are set by the domain expert.

After specifying the constants $g_1, \ldots, g_n$ we can consider a more complicated forms by eliciting the need for non-diagonal entries in g representing the coupling of different dimensions. Finally, extending the elicitation to non-constant metrics, we can start with a base metric form $g'(u, v)$ and modulate it as necessary in different dimensions according to its position e.g., $g_x(u, v) = \prod_{j=1}^{n} h_j(x_j) g'(u, v)$. The choice of simple modulation functions facilitate their characterisation by the domain expert. For example, modulation functions such as $h_j(z) = \exp(c_j z)$ represent monotonic increase or decrease and can be characterised by eliciting the constants $c_1, \ldots, c_n$. Note that the elicitation process described here results in a well-defined metric i.e. symmetric bilinear positive definite $g_x(u, v)$ that is smooth in x.

It is important to ensure that the elicited geometry lead to efficient computation of (17.3) and (17.4). This can be achieved by limiting the classes of metrics under consideration to include only metrics g leading to closed form expressions (17.3), (17.4). We examine such classes of metrics in Section 17.5.

17.3.2 Estimating geometry from data

An alternative approach to elicitation is to estimate the geometry from data. We start by discussing first the unsupervised learning scenario where the available data $\{x_i : i = 1, \ldots, m\} \subset X$ is unlabelled and then proceed to the case of supervised learning where the available data is labelled $\{(x_i, y_i) : i = 1, \ldots, m\} \subset X \times \{-1, +1\}$.

It is often the case that while $X \subset \mathbb{R}^N$, the space X itself is of much lower dimensionality. For example, images of size 100×100 are embedded in $\mathbb{R}^{10^5}$ by vectorising the array of 100×100 pixels. Assuming that the images share a certain characteristic such as describing natural scenes or faces, the set X of possible data is a relatively small subset of $\mathbb{R}^{10^5}$. In fact, for many classes of images such as face images or handwritten digits X can be shown to be a smooth low-dimensional subset of $\mathbb{R}^{10^5}$.

Manifold learning is the task of separating the lower-dimensional X from the higher-dimensional embedding space $\mathbb{R}^N$ (Saul and Roweis 2003). Assuming no further information it is customary to consider the metric $g = \delta$ on X that is inherited from the embedding Euclidean space i.e. $g_x(u, v) = \sum u_i v_i$ where $u, v \in$

$T_x X$ are expressed in coordinates of the embedding tangent space $T_p \mathbb{R}^N \cong \mathbb{R}^N$. The resulting distance $d(x, y)$ between two points $x, y \in X$ is the Euclidean length of the shortest curve α connecting x, y that lies completely within X. In contrast to the Euclidean distance, this distance is customised to the submanifold $X \subset \mathbb{R}^N$ and does not consider curves passing through $\mathbb{R}^N \setminus X$ in (17.4).

An alternative approach is to select a metric g from a parametric family $\mathcal{G} = \{g^\theta : \theta \in \Theta\}$ based on maximisation of normalised data volume (Lebanon 2006)

$$\prod_{i=1}^{m} \frac{\mathrm{dvol}(g_{x_i})}{\int_X \mathrm{dvol}(g_x)\, dx} \quad \text{where} \quad \mathrm{dvol}(g_x) = \sqrt{\det g_x}.$$

In contrast to manifold learning, this approach has the advantage that by carefully selecting the metric family $\mathcal{G}$ it is possible to ensure that the obtained metric leads to efficient computation of the distance function and other related quantities (Lebanon 2006).

In the supervised case, the presence of data labels y_i can be used to obtain a geometry that emphasises certain aspects of the data. For example, since the task of classifying or estimating $p(y|x)$ requires significant geometric separation between the two classes $U = \{x_j : y_j = 1\}$, $V = \{x_j : y_j = -1\}$, it makes sense to obtain a metric $g \in \mathcal{G}$ that realises a high degree of separation between u and V. The selected metric g can then be used in constructing a conditional model $p(y|x)$ or a classifier. As in the previous case, careful selection of $\mathcal{G}$ can ensure efficient computing of the distances and other geometric quantities.

17.3.3 Axiomatic characterisations

Axiomatic characterisations employ geometrical tools to single out a single metric, or a family of metrics that enjoy certain desirable properties or axioms. It is remarkable that the characterised geometries are often related to well-known statistical procedures and distances. As a result, the axiomatic characterisation may be used to motivate these procedures from a geometrical perspective. On the other hand, modifying or augmenting the axioms results in new geometries that may be more appropriate for the specific space X under consideration. The next section contains more details on this topic and Section 17.5 discusses it in the context of text data.

17.4 Congruent embeddings and simplicial geometries

The n-dimensional simplex

$$\mathbb{P}_n = \left\{ x \in \mathbb{R}^{n+1} : \forall i \ x_i > 0, \sum_{i=1}^{n} x_i = 1 \right\} \subset \mathbb{R}^{n+1}$$

represents the set of all positive probability distributions, or alternatively multinomial parameters, over $n + 1$ items. In the case of $n = 2$ it is easy to visualise the simplex $\mathbb{P}_2$ as a 2-D triangle shaped surface in $\mathbb{R}^3$ or $\mathbb{R}^2$ (see Figure 17.1). Closely

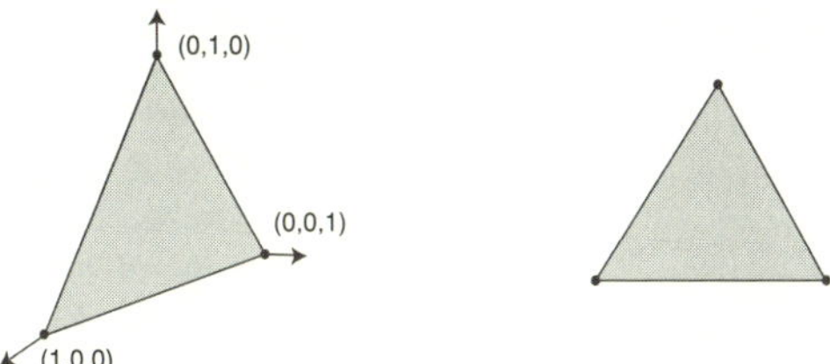

Fig. 17.1 The 2-simplex $\mathbb{P}_2$ may be visualised as a surface in $\mathbb{R}^3$ (left) or as a triangle in $\mathbb{R}^2$ (right).

related to the simplex is

$$\mathbb{R}^n_{>0} = \{x \in \mathbb{R}^n : \forall i \ x_i > 0\}$$

representing non-normalised non-negative measures.

We consider the simplex $\mathbb{P}_n$ and $\mathbb{R}^n_{>0}$ rather than their closures $\overline{\mathbb{P}_n}, \overline{\mathbb{R}^n_{>0}}$ which contain zero components to ensure that they are smooth manifolds. The discussion concerning $\mathbb{P}_n$ and $\mathbb{R}^n_{>0}$ presented here applies to the above closures in their entirety through the use of limiting arguments such as the ones described in (Lafferty and Lebanon 2005).

At first glance, the simplex seems to describe probabilities or statistical models rather than the data space itself. However, many types of structured data can be represented as points in $\overline{\mathbb{P}_n}, \overline{\mathbb{R}^n_{>0}}$ or their products $\overline{\mathbb{P}^k_n}, \overline{\mathbb{R}^{nk}_{>0}}$ using embedding arguments (Lebanon 2005b). We elaborate on this in the next section where we demonstrate various embedding techniques of text documents as distributions and conditional distributions. As a result, an axiomatic characterisation of the geometry underlying the above spaces is directly applicable to modelling the embedded data using the ideas mentioned in Section 17.2.

Čencov's characterisation of the simplex geometry makes use of invariance under congruent embedding by Markov morphisms. We start by informally defining the necessary geometric concepts. Our presentation is based on the relatively simple exposition given by Campbell (Campbell 1986) rather than Čencov's original formulation (Čencov 1982).

A bijective and smooth mapping between two Riemannian manifolds $f : (M, g) \to (N, h)$, defines the push-forward transformation $f_* : T_x M \to T_{f(x)} N$ which maps tangent vectors in M to the corresponding tangent vectors in N. Since tangent vectors correspond to differentiation operators, f_* generalises the well-known Jacobian mapping from real analysis. Using the push-forward map we define the pull-back metric $f^* h$ on M defined as

$$(f^* h)_x(u, v) = h_{f(x)}(f_* u, f_* v).$$

If $f^* h = g$ we say that the mapping f is an isometry between (M, g) and (N, h). In this case, the two manifolds may be considered geometrically equivalent as all their geometrical content including distances, volume, angles and curvature are in perfect agreement.

Definition 17.1 A Markov morphism is a matrix $Q \in \mathbb{R}^{n \times l}$, with $n \leq l$, having non-negative entries such that every row sums to 1 and every column has precisely one non-zero element.

Markov morphisms $Q \in \mathbb{R}^{n \times l}$ are linear transformations which map $\mathbb{P}_{n-1}$ injectively into $\mathbb{P}_{l-1}$. Referred to as congruent embeddings by Markov morphism, these mappings are realised by $x \mapsto xQ$ where $x \in \mathbb{P}_{n-1}$ and $xQ \in \mathbb{P}_{l-1}$ are considered as row vectors. A close examination of the mapping $x \mapsto xQ$ shows that it corresponds to probabilistic refining of the event space $\{1, \ldots, n\} \mapsto \{1, \ldots, l\}$ where the refinement $i \to j$ occurs with probability Q_{ij}.

Proposition 17.1 ((Čencov 1982)) *Let $\{(\mathbb{P}_n, g^{(n)}) : n = 1, 2, 3, \ldots\}$ be a sequence of Riemannian manifolds. Then, any congruent embedding by a Markov morphism acting on this sequence $Q : (\mathbb{P}_k, g^{(k)}) \to (\mathbb{P}_l, g^{(l)})$ is an isometry onto its image if and only if*

$$g_x^{(n)}(u, v) \propto \sum_{i=1}^{n+1} \frac{u_i v_i}{x_i}, \quad x \in \mathbb{P}_n \tag{17.8}$$

where $u, v \in T_x \mathbb{P}_n$ are expressed in coordinates of the embedding $T_p \mathbb{R}^{n+1}$ i.e. $u, v \in \{z \in \mathbb{R}^{n+1} : \sum_{i=1}^{n+1} z_i = 0\}$.

The metric (17.8) coincides with the Fisher information

$$g_\theta^{(n)}(u, v) = \mathrm{E}_{p_\theta}\left\{ (D_u \log p_\theta)[D_v \log p_\theta] \right\}$$

where p_θ is the multinomial distribution parametrised by θ. D_u, D_v are the partial differentiation operators corresponding to the tangent vectors u, v. As a result, the metric (17.8) is commonly referred to as the Fisher metric for $\mathbb{P}_n$ and the resulting geometry is certainly the most important example of information geometry.

The axioms underlying the characterisation theorem are sometimes referred to as invariance under sufficient statistics transformations (Amari and Nagaoka 2000). The name comes from the fact that the inverses of Markov morphisms correspond to extracting statistics which are sufficient by definition under the multinomial associated with the rougher event space $\mathbb{P}_n$. The above proposition implies that under Markov morphisms any Riemannian metric different from (17.8) will necessarily transform to a different functional form. This makes it difficult to know that a metric different from (17.8) is the appropriate one since its precise shape depends on the granularity of the event space.

An interesting way to visualise the Fisher metric is to consider the isometry ν between the Fisher information inner product on the simplex and the Euclidean inner product on the positive orthant of the sphere. This isometry $\nu : (\mathbb{P}_n, g^{(n)}) \to (\mathbb{S}_m^+, \delta)$ is defined by $\nu(x_1, \ldots, x_{n+1}) = (\sqrt{x_1}, \ldots, \sqrt{x_{n+1}})$ where

$$\mathbb{S}_n^+ = \left\{ x \in \mathbb{R}^{n+1} : \forall i \; x_i > 0, \; \sum_i x_i^2 = 1 \right\}$$

and δ is as before the metric inherited from the embedding Euclidean space $\delta_x(u, v)$

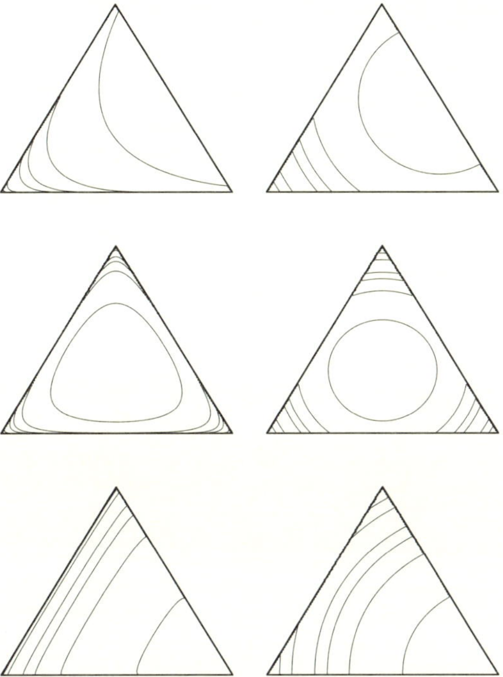

Fig. 17.2 Equal distance contours on $\mathbb{P}_2$ from the upper right edge (top row), the centre (centre row), and lower right corner (bottom row). The distances are computed using the Fisher information (left) and Euclidean (right) metrics.

$= \langle u, v \rangle = \sum_i u_i v_i$. In other words, transforming the probability vector by taking square roots maps the simplex to the positive portion of the sphere where the Fisher metric $g^{(n)}$ becomes the standard Euclidean inner product. As a result, the distance function $d(x, y), x, y \in \mathbb{P}_n$ corresponding to the Fisher metric may be computed as the length of the shortest curve connecting $\nu(x), \nu(y)$ on the sphere

$$d(x, y) = \arccos \left(\sum_i \sqrt{x_i y_i} \right). \tag{17.9}$$

Figure 17.2 illustrates (17.9) on the simplex $\mathbb{P}_2$ and contrasts it with the Euclidean distance function $\|x - y\|_2$ resulting from $(\mathbb{P}_n, \delta)$.

As mentioned in Section 17.2, the metric contains additional information besides the distance function that may be used for statistical modelling. For example, flat surfaces in $(\mathbb{P}_n, g^{(n)})$ are curved in $(\mathbb{P}_n, \delta)$ and vice versa. An interesting visualisation of this can be found in Figure 17.3 which contrasts the standard definition of logistic regression (17.1) which assumes Euclidean geometry with its Fisher information analogue (17.7). The decision boundaries in the non Euclidean case correspond to flat surfaces in the Fisher geometry which are the correct geometric analogue of linear hyperplanes. A similar demonstration may be found in Figure 17.4 which contrasts the decision boundaries obtained by support vector machines (SVM) with

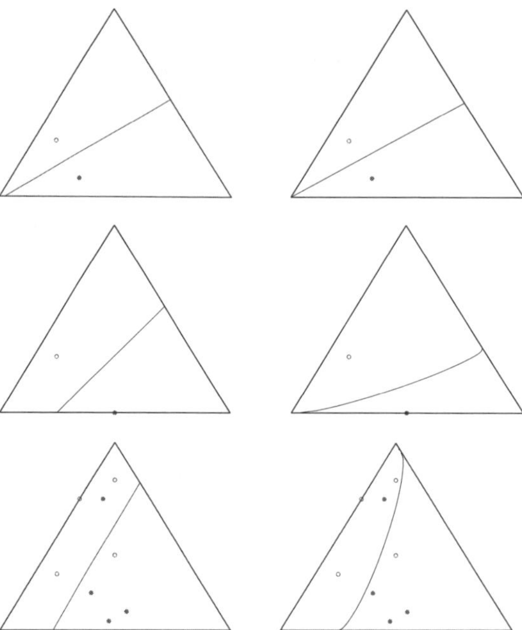

Fig. 17.3 Experiments contrasting flat decision boundaries obtained by the maximum likelihood estimator (MLE) for Euclidean logistic regression (left column) with multinomial logistic regression (right column) for toy data in $\mathbb{P}_2$.

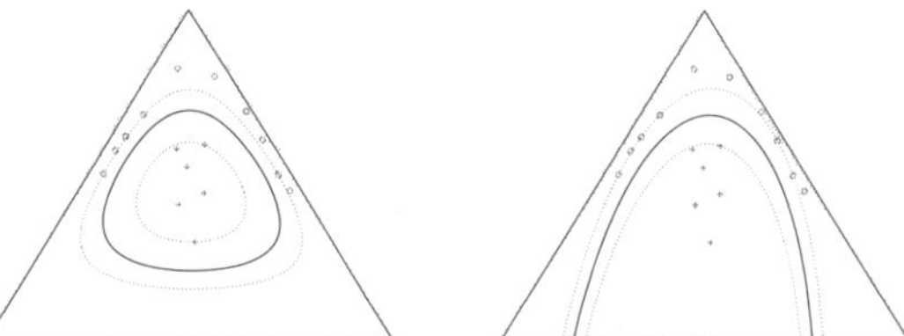

Fig. 17.4 Decision boundaries obtained by SVM trained on synthetic data using the Euclidean heat kernel (right) and the information geometry heat kernel (left).

the Euclidean diffusion kernel (also known as radial basis function or RBF kernel) and with the Fisher geometry diffusion kernel (Lafferty and Lebanon 2005).

Proposition 17.2 ((Campbell 1986)) *Let* $\{(\mathbb{R}^n_{>0}, g^{(n)}) : n = 2, 3, \ldots\}$ *be a sequence of Riemannian manifolds. Then, any congruent embedding by a Markov morphism* $Q : (\mathbb{R}^n_{>0}, g^{(n)}) \to (\mathbb{R}^l_{>0}, g^{(l)})$ *is an isometry onto its image if, and only if,*

$$g_x^{(n)}(u, v) = A(|x|) \sum_i \sum_j u_i v_j + |x| B(|x|) \sum_i \frac{u_i v_i}{x_i} \tag{17.10}$$

where $|x| = \sum x_i$ *and* $A, B : \mathbb{R} \to \mathbb{R}$ *are smooth functions.*

The restriction of $\mathbb{R}^n_{>0}$ to the simplex results in $x \in \mathbb{P}_{n-1} \subset \mathbb{R}^n_{>0}$, $\sum u_i = \sum v_i = 0$

making the choice of A immaterial as the first term in (17.10) zeros out. Similarly, in this case, $|x| = 1$ making the choice of B immaterial as well and reducing Proposition 17.2 to Proposition 17.1.

The extension of Proposition 17.1 to products $\mathbb{P}_n^k$ and $\mathbb{R}_{>0}^{nk}$ corresponding to spaces of conditional distributions and non-negative measures is somewhat more complicated as the definition of Markov morphisms need to be carefully formulated. The appropriate extension characterises the invariant metric on $\mathbb{R}_{>0}^{kn}$ as

$$g_x^{(k,n)}(u,v) = A(|x|) \sum_{a,b,c,d} u_{ab}v_{cd} + |x|B(|x|) \sum_{a,b,d} \frac{u_{ab}v_{ad}}{|x_a|}$$

$$+ |x|C(|x|) \sum_{a,b} \frac{u_{ab}v_{ab}}{x_{ab}}, \quad x \in \mathbb{R}_{>0}^{kn} \tag{17.11}$$

where $u, v \in T_x \mathbb{R}_{>0}^{kn} \cong \mathbb{R}^{k \times n}$, $|x| \stackrel{\text{def}}{=} \sum_a |x_a| \stackrel{\text{def}}{=} \sum_{a,b} x_{ab}$, and $A, B, C : \mathbb{R} \to \mathbb{R}$ are smooth functions. See (Lebanon 2005a) for further details and for the analogue expression corresponding to spaces of conditional distributions $\mathbb{P}_n^k$.

17.5 Text documents

Documents are most accurately described as time series $y = \langle y_1, \ldots, y_N \rangle$ containing words or categorical variables $y_i \in V$. We assume that the vocabulary or set of possible words V is finite and with no loss of generality, define it to consist of integers $V = \{1, \ldots, |V|\}$.

The representation of y as categorical time series is problematic due to its high dimensionality and since the representation depends on the document length which makes it hard to compare documents of varying lengths. A popular alternative is to represent the document using its word histogram, also known as the bag of words (bow) representation

$$\gamma^{\text{hist}}(y) \stackrel{\text{def}}{=} \left(\frac{1}{N} \sum_{j=1}^N \delta_{1,y_j}, \ldots, \frac{1}{N} \sum_{j=1}^N \delta_{k,y_j} \right) \in \mathbb{R}^{|V|}. \tag{17.12}$$

For example, assuming $V = \{1, \ldots, 5\}$ we have

$$\gamma^{\text{hist}}(\langle 1, 4, 3, 1, 4 \rangle) = \gamma^{\text{hist}}(\langle 4, 4, 3, 1, 1 \rangle) = \left(\frac{2}{5}, 0, \frac{1}{5}, \frac{2}{5}, 0 \right). \tag{17.13}$$

The histogram representation (17.12) maps documents to the simplex $\mathbb{P}_{V-1}$ but the embedding γ^{hist} is neither injective nor onto. The lack of injectivity is not a serious problem since by definition the histogram representation ignores word order and identifies two documents with the same word contents as the same document. The image of the histogram representation is a strict subset of the simplex containing only vectors with rational coefficients $\text{image}(\gamma^{\text{hist}}) = \mathbb{P}_{V-1} \cap \mathbb{Q}^V$. However, since the $\text{image}(\gamma^{\text{hist}})$ is a discrete set, it makes sense to consider instead the interior of its completion $\text{int}(\overline{\text{image}(\gamma^{\text{hist}})})$ which coincides with the simplex $\mathbb{P}_{V-1}$.

The histogram embedding of text in $\mathbb{P}_{V-1}$ has a clear statistical interpretation. Assuming that text documents are generated by unknown multinomial distributions

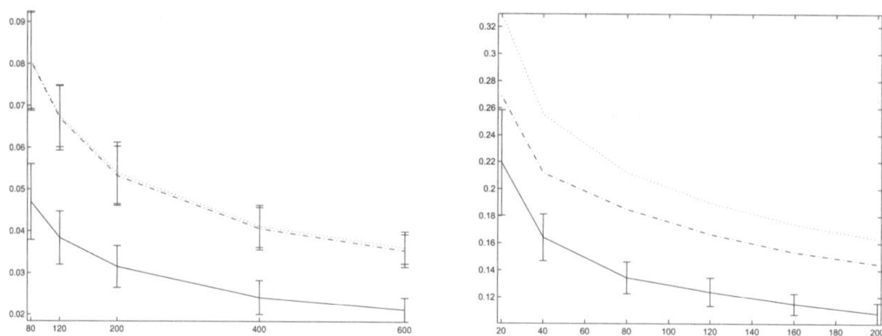

Fig. 17.5 Error rate over a held out set as a function of the training set size (WebKB data). Left: SVM using Fisher diffusion kernels (solid), Euclidean diffusion kernel (dashed), and linear kernel (dotted). Right: Logistic regression using the Fisher geometry (solid), Euclidean geometry (dashed) and Euclidean geometry following L_2 normalisation (dotted). Error bars represent one standard deviation.

$y \sim \mathrm{Mult}(\theta_y)$, we have that the histogram representation is the maximum likelihood estimator for the multinomial parameter $\gamma^{\mathrm{hist}}(y) = \hat{\theta}_y^{\mathrm{mle}}$. Viewed in this way, $\gamma^{\mathrm{hist}}(y)$ is but one possible embedding in the simplex. Other estimators such as the maximum posterior under a Dirichlet prior $\gamma^{\mathrm{map}}(y) \propto \gamma^{\mathrm{hist}}(y) + \alpha$ and empirical Bayes would result in other embeddings in the simplex.

Since the embedded documents represent multinomial distributions, it is natural to invoke Čencov's theorem and to use the Fisher geometry in modelling them. Experiments on a number of real-world text classification datasets indicate that the resulting classifiers perform significantly better than their Euclidean versions. Figure 17.5 contrasts the error rates for the Fisher and Euclidean based SVM (using diffusion kernels) and logistic regression. Further details and additional results may be found in (Lebanon and Lafferty 2004, Lafferty and Lebanon 2005, Lebanon 2005b).

In the case of embedded text documents it is beneficial to consider metrics g that are not symmetric i.e. $g_x(u,v) \neq g_{\pi(x)}(\pi(u), \pi(v))$ where $\pi(z)$ permutes the components of the vector z. Intuitively, the different components correspond to different vocabulary words which carry a non-exchangeable semantic meaning. For example, stop words such as 'the' or 'a' usually carry less meaning than other content words and their corresponding component should influence the metric $g_x(u,v)$ less than other components. Similarly, some words are closely related to each other such as 'often' and 'frequently' and should not be treated in the same manner as two semantically unrelated words. Some progress along these lines is described in (Lebanon 2006) where the invariance axioms in Proposition 17.1 are extended in a way that leads to characterisation of non-symmetric metrics.

While the histogram embedding provides a convenient document representation and achieves reasonable accuracy in text classification, it is less suitable for more sequential tasks. Since it completely ignores word ordering e.g., (17.13), it is not suitable for modelling the sequential progression of semantics throughout documents. A reasonable alternative is to assume that different words y_s, y_t in the document y are generated by different multinomials θ_s, θ_t, where $\theta_s \to \theta_t$ as $s \to t$ i.e., close

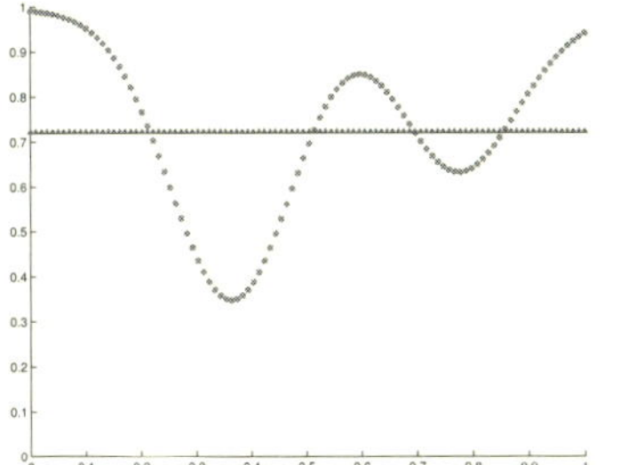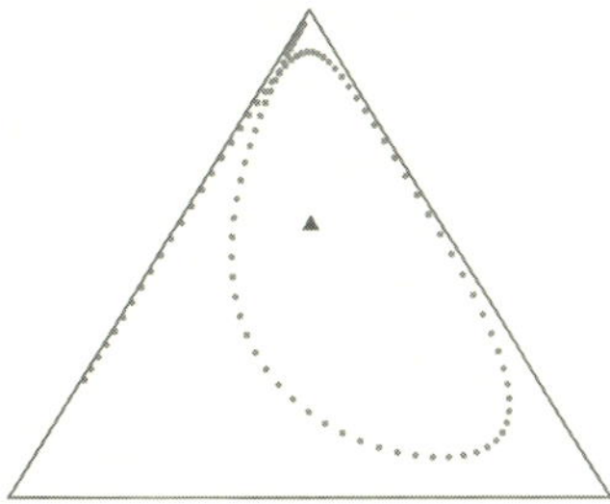

Fig. 17.6 Documents over $V = \{1, 2\}$ and $V = \{1, 2, 3\}$ can be represented as curves in the simplex $\mathbb{P}_1$ (left) and $\mathbb{P}_2$ (right). The horizontal line (left) and triangle (right) represent the global histogram representation of the document.

words are generated by similar multinomials. The local likelihood estimator for this semi-parametric model uses local smoothing to estimate the locally weighted bag of words (lowbow) or multinomial models (Lebanon *et al.* 2007). Replacing the discrete location parameter $1, \ldots, N$ within documents by a continuous interval I the local estimator provides a smooth curve in the simplex $\mathbb{P}_{V-1}$ representing the smooth transition of the local multinomial models $\{\theta_t : t \in I\}$ throughout the document. For example, the curves corresponding to the documents $z = \langle 1,1,1,2,2,1,1,1,2,1,1 \rangle$ and $w = \langle 1,3,3,3,2,2,1,3,3 \rangle$ over $V = \{1, 2\}$ and $V = \{1, 2, 3\}$ (respectively) are illustrated in Figure 17.6.

The resulting curve $\gamma(y) : I \to \mathbb{P}_{V-1}$ embeds documents in an infinite product of simplices $\mathbb{P}_{V-1}^I$. Probabilistically, the curve $\gamma(y) \in \mathbb{P}_{V-1}^I$ represents a conditional multinomial distribution. Using the characterisation (17.11) we obtain a geometry for use in sequential modelling of the curves $\gamma(y)$ (Lebanon *et al.* 2007, Mao *et al.* 2007). Experiments reported in (Lebanon *et al.* 2007) confirm the practical benefit of using the sequential embedding in $\mathbb{P}_{V-1}^I$ using the characterised geometry.

17.6 Discussion

Modelling high-dimensional structured data is often poorly understood. The standard approach is to use existing models or classifiers as black boxes without considering whether the underlying assumptions are appropriate for the data. In particular many existing popular models assume, either explicitly or implicitly, that the data space X is well characterised by the Euclidean geometry. Explicitly obtaining a geometry for the data space X through elicitation, learning from data, or axiomatic characterisation, enables the construction of more accurate and data-specific models.

In this chapter, we discussed the role of data geometry in statistical modelling and described several approaches to obtaining a geometry for the data space. Using the embedding principle and Čencov's theorem we describe several axiomatic characterisations of the geometry of X. These geometries are closely related to the Fisher information and provide an interesting connection to the theory of information geometry and its relation to asymptotic inference. Furthermore, experimental evidence demonstrates that the characterised geometries lead to geometric generalisations of popular classifiers which provide state-of-the-art performance in modelling text documents.

References

Amari, S.-I. and Nagaoka, H. (2000). *Methods of Information Geometry* (American Mathematical Society, Oxford University Press).

Campbell, L. L. (1986). An extended Čencov characterization of the information metric. In *Proc. of the American Mathematical Society* **98**(1), 135–41.

Čencov, N. N. (1982). *Statistical Decision Rules and Optimal Inference* (Providence, RI, American Mathematical Society).

Kass, R. E. and Voss, P. W. (1997). *Geometrical Foundation of Asymptotic Inference* (New York, John Wiley & Sons).

Lafferty, J. and Lebanon, G. (2005). Diffusion kernels on statistical manifolds, *Journal of Machine Learning Research* **6**, 129–63.

Lebanon, G. (2005a). Axiomatic geometry of conditional models, *IEEE Transactions on Information Theory* **51**(4), 1283–94.

Lebanon, G. (2005b). Riemannian geometry and statistical machine learning. PhD thesis, School of Computer Science, Carnegie Mellon University.

Lebanon, G. (2006). Metric learning for text documents, *IEEE Transactions on Pattern Analysis and Machine Intelligence* **28**(4), 497–508.

Lebanon, G. and Lafferty, J. (2004). Hyperplane margin classifiers on the multinomial manifold. In *Proc. of the 21st International Conference on Machine Learning* (ACM press).

Lebanon, G., Mao, Y. and Dillon, J. (2007). The locally weighted bag of words framework for document representation, *Journal of Machine Learning Research* **8**, 2405–41.

Mao, Y., Dillon, J. and Lebanon, G. (2007). Sequential document visualization, *IEEE Transactions on Visualization and Computer Graphics* **13**(6), 1208–15.

Saul, L. and Roweis, S. (2003). Think globally, fit locally: Unsupervised learning of low dimensional manifolds, *Journal of Machine Learning Research* **4**(2), 119–55.

Spivak, M. (1975). *A Comprehensive Introduction to Differential Geometry*, Vol 1–5, (Publish or Perish).

Wand, M. P. and Jones, M. C. (1995). *Kernel Smoothing* (Boca Raton, Chapman & Hall/CRC).

18

Exponential manifold by reproducing
kernel Hilbert spaces

Kenji Fukumizu

Abstract

The purpose of this chapter is to propose a method of constructing exponential families of Hilbert manifolds, on which estimation theory can be built. Although there have been works on infinite-dimensional exponential families of Banach manifolds (Pistone and Sempi 1995, Gibilisco and Pistone 1998, Pistone and Rogantin 1999), they are not appropriate for discussing statistical estimation with a finite sample; the likelihood function with a finite sample is not realised as a continuous function on the manifold.

The proposed exponential manifold uses a reproducing kernel Hilbert space (RKHS) as a functional space in the construction. A RKHS is defined as a Hilbert space of functions such that evaluation of a function at an arbitrary point is a continuous functional on the Hilbert space. Since evaluation of the likelihood function is necessary for the estimation theory, it is very natural to use a manifold associated with a RKHS. Such a manifold can be either finite or infinite dimensional depending of the choice of RKHS.

This chapter focuses on the maximum likelihood estimation (MLE) with the exponential manifold associated with a RKHS. As in many non-parametric estimation methods, straightforward extension of MLE to an infinite-dimensional exponential manifold can be an ill-posed problem; the estimator is chosen from the infinite-dimensional space, while only a finite number of constraints is given by the sample. To solve this problem, a pseudo-maximum likelihood method is proposed by restricting the infinite-dimensional manifold to a series of finite-dimensional sub-manifolds, which enlarge as the sample size increases. Some asymptotic results in the limit of infinite sample are shown, including the consistency of the pseudo-MLE.

18.1 Exponential family associated with a reproducing kernel Hilbert space

18.1.1 Reproducing kernel Hilbert space

This subsection provides a brief review of reproducing kernel Hilbert spaces. Only real Hilbert spaces are discussed in this chapter, while a RKHS is

Algebraic and Geometric Methods in Statistics, ed. Paolo Gibilisco, Eva Riccomagno, Maria Piera Rogantin and Henry P. Wynn. Published by Cambridge University Press. © Cambridge University Press 2010.

291

defined as a complex Hilbert space in general. For the details on RKHS, see (Aronszajn 1950).

Let Ω be a set, and $\mathcal{H}$ be a Hilbert space included in the set of all real-valued functions on Ω. The inner product of $\mathcal{H}$ is denoted by $\langle\,,\,\rangle_{\mathcal{H}}$. The Hilbert space $\mathcal{H}$ is called a *reproducing kernel Hilbert space* (RKHS) if there is a function

$$k : \Omega \times \Omega \to \mathbb{R}$$

such that (i) $k(\cdot, x) \in \mathcal{H}$ for all $x \in \Omega$, and (ii) for any $f \in \mathcal{H}$ and $x \in \Omega$, $\langle f, k(\cdot, x)\rangle_{\mathcal{H}} = f(x)$. The condition (ii) is called the *reproducing property* and k is called a *reproducing kernel*.

As $k(x, y) = \langle k(\cdot, y), k(\cdot, x)\rangle_{\mathcal{H}} = \langle k(\cdot, x), k(\cdot, y)\rangle_{\mathcal{H}} = k(y, x)$, a reproducing kernel is symmetric. It is easy to see that a reproducing kernel is unique if it exists. The following proposition is a characterisation of RKHS.

Proposition 18.1 *A Hilbert space of functions on Ω is a RKHS if and only if the evaluation mapping $e_x : \mathcal{H} \to \mathbb{R}$, $f \mapsto f(x)$, is a continuous linear functional on $\mathcal{H}$ for any $x \in \Omega$.*

Proof Suppose $k : \Omega \times \Omega \to \mathbb{R}$ is a reproducing kernel of $\mathcal{H}$. For any $x \in \Omega$ and $f \in \mathcal{H}$, we have $|e_x(f)| = |f(x)| = |\langle f, k(\cdot, x)\rangle_{\mathcal{H}}| \leq \|f\|_{\mathcal{H}}\|k(\cdot, x)\|_{\mathcal{H}} = \|f\|_{\mathcal{H}}\sqrt{k(x, x)}$, which shows e_x is bounded. Conversely, if the evaluation mapping e_x is bounded, by Riesz's representation theorem, there exists $\phi_x \in \mathcal{H}$ such that $f(x) = e_x(f) = \langle f, \phi_x\rangle_{\mathcal{H}}$. The function $k(y, x) = \phi_x(y)$ is then a reproducing kernel on $\mathcal{H}$. $\square$

A function $k : \Omega \times \Omega \to \mathbb{R}$ is said to be *positive definite* if it is symmetric, $k(x, y) = k(y, x)$ for $x, y \in \Omega$, and for any points $x_1, \ldots, x_n \in \Omega$ the symmetric matrix $(k(x_i, x_j))_{i,j}$ is positive semidefinite, i.e., for any real numbers $c_1, \ldots, c_n$ the inequality $\sum_{i,j=1}^{n} c_i c_j k(x_i, x_j) \geq 0$ holds.

A RKHS and a positive definite kernel have a one-to-one correspondence. If $\mathcal{H}$ is a RKHS on Ω, the reproducing kernel $k(x, y)$ is positive definite, because $\sum_{i,j} c_i c_j k(x_i, x_j) = \|\sum_i c_i k(\cdot, x_i)\|_{\mathcal{H}}^2 \geq 0$. It is also known (Aronszajn 1950) that for a positive definite kernel k on Ω there uniquely exists a RKHS $\mathcal{H}_k$ such that $\mathcal{H}_k$ consists of functions on Ω, the class of functions $\sum_{i=1}^{m} a_i k(\cdot, x_i)$ $(m \in \mathbb{N}, x_i \in \Omega, a_i \in \mathbb{R})$ is dense in $\mathcal{H}_k$, and $\langle f, k(\cdot, x)\rangle_{\mathcal{H}_k} = f(x)$ holds for any $f \in \mathcal{H}_k$ and $x \in \Omega$. Thus, a Hilbert space $\mathcal{H}$ of functions on Ω is a RKHS if and only if $\mathcal{H} = \mathcal{H}_k$ for some positive definite kernel k. In the following, a RKHS is usually given by a positive definite kernel.

If Ω is a topological space and k is a continuous positive definite kernel, the corresponding RKHS $\mathcal{H}$ consists of continuous functions on Ω. In fact, from $|f(x) - f(y)| = |\langle f, k(\cdot, x) - k(\cdot, y)\rangle_{\mathcal{H}}| \leq \|f\|_{\mathcal{H}}\|k(\cdot, x) - k(\cdot, y)\|_{\mathcal{H}}$ for $f \in \mathcal{H}$, the assertion follows from $\|k(\cdot, x) - k(\cdot, y)\|_{\mathcal{H}}^2 = k(x, x) - 2k(x, y) + k(y, y)$.

The following functions are known to be positive definite on $\mathbb{R}^n$: (i) linear kernel $k(x, y) = x^\top y$; (ii) Gaussian kernel $k(x, y) = \exp(-\|x - y\|^2/\sigma^2)$ $(\sigma > 0)$; (iii) polynomial kernel $k(x, y) = (x^\top y + c)^d$ $(c \geq 0, d \in \mathbb{N})$.

The linear kernel provides the n-dimensional Euclidean space. The RKHS given by the polynomial kernel with degree d and $c > 0$ consists of all the polynomials of degree d. It is known that the Gaussian kernel gives an infinite-dimensional RKHS.

18.1.2 Exponential manifold associated with a RKHS

In this chapter, it is assumed that Ω is a topological space, and μ is a Borel probability measure. The *support* of μ is defined by the smallest closed set F such that $\mu(\Omega \backslash F) = 0$. It is also assumed that the support of μ is Ω. The set of continuous positive probability density functions with respect to μ is denoted by

$$\mathcal{M}_\mu = \left\{ f : \Omega \to \mathbb{R} ,\ f \text{ is continuous},\ f > 0,\ \text{and}\ \int_\Omega f\, d\mu = 1 \right\}.$$

Hereafter, the probability given by a density $f \in \mathcal{M}_\mu$ is denoted by $f\mu$, and the expectation of a measurable function on Ω with respect to $f\mu$ is denoted by $\mathrm{E}_f[u]$ or $\mathrm{E}_f[u(X)]$.

For estimating a probability density function from a sample, it is required that the probabilities and the density functions are in one-to-one correspondence. The class $\mathcal{M}_\mu$ of density functions guarantees it; for $f, g \in \mathcal{M}_\mu$ the probabilities $f\mu$ and $g\mu$ coincide if and only if $f = g$. In fact, if $f \neq g$, there is a non-empty open set U and $\varepsilon > 0$ such that $f(x) - g(x) > \varepsilon$ on U. From $\mu(U) > 0$, $f\mu(U)$ and $g\mu(U)$ must differ. For the probability $\mu = w(x)dx$ on the Euclidean space $\mathbb{R}^m$, where w is a positive continuous density function with respect to Lebesgue measure dx, the class $\mathcal{M}_\mu$ consists of all the positive continuous density functions with respect to the Lebesgue measure.

Let $k : \Omega \times \Omega \to \mathbb{R}$ be a continuous positive definite kernel on Ω. Define a subclass of $\mathcal{M}_\mu$ by

$$\mathcal{M}_\mu(k) = \left\{ f \in \mathcal{M}_\mu \ :\ \text{there exists } \delta > 0 \text{ such that } \int e^{\delta \sqrt{k(x,x)}} f(x)\, d\mu(x) < \infty \right\}.$$

A positive definite kernel k is bounded if and only if the function $k(x,x)$ on Ω is bounded, since $|k(x,y)| \leq k(x,x)k(y,y)$ by the positive semidefiniteness. For a bounded k, we have $\mathcal{M}_\mu(k) = \mathcal{M}_\mu$.

Throughout this chapter, the following assumption is made unless otherwise mentioned.

(A-0)   The RKHS $\mathcal{H}_k$ contains the constant functions.

This is a mild assumption, because for any RKHS $\mathcal{H}_k$ the direct sum $\mathcal{H}_k + \mathbb{R}$, where $\mathbb{R}$ denotes the RKHS associated with the positive definite kernel 1 on Ω, is again a RKHS with reproducing kernel $k(x,y) + 1$, see (Aronszajn 1950).

For any $f \in \mathcal{M}_\mu(k)$, $\mathrm{E}_f[\sqrt{k(X,X)}]$ is finite, because $\delta\, \mathrm{E}_f[\sqrt{k(X,X)}] \leq \mathrm{E}_f[e^{\delta \sqrt{k(X,X)}}] < \infty$. From $|u(x)| = |\langle u, k(\cdot, x)\rangle_{\mathcal{H}_k}| \leq \sqrt{k(x,x)}\|u\|_{\mathcal{H}_k}$, the mapping $u \mapsto \mathrm{E}_f[u(X)]$ is a bounded functional on $\mathcal{H}_k$ for any $f \in \mathcal{M}_\mu(k)$. We define a closed subspace T_f of $\mathcal{H}_k$ by

$$T_f := \{ u \in \mathcal{H}_k \mid \mathrm{E}_f[u(X)] = 0 \},$$

which works as a tangent space at f, as we will see later. Note that, by the assumption (A-0), $u - \mathrm{E}_f[u]$ is included in T_f for any $u \in \mathcal{H}_k$.

For $f \in \mathcal{M}_\mu(k)$, let $\mathcal{W}_f$ be a subset of T_f defined by

$$\mathcal{W}_f = \left\{ u \in T_f \ : \ \text{there exists } \delta > 0 \text{ such that } \mathrm{E}_f[e^{\delta\sqrt{k(X,X)}+u(X)}] < \infty \right\}.$$

The cumulant generating function Ψ_f on $\mathcal{W}_f$ is defined by

$$\Psi_f(u) = \log \mathrm{E}_f[e^{u(X)}].$$

Lemma 18.1 *For any $u \in \mathcal{W}_f$, the probability density function*

$$e^{u - \Psi_f(u)} f$$

belongs to $\mathcal{M}_\mu(k)$.

Proof It is obvious that $\Psi(u)$ is finite for any $u \in \mathcal{W}_f$, so that the above probability density function is well-defined. By the definition of $\mathcal{W}_f$, there is $\delta > 0$ such that $\mathrm{E}_f[e^{\delta\sqrt{k(X,X)}+u(X)}] < \infty$, which derives

$$\int e^{\delta\sqrt{k(x,x)}} e^{u(x) - \Psi_f(u)} f(x)\, d\mu(x) = e^{-\Psi_f(u)}\, \mathrm{E}_f\left[e^{\delta\sqrt{k(X,X)}+u(X)}\right]$$

which is smaller than infinity. $\qquad\square$

From Lemma 18.1, the mapping

$$\xi_f : \mathcal{W}_f \to \mathcal{M}_\mu(k), \qquad u \mapsto e^{u - \Psi_f(u)} f$$

is defined. The map ξ_f is one-to-one, because $\xi_f(u) = \xi_f(v)$ implies $u - v$ is constant, which is necessarily zero from $\mathrm{E}_f[u] = \mathrm{E}_f[v] = 0$. Let $\mathcal{S}_f = \xi_f(\mathcal{W}_f)$, and φ_f be the inverse of ξ_f, that is,

$$\varphi_f : \mathcal{S}_f \to \mathcal{W}_f, \qquad g \mapsto \log\frac{g}{f} - \mathrm{E}_f\left[\log\frac{g}{f}\right].$$

It will be shown that φ_f works as a local coordinate that makes $\mathcal{M}_\mu(k)$ a Hilbert manifold. The following facts are basic;

Lemma 18.2 *Let f and g be arbitrary elements in $\mathcal{M}_\mu(k)$. Then,*

(i) $\mathcal{W}_f$ is an open subset of T_f, and
(ii) $g \in \mathcal{S}_f$ if and only if $\mathcal{S}_g = \mathcal{S}_f$.

Proof (i). For an arbitrary $u \in \mathcal{W}_f$, take $\delta > 0$ so that

$$\mathrm{E}_f[e^{u(X)+\delta\sqrt{k(X,X)}}] < +\infty.$$

Define an open neighborhood V_u of u in T_f by $V_u = \{v \in T_f \,\|v - u\|_{\mathcal{H}_k} < \delta/2\}$.

Then, for any $v \in V_u$,

$$
\begin{aligned}
\mathrm{E}_f\left[e^{(\delta/2)\sqrt{k(X,X)}+v(X)}\right] &= \mathrm{E}_f\left[e^{(\delta/2)\sqrt{k(X,X)}+\langle v-u,k(\cdot,X)\rangle_{\mathcal{H}_k}+u(X)}\right] \\
&\leq \mathrm{E}_f\left[e^{(\delta/2)\sqrt{k(X,X)}+\|v-u\|_{\mathcal{H}_k}\sqrt{k(X,X)}+u(X)}\right] \\
&\leq \mathrm{E}_f\left[e^{\delta\sqrt{k(X,X)}+u(X)}\right] \quad < \infty,
\end{aligned}
$$

which implies $\mathcal{W}_f$ is open.

(ii). 'If' part is obvious. For the 'only if' part, we first prove $\mathcal{S}_g \subset \mathcal{S}_f$ on condition $g \in \mathcal{S}_f$. Let h be an arbitrary element in $\mathcal{S}_g$, and take $u \in \mathcal{W}_f$ and $v \in \mathcal{W}_g$ such that $g = e^{u-\Psi_f(u)}f$ and $h = e^{v-\Psi_g(v)}g$. From the fact $g \in \mathcal{W}_f$, there is $\delta > 0$ such that $\mathrm{E}_g[e^{v(X)+\delta\sqrt{k(X,X)}}] < \infty$. We have $\int e^{v(x)+u(x)+\delta\sqrt{k(x,x)}-\Psi_f(u)}f(x)\,d\mu(x) < \infty$, which means $v + u - \mathrm{E}_f[v] \in \mathcal{W}_f$. From $h = e^{(v+u-\mathrm{E}_f[v])-(\Psi_f(u)+\Psi_g(v)-\mathrm{E}_f[v])}f$, we have $\Psi_f(v+u-\mathrm{E}_f[v]) = \Psi_f(u) + \Psi_g(v) - \mathrm{E}_f[v]$ and $h = \xi_f(v+u-\mathrm{E}_f[v]) \in \mathcal{S}_f$.

For the opposite inclusion, it suffices to show $f \in \mathcal{S}_g$. Let $\gamma > 0$ be a constant so that $\mathrm{E}_f[e^{\gamma\sqrt{k(X,X)}}] < \infty$.

From $e^{-u}g = e^{-\Psi_f(u)}f$, we see $\int e^{\gamma\sqrt{k(x,x)}-u(x)}g(x)\,d\mu(x) < \infty$, which means $-u + \mathrm{E}_g[u] \in \mathcal{W}_g$. It follows that $f = e^{-u+\Psi_f(u)}g = e^{(-u+\mathrm{E}_g[u])-(-\Psi_f(u)+\mathrm{E}_g[u])}g$ means $f = \xi_g(-u+\mathrm{E}_g[u]) \in \mathcal{S}_g$. $\qquad\square$

The map φ_f defines a structure of Hilbert Manifold on $\mathcal{M}_\mu(k)$, which we call *reproducing kernel exponential manifold*.

Theorem 18.1 *The system* $\{(\mathcal{S}_f, \varphi_f)\}_{f \in \mathcal{M}_\mu(k)}$ *is a* C^∞*-atlas of* $\mathcal{M}_\mu(k)$*; that is,*

(i) if $\mathcal{S}_f \cap \mathcal{S}_g \neq \emptyset$*, then* $\varphi_f(\mathcal{S}_f \cap \mathcal{S}_g)$ *is an open set in* T_f*, and*

(ii) if $\mathcal{S}_f \cap \mathcal{S}_g \neq \emptyset$*, then*

$$
\varphi_g \circ \varphi_f^{-1}\big|_{\varphi_f(\mathcal{S}_f \cap \mathcal{S}_g)} : \varphi_f(\mathcal{S}_f \cap \mathcal{S}_g) \to \varphi_g(\mathcal{S}_f \cap \mathcal{S}_g)
$$

is a C^∞ *map.*

Thus, $\mathcal{M}_\mu(k)$ *admits a structure of* C^∞*-Hilbert manifold.*

Proof The assertion (i) is obvious, because $\mathcal{S}_f \cap \mathcal{S}_g \neq \emptyset$ means $\mathcal{S}_f = \mathcal{S}_g$ from Lemma 18.2. Suppose $\mathcal{S}_f \cap \mathcal{S}_g \neq \emptyset$, that is, $\mathcal{S}_f = \mathcal{S}_g$. For any $u \in \mathcal{W}_f$,

$$
\begin{aligned}
\varphi_g \circ \varphi_f^{-1}(u) = \varphi_g\big(e^{u-\Psi_f(u)}f\big) &= \log\frac{e^{u-\Psi_f(u)}f}{g} - \mathrm{E}_g\left[\log\frac{e^{u-\Psi_f(u)}f}{g}\right] \\
&= u + \log(f/g) - \mathrm{E}_g\big[u + \log(f/g)\big],
\end{aligned}
$$

from which the assertion (ii) is obtained, because $u \mapsto \mathrm{E}_g[u]$ is of C^∞ on $\mathcal{W}_f$. It is known that with the assertions (i) and (ii) a topology is introduced on $\mathcal{M}_\mu(k)$ so that all $\mathcal{S}_f$ are open, and $\mathcal{M}_\mu(k)$ is equipped with the structure of a C^∞-Hilbert manifold, see (Lang 1985). $\qquad\square$

The open set $\mathcal{S}_f$ is regarded as a maximal exponential family in $\mathcal{M}_\mu(k)$. In fact, we have the following

Theorem 18.2 *For any $f \in \mathcal{M}_\mu(k)$,*

$$\mathcal{S}_f = \{g \in \mathcal{M}_\mu(k) : \text{ there exists } u \in T_f \text{ such that } g = e^{u - \Psi_f(u)} f\}.$$

Proof It suffices to show that $g = e^{u - \Psi_f(u)} f$ in the right-hand side is included in the left-hand side, as the opposite inclusion is obvious. From $g \in \mathcal{M}_\mu(k)$, there is $\delta > 0$ such that $E_g[e^{\delta \sqrt{k(X,X)}}] < \infty$, which means $E_f[e^{\delta \sqrt{k(X,X)} + u(X)}] < \infty$. Therefore, $u \in \mathcal{W}_f$ and $g = \xi_f(u) \in \mathcal{S}_f$. $\square$

From Lemma 18.2 (ii), we can define an equivalence relation such that f and g are equivalent if and only if they are in the same local maximal exponential family, that is, if and only if $\mathcal{S}_f \cap \mathcal{S}_g \neq \emptyset$. Let $\{\mathcal{S}^{(\lambda)}\}_{\lambda \in \Lambda}$ be the equivalence class. Then, they are equal to the set of connected components.

Theorem 18.3 *Let $\{\mathcal{S}^{(\lambda)}\}_{\lambda \in \Lambda}$ be the equivalence class of the maximum local exponential families. Then, $\mathcal{S}^{(\lambda)}$, $\lambda \in \Lambda$ are the connected components of $\mathcal{M}_\mu(k)$. Moreover, each component $\mathcal{S}^{(\lambda)}$ is simply connected.*

Proof From Lemma 18.2 and Theorem 18.1, $\{\mathcal{S}^{(\lambda)}\}_{\lambda \in \Lambda}$ are disjoint open covering of $\mathcal{M}_\mu(k)$. The proof is completed if every $\mathcal{W}_f$ is shown to be convex. Let u_0 and u_1 be arbitrary elements in $\mathcal{W}_f$. Then, there exists $\delta > 0$ such that $E_f[e^{\delta \sqrt{k(X,X)} + u_0(X)}] < \infty$ and $E_f[e^{\delta \sqrt{k(X,X)} + u_1(X)}] < \infty$. For $u_t = t u_1 + (1 - t)u_0 \in T_f$ $(t \in [0,1])$, we have $e^{u_t(x)} \leq t e^{u_1(x)} + (1 - t)e^{u_0(x)}$ by the convexity of $z \mapsto e^z$. It leads to

$$E_f\left[e^{\delta \sqrt{k(X,X)} + u_t(X)}\right] \leq t E_f\left[e^{\delta \sqrt{k(X,X)} + u_1(X)}\right] + (1-t)\,E_f\left[e^{\delta \sqrt{k(X,X)} + u_0(X)}\right] < \infty,$$

which means $u_t \in \mathcal{W}_f$. $\square$

The Hilbert space $\mathcal{H}_k$, which is used for giving a manifold structure to $\mathcal{M}_\mu(k)$, has stronger topology than the Orlicz space used for the exponential manifold by (Pistone and Sempi 1995). Recall that a function u is an element of the Orlicz space $L^{\cosh - 1}(f)$ if and only if there is $\alpha > 0$ such that

$$E_f\left[\cosh\left(\frac{u}{\alpha}\right) - 1\right] < \infty.$$

The space $u \in L^{\cosh - 1}(f)$ is a Banach space with the norm

$$\|u\|_{L^{\cosh - 1}(f)} = \inf\left\{\alpha > 0 \,\Big|\, E_f\left[\cosh\left(\frac{u}{\alpha}\right) - 1\right] \leq 1\right\}.$$

For details on this space, see (Pistone and Sempi 1995).

Proposition 18.2 *For any $f \in \mathcal{M}_\mu(k)$, the RKHS $\mathcal{H}_k$ is continuously included in $L^{\cosh - 1}(f)$. Moreover, if a positive number A_f is defined by*

$$A_f = \inf\left\{\alpha > 0 : \int e^{\frac{\sqrt{k(x,x)}}{\alpha}} f(x)\, d\mu(x) \leq 2\right\},$$

then for any $u \in \mathcal{H}_k$

$$\|u\|_{L^{\cosh -1}(f)} \leq A_f \|u\|_{\mathcal{H}_k}.$$

Proof From the inequality

$$\mathrm{E}_f\left[\cosh(u(X)/\alpha) - 1\right] \leq \mathrm{E}_f\left[e^{|u(X)|/\alpha}\right] - 1 \leq \mathrm{E}_f\left[e^{\frac{1}{\alpha}\|u\|_{\mathcal{H}_k}\sqrt{k(X,X)}}\right] - 1,$$

if $\|u\|_{\mathcal{H}_k}/\alpha < 1/A_f$, then $\mathrm{E}_f[\cosh(u/\alpha)-1] \leq 1$. This means $A_f\|u\|_{\mathcal{H}_k} \geq \|u\|_{L^{\cosh -1}(f)}$.
$\square$

Proposition 18.2 states that the manifold $\mathcal{M}_\mu(k)$ is a subset of the maximum exponential manifold. However, the former is not necessarily a submanifold of the latter, because $\mathcal{H}_k$ is not a closed subspace of $L^{\cosh -1}(f)$ in general. Note also that $L^{\cosh -1}(f)$ is continuously embedded in $L^p(f)$ for all $p \geq 1$. Thus, $\mathrm{E}_f |u|^p$ is finite for any $f \in \mathcal{M}_\mu(k)$, $u \in \mathcal{H}_k$, and $p \geq 1$.

The reproducing kernel exponential manifold and its connected components depend on the underlying RKHS. It may be either finite or infinite dimensional. A different choice of the positive definite kernel results in a different exponential manifold. A connected component of $\mathcal{M}_\mu(k)$ in Theorem 18.3 is in general smaller than the maximal exponential model discussed in (Pistone and Sempi 1995).

18.1.3 Mean and covariance on reproducing kernel exponential manifolds

As in the case of finite-dimensional exponential families and the exponential manifold by (Pistone and Sempi 1995), the derivatives of the cumulant generating function provide the cumulants or moments of the random variables given by tangent vectors. Let $f \in \mathcal{M}_\mu(k)$ and $v_1, \ldots, v_d \in T_f$. The d-th derivative of Ψ_f in the directions $v_1, \ldots, v_d$ at $f_u = e^{u - \Psi_f(u)} f$ is denoted by $D_u^d \Psi_f(v_1, \ldots, v_d)$. We have

$$D_u \Psi_f(v) = \mathrm{E}_{f_u}[v], \qquad D_u^2 \Psi_f(v_1, v_2) = \mathrm{Cov}_{f_u}[v_1(X), v_2(X)],$$

where $\mathrm{Cov}_g[v_1, v_2] = \mathrm{E}_g[v_1(X)v_2(X)] - \mathrm{E}_g[v_1(X)]\,\mathrm{E}_g[v_2(X)]$ is the covariance of v_1 and v_2 under the probability $g\mu$.

The first and second moments are expressed also by an element and an operator of the Hilbert space. Let P be a probability on Ω such that $\mathrm{E}_P[\sqrt{k(X,X)}] < \infty$. Because the functional $\mathcal{H}_k \ni u \mapsto \mathrm{E}_P[u(X)]$ is bounded, there exists $m_P \in \mathcal{H}_k$ such that

$$\mathrm{E}_P[u(X)] = \langle u, m_P \rangle_{\mathcal{H}_k}$$

for all $u \in \mathcal{H}_k$. We call m_P the *mean element* for P. Noticing that the mapping $\mathcal{H}_k \times \mathcal{H}_k \ni (v_1, v_2) \mapsto \mathrm{Cov}_P[v_1(X), v_2(X)]$ is a bounded bilinear form, we see that there exists a bounded operator Σ_P on $\mathcal{H}_k$ such that

$$\mathrm{Cov}_P[v_1(X), v_2(X)] = \langle v_1, \Sigma_P v_2 \rangle_{\mathcal{H}_k}$$

holds for all $v_1, v_2 \in \mathcal{H}_k$. The operator Σ_P is called the *covariance operator* for P. For the details about covariance operators on a RKHS, see (Fukumizu *et al.* 2007).

When a local coordinate $(\varphi_{f_0}, \mathcal{S}_{f_0})$ in a reproducing kernel exponential manifold $\mathcal{M}_\mu(k)$ is assumed, the notations m_u and Σ_u are also used for the mean element and covariance operator, respectively, with respect to the probability density $f_u = e^{u - \Psi_{f_0}(u)} f_0$. The mapping $\mathcal{W}_f \ni u \mapsto m_u \in \mathcal{H}_k$ is locally one-to-one, because the derivative $\Sigma_u|_{T_{f_0}}$ is injective for non-degenerate μ. We call m_u the *mean parameter* for the density f_u. We have

$$D_u \Psi_f(v) = \langle m_u, v \rangle_{\mathcal{H}_k}, \qquad D_u^2 \Psi_f(v_1, v_2) = \langle v_1, \Sigma_u v_2 \rangle_{\mathcal{H}_k}.$$

The mean element $m_P(y)$ as a function is explicitly expressed by

$$m_P(y) = \mathrm{E}_P[k(X, y)]$$

from $m_P(y) = \langle m_P, k(\cdot, y) \rangle_{\mathcal{H}_k} = \mathrm{E}_P[k(X, y)]$. The operator Σ_u is an extension of the Fisher information matrix. It is interesting to ask when the mean element specifies a probability.

Definition 18.1 Let $(\Omega, \mathscr{B})$ be a measurable space, and k be a measurable positive definite kernel on Ω such that $\int k(x, x) \, dP(x)$ is finite for any probability P on $(\Omega, \mathscr{B})$. The kernel k is called *characteristic* if the mapping $P \mapsto m_P$ uniquely determines a probability.

It is known that a Gaussian kernel is characteristic on R^n equipped with the Borel σ-field (Fukumizu *et al.* 2008). If $k(x, y) = \exp(-\|x - y\|^2 / \sigma^2) + 1 \ (\sigma > 0)$ is used for defining $\mathcal{M}_k(\mu)$, the mean parameter m_u uniquely determines a probability on $\mathcal{M}(\mu)$.

18.1.4 Kullback–Leibler divergence

Let $f_0 \in \mathcal{M}_\mu(k)$ and $u, v \in \mathcal{W}_{f_0}$. With the local coordinate $(\varphi_{f_0}, \mathcal{S}_{f_0})$, it is easy to see that the Kullback–Leibler divergence from $f_u = e^{u - \Psi_{f_0}(u)} f_0$ to $f_v = e^{v - \Psi_{f_0}(v)} f_0$ is given by

$$\mathrm{KL}(f_u \| f_v) = \Psi_{f_0}(v) - \Psi_{f_0}(u) - \langle v - u, m_u \rangle_{\mathcal{H}_k}. \tag{18.1}$$

Let f_u, f_v and f_w be points in $\mathcal{S}_{f_0}$. It is straightforward to see

$$\mathrm{KL}(f_u \| f_w) = \mathrm{KL}(f_u \| f_v) + \mathrm{KL}(f_v \| f_w) - \langle w - v, m_u - m_v \rangle_{\mathcal{H}_k}. \tag{18.2}$$

Let U be a closed subspace of T_{f_0} and $\mathcal{V} = U \cap \mathcal{W}_{f_0}$. The subset $\mathcal{N} = \varphi_{f_0}^{-1}(\mathcal{V})$ is a submanifold of $\mathcal{S}_{f_0}$, which is also an exponential family. Let $f_* = e^{u_* - \Psi_{f_0}(u_*)}$ be a point in $\mathcal{S}_{f_0}$, and consider the minimiser of the KL divergence from f_* to a point in $\mathcal{N}$

$$u_{opt} = \arg\min_{u \in \mathcal{V}} \mathrm{KL}(f_* \| f_u). \tag{18.3}$$

Theorem 18.4 *Under the assumption that the minimiser u_{opt} in Equation 18.3 exists, the orthogonal relation*

$$\langle u - u_{opt}, m_{u_*} - m_{u_{opt}} \rangle_{\mathcal{H}_k} = 0 \tag{18.4}$$

and the Pythagorean equation

$$\mathrm{KL}(f_*\|f_u) = \mathrm{KL}(f_*\|f_{u_{opt}}) + \mathrm{KL}(f_{u_{opt}}\|f_u) \tag{18.5}$$

hold for any $u \in \mathcal{V}$.

Proof Since $\mathcal{W}_{f_0}$ is an open convex set, $u_t = t(u - u_{opt}) + u_{opt}$ lies in $\mathcal{W}_{f_0}$ for all $t \in (-\delta, \delta)$ with sufficiently small $\delta > 0$. From Equation 18.2, $\mathrm{KL}(f_*\|f_{u_t})$ is differentiable with respective to t, and $\frac{d}{dt}\mathrm{KL}(f_*\|f_{u_t})|_{t=0} = 0$ by the minimality. This derives

$$\langle u - u_{opt}, m_{u_{opt}}\rangle_{\mathcal{H}_k} - \langle u - u_{opt}, m_{u_*}\rangle_{\mathcal{H}_k} = 0,$$

which is the orthogonal relation. The Pythagorean relation is obvious from Equations (18.2) and (18.4). $\qquad\square$

18.2 Pseudo maximum likelihood estimation with $\mathcal{M}_\mu(k)$

In this section, statistical estimation with a reproducing kernel exponential manifold is discussed. Throughout this section, a continuous positive definite kernel k with the assumption (A-0) and a connected component $\mathcal{S}$ of $\mathcal{M}_\mu(k)$ are fixed.

From Lemma 18.2 and Theorem 18.2, for any $f_0 \in \mathcal{S}$ the component $\mathcal{S}$ can be expressed by

$$\mathcal{S} = \{f \in \mathcal{M}_\mu(k) : f = e^{u - \Psi_0(u)} f_0 \text{ for some } u \in T_{f_0}\},$$

where Ψ_0 is an abbreviation of Ψ_{f_0}. For notational simplicity, $\mathcal{W}_0 = \mathcal{W}_{f_0}$ and $f_u = e^{u - \Psi_0(u)} f_0$ for $u \in \mathcal{W}_0$ are used.

It is assumed that $(X_1, X_2, \ldots, X_n)$ is an independent and identically distributed (i.i.d.) sample with probability $f_*\mu$ with $f_* \in \mathcal{S}$, which is called a true probability density. We discuss the problem of estimating f_* with the statistical model $\mathcal{S}$ given the finite sample.

18.2.1 Likelihood equation on a reproducing kernel exponential manifold

The maximum likelihood estimation (MLE) is the most popular estimation method for finite-dimensional exponential families. In the following, we consider the MLE approach with the reproducing kernel exponential manifold $\mathcal{S}$, which may not be finite dimensional. The objective function of MLE with $\mathcal{S}$ is given by

$$\sup_{u \in \mathcal{W}_0} L_n(u), \qquad L_n(u) = \frac{1}{n}\sum_{i=1}^{n} u(X_i) - \Psi_0(u),$$

where $L_n(u)$ is called the log likelihood function. By introducing the empirical mean element

$$\widehat{m}^{(n)} = \frac{1}{n}\sum_{i=1}^{n} k(\cdot, X_i),$$

the log likelihood function is rewritten by

$$L_n(u) = \langle \widehat{m}^{(n)}, u \rangle_{\mathcal{H}_k} - \Psi_0(u).$$

Taking the partial derivative of $L_n(u)$, we obtain the likelihood equation,

$$\langle \widehat{m}^{(n)}, v \rangle_{\mathcal{H}_k} = \langle m_u, v \rangle_{\mathcal{H}_k} \qquad (\forall v \in \mathcal{H}_k),$$

where m_u is the mean parameter corresponding to the density f_u. Note that the above equation holds not only for $v \in T_{f_0}$ but for all $v \in \mathcal{H}_k$, since $\langle \widehat{m}^{(n)} - m_u, 1 \rangle_{\mathcal{H}_k}$ always vanishes. The log likelihood equation is thus reduced to

$$m_u = \widehat{m}^{(n)}, \tag{18.6}$$

that is, the mean parameter for the maximum likelihood estimator is the empirical mean element $\widehat{m}^{(n)}$.

If $\mathcal{H}_k$ is finite dimensional and $(\phi_1, \dots, \phi_d)$ is a basis of T_{f_0}, Equation (18.6) is equivalent to

$$m_u^j = \frac{1}{n} \sum_{i=1}^n \phi_j(X_i) \qquad (j = 1, \dots, d),$$

where $(m_u^1, \dots, m_u^d)$ is the component of m_u with respect to the basis $(\phi_1, \dots, \phi_d)$. If the mapping $u \mapsto m_u$ is invertible, which is often the case with ordinary finite-dimensional exponential families, the MLE $\widehat{u}$ is given by the inverse image of $\widehat{m}^{(n)}$.

Unlike the finite-dimensional exponential family, the likelihood Equation (18.6) does not necessarily have a solution in the canonical parameter u. As (Pistone and Rogantin 1999) point out for their exponential manifold, the inverse mapping from the mean parameter to the canonical parameter u is not bounded in general. For reproducing kernel exponential manifolds, the unboundedness of the inverse of $u \mapsto m_u$ can been seen by investigating its derivative. In fact, the derivative of the map $u \mapsto m_u$ is given by the covariance operator Σ_u, which is known to be of trace class by $\mathrm{E}_{f_0}[k(X, X)] < \infty$, see (Fukumizu *et al.* 2007). If $\mathcal{H}_k$ is infinite dimensional, Σ_u has arbitrary small positive eigenvalues, which implies Σ_u does not have a bounded inverse. Thus, the mean parameter does not give a coordinate system for infinite-dimensional manifolds.

Another explanation for the fact that the likelihood equation does not have a solution is given by the interpretation as moment matching; the empirical distribution $\frac{1}{n} \sum_{i=1}^n \delta_{X_i}$ and the probability $e^{u - \Psi_0(u)} f_0 \mu$ must have the same mean element. If k is characteristic (see Definition 18.1), these two probabilities must be the same; this is impossible if the support of μ is uncountable.

To solve this problem, a method of pseudo maximum likelihood estimation will be proposed in Section 18.2.3, in which asymptotic properties of the mean parameter yet play an important role.

18.2.2 $\sqrt{n}$-*consistency of the mean parameter*

The next theorem establishes $\sqrt{n}$-consistency of the mean parameter in a general form.

Theorem 18.5 *Let $(\Omega, \mathcal{B}, P)$ be a probability space, $k : \Omega \times \Omega \to \mathbb{R}$ be a positive definite kernel so that $\mathrm{E}_P[k(X, X)] < \infty$, and $m_P \in \mathcal{H}_k$ be the mean element with respect to P. Suppose $X_1, \ldots, X_n$ are i.i.d. sample from P, and define the empirical mean element $\widehat{m}^{(n)}$ by $\widehat{m}^{(n)} = \frac{1}{n} \sum_{i=1}^n k(\cdot, X_i)$. Then, we have*

$$\|\widehat{m}^{(n)} - m_P\|_{\mathcal{H}_k} = O_p\big(1/\sqrt{n}\big) \quad (n \mapsto \infty).$$

Proof Let $\mathrm{E}_X[\cdot]$ denote the expectation with respect to the random variable X which follows P. Suppose $X, \tilde{X}, X_1, \ldots, X_n$ are i.i.d. We have

$$E\|\widehat{m}^{(n)} - m_P\|_{\mathcal{H}_k}^2$$

$$= \frac{1}{n^2} \sum_{i=1}^n \sum_{j=1}^n \mathrm{E}_{X_i} \, \mathrm{E}_{X_j}[k(X_i, X_j)] - \frac{2}{n} \sum_{i=1}^n \mathrm{E}_{X_i} \, \mathrm{E}_X[k(X_i, X)] + \mathrm{E}_X \, \mathrm{E}_{\tilde{X}}[k(X, \tilde{X})]$$

$$= \frac{1}{n^2} \sum_{i=1}^n \sum_{j \neq i} E[k(X_i, X_j)] + \frac{1}{n} \mathrm{E}_X[k(X, X)] - \mathrm{E}_X \, \mathrm{E}_{\tilde{X}}[k(X, \tilde{X})]$$

$$= \frac{1}{n}\{\mathrm{E}_X[k(X, X)] - \mathrm{E}_X \, \mathrm{E}_{\tilde{X}}[k(X, \tilde{X})]\} = O(1/n).$$

The assertion is obtained by Chebyshev's inequality. $\qquad\square$

By a similar argument to (Gretton *et al.* 2008), it is further possible to see that $n\|\widehat{m}^{(n)} - m_P\|_{\mathcal{H}_k}^2$ converges in law to a normal distribution.

18.2.3 *Pseudo maximum likelihood estimation*

This subsection proposes the pseudo maximum likelihood estimation using a series of finite-dimensional subspaces in $\mathcal{H}_k$ to make the inversion from the mean parameter to the canonical parameter possible. With an infinite-dimensional reproducing kernel exponential manifold, the estimation of the true density with a finite sample is an ill-posed problem, because it attempts to find a function from the infinite-dimensional space with only a finite number of constraints made by the sample. Among many methods of regularisation to solve such ill-posed problems, one of the most well-known methods is Tikhonov regularisation (Groetsch 1984), which adds a regularisation term to the objective function for making inversion stable. (Canu and Smola 2006) have proposed a kernel method for density estimation using an exponential family defined by a positive definite kernel, while they do not formulate it rigorously. They discuss Tikhonov-type regularisation for estimation. Another major approach to regularisation is to approximate the original infinite-dimensional space by finite-dimensional subspaces (Groetsch 1984). This chapter uses the latter approach, because it matches better the geometrical apparatus developed in the previous sections.

Let $\{\mathcal{H}^{(\ell)}\}_{\ell=1}^\infty$ be a series of finite-dimensional subspaces of $\mathcal{H}_k$ such that $\mathcal{H}^{(\ell)} \subset \mathcal{H}^{(\ell+1)}$ for all $\ell \in \mathbb{N}$. For any $f \in \mathcal{M}_\mu(k)$, a subspace $T_f^{(\ell)}$ of T_f is defined by $T_f^{(\ell)} = T_f \cap \mathcal{H}^{(\ell)}$, and an open set $\mathcal{W}_f^{(\ell)}$ of $T_f^{(\ell)}$ is defined by $\mathcal{W}_f^{(\ell)} = \mathcal{W}_f \cap \mathcal{H}^{(\ell)}$.

Also, the notations $\mathcal{W}^{(\ell)}$ and $\mathcal{S}^{(\ell)}$ are used for $\mathcal{W}_{f_0}^{(\ell)}$ and $\{f_u \in \mathcal{S} : u \in \mathcal{W}^{(\ell)}\}$, respectively.

For each $\ell \in \mathbb{N}$, the pseudo maximum likelihood estimator $\widehat{u}^{(\ell)}$ in $\mathcal{W}^{(\ell)}$ is defined by

$$\widehat{u}^{(\ell)} = \arg\max_{u \in \mathcal{W}^{(\ell)}} \langle \widehat{m}^{(n)}, u \rangle_{\mathcal{H}_k} - \Psi_0(u).$$

In the following discussion, it is assumed that the maximiser $\widehat{u}^{(\ell)}$ exists in $\mathcal{W}^{(\ell)}$, and further the following two assumptions are made:

(A-1) For all $u \in \mathcal{W}_0$, let $u_*^{(\ell)} \in \mathcal{W}^{(\ell)}$ ($\ell \in \mathbb{N}$) be the minimiser of $\min_{u^{(\ell)} \in \mathcal{W}^{(\ell)}} \mathrm{KL}(f_u \| f_{u^{(\ell)}})$. Then

$$\| u - u_*^{(\ell)} \|_{\mathcal{H}_k} \to 0 \qquad (\ell \to \infty).$$

(A-2) For $u \in \mathcal{W}_0$, let $\lambda^{(\ell)}(u)$ be the least eigenvalue of the covariance operator Σ_u restricted on $T_{f_u}^{(\ell)}$, that is,

$$\lambda^{(\ell)}(u) = \inf_{v \in T_{f_u}^{(\ell)},\, \|v\|_{\mathcal{H}_k}=1} \langle v, \Sigma_u v \rangle_{\mathcal{H}_k}.$$

Then, there exists a sub-sequence $(\ell_n)_{n=1}^{\infty}$ of $\mathbb{N}$ such that for all $u \in \mathcal{W}_0$ we can find $\delta > 0$ for which

$$\tilde{\lambda}_u^{(\ell)} = \inf_{u' \in \mathcal{W}_0,\, \|u'-u\|_{\mathcal{H}_k} \leq \delta} \lambda^{(\ell)}(u')$$

satisfies

$$\lim_{n \to \infty} \sqrt{n}\,\tilde{\lambda}_u^{(\ell_n)} = +\infty.$$

The assumption (A-1) means $\mathcal{S}^{(\ell)}$ can approximate a function in $\mathcal{S}$ at any precision as ℓ goes to infinity. The assumption (A-2) provides a stable MLE in the sub-model $\mathcal{S}^{(\ell)}$ by lower-bounding the least eigenvalue of the derivative of the map $u \mapsto m_u$.

Theorem 18.6 *Under the assumptions (A-1) and (A-2),*

$$\mathrm{KL}(f_* \| f_{\widehat{u}^{(\ell_n)}}) \to 0 \qquad (n \mapsto \infty)$$

in probability. Moreover, let $u_* \in \mathcal{W}_0$ be the element which gives $f_{u_*} = f_*$, and $u_*^{(\ell)}$ be the element in (A-1) with respect to u_*. If positive constants γ_n and ε_n satisfy

$$\| u_* - u_*^{(\ell_n)} \|_{\mathcal{H}_k} = o(\gamma_n) \qquad (n \mapsto \infty) \tag{18.7}$$

and

$$\frac{1}{\sqrt{n}\,\tilde{\lambda}_{u_*}^{(\ell_n)}} = o(\varepsilon_n) \qquad (n \mapsto \infty), \tag{18.8}$$

then we have

$$\mathrm{KL}(f_* \| f_{\widehat{u}^{(\ell_n)}}) = o_p(\max\{\gamma_n, \varepsilon_n\}) \qquad (n \mapsto \infty).$$

Proof We prove the second assertion of the theorem. The first one is similar. Let m_* and $m_*^{(\ell)}$ be the mean parameters corresponding to u_* and $u_*^{(\ell)}$, respectively. From Equations (18.4) and (18.5), we have

$$\langle u - u_*^{(\ell)}, m_*^{(\ell)}\rangle_{\mathcal{H}_k} = \langle u - u_*^{(\ell)}, m_*\rangle_{\mathcal{H}_k} \tag{18.9}$$

for all $u \in \mathcal{W}^{(\ell)}$, and $\mathrm{KL}(f_*\|f_{\widehat{u}^{(\ell_n)}}) = \mathrm{KL}(f_*\|f_{u_*^{(\ell_n)}}) + \mathrm{KL}(f_{u_*^{(\ell_n)}}\|f_{\widehat{u}^{(\ell_n)}})$.
Equations (18.1) and (18.7) imply

$$\mathrm{KL}(f_*\|f_{u_*^{(\ell_n)}}) = o(\gamma_{\ell_n}) \qquad (n \mapsto \infty).$$

Thus, the proof is done if we show

$$\mathrm{Pr}\big(\|\widehat{u}^{(\ell_n)} - u_*^{(\ell_n)}\|_{\mathcal{H}_k} \geq \varepsilon_n\big) \to 0 \qquad (n \mapsto \infty). \tag{18.10}$$

In fact, since Equations (18.1) and (18.9) give

$$\mathrm{KL}(f_{u_*^{(\ell_n)}}\|f_{\widehat{u}^{(\ell_n)}}) = \Psi_0(\widehat{u}^{(\ell_n)}) - \Psi_0(u_*^{(\ell_n)}) - \langle m_*, \widehat{u}^{(\ell_n)} - u_*^{(\ell_n)}\rangle_{\mathcal{H}_k},$$

Equation 18.10 means $\mathrm{KL}(f_{u_*^{(\ell_n)}}\|f_{\widehat{u}^{(\ell_n)}}) = o_p(\varepsilon_n)$ $(n \mapsto \infty)$. Let $\delta > 0$ be the constant in the assumption (A-2) with respect to u_*. If the event of the probability in Equation 18.10 holds, we have

$$\sup_{\substack{u \in \mathcal{W}^{(\ell_n)} \\ \|u - u_*^{(\ell_n)}\|_{\mathcal{H}_k} \geq \varepsilon_n}} L_n(u) - L_n(u_*^{(\ell_n)}) \geq 0, \tag{18.11}$$

where $L_n(u) = \langle u, \widehat{m}^{(n)}\rangle_{\mathcal{H}_k} - \Psi_0(u)$. On the other hand, it follows from Equation 18.9 and Taylor expansion that for any $u \in \mathcal{W}^{(\ell_n)}$

$$\begin{aligned}
L_n(u) &- L_n(u_*^{(\ell_n)})\\
&= \langle u - u_*^{(\ell_n)}, \widehat{m}^{(n)} - m_*\rangle_{\mathcal{H}_k} - \big\{\Psi_0(u) - \Psi_0(u_*^{(\ell_n)}) - \langle u - u_*^{(\ell_n)}, m_*^{(\ell_n)}\rangle_{\mathcal{H}_k}\big\}\\
&= \langle u - u_*^{(\ell_n)}, \widehat{m}^{(n)} - m_*\rangle_{\mathcal{H}_k} - \frac{1}{2}\langle u - u_*^{(\ell_n)}, \Sigma_{\widetilde{u}}(u - u_*^{(\ell_n)})\rangle_{\mathcal{H}_k},
\end{aligned}$$

where $\widetilde{u}$ is a point in the line segment between u and $u_*^{(\ell_n)}$. By the definition of $\widetilde{\lambda}^{(\ell)}$, for sufficiently large n so that $\|u_*^{(\ell_n)} - u_*\|_{\mathcal{H}_k} \leq \delta$, we obtain

$$\sup_{\substack{u \in \mathcal{W}^{(\ell_n)} \\ \|u - u_*^{(\ell_n)}\|_{\mathcal{H}_k} \geq \varepsilon_n}} L_n(u) - L_n(u_*^{(\ell_n)})$$

$$\leq \sup_{\substack{u \in \mathcal{W}^{(\ell_n)} \\ \|u - u_*^{(\ell_n)}\|_{\mathcal{H}_k} \geq \varepsilon_n}} \|u - u_*^{(\ell_n)}\|_{\mathcal{H}_k}\|\widehat{m}^{(n)} - m_*\|_{\mathcal{H}_k} - \frac{1}{2}\widetilde{\lambda}^{(\ell_n)}\|u - u_*^{(\ell_n)}\|_{\mathcal{H}_k}^2$$

$$\leq \sup_{\substack{u \in \mathcal{W}^{(\ell_n)} \\ \|u - u_*^{(\ell_n)}\|_{\mathcal{H}_k} \geq \varepsilon_n}} \|u - u_*^{(\ell_n)}\|_{\mathcal{H}_k}\Big\{\|\widehat{m}^{(n)} - m_*\|_{\mathcal{H}_k} - \frac{1}{2}\widetilde{\lambda}^{(\ell_n)}\varepsilon_n\Big\}. \tag{18.12}$$

Equations (18.11) and (18.12) show that the probability in Equation 18.10 is upper bounded by

$$\mathrm{Pr}\big(\|\widehat{m}^{(n)} - m_*\|_{\mathcal{H}_k} \geq \tfrac{1}{2}\widetilde{\lambda}^{(\ell_n)}\varepsilon_n\big),$$

which converges to zero by Theorem 18.5 and Equation 18.8. $\qquad\square$

There is a trade-off between the decay rates of ε_n and γ_n; if the subspace $\mathcal{W}^{(\ell_n)}$ enlarges rapidly, the approximation accuracy γ_n decreases fast, while a small value for $\tilde{\lambda}_{u_*}^{(\ell_n)}$ results in a slow rate of ε_n.

18.3 Concluding remarks

This chapter has proposed a new family of statistical models, the reproducing kernel exponential manifold, which includes infinite-dimensional exponential families. The most significant property of this exponential manifold is that the empirical mean parameter is included in the Hilbert space. Thus, estimation of the density function with a finite sample can be discussed based on this exponential manifold, while many other formulations of exponential manifold cannot provide a basis for estimation with a finite sample. Using the reproducing kernel exponential manifold, a method of pseudo maximum likelihood estimation has been proposed with a series of finite-dimensional submanifolds, and consistency of the estimator has been shown.

Many problems remain unsolved, however. One of them is a practical method for constructing a sequence of subspaces used for the pseudo maximum likelihood estimation. A possible way of defining the sequence is to use the subspace spanned by $k(\cdot, X_1), \ldots, k(\cdot, X_\ell)$. However, with this construction the subspaces are also random depending on the sample, and the results in this chapter should be extended to the case of random subspaces to guarantee the consistency. Another practical issue is how to choose the sub-sequence ℓ_n so that the assumption (A-2) is satisfied. We need to elucidate the properties of the least eigenvalue of the covariance operator restricted to finite-dimensional subspaces, which is not necessarily obvious. Also, providing examples of the estimator for specific kernels is practically important. Investigation of these problems will be among our future works.

Acknowledgements

This work has been partially supported by JSPS KAKENHI 19500249, Japan, and the Alexander-von-Humboldt fellowship, Germany.

References

Aronszajn, N. (1950). Theory of reproducing kernels, *Transactions of the American Mathematical Society* **69**(3), 337–404.

Canu, S. and Smola, A. J. (2006). Kernel methods and the exponential family, *Neurocomputing*, **69**(7-9), 714–20.

Fukumizu, K., Bach, F. R. and Gretton, A. (2007). Statistical consistency of kernel canonical correlation analysis, *Journal of Machine Learning Research*, **8**, 361–83.

Fukumizu, K., Gretton, A., Sun, X. and Schölkopf, B. (2008). Kernel measures of conditional dependence, *Advances in Neural Information Processing Systems* **20**, 489–96.

Gibilisco, P. and Pistone, G. (1998). Connections on non-parametric statistical manifolds by Orlicz space geometry, *Infinite Dimensional Analysis, Quantum Probability and Related Topics* **1**(2), 325–47.

Gretton, A., Fukumizu, K., Teo, C. H., Song, L., Schölkopf, B. and Smola, A. (2008). A Kernel Statistical Test of Independence, *Advances in Neural Information Processing Systems* **20**, 585–92.

Groetsch, C. W. (1984). *The Theory of Tikhonov Regularization for Fredholm Equations of the First Kind* (London, Pitman).

Lang, S. (1985). *Differential Manifolds* 2nd edn (New York, Springer-Verlag).

Pistone, G. and Rogantin, M. P. (1999). The exponential statistical manifold, mean parameters, orthogonality and space transformations, *Bernoulli* **5**, 721–60.

Pistone, G. and Sempi, C. (1995). An infinite-dimensional geometric structure on the space of all the probability measures equivalent to the given one, *Annals of Statistics* **23**, 1543–61.

19

Geometry of extended exponential models

Daniele Imparato

Barbara Trivellato

Abstract

We discuss the extended exponential models obtained by extending a canonical exponential model with its limits. We aim to clarify the geometry of one of the possible definitions of the extended exponential model from the differential geometry point of view. New technical results and examples of applications will be given in later sections. The properties of the Kullback–Leibler divergence are shown in the last section and its relations with exponential models are discussed. Reference should be made to Chapter 21 for the algebraic aspects of the exponential models.

19.1 A general framework

A general notion of exponential model which includes all strictly positive densities is hard to define, unless the reference sample space is finite. Because of this, we define a notion of a maximal exponential model which is mathematically rich enough to handle interesting statistical problems. The maximal exponential model was constructed on the set of strictly positive densities $\mathcal{M}_>$ on the probability space $(\Omega, \mathcal{F}, \mu)$ in (Pistone and Sempi 1995, Gibilisco and Pistone 1998, Pistone and Rogantin 1999, Cena and Pistone 2007), to which we refer for details.

Let $\mathcal{M}_\geq$ be the set of all μ-densities. Both $\mathcal{M}_>$ and $\mathcal{M}_\geq$ are convex subsets of $L^1(\mu)$, the second one being the closure of the first one. In order to ensure the integrability of positive random variables of the form $q = e^v \cdot p$ in a suitably defined neighbourhood of v, we introduce the convex function $\Phi_1(x) = \cosh(x) - 1$ and denote with $L^{\Phi_1}(p)$ the vector space of the random variables v so that $E_p[\Phi_1(\alpha v)] < +\infty$ for some $\alpha > 0$. This is a Banach space of the Orlicz type based on the Young function Φ_1, see (Rao and Ren 2002). We use the norm whose unit open ball is the set $\{v : E_p[\Phi_1(v)] < 1\}$. The Orlicz space $L^{\Phi_1}(p)$ is contained in $L^1(p)$ and we denote the subspace of the centred random variables with $L_0^{\Phi_1}(p)$ or with B_p .

Our aim is to represent each density $q \in \mathcal{M}_\geq$ in exponential form, $q = e^v \cdot p$, where v is a random variable with values in the left-extended real line $[-\infty, +\infty[$,

Algebraic and Geometric Methods in Statistics, ed. Paolo Gibilisco, Eva Riccomagno, Maria Piera Rogantin and Henry P. Wynn. Published by Cambridge University Press. © Cambridge University Press 2010.

and $p \in \mathcal{M}_>$. As $E_p[e^v] = 1$, it follows from

$$E_p\left[e^{v^+}\right] = E_p[e^v(v \geq 0)] + E_p[(v < 0] \leq 2,$$

where v^+ is the positive part of v, that

$$E_p\left[\Phi_1(v^+)\right] = \frac{1}{2}\left(E_p\left[e^{v^+}\right] + E_p\left[e^{-v^+}\right]\right) - 1 \leq \frac{1}{2},$$

which in turn implies $v^+ \in L^{\Phi_1}(p)$ with a norm smaller than 1. On the other side, there are no restrictions on the negative part v^-.

In order to force a linear structure, in what follows it is assumed that $v^- \in L^{\Phi_1}(p)$. All the other cases belong to the closure in $L^1(p)$ of such a class. Indeed, let $v_n = \max(v, -n)$, $n = 0, 1, \ldots$ The sequence $f_n = e^{v_n} \cdot p$ is a.s decreasing to q, and

$$E_p[f_0] = E_p[e^v(v \geq 0)] + E_p[(v < 0)] \leq 2.$$

By dominated convergence, we obtain that the sequence of densities $q_n = f_n/E_p[f_n]$ is convergent to q in $L^1(p)$.

19.1.1 Cumulant generating functional

The general setting requires a formal presentation. For a given $p \in \mathcal{M}_>$, the *moment generating functional* is the functional $M_p(v) = E_p[e^v]$, $v \in L^{\Phi_1}(p)$.

Proposition 19.1 *M_p is convex and Gâteaux-differentiable at any point v of its proper domain, $\{M_p < +\infty\}$. Its n-th derivative in the direction v is given by $v \to E_p[v^n e^v]$. Furthermore, M_p is Fréchet-differentiable and analytic in the open unit ball of $L^{\Phi_1}(p)$.*

Proof See (Cena and Pistone 2007) for the Fréchet-differentiability and analyticity; for the rest, see (Pistone and Rogantin 1999). □

The *cumulant generating functional* is $K_p(u) = \log E_p[e^u]$, $u \in L_0^{\Phi_1}(p)$. In the sequel, we shall denote the proper domain of K_p with $\mathcal{K}_p$ and its topological interior with $\mathcal{S}_p$. Furthermore, let $C_p = \{u \in L_0^{\Phi_1}(p) : \text{for all } t \in \mathbb{R}, K_p(tu) < \infty\} \subset \mathcal{S}_p$. When μ is an atomic measure with a finite number of atoms, $\mathcal{S}_p = C_p = L_0^{\Phi_1}(p)$ and $L_0^{\Phi_1}(p)$ is a separable space. The following proposition summarises some properties of the domain of K_p. For a sketch of the proof see the on-line supplement.

Proposition 19.2 *Suppose that $(\Omega, \mathcal{F}, \mu)$ is not atomic with a finite number of atoms. Then*

(A-1) $L_0^{\Phi_1}(p)$ is a non-separable space.

(A-2) $C_p = \overline{L^\infty \cap L_0^{\Phi_1}(p)} \neq L_0^{\Phi_1}(p)$.

(A-3) $\mathcal{K}_p$ is neither a closed nor an open set.

(A-4) $\mathcal{S}_p$ satisfies a cylindrical property, that is, $v + C_p \in \mathcal{S}_p$ if $v \in \mathcal{S}_p$.

Proposition 19.3 *K_p is a convex functional, which is infinitely Gâteaux-differentiable in $\mathcal{S}_p$ and Fréchet-differentiable in the unit open ball of $L_0^{\Phi_1}(p)$.*

Proof It follows directly from Proposition 19.1. $\qquad\square$

The first three derivatives of K_p are listed below, where we assume $q = \exp(u - K_p(u)) \cdot p$

$$\mathrm{D}\,K_p\,(u)\,v = \mathrm{E}_q\,[v]\,, \quad v \in L_0^{\Phi_1}(p),$$
$$\mathrm{D}^2\,K_p\,(u)\,v_1 \otimes v_2 = \mathrm{Cov}_q\,(v_1, v_2)\,, \quad v_1, v_2 \in L_0^{\Phi_1}(p),$$
$$\mathrm{D}^3\,K_p(u)v_1 \otimes v_2 \otimes v_3 = \mathrm{E}_q\,[(v_1 - \mathrm{E}_q\,[v_1])(v_2 - \mathrm{E}_q\,[v_2])(v_3 - \mathrm{E}_q\,[v_3])]$$

for $v_1, v_2, v_3 \in L_0^{\Phi_1}(p)$.

19.2 The maximal exponential model

Two densities $q, r \in \mathcal{M}_\geq$ are always connected by the Hellinger arc $p(t) \propto q^{1-\lambda} r^\lambda$, $\lambda \in [0, 1]$. However, if $\mathrm{Supp}\,q \neq \mathrm{Supp}\,r$, such an arc is $L^1(\mu)$-discontinuous at least at one of the endpoints. Here Supp indicates support. Because of the issue represented by different supports, we cannot say that Hellinger arcs are 'bona fide' exponential models. We use the following more restrictive definitions of an exponential arc connecting two densities and of a general exponential model.

Definition 19.1 *Two densities $p, q \in \mathcal{M}_>$ are connected by an open exponential arc if $r \in \mathcal{M}_>$, a random variable u and an open interval $I = (t_0 - \varepsilon, t_1 + \varepsilon)$ exist so that $p(t) \propto e^{tu} \cdot r$, $t \in I$, is a one-dimensional exponential model containing both p and q at t_0, t_1 respectively. The exponential arc is left open at p if $I = (t_0 - \varepsilon, t_1]$.*

Definition 19.2 *For $p \in \mathcal{M}_>$, the maximal exponential model at p is the set of densities $\mathcal{E}(p) = \{\exp(u - K_p(u)) \cdot p : u \in \mathcal{S}_p\}$.*

It was proved, in (Pistone and Rogantin 1999), that $\mathcal{E}(p)$ coincides with the set of densities which can be connected to p by an open exponential arc. Moreover, in (Pistone and Sempi 1995), the maximal exponential model was endowed with an infinite-dimensional differentiable manifold structure, that was investigated more deeply in (Pistone and Rogantin 1999) and (Cena and Pistone 2007).

19.3 Enlarged exponential models

The idea of the maximal exponential model is extended by weakening the requirement of the connection by arcs.

Definition 19.3 *The set of densities $\widehat{\mathcal{E}}(p)$ that can be connected to p by an exponential arc which is left open at p is called the enlarged maximal exponential model.*

Proposition 19.4 *The following statements are equivalent*

(A-1) $q \in \widehat{\mathcal{E}}(p)$.

(A-2) $\log(q/p) \in L^{\Phi_1}(p)$.

(A-3) $p/q \in L^a(p)$ *for some* $a > 0$.

(A-4) $q = e^{u - K_p(u)} \cdot p$ *for some* $u \in \mathcal{K}_p$.

(A-5) A sequence $q_n = e_p(u_n)$, $u_n \in \mathcal{S}_p$, $n = 1, 2, \ldots$, *exists so that* $\lim_{n \to \infty} u_n = u$
     μ*-almost surely and in* $L^{\Phi_1}(p)$, $\lim K_p(u_n) = K_p(u)$, *and* $q = e^{u - K_p(u)} \cdot p$.

For the proof see the on-line supplement. Proposition 19.4 applies, in particular, to maximal exponential models because the connection with open arcs implies that there are both types of half-closed arcs. For example, from Item (A-2), we have that if q belongs to $\mathcal{E}(p)$ then $\log(q/p)$ belongs to both $L^{\Phi_1}(p)$ and $L^{\Phi_1}(q)$, see (Cena and Pistone 2007).

The following corollary shows that the definition of the enlarged maximal exponential model is consistent with a geometric approach, since $\widehat{\mathcal{E}}(p)$ does not depend on the reference measure.

Corollary 19.1 *Let* $p_1, p_2 \in \mathcal{E}(p)$ *for some* $p \in \mathcal{M}_>$. *Then* $\widehat{\mathcal{E}}(p_1) = \widehat{\mathcal{E}}(p_2)$.

Proof Let $q \in \widehat{\mathcal{E}}(p_1)$. From $\log q/p_2 = \log q/p_1 + \log p_1/p_2$, it follows that $\log q/p_2 \in L^{\Phi_1}(p_2)$. Both $\log q/p_1$ and $\log p_1/p_2$ belong to $L^{\Phi_1}(p_2)$, the first one by assumption and the second one because p_1 and p_2 can be connected by an open exponential arc. Therefore $q \in \widehat{\mathcal{E}}(p_1)$. The reverse inclusion can be proved analogously. $\qquad \square$

The following is a sufficient condition for a limit point of a maximal exponential model to be in the corresponding extended model.

Corollary 19.2 *Let* $q \in \overline{\mathcal{E}(p)}$, *i.e let* (q_n), $q_n \in \mathcal{E}(p)$, *be a sequence so that* $q_n \to q$ *in* $L^1(\mu)$. *Assume that* $p^{\alpha+1}/q_n^\alpha$ *converges in* $L^1(\mu)$ *for some* $\alpha > 0$. *Then* $q \in \widehat{\mathcal{E}}(p)$.

Proof Possibly for a sub-sequence, $q_n \to q$ almost everywhere, so that $(p/q_n)^\alpha \to (p/q)^\alpha$ in $L^1(p)$. Hence, from Proposition 19.4, $q \in \widehat{\mathcal{E}}(p)$. $\qquad \square$

While $\mathcal{K}_p$ is the topological closure of $\mathcal{S}_p$, it is not obvious how $\widehat{\mathcal{E}}(p)$ is the closure of $\mathcal{E}(p)$. It is not the closure in $L^1(\mu)$ nor the closure in the sense of the topology induced by the manifold structure.

Let $p \in \mathcal{M}_>$; the model $\partial \mathcal{E}(p) = \widehat{\mathcal{E}}(p) \setminus \mathcal{E}(p)$ is called the *border maximal exponential model at* p. If $q \propto e^u \cdot p$ so that $q \in \partial \mathcal{E}(p)$, then $\|u\|_{\Phi_1, p} \geq 1$. This property reveals the counter-intuitive structure of $\partial \mathcal{E}(p)$. In fact, if $p_1 \in \mathcal{E}(p)$, q can be represented as $q = \exp(u_1 - K_{p_1}(u_1))p_1$, where $u_1 = u - \mathbb{E}_{p_1}[u]$. Hence, for each $\tilde{p} \in \mathcal{E}(p)$ it holds that $\|u - \mathbb{E}_{\tilde{p}}[u]\|_{\Phi_1, \tilde{p}} \geq 1$.

Proposition 19.5 *Let* $p \in \mathcal{M}_>$; *then*

(A-1) $q \in \widehat{\mathcal{E}}(p)$ *if, and only if, a left open right closed exponential arc exists that*
     connects p *to* q. *In particular,* $q \in \partial \mathcal{E}(p)$ *if, and only if, such an arc cannot*
     be right open.

(A-2) $\widehat{\mathcal{E}}(p)$ *is a convex set.*

For the proof see the on-line supplement. The convexity property shown in Proposition 19.5 states in particular that, given $q \in \widehat{\mathcal{E}}(p)$, the mixture arc connecting p to q is completely contained in $\widehat{\mathcal{E}}(p)$.

One might want to further extend $\widehat{\mathcal{E}}(p)$ by requiring even fewer assumptions. To this end, it is possible to introduce the model

$$\widetilde{\mathcal{E}}(p) = \{q = e^{u - K_p(u)} p : u \in L^1(p) \text{ s.t. } \mathrm{E}_p\left[e^u\right] < \infty\};$$

equivalently, if needed by subtracting their expectation, $\widetilde{\mathcal{E}}(p)$ is the set of densities which can be parametrised by centred random variables. However, such an extension leads to less regularity than the corresponding exponential arcs. Let $q_1, q_2 \in \widetilde{\mathcal{E}}(p) \setminus \widehat{\mathcal{E}}(p)$; then q_1 and q_2 can be connected by a left open right closed exponential arc. Such an arc is discontinuous at its end-points and continuous in its interior. In fact, none of the internal points can belong to either $\mathcal{E}(q_1)$ or $\mathcal{E}(q_2)$. On the other hand, any internal point must belong to the same connected component $\mathcal{E}(r)$, for some $r \in \mathcal{M}_>$.

Therefore, we feel that the introduction of exponential Orlicz spaces is justified because both the maximal exponential model $\mathcal{E}(p)$ and its enlargement $\widehat{\mathcal{E}}(p)$ are models with enough regularity for statistical purposes.

19.3.1 Non-maximal exponential models

As in the parametric case, it is possible to define a generic exponential model at p as a special subset of $\mathcal{E}(p)$. More precisely, we give the following definition.

Definition 19.4 An *exponential model at p* is a subset of $\mathcal{E}(p)$ of the type

$$\mathcal{E}_V(p) = \{q = e^{u - K_p(u)} \cdot p, \quad u \in \mathcal{S}_p \cap V\},$$

for a subspace V of B_p. Analogously, we define

$$\widehat{\mathcal{E}}_V(p) = \{q = e^{u - K_p(u)} \cdot p, \quad u \in \mathcal{K}_p \cap V\}.$$

Example 19.1 (Parametric exponential model) Consider the vector space V defined as $V = \mathrm{Span}\,(u_1, \dots, u_n)$, $u_i \in B_p$, $i = 1, \dots n$. Then

$$\widehat{\mathcal{E}}_V(p) = \left\{ \exp\left(\sum_{i=1}^{n} \theta_i u_i - \psi(\theta)\right) \cdot p, \quad \psi(\theta) = K_p\left(\sum_{i=1}^{n} \theta_i u_i\right) < \infty \right\},$$

that is, $\widehat{\mathcal{E}}_V(p)$ is the classical parametric exponential model with canonical statistics u_i.

According to classical definitions, e.g. (Lang 1995), $\mathcal{E}_V(p)$ is a submanifold of $\mathcal{E}(p)$ if, and only if, the subspace V splits in B_p. That is, a second subspace W of B_p exists so that $B_p = V \oplus W$. This is true in the finite-dimensional case of Example 19.1, but in general it is false. This issue is discussed briefly next.

The statistical connection theory by (Amari and Nagaoka 2000) and the idea of mixed parametrisation of exponential models, see e.g. (Barndorff-Nielsen 1978), suggest a relaxation of the splitting condition. Let Ψ_1 be the conjugate Young function of Φ_1. Then, the Orlicz space $L_0^{\Psi_1}(p)$ is the pre-dual of B_p, see (Rao and Ren 2002). Let V be a closed subspace of B_p and let $V^\perp$ be its orthogonal in $L_0^{\Psi_1}(p)$. It is possible to represent each $q \in \mathcal{E}(p)$ uniquely as $q = e^{u-K_p(u)} \cdot (1+v) \cdot p$, $u \in V$ and $v \in V^\perp$. See (Cena and Pistone 2007) for some developments in this direction.

The same weak splitting can be used to derive a necessary implicit condition for the membership problem for $\widehat{\mathcal{E}}_V(p)$. If $q \in \widehat{\mathcal{E}}_V(p)$, then $u - K_p(u) = \log(q/p)$ is orthogonal to $V^\perp$. If $v \in V^\perp$ then densities $r^+, r^- \in \mathcal{M}_\geq$ exist so that $v \propto r^+ - r^-$, $r^+ r^- = 0$, and $\mathrm{E}_{r^+}[\log(q/p)] = \mathrm{E}_{r^-}[\log(q/p)]$. See Chapter 21 for further developments.

In general, in the infinite-dimensional case, the concept of the canonical statistics of an exponential model becomes meaningless if the space V does not admit a numerable basis. However, in some cases a similar definition can be given. Let $\sum_{i=1}^{\infty} \theta_i u_i$, $\theta_i \in \mathbb{R}$, where (u_i), $u_i \in \mathcal{S}_p \cap V$, is a converging sequence. The coefficients (θ_i) could be considered the canonical parameters of the infinite-dimensional exponential model

$$\mathcal{E}_V(p) = \exp\left(\sum_i \theta_i u_i - K_p\left(\sum_i \theta_i u_i\right)\right).$$

Example 19.2 (Infinite-dimensional model) Let us consider the sample space $[0,1]$ equipped with the Lebesgue measure. The space $V = \widetilde{L_0^{\Phi_1}}(1)$ of the centered random variables so that, for each $\alpha \in \mathbb{R}$, $\alpha u \in L^{\Phi_1}(1)$ is a closed and separable subspace of B_p and $V \subset \mathcal{S}_1$. The system of the Haar functions $(H_{n,k})$ is a basis for V, see e.g. (Rao and Ren 1990). Hence, the corresponding exponential model is

$$\mathcal{E}_V(p) = \left\{ \exp\left(\sum_{m,k} \theta_{m.k} H_{m,k} - \psi((\theta)) \right) \right\},$$

i.e. $\mathcal{E}_V(p)$ is an infinite-dimensional parametric exponential model whose canonical statistics are the Haar functions.

19.3.2 MLE for exponential models

In the finite case, the likelihood function for $\mathcal{E}_V(p)$ can be expressed as $q^{\otimes n} = \exp\left(\sum_{i=1}^n u(x_i) - nK_p(u)\right) \cdot p^{\otimes n}$, where $x_1, \ldots, x_n$ are n given observed data. The maximum likelihood estimator (MLE) maximises

$$\frac{1}{n} \sum_{i=1}^n u(x_i) - K_p(u) = \langle u, \widehat{F}_n \rangle - K_p(u),$$

and is always well defined.

In analogy with the finite case, it is possible to generalise the concept of MLE for $\mathcal{E}_V(p)$. Let $F \in (L_0^{\Phi_1}(p))^*$ and $u \in L^1(F) \cap \mathcal{S}_p \cap V$; the log-likelihood becomes

$$l(u) = \int u\, dF - K_p(u). \tag{19.1}$$

In order to minimise $l(u)$, the extremal point of (19.1) should be found. Therefore, the ML estimator $\widehat{u}$ satisfies

$$\langle v, F \rangle - DK_p(\widehat{u}) \cdot v = \langle v, F \rangle - \mathrm{E}_{\widehat{q}}\,[v] = 0, \ \text{for all } v \in V,$$

where $\widehat{q} = e_p(\widehat{u})$; that is, $F - \dfrac{\widehat{q}}{p} \in V^{\perp}$.

19.3.3 The Cameron–Martin densities on the Wiener space

Let $\Omega = \mathcal{C}([0,1])$, μ the Wiener measure on Ω, and $(\mathcal{F}_t)$, $t \in [0,1]$, the canonical filtration.

Proposition 19.6 *Consider the set of Cameron–Martin densities*

$$\mathcal{G} = \left\{ \exp\left(\int_0^1 f_s\, dW_s - 1/2 \int_0^1 f^2\, ds \right), \ f \in L^2([0,1]) \right\}.$$

(A-1) $\mathcal{G}$ *is an exponential model* $\mathcal{G} = \mathcal{E}_V$, *where*

$$V = \left\{ u = \int_0^1 f_s\, dW_s, \ f \in L^2([0,1]) \right\}. \tag{19.2}$$

(A-2) The border model is empty: $\mathcal{E}_V = \widehat{\mathcal{E}}_V$.

(A-3) The orthogonal space is generated by the random variables of the form $\int_0^1 F_t\, dW_t$ with $\mathrm{E}\,(F_t) = 0$ almost everywhere.

(A-4) $\mathcal{G}$ *is an infinite-dimensional parametric exponential model, with canonical statistics* $T_i = \int_0^1 H_i(s)\, dW_s$, *where* (H_i), $i \in \mathbb{N}$, *is a basis of* $L^2([0,1])$.

Proof It was remarked in (Pistone and Sempi 1995) that the Orlicz norm of the exponential space $L^{\Phi_1}(\mu)$ is equivalent to the usual $L^2(\mu)$-norm for all Gaussian random variables. More precisely, we have

$$\left\| \int_0^1 f(s)\, dW_s \right\|_{\Phi_1} = \frac{1}{\log 2}\, \|f\|_2\,.$$

Therefore, since for each $f \in L^2([0,1])$, it holds that $\int_0^1 f(s)\, dW_s \in L^2(\mu)$, the space V of Equation (19.2) is a closed subspace of $L^{\Phi_1}(\mu)$. Moreover, as

$$\mathrm{E}_\mu\left[\exp\left(t \int_0^1 f(s)\, dW_s \right) \right] = \exp\left(-\frac{t}{2} \|f\|_2 \right),$$

the space V is a subspace of C_μ, see Proposition 19.2. By comparing the definition of $\mathcal{G}$ with the definition of the exponential model, i.e.

$$\exp\left(\int_0^1 f_s\, dW_s - 1/2 \int_0^1 f^2(s)\, ds \right) = \exp\left(u - K_p(u) \right),$$

we obtain $u = \int_0^1 f(s)\, dW_s \in \mathcal{S}_p \cap V$ and $K_p(u) = 1/2 \int_0^1 f^2(s)\, ds$. It should be noted that $\mathcal{E}_V(p) = \widehat{\mathcal{E}}_V(p)$, since $\mathcal{S}_p \equiv K_p$. Given a basis (H_i) of $L^2[0,1]$ and $f(s) = \sum_i \theta_i H_i(s)$, if $T_i = \int_0^1 H_i(s)\, dW_s$, the series $\sum_i \theta_i T_i$ converges in L^2, hence in

$L_0^{\Phi_1}(p)$, and $\int_0^1 f(s)\,dW_s = \sum_i \theta_i T_i$. Thus, a Girsanov density in $\mathcal{G}$ can be expressed as $\exp\left(\sum_i \theta_i T_i - \psi(\theta)\right)$, where the canonical statistics are $T_i = \int_0^1 H_i(s)\,dW_s$. $\qquad \square$

Example 19.3 Example 19.2 can be extended to a larger subset of Girsanov densities, for example by defining

$$V_G = \left\{ u = \int_0^1 f_s\,dW_s, \quad f_s \in \widetilde{L^{2,\Phi_1}}(\mu) \text{ for all } s \in [0,1] \right\},$$

where L^{2,Φ_1} is the Orlicz space of the Young function $\Phi_1(x^2)$, see (Imparato 2008). The set

$$\mathcal{E}_{V_G}(p) = \{ e^{u - K_p(u)}, \ u \in V_G \},$$

where, as above, $K_p(u) = 1/2 \int_0^1 f^2(s)\,ds$, is an exponential model, since $u \in L_0^{\Phi_1}(p)$, because

$$\mathrm{E}_p[e^{\pm \alpha(W_t - W_s)f_s}] = \mathrm{E}_p[e^{\alpha^2(t-s)f_s^2}] \leq \infty.$$

19.3.4 The compound Poisson density model

Let $(\Omega, \mathcal{F}, \mu)$ be a probability space, X_n, $n = 1, 2, \ldots$, independent and identically distributed (i.i.d.) random variables with a uniform distribution on $[0,1]$, and $\{N(t),\ t \geq 0\}$ a homogeneous Poisson process of rate λ, independent of the sequence (X_n). For each $f \in L^1[0,1]$, the i.i.d. sequence $f(X_i)$, $i = 1, 2, \ldots$, is independent of the process $N(t)$ and the compound Poisson process $Y(f)$ is defined as $Y_t(f) = \sum_{i=1}^{N(t)} f(X_i)$, $t \geq 0$. The set

$$V_t = \left\{ Y_t(f) - \lambda t \int_0^1 f(x)\,dx, \quad f \in L_0^{\Phi_1}[0,1] \right\}$$

is a subspace of $L_0^{\Phi_1}(\mu)$. In fact, $\mathrm{E}(Y_t(f)) = \lambda t \int_0^1 f(s)\,ds$ and, for each $\alpha > 0$,

$$\mathrm{E}\left(e^{\pm \alpha Y_t(f)}\right) = e^{-\lambda t} \sum_{k=1}^{\infty} \frac{(\lambda t)^k}{k!} \mathrm{E}\left(\exp(\pm \alpha \sum_{i=1}^{k} f(X_i))\right)$$

$$= e^{-\lambda t} \sum_{k=1}^{\infty} \frac{(\lambda t)^k}{k!} \left(\int_0^1 \exp(\pm \alpha f(x)\,dx)\right)^k$$

$$= e^{-\lambda t}\left(\exp\left(\lambda t \int_0^1 \exp(\pm \alpha f(x))\,dx\right) - 1\right),$$

which is finite for a suitable α if, and only if, $f \in L_0^{\Phi_1}[0,1]$. Therefore, it is possible to define the exponential model

$$\mathcal{E}_{V_t} = \left\{ e^{u_t - K_\mu(u_t)}, \quad u_t \in V_t \cap \mathcal{S} \right\},$$

where $u_t = Y_t(f) - \lambda t \int_0^1 f(x)\, dx$ and

$$e^{K_\mu(u_t)} = \mathrm{E}\left(\exp\left(Y_t(f) - \lambda t \int_0^1 f(x)\, dx \right) \right)$$

$$= e^{-\lambda t} \left(\exp\left(\lambda t \int_0^1 e^{f(x)}\, dx \right) - 1 \right) \exp\left(-\lambda t \int_0^1 f(x)\, dx \right).$$

Note that $K_\mu(u_t)$ is finite if, and only if, $K(f) = \log\left(\int_0^1 e^{f(x)}\, dx \right) < \infty$. The enlarged model $\widehat{\mathcal{E}}_{V_t}(p)$ is an infinite-dimensional model, whose parameters are identified with the f's and the parameter space with $\mathcal{K} = \left\{ f : \int_0^1 e^{f(x)}\, dx < \infty \right\}$. In this example, $\mathcal{E}_{V_t}(p) \subsetneq \widehat{\mathcal{E}}_{V_t}(p)$, since here $V_t \cap \mathcal{S}_p \subsetneq V_t \cap \mathcal{K}_p$.

19.4 Extended exponential models

This section is devoted to the discussion of some partial results on the $L^1(\mu)$-closure of an exponential model. The $L^1(\mu)$ convergence, when restricted to the set of non-negative densities $\mathcal{M}_\geq$, is equivalent to the convergence in μ-probability. If $q, q_n \in \mathcal{M}_\geq$, $n = 1, 2, \ldots$, and $q_n \to q$ in μ probability, then the sequence q_n is uniformly integrable because $\int q_n$ converges trivially to $\int q$, see e.g. (Dellacherie and Mayer 1975). This property is called the Scheffé lemma. The $L^1(\mu)$-convergence of parametric exponential models was first introduced in (Čencov 1972) and later discussed in great detail by (Csiszár and Matúš 2005).

Definition 19.5 The closure in L^1 topology of the exponential model $\mathcal{E}_V(p)$ is called the *extended exponential model at p in $\mathcal{M}_\geq$*. Such a closure is denoted by $\overline{\mathcal{E}_V(p)}$.

First we prove that the closure $\overline{\mathcal{E}(p)}$ of the maximal exponential $\mathcal{E}(p)$ consists trivially of all non-negative densities $\mathcal{M}_\geq$. It should be noticed that, as the closure is always the same, irrespective of the maximal exponential model, each non-negative density can be approximated by a sequence of densities belonging to any maximal exponential model of choice. Maximal exponential models are open in the exponential topology, while each of them is dense in the $L^1(\mu)$ topology of the set of all densities $\mathcal{M}_\geq$. The most interesting case of the closure of a non-maximal exponential model will be discussed next in Theorem 19.1 giving the main result of the chapter.

Theorem 19.1 *For any $p \in \mathcal{M}$, the maximal exponential model at p is dense in the non-negative densities $\mathcal{M}_\geq$, i.e. $\overline{\mathcal{E}(p)} = \mathcal{M}_\geq$.*

Proof First, let us show that for any $p \in \mathcal{M}$ the set of the simple densities is included in the extended maximal exponential model $\overline{\mathcal{E}(p)}$. Let q be such a density, and let $\Omega' = \mathrm{Supp}\, q$, $\Omega'' = (\mathrm{Supp}\, q)^c$. Let us consider the increasing sequence of truncated values of p,

$$p_n = p(p \leq n) + n(p > n) \quad n \in \mathbb{Z}_{>0}$$

and let $\Omega_n = \{\omega \in \Omega : p(\omega) > n\}$ be the event where $p \neq p_n$; observe that $p_n \chi_{\Omega'} \to p\chi_{\Omega'}$ point-wise, where χ_A is the indicator function of A. For each $n \in \mathbb{N}$, let us define the densities $q_n = e^{v_n} \cdot p$, where

$$v_n = \begin{cases} \log(q/p_n) - \log\left(\int (q/p_n)\,pd\mu + e^{-n}\,\mathrm{P}\left(\Omega''\right)\right) & \text{if } \omega \in \Omega', \\ -n - \log\left(\int (q/p_n)\,pd\mu + e^{-n}\,P(\Omega'')\right) & \text{if } \omega \in \Omega''. \end{cases}$$

As q is a simple density whose support is Ω', then a and A exist so that $0 < a \leq q(x) \leq A < \infty$, $x \in \Omega'$. It follows that the integral $\int q/p_n pd\mu$ is finite because

$$\int q/p_n\, pd\mu = \int_{\Omega_n} (q/p_n)\,pd\mu + \int_{\complement\Omega_n} (q/p_n)\,pd\mu$$
$$= \frac{1}{n}\int_{\Omega_n} q\,pd\mu + \int_{\complement\Omega_n} qd\mu < \frac{A}{n} + 1.$$

It also follows that $q_n \in \mathcal{E}(p)$. In fact, $\int e^{\pm v_n}\,pd\mu$ is dominated by $\int_{\Omega'} e^{\pm \log(q/p_n)}\,pd\mu$, which in turn is finite:

$$\int_{\Omega'} e^{+\log(q/p_n)}\,pd\mu = \int_{\Omega'} q(p/p_n)\,d\mu \leq A\int_{\Omega'}(p/p_n)\,d\mu < A\left(\frac{1}{n}+1\right),$$
$$\int_{\Omega'} e^{-\log(q/p_n)}\,pd\mu = \int_{\Omega'}(1/q)p_n\,pd\mu \leq \frac{1}{a}\int_{\Omega'} p_n\,pd\mu < \frac{n}{a}.$$

Finally, it is easy to verify that $q_n \to q$ almost everywhere, since, by the monotone convergence theorem, $\int_{\Omega'} q/p_n\,pd\mu \to 1$. Therefore, by Scheffé's Lemma, $q_n \to q$ in $L^1(\mu)$. Since any density can be written as the limit of simple densities in $L^1(\mu)$, we have proved that $\mathcal{M}_\geq \subset \overline{\mathcal{E}(p)}$. The converse inclusion follows from the definition of $\overline{\mathcal{E}(p)}$. $\qquad\qquad\square$

Weaker notions of extension of an exponential model have been introduced in Section 19.3. Since $\mathcal{E}(p) \subset \widehat{\mathcal{E}}(p) \subset \widetilde{\mathcal{E}}(p) \subset \overline{\mathcal{E}(p)} = \mathcal{M}_\geq$, it follows that $\overline{\mathcal{E}(p)} = \overline{\widehat{\mathcal{E}}(p)} = \overline{\widetilde{\mathcal{E}}(p)} = \mathcal{M}_\geq$.

In the finite state space case, since $\mathcal{E}(p) = \widehat{\mathcal{E}}(p)$ contains all the densities q so that $\operatorname{Supp} q$ is full, it follows that $\overline{\mathcal{E}(p)} \setminus \mathcal{E}(p) = \mathcal{M}_\geq \setminus \mathcal{E}(p)$ coincides with the set of densities q so that $\operatorname{Supp} q$ is not full.

Theorem 19.2 analyses the convergence of the u's and the K's in sequences of densities in $\mathcal{E}(p)$ which are convergent in the L^1 sense. Let us first consider an example.

Example 19.4 Let I_n be a sequence of intervals on the space $\Omega = [0,1]$ with the Lebesgue measure m, so that $m\left(\limsup_{n\to\infty} I_n\right) = 1$ and $\lim_{n\to\infty} m(I_n) = 0$. The random variables

$$u_n(x) = \begin{cases} 1 & \text{if } x \in \complement I_n, \\ 1 - \frac{1}{m(I_n)} & \text{if } x \in I_n, \end{cases}$$

where $\complement I_n$ is the complementary set of I_n in $[0,1]$, have the expected value $(1 - m(I_n)) + \left(1 - \frac{1}{m(I_n)}\right) m(I_n) = 0$. The sequence u_n is convergent in m-probability to 1, while $\liminf_{n\to\infty} u_n = -\infty$. The cumulant functionals are

$$K(u_n) = \log\left(e^1(1 - m(I_n)) + e^{1 - \frac{1}{m(I_n)}}m(I_n)\right)$$

and $\lim_{n\to\infty} K(u_n) = 1$. The sequence of densities $q_n = e^{u_n - K(u_n)}$ is convergent to the uniform density e^{1-1} where $u = 0$. Note that the sequence (u_n) has limit 1, which is different from the centred random variable that represents the centered exponent of the limit density.

Theorem 19.2 *Let $p \in \mathcal{M}_>$ and $q \in \mathcal{M}_\geq = \overline{\mathcal{E}(p)}$. Consider sequences $u_n \in \mathcal{S}_p$ and $q_n = e^{u_n - K_p(u_n)} \cdot p \in \mathcal{E}(p)$, $n = 1, 2, \ldots$, such that $q_n \to q$ in $L^1(\mu)$ as $n \to \infty$.*

(A-1) The sequence $v_n = u_n - K_p(u_n)$ converges in $p \cdot \mu$-probability, as $n \to \infty$, to a $[-\infty, +\infty[$-valued random variable v and $\operatorname{Supp} q = \{v \neq -\infty\}$.

(A-2) $\liminf_{n\to\infty} v_n \leq \liminf_{n\to\infty} u_n$. If the sequence $(v_n)_n$ is μ-almost surely convergent, then $v \leq \liminf_{n\to\infty} u_n$.

(A-3) If $\operatorname{Supp} q = \Omega$, then either

(a) $\limsup_{n\to\infty} K_p(u_n) < +\infty$ *and for each sub-sequence $n(k)$ such that $u_{n(k)}$ is $p \cdot \mu$-convergent, it holds that*

$$-\infty < v + \liminf_{n\to\infty} K_p(u_n) \leq \lim_{k\to\infty} u_{n(k)}$$
$$\leq v + \limsup_{n\to\infty} K_p(u_n) < +\infty,$$

μ-almost surely, or

(b) $\limsup_{n\to\infty} K_p(u_n) = +\infty$ *and for each sub-sequence $n(k)$ such that $u_{n(k)}$ is $p \cdot \mu$-convergent, it holds that $\lim_{k\to\infty} u_{n(k)} = +\infty$.*

(A-4) If $\operatorname{Supp} q \neq \Omega$, then $\lim_{n\to\infty} K_p(u_n) = +\infty$ and $\lim_{n\to\infty} u_n = +\infty$ $p \cdot \mu$-a.s on $\operatorname{Supp} q$. Moreover, $\lim_{n\to\infty} u_n - K_p(u_n) = -\infty$ on $\{q = 0\}$.

Theorem 19.2, for whose proof we refer to the on-line supplement, gives necessary conditions on the u's to ensure L^1-convergence of the densities. Proposition 19.7 below considers sufficient conditions. It gives a re-interpretation in terms of weak convergence of the non-parametric version of (Csiszár and Matúš 2005, Lemma 1.2), see also (Čencov 1972). For sake of simplicity, we present our result for the reference density $p \equiv 1$.

Proposition 19.7 *Let $q_n = e^{u_n - K(u_n)} \in \widehat{\mathcal{E}}(\mu)$ and suppose that, in the weak topology, the image measures $u_n^*(\mu)$ and $u_n^*(q_n)$ converge to $u^*(\mu)$ and to some measure P, respectively. Then $\lim_{n\to\infty} K_p(u_n) = K_p(u) < \infty$ and $P = u^*(q)$, where $q = e^{u - K_p(u)} \cdot p$.*

Proof First of all, observe that

$$u_n^*(q_n)(dy) = e^{y - K_p(u_n)} u_n^*(\mu)(dy) \tag{19.3}$$

Let B a compact continuity set of P. Then,

$$P(B) = \lim_{n \to \infty} e^{-K_p(u_n)} \int_B e^y u_n^*(\mu)\,(dy) = \lim_{n \to \infty} e^{-K_p(u_n)} \int_B e^y u^*(\mu)\,(dy), \quad (19.4)$$

which in turn shows that $\lim_{n \to \infty} K_p(u_n) = \kappa$ exists and is finite. Therefore $P(A) = e^{-\kappa} \int_A e^y\,du^*(\mu)\,(dy)$ for all Borel sets A. In particular, $1 = e^{-\kappa} \int e^u\,d\mu$, so that $\lim_{n \to \infty} K(u_n) = K(u)$. In reality, (19.4) can be extended to each Borel set, so that by taking $B = \mathbb{R}$ we obtain $\lim K_p(u_n) = K_p(u) < \infty$. In fact, if $u \ni K_p$, then $P(B)$ would be equal to zero for each Borel set B. Hence, again from (19.4), for each Borel set B we have

$$P(B) = \lim_n \int_B e^{y - K_p(u_n)} u_n^*(\mu) = \int_B e^{y - K_p(u)} u^*(\mu),$$

i.e., $P = u^*(q)$, where $q = e^{u - K_p(u)}p$. □

Theorem 19.3 is similar to (Csiszár and Matúš 2005, Corollary 1, p. 590), adapted to the non-parametric case, and follows from Proposition 19.7. For the proof see the on-line supplement.

Theorem 19.3 *Let $q_n = e_p(u_n) \in \mathcal{E}(p)$, and suppose that $u_n \to u$ in μ-probability. Then, possibly for a sub-sequence, the following statements are equivalent*

(A-1) $u_n^*(q_n) \to u^*(q)$ weakly, where $q = e^{u - k_p(u)}p$.
(A-2) $u_n \to u$ almost everywhere and $K_p(u_n) \to K_p(u) < \infty$.
(A-3) $q_n \to q$ in $L^1(\mu)$, where $q = e^{u - k_p(u)}p$.

Corollary 19.3 *Let $q \in \mathcal{M}_\geq = \overline{\mathcal{E}(p)}$, i.e. sequences $(u_n)_n$, $u_n \in \mathcal{S}_p$ and $q_n = e_p(u_n)$, $q_n \to q$ in $L^1(\mu)$, exist and suppose that $u_n \to u$ in μ-probability. Then, $q = e^{u - K_p(u)}p$ and, possibly for a sub-sequence, $K_p(u_n) \to K_p(u)$.*

For the proof see the on-line supplement. From Corollary 19.3 we cannot conclude that $q \in \widetilde{\mathcal{E}}(p)$, since u does not necessarily belong to $L^1(p)$. In order to ensure this, stronger conditions are needed, see e.g. Proposition 19.13. However, Corollary 19.3 implies that for any $q \in \mathcal{M}_\geq$, either $q = e^{u - K_p(u)}$ or u is not a limit in μ-probability. It is in fact possible to find examples of densities whose proper domain is Ω and which can be represented as a limit of densities $q_n = e_p(u_n)$ so that both u_n and $K_p(u_n)$ diverge.

Example 19.5 Let μ be the Lebesgue measure on $[0, 1]$ and define

$$u_n(x) = \begin{cases} \dfrac{n^2}{n-1} & \text{if } x \in [0, 1 - 1/n] \\ -n^2 & \text{if } x \in (1 - 1/n, 1]. \end{cases}$$

It is possible to verify that $u_n \in \mathcal{S}_p$ and that both u_n and $K_p(u_n)$ diverge. However, the densities $q_n = e_p(u_n)$ converge to the constant density $q \equiv 1$.

However, the following proposition shows that such a degenerate case is a peculiarity of the infinite-dimensional case.

Proposition 19.8 *Let Ω be finite, $q \in \mathcal{M}_{\geq} = \overline{\mathcal{E}(p)}$ and suppose that $\operatorname{Supp} q = \Omega$, i.e. the sequences $(u_n)_n$, $u_n \in \mathcal{S}_p$ and $q_n = e_p(u_n)$, $q_n \to q$ in $L^1(\mu)$, exist. Then, possibly for a sub-sequence,*

$$-\infty < \lim u_n < \infty \ \text{almost everywhere} \quad \text{and} \quad \lim K_p(u_n) < \infty.$$

Proof First, we show that u_n cannot diverge. To this end, let us suppose that $|\Omega| = k$ and let $p = (p_1, \ldots, p_k)$ be a distribution on Ω. Next, we consider the sequence

$$q_n(i) = e^{v_n(i)} = e^{u_n(i) - K_p(u_n)}, \quad i = 1, \ldots k,$$

where $u_n = v_n - \mathrm{E}_p[u_n]$, converging to q in $L^1(\mu)$. A straightforward computation shows then that

$$u_n(i) = \log \left[\frac{(q_n(i))^{1 - p_i}}{\prod_{j \neq i} (q_n(j))^{p_j}} \right], \quad i = 1, \ldots k.$$

Since, possibly for a sub-sequence, $q_n \to q$ almost everywhere, u_n cannot diverge, as $q(i) > 0$ for each $i = 1, \ldots k$. $\qquad\square$

19.4.1 Extension of the Cameron–Martin model

Proposition 19.9 *Let $\mathcal{E}_V(p)$ be the Cameron–Martin model and let $q \in \overline{\mathcal{E}_V(p)}$, i.e. a sequence $q_n = e_p(u_n) \in \mathcal{E}_V(p)$ exists that converges to q in $L^1(\mu)$. Then, $\operatorname{Supp} q = \Omega$ and, possibly for a sub-sequence, $\lim_n \int_0^1 f_n^2(s) \, ds < \infty$ and $\lim \int_0^1 f_n(s) \, dW_s$ is finite almost everywhere.*

Proof Let q_n be a Cameron–Martin density; then $q_n = e_p(u_n) \in \mathcal{E}_V(p)$, where $u_n = \int_0^1 f_n(s) \, dW_s$ is a Gaussian variable. Hence, u_n and $-u_n$ have the same joint distributions, so that

$$\limsup_n u_n \sim \limsup(-u_n) = -\liminf_n u_n.$$

Hence, $\limsup u_n = +\infty$ almost everywhere implies $\liminf u_n = -\infty$ almost everywhere, since $P(\limsup u_n = +\infty) = 1$ implies

$$P(\liminf_n u_n = -\infty) = P(\limsup_n u_n = +\infty) = 1.$$

This means that, from Theorem 19.2, $\operatorname{Supp} q = \Omega$ and, possibly for a sub-sequence, $\lim_n K_p(u_n) < \infty$ and $\lim_n u_n = u$ almost everywhere. $\qquad\square$

Theorem 19.4 *The Cameron–Martin density model $\mathcal{E}_V(p)$ is closed in L^1-topology, i.e., $\mathcal{E}_V(p) = \overline{\mathcal{E}_V(p)}$.*

Proof Let $q_n = e^{v_n} = e^{u_n - K_p(u_n)} p \in \mathcal{E}_V(p)$, where $u_n = \int_0^1 f_n(s) \, dW_s$ and $K_p(u_n) = 1/2 \int_0^1 f_n^2(s) \, ds$ be a sequence converging to q in L^1. From Proposition 19.9, possibly for a sub-sequence, $\lim_n K_p(u_n) = c \in \mathbb{R}$ and $\lim u_n = u$

almost everywhere, where $u = \int_0^1 f_s(\omega)\,dW_s$ is Gaussian, as the limit of Gaussian variables whose mean and variance converge. More precisely, it is possible to show that $u = \int_0^1 f(s)\,dW_s$. Since u_n is a martingale, in fact, for any martingale $M_n = \int_0^{\cdot} G_s(\omega)\,dW_s$, where (G_s) is a predictable process in $L^2([0,1] \times \Omega)$, it holds that

$$\mathrm{E}_p\left[u_n \int_0^1 G_s(\omega)\,dW_s\right] = \mathrm{E}_p\left[\int_0^1 f_n(s)G_s\,ds\right] = \int_0^1 f_n(s)\mathbb{E}_p[G_s]\,ds. \qquad (19.5)$$

Since $\lim_n \int_0^1 f_n^2(s)\,ds < \infty$ exists, the sequence (f_n) converges in the weak topology of $L^2[0,1]$ to f; hence, by taking the limits in (19.5) one obtains

$$\mathrm{E}_p\left[u \int_0^1 G_s\,dW_s\right] = \int_0^1 f(s)\mathrm{E}_p[G_s]\,ds,$$

that is, $u = \int_0^1 f(s)\,dW_s$. Finally, it can be observed that $c = 1/2 \int_0^1 f^2(s)\,ds$, since possibly for a sub-sequence

$$1 = \mathrm{E}_p[q] = e^{-c}\mathbb{E}_p[e^u]$$
$$= e^{-c}\mathrm{E}_p\left[\exp\left(\int_0^1 f(s)\,dW_s - 1/2 \int_0^1 f^2(s)\,ds + 1/2 \int_0^1 f^2(s)\,ds\right)\right]$$
$$= \exp\left(-c + 1/2 \int_0^1 f^2(s)\,ds\right).$$

Hence, possibly for a sub-sequence

$$q = \exp\left(\int_0^1 f(s)\,dW_s - 1/2 \int_0^1 f^2(s)\,ds\right)p = \exp\left(u - K_p(u)\right)p \in \mathcal{E}_V(p),$$

and we can conclude. $\square$

19.5 The KL-divergence

Classes of exponential models, based on the domain and regularity of the functional $u \mapsto \log(\mathrm{E}_p[e^u])$, were introduced in Section 19.3. If $q \propto e^u \cdot p$, then $\mathrm{E}_p[e^u] < \infty$, and

$$q \in \mathcal{E}(p) \text{ if and only if } u \in \mathcal{S}_p,$$
$$q \in \widehat{\mathcal{E}}(p) \text{ if and only if } u \in \mathcal{K}_p,$$
$$q \in \widetilde{\mathcal{E}}(p) \text{ if and only if } u \in L^1(p).$$

Next, an attempt is made to relate such models to the finiteness of the *KL*-divergence, as defined in the next section, cf. (Csiszár 1975).

19.5.1 KL-divergence and exponential model

Definition 19.6 The *Kullback–Leibler (KL) divergence* or *relative entropy* is defined as $D(q\|p) = \int_{\{q>0\}} \log(q/p)\,qd\mu$, $q \in \mathcal{M}_{\geq}$, $p \in \mathcal{M}_{>}$.

It is well known, see e.g (Cover and Thomas 2006), that $D(\cdot\|p)$ is a non-negative convex map, which is zero if, and only if, $q = p$. Furthermore, it dominates the L^1-distance, as stated by Pinsker's inequality, see e.g. (Cover and Thomas 2006); namely,

$$\|p - q\|_{L^1(\mu)}^2 \leq 2D(p\|q).$$

However, it is not a distance, since it does not satisfy the triangular inequality in general.

(Cena and Pistone 2007) proved that $\mathrm{KL}(q\|p)$ is finite whenever $q \in \mathcal{E}(p)$. General conditions can be required for the KL-divergence $\mathrm{KL}(q\|p)$ to be finite, when $q, p \in \mathcal{M}_>$. To this end, we need to introduce the class of Young functions

$$\Psi_r(x) = (1 + |x^r|)\log(1 + |x^r|) - |x^r|, \quad r \geq 1.$$

It should be noticed that Ψ_1 is equivalent to the conjugate function of Φ_1. In (Cena and Pistone 2007, p. 15), it was proved that $\mathrm{KL}(q\|p) < \infty$ if, and only if, $q/p \in L^{\Psi_1}(p)$. The duality between the Orlicz spaces L^{Φ_1} and L^{Ψ_1} and the consequences for the geometry of statistical models are not discussed here, but reference can be made to (Pistone and Rogantin 1999), (Cena 2002).

Our aim, now, is to investigate conditions for the finiteness of $\mathrm{KL}(q\|p)$ and $D(p\|q)$ when $q \in \widehat{\mathcal{E}}(p)$.

Lemma 19.1 *If $q \in \widehat{\mathcal{E}}(p)$, then for each $v \in L^{\Psi_1}(p)$, $vp/q \in L^{\Psi_1}(q)$.*

Proof If $q \in \widehat{\mathcal{E}}(p)$, $L^{\Psi_1}(q) \subset L^{\Psi_1}(p)$. Then the map $v \mapsto vp/q$ is an injection from $L^{\Psi_1}(p)$ to $L^{\Psi_1}(q)$, see Cena and Pistone (2007). $\qquad\square$

Proposition 19.10 *If $q \in \widehat{\mathcal{E}}(p)$, then $p/q \in L^{\Psi_1}(q)$, that is, $D(p\|q) < \infty$.*

Proposition 19.11 is a first attempt at discussing the reverse inclusion. The proof follows directly from Lemma 19.2.

Lemma 19.2 *Let $p, q \in \mathcal{M}_>$ be two densities so that $p/q \in L^{\Psi_{1+\alpha}}(q)$ for some $\alpha > 0$, where for $r > 0$ $\Psi_r(x) = \Psi_1(x^r)$. Then $q \in \widehat{\mathcal{E}}(p)$.*

Proof In order to prove the thesis, we first observe that $p/q \in L^\alpha(p)$, for some $\alpha > 0$. In fact, by using Young's inequality $|xy| \leq \Psi_1(x) + \Phi_1(y)$ with $x = (p/q)^{1+\alpha}$ and $y = 1$, we obtain, for $\alpha > 0$,

$$\mathrm{E}_p\left[(p/q)^\alpha\right] = \mathrm{E}_q\left[(p/q)^{\alpha+1}\right] \leq \mathrm{E}_q\left[\Psi_1(p/q)^{\alpha+1}\right] < \infty.$$

Therefore, the thesis follows from Proposition 19.4. $\qquad\square$

Proposition 19.11 *Let $p/q \in L^{\Psi_{1+\alpha}}(q)$ for some $\alpha > 0$. Then, equivalently, (i) $D(p\|q) < \infty$, that is, $p/q \in L^{\Psi_1}(q)$ and (ii) $q \in \widehat{\mathcal{E}}(p)$ hold.*

In Lemma 19.2 we assume that $p/q \in L^{\Psi_{1+\alpha}}(q)$. A direct computation shows that if $p/q \in L^{\Psi_1}(p)$, then $p/q \in L^{\Psi_2}(q)$, so that the hypotheses of Lemma 19.2 are satisfied.

Corollary 19.4 *If $L^{\Psi_1}(p) = L^{\Psi_1}(q)$, then $q \in \widehat{\mathcal{E}}(p)$ if, and only if, $D(p\|q) < \infty$. In particular, it holds if q and p are connected by an open mixture arc.*

Proof From Proposition 19.10, if $q \in \widehat{\mathcal{E}}(p)$ then $p/q \in L^{\Psi_1}(q)$. Conversely, as noted after Proposition 19.11 and from Proposition 19.10, if $p/q \in L^{\Psi_1}(q) = L^{\Psi_1}(p)$, then $p/q \in L^{\Psi_2}(q)$ and therefore $q \in \widehat{\mathcal{E}}(p)$. In particular, it was proved in (Cena and Pistone 2007) that, if q and p are connected by an open mixture arc, then $L^{\Phi}(p) = L^{\Phi}(q)$ for any Φ. $\qquad\square$

Proposition 19.12 *Let $q \in \widehat{\mathcal{E}}(p)$; then $\mathrm{KL}(q\|p) < \infty$ if, and only if, $r \in \mathcal{E}(p)$ exists so that $D(q\|q + r/2) < \infty$.*

Proof Due to the parallelogram identity for the relative entropy, see (Csiszár 1975), given $r \in \mathcal{E}(p)$

$$\mathrm{KL}(q\|p) = 2D\left(\frac{r+q}{2}\Big\|p\right) + D\left(r\Big\|\frac{r+q}{2}\right) + D\left(q\Big\|\frac{r+q}{2}\right) - D(r\|p).$$

Observe that $r \in \mathcal{E}(p)$ and $q \in \widehat{\mathcal{E}}(p)$ imply that $(r+q)/2 \in \mathcal{E}(p)$, so that $\mathrm{KL}(q\|p) < \infty$ if, and only if, $D(q\|q + r/2)$ is finite. $\qquad\square$

Propositions 19.13 and 19.14 relate the convergence of the KL-divergence to the analytical framework of the exponential models. In particular, Lemma 19.3 and Corollary 19.5 are reinterpretations of classical propositions which were stated and proved in (Csiszár 1975, p. 157).

Proposition 19.13 *Let $p, q \in \mathcal{M}_>$. From Theorem 19.1, a sequence $(q_n) \in \mathcal{E}(p)$ exists so that $q_n \to q$ in $L^1(\mu)$. Suppose that $D(p\|q_n) \to D(p\|q) < \infty$. Then $q \in \widetilde{\mathcal{E}}(p)$.*

Proof Let $q = e^v p$. Since $q_n \to q$, possibly for a sub-sequence $q_n \to q$ almost everywhere, i.e. $\lim_n (u_n - K_p(u_n)) = \log(q/p) = v$. A direct computation shows that $D(p\|q_n) = K_p(u_n)$, so that the hypothesis of convergence of the relative entropy implies that $\lim_n K_p(u_n) = D(p\|q) = \mathrm{E}_p(v) < \infty$. Hence, $\lim u_n = u = v - \mathrm{E}_p[v]$. This implies that $q = e^{u - K_p(u)} p \in \widetilde{\mathcal{E}}(p)$, where $K_p(u) = \mathrm{E}_p[v] = \lim K_p(u_n)$. $\qquad\square$

Lemma 19.3 *If $D(q_n\|p) \to 0$, then $\mathrm{E}_{q_n}[u] \to \mathrm{E}_p[u]$ for all $u \in L^{\Phi_1}(p)$.*

Corollary 19.5 *Let $u \in B_p$ and $q_n = e_p(u_n)$. Then $D(q_n\|p) \to 0$ implies $\mathrm{E}_{q_n}[u] \to 0$, i.e. $DK_p(u_n) \cdot u \to 0$.*

Equivalently, Corollary 19.5 states that $D(q_n\|p) \to 0$ implies that $q_n/p \to 1$ in the weak topology with respect to the dual space of B_p.

Lemma 19.4 *Let $q \in \mathcal{M}_\geq$ such that $\mathrm{KL}(q\|p) < \infty$. Then $L^{\Phi_1}(p) \subset L^1(q)$.*

Proof See (Biagini and Frittelli 2008). $\qquad\square$

Proposition 19.14 *Let $q = \exp(u - K_p(u)) \cdot p \in \widehat{\mathcal{E}}(p)$ such that $\mathrm{KL}(q\|p) < \infty$; consider $v \in L_0^{\Phi_1}(p)$ such that $u + v \in \mathcal{S}_p$ and define $r = e^{u+v-K_p(u+v)} \cdot p \in \mathcal{E}(p)$. If $\mathrm{KL}(q\|p) \leq D(q\|r)$, then K_p admits a sub-differential in u along v equal to $\partial K_p(u) \cdot v = \mathrm{E}_q[v]$.*

Proof Firstly, observe that $\mathrm{KL}(q\|p)$ is finite since $q \in \mathcal{E}(p)$. This implies, using Lemma 19.4, that both $\mathrm{E}_q[v]$ and $\mathrm{E}_q[u]$ are finite. From the definition of the sub-differential, it is necessary to prove that $\mathrm{E}_q[v] + K_p(u) \leq K_p(u + v)$, that is,

$$\frac{\mathrm{E}_p[ve^u] + \mathrm{E}_p[e^u]\log(\mathrm{E}_p[e^u])}{\mathrm{E}_p[e^u]} \leq K_p(u + v).$$

An application of the Jensen inequality to the convex function $f(x) = x \log x$ leads to

$$\frac{\mathrm{E}_p[ve^u] + \mathrm{E}_p[e]^u\log(\mathrm{E}_p[e^u])}{\mathrm{E}_p[e^u]} \leq \frac{\mathrm{E}_p[ve^u] + \mathrm{E}_p[ue^u]}{\mathrm{E}_p[e^u]} = \mathrm{E}_q[u + v].$$

A direct computation shows that $\mathrm{KL}(q\|p) \leq D(q\|r)$ is equivalent to stating that $\mathrm{E}_q[u + v] \leq K_p(u + v)$, so that the thesis follows. $\qquad\square$

Example 19.6 (Cameron–Martin) Let $\mathcal{F}_t$, $t \in [0, 1]$, be a filtration on $\Omega = \mathcal{C}([0, 1])$ and μ be the Wiener measure. Let q be a Cameron–Martin density with respect to μ, i.e. $q = \exp(\int_0^1 f(s)\,dW_s - 1/2\int_0^1 f^2(s)\,ds) \in \mathcal{E}_V(p)$, where $V = \{u = \int_0^1 g_s\,dW_s,\ g \in L^2([0, 1])\}$ and $p \equiv 1$. Accordingly, the divergence $\mathrm{KL}(q\|p)$ is expected to be finite. In fact, by definition and by an application of the Girsanov theorem, it holds that

$$\mathrm{KL}(q\|p) = \mathrm{E}_q[\log q] = \mathrm{E}_q\left[\int_0^1 f(s)\,dW_s - 1/2\int_0^1 f^2(s)\,ds\right]$$

$$= \mathrm{E}_q\left[1/2\int_0^1 f^2(s)\,ds\right] = 1/2\,\|f\|_{L^2[0,1]}.$$

It should be noted that for the Cameron–Martin model, the divergence $\mathrm{KL}(q\|p)$ coincides with the reverse $D(p\|q)$:

$$D(p\|q) = -\mathrm{E}(\log q) = -\mathrm{E}\left(\int_0^1 f_s\,dW_s - 1/2\int_0^1 f^2(s)\,ds\right) = 1/2\,\|f\|_{L^2[0,1]},$$

which is also equal to the cumulant generating functional $K_p(u)$.

19.5.2 I-closure and rI-closure of the maximal exponential models

We recall the notion of *I-closure* and *rI-closure* of $\mathcal{E}(p)$ as introduced in (Csiszár and Matúš 2005).

Definition 19.7 Let $p \in \mathcal{M}_>$; we denote with $\overline{\mathcal{E}(p)}^I$ the *closure of $\mathcal{E}(p)$ with respect to the relative entropy*. Similarly, $\overline{\mathcal{E}(p)}^{rI}$ denotes the *closure of $\mathcal{E}(p)$ with respect to the reverse relative entropy.*

Clearly Definition 19.7 means that $\overline{\mathcal{E}(p)}^{I}$ is the set of densities $q \in \mathcal{M}_>$ so that a sequence $q_n \in \mathcal{E}(p)$, $n = 1, 2, \ldots$, exists that satisfies the condition $\lim_{n\to\infty} D(q_n\|q) = 0$. Furthermore, $\overline{\mathcal{E}(p)}^{rI}$ the set of densities $q \in \mathcal{M}_\geq$ so that a sequence $q_n \in \mathcal{E}(p)$, $n = 1, 2, \ldots$ exists that satisfies $\lim_{n\to\infty} D(q\|q_n) = 0$.

In their work, (Csiszár and Matúš 2005) observed that, due to Pinsker's inequality, for any $p, q \in \mathcal{M}_\geq$, both the I-closure and the rI-closure of an exponential model is contained in the corresponding variation closure. Hence, their aim was to investigate conditions so that the reverse inclusion could be satisfied. Their results are in particular concerned with the concept of *partial mean*.

In the framework of the non-parametric exponential model, without considering Pinsker's inequality, it is known that trivially both $\overline{\mathcal{E}(p)^I}$ and $\overline{\mathcal{E}(p)^{rI}}$ are contained in $\overline{\mathcal{E}(p)} = \mathcal{M}_\geq$. On the other hand, the property of partial mean is closely related to the finite-dimensional structure of the exponential model they considered and becomes meaningless in our context. Theorem 19.5 shows how the concepts of I-closure and rI-closure are related to the previously introduced density sets.

Theorem 19.5 *For any $p \in \mathcal{M}_>$, (i) $\overline{\mathcal{E}(p)}^{rI} = \overline{\mathcal{E}(p)} = \mathcal{M}_\geq$, and (ii) $\overline{\mathcal{E}(p)}^{I} = \mathcal{M}_>$ holds.*

Proof (i) Let $p \in \mathcal{M}_>$ and $q \in \mathcal{M}_\geq$. Given $\Omega' = \operatorname{Supp} q$ and $\Omega'' = (\operatorname{Supp} q)^c$ (where ^{c} stands for complement), let us consider the increasing truncated sequence

$$\left(\frac{p}{q}\right)_n (\omega) = \frac{p}{q}(\omega) \left(\frac{p}{q}(\omega) \leq n\right) + n \left(\frac{p}{q}(\omega) > n\right), \quad \omega \in \Omega',$$

which converges to p/q a.e in Ω', and let $(q/p)_n = (p/q)_n^{-1}$ and $\Omega_n = \{\omega \in \Omega : p/q > n\}$. Let

$$v_n = \begin{cases} -n - \log c_n & \text{if } \omega \in \Omega'' \\ -\log(c_n (p/q)_n) & \text{if } \omega \in \Omega', \end{cases}$$

where $c_n = \int_{\Omega'} (q/p)_n \, pd\mu + e^{-n} P(\Omega'')$ is well defined since

$$\int_{\Omega'} (q/p)_n \, pd\mu = \int_{\Omega'\cap\Omega_n} (q/p)_n \, pd\mu + \int_{\Omega'\cap\Omega_n^c} (q/p)_n \, pd\mu$$
$$= \frac{1}{n} \int_{\Omega'\cap\Omega_n} pd\mu + \int_{\Omega'\cap\Omega_n^c} qd\mu < \infty.$$

Next, let $\widetilde{q}_n = e^{v_n} p$. It should be observed that $\widetilde{q}_n \in \mathcal{E}(p)$; in fact, it suffices to prove that $\int_{\Omega'} e^{\pm \log(q/p)_n} \, pd\mu < \infty$, which is true, since

$$\int_{\Omega'} (p/q)_n \, pd\mu \leq n \int_{\Omega'} pd\mu < \infty.$$

Next, a straightforward computation shows that

$$D(q\|\widetilde{q}_n) = E_q \left[\log q/\widetilde{q}_n\right] = E_q \left[\log((p/q)_n / (p/q))\right] + \log c_n,$$

which converges to zero, due to an application of the monotone convergence theorem to the sequence $(p/q)_n$.

(ii) By definition of the I-closure, $\overline{\mathcal{E}(p)}^I \subset \mathcal{M}_>$. Conversely, if $p, q \in \mathcal{M}_>$, it is possible to define $(p/q)_n$, v_n and $\widetilde{q}_n = \mathrm{e}^{v_n} p$ as before. Hence,

$$
\begin{aligned}
D(\widetilde{q}_n \| q) = \mathrm{E}_{\widetilde{q}_n} \left[\log \widetilde{q}_n / q \right] &= \int_\Omega \frac{q_n p}{p_n c_n} \log \left(\frac{q_n p}{q p_n c_n} \right) d\mu \\
&= 1/c_n \left[\mathrm{E}_p \left[\frac{q_n}{p_n} \log \left(\frac{q_n/p_n}{q/p} \right) \right] - (\log c_n) \mathrm{E}_p \left[q_n/p_n \right] \right].
\end{aligned}
$$

It should be observed that $(q/p)_n$ and $(q/p)_n \log \left((q/p)_n / (q/p) \right)$ are decreasing sequences. Hence, again through the monotone convergence theorem, it can be concluded that $D(\widetilde{q}_n \| q)$ converges to zero, since $\lim_n c_n = 1$, so that $\mathcal{M} \subset \overline{\mathcal{E}(p)}^I$.

$\square$

19.6 Conclusion

The geometry of non-parametric exponential models and its analytical properties in the topology of the (exponential) Orlicz space were studied in previous works, see (Pistone and Sempi 1995), (Pistone and Rogantin 1999), (Gibilisco and Pistone 1998), (Cena and Pistone 2007). Inspired by some results in (Csiszár and Matúš 2005) in the parametric case, that framework has been extended to include the closure of these models in the L^1-topology. Examples have been presented and the use of relative entropy discussed.

The analytical framework of the Orlicz spaces is difficult to handle in the general case. However, as shown in some examples, this topology often reduces to a Hilbert space and the technical results become friendly. This suggests investigating the extension of the exponential model in Hilbert topologies. Chapter 18 by K. Fukumizu develops this.

Practical applications of the extension procedure of exponential models can arise in several fields. In optimisation problems, for instance, the maximum of a given function, called *fitness* function, is obtained using a convergence algorithm of densities towards a density with reduced support. The support of the limit density consists of the points where a maximum is reached. Densities with reduced support with respect to the original model are obtained by considering extended exponential models. Optimisation procedures are considered in Chapter 21.

The connection between exponential models and mixture models may have applications in Finance. We motivate this as follows. Proposition 19.5 states that the left open right closed mixture arc connecting $p, q \in \mathcal{M}_>$ is contained in $\widehat{\mathcal{E}}(p)$. However, the connection of p to q by a left open right closed mixture arc is equivalent to the boundedness of q/p. This result is a slight modification of (Cena and Pistone 2007, Prop. 15(1)). Now, if p is the objective measure in a discrete market, the non-arbitrage condition is equivalent to the existence of a martingale q so that q/p is bounded. Therefore, the martingale measure is contained in $\widehat{\mathcal{E}}(p)$.

Acknowledgements

The authors are grateful to Professor Giovanni Pistone for his helpful comments and fruitful discussions.

References

Amari, S. and Nagaoka, H. (2000). *Methods of Information Geometry* (American Mathematical Society, Oxford University Press) translated from the 1993 Japanese original by Daishi Harada.

Barndorff-Nielsen, O. E. (1978). *Information and Exponential Families in Statistical Theory* (New York, John Wiley & Sons).

Biagini, S. and Frittelli, M. (2008). A unifying framework for utility maximization problems: an Orlicz space approach, *Annals of Applied Probability* **18**(3), 929–66.

Cena, A. (2002). Geometric structures on the non-parametric statistical manifold. PhD thesis, Dipartimento di Matematica, Università di Milano.

Cena, A. and Pistone, G. (2007). Exponential statistical manifold, *Annals of the Institute of Statistical Mathematics* **59**, 27–56.

Čencov, N. N. (1972). *Statistical Decision Rules and Optimal Inference* (Providence, RI, American Mathematical Society), translation 1982.

Cover, T. M. and Thomas, J. A. (2006). *Elements of information theory* 2edn (Hoboken, NJ, John Wiley & Sons).

Csiszár, I. (1975). I-divergence geometry of probability distributions and minimization problems, *Annals of Probability* **3**, 146–58.

Csiszár, I. and Matúš, F. (2005). Closures of exponential families, *Annals of Probability* **33**(2), 582–600.

Dellacherie, C. and Mayer, P.-A. (1975). *Probabilités et potentiel. Chapitres I à IV. Édition entièrment refondue* (Paris, Hermann).

Gibilisco, P. and Pistone, G. (1998). Connections on non-parametric statistical manifolds by Orlicz space geometry, *Infinite Dimensional Analysis, Quantum Probability and Related Topics* **1**(2), 325–47.

Imparato, D. (2008). Exponential models and Fisher information. Geometry and applications. PhD thesis, Dipartimento di Matematica, Politecnico di Torino.

Lang, S. (1995). *Differential and Riemannian manifolds* 3rd edn (New York, Springer-Verlag).

Pistone, G. and Rogantin, M. P. (1999). The exponential statistical manifold: mean parameters, orthogonality and space transformations, *Bernoulli* **5**(4), 721–60.

Pistone, G. and Sempi, C. (1995). An infinite-dimensional geometric structure on the space of all the probability measures equivalent to a given one, *Annals of Statistics* **23**(5), 1543–61.

Rao, M. M. and Ren, Z. D. (1990). *Theory of Orlicz spaces* (New York, Marcel Dekker).

Rao, M. M. and Ren, Z. D. (2002). *Applications of Orlicz spaces* (New York, Marcel Dekker).

20

Quantum statistics and measures of quantum information

Frank Hansen

Abstract

The geometrical formulation of quantum statistics and the theory of measures of quantum information are intimately connected by the introduction of the notion of metric adjusted skew information. We survey the area with a focus on the 'representing operator monotone functions'. In particular, we exhibit a new order structure that renders the set of metric adjusted skew informations into a lattice with a maximal element. The Wigner–Yanase–Dyson skew informations (with parameter p) are increasing with respect to this order structure for $0 < p \le 1/2$ and decreasing for $1/2 \le p < 1$ with maximum in the Wigner–Yanase skew information.

20.1 Introduction

The geometrical formulation of quantum statistics and the theory of measures of quantum information are two distinct theories with separate motivations and histories, and for a long time they did not seem to be related in any way. Early contributions which view quantum information in statistical terms were given by Hasegawa and Petz (Hasegawa and Petz 1996), and by Luo (Luo 2003a, Luo 2003b, Luo 2005). Today the two theories are largely fused with the notion of metric (or metrically) adjusted skew information and a common set of tools and techniques involving certain operator monotone functions and their representations.

The geometrical approach to statistics reveals its fundamental nature when we try to generalise classical statistics to the quantum setting. The key to obtaining quantisation of the Fisher information is to consider it as a Riemannian metric with a certain behaviour in the presence of noise.

20.1.1 Aspects of classical Fisher information

Consider the (open) probability simplex

$$\mathcal{P}_n = \{p = (p_1, \ldots, p_n) \mid p_i > 0, \ \sum_i p_i = 1\}$$

Algebraic and Geometric Methods in Statistics, ed. Paolo Gibilisco, Eva Riccomagno, Maria Piera Rogantin and Henry P. Wynn. Published by Cambridge University Press. © Cambridge University Press 2010.

with tangent space $T\mathcal{P}_n = \{u \in \mathbb{R}^n \mid \sum_i u_i = 0\}$. The Fisher–Rao metric is given by

$$g_{p,F}(u,v) = \sum_{i=1}^{n} \frac{u_i v_i}{p_i}, \qquad u,v \in T\mathcal{P}_n\,.$$

The geometry defined in this way is 'spherical' in the following sense. Let M be a differentiable manifold, and let (N,g) be a Riemannian manifold. Suppose $\varphi : M \to N$ is an immersion, that is a differentiable map such that its differential $D_p\varphi : T_pM \to T_pN$ is injective for any $p \in M$. Then there exists a unique Riemannian scalar product g^φ on M such that $\varphi : (M,g^\varphi) \to (N,g)$ is a Riemannian isometry. The scalar product g^φ is called the *pull-back* metric induced by φ and by its very definition one has

$$g_p^\varphi(u,v) = g_{\varphi(p)}(D_p\varphi(u), D_p\varphi(v)).$$

The various geometrical aspects of N are in this way 'pulled back' to M. If for example $\gamma : [0,1] \to M$ is a curve and $L(\gamma)$ denotes its length then $L(\gamma) = L(\varphi \circ \gamma)$.

Consider now $\mathcal{P}_n$ as a differentiable manifold (with no Riemannian structure) and let S_n^2 be the sphere of radius two in $\mathbb{R}^n$ considered as a Riemannian submanifold of $\mathbb{R}^n$. Let $\varphi : \mathcal{P}_n \to S_n^2$ be the map

$$\varphi(p) = \varphi(p_1,\dots,p_n) = 2(\sqrt{p_1},\dots,\sqrt{p_n})$$

with differential $D_p\varphi = M_{p^{-1/2}}$ where $M_p(u) = (p_1 u_1,\dots,p_n u_n)$. Then

$$g_p^\varphi(u,v) = g_{\varphi(p)}(D_p\varphi(u), D_p\varphi(v)) = \langle M_{p^{-1/2}}(u), M_{p^{-1/2}}(v)\rangle$$

$$= \sum_{i=1}^{n} \frac{u_i v_i}{p_i} = g_{p,F}(u,v). \tag{20.1}$$

The Fisher information is thus the pull-back by the square root mapping of the standard spherical geometry defined on the simplex of probability vectors.

There is another important approach to Fisher information. Consider the Kullback–Leibler relative entropy given by

$$K(p,q) = \sum_{i=1}^{n} p_i(\log p_i - \log q_i) \tag{20.2}$$

and the calculation

$$-\frac{\partial^2}{\partial t \partial s} K(p + tu, p + sv)\Big|_{t=s=0}$$

$$= -\frac{\partial}{\partial t} \sum_{i=1}^{n} (p_i + tu_i)\left(-\frac{1}{p_i + sv_i}\right) \cdot v_i \Big|_{t=s=0}$$

$$= \sum_{i=1}^{n} \frac{u_i v_i}{p_i + sv_i}\Big|_{t=s=0} = \sum_{i=1}^{n} \frac{u_i v_i}{p_i} = g_{p,F}(u,v).$$

The Fisher information may therefore be obtained also as the Hessian geometry associated with the relative entropy.

In the next section we shall introduce analogues to these two approaches in the quantum setting.

20.1.2 Quantum counterparts

Let M_n denote the set of $n \times n$ complex matrices equipped with the Hilbert–Schmidt scalar product $\langle A, B \rangle = \operatorname{Tr} A^* B$. Let $\mathcal{P}_n$ be the set of strictly positive (self-adjoint) elements of M_n with unit trace, that is

$$\mathcal{P}_n = \{\rho \in M_n \mid \operatorname{Tr} \rho = 1, \, \rho > 0\}.$$

The tangent space to $\mathcal{P}_n$ at ρ is given by

$$T_\rho \mathcal{P}_n = \{A \in M_n \mid A = A^*, \, \operatorname{Tr} A = 0\}.$$

It is useful to decompose $T_\rho \mathcal{P}_n$ as the direct sum of a 'commuting' and a 'non-commuting part' with respect to ρ. More precisely, we set

$$(T_\rho \mathcal{P}_n)^c = \{A \in T_\rho \mathcal{P}_n \mid [A, \rho] = 0\}$$

and define $(T_\rho \mathcal{P}_n)^o$ as the orthogonal complement of $(T_\rho \mathcal{P}_n)^c$ with respect to the Hilbert–Schmidt scalar product. Obviously we then have

$$T_\rho \mathcal{P}_n = (T_\rho \mathcal{P}_n)^c \oplus (T_\rho \mathcal{P}_n)^o.$$

It is easy to derive that any commutator $i[\rho, A]$ belongs to the non-commuting part $(T_\rho \mathcal{P}_n)^o$ of the tangent space. Let S_n^2 denote the sphere in M_n of radius two and consider the Riemannian manifold structure induced by the Hilbert–Schmidt scalar product. Let us also consider the map $\varphi : \mathcal{P}_n \to S_n^2$ given by

$$\varphi(\rho) = 2\sqrt{\rho}.$$

We may pull-back the spherical structure on S_n^2 and study the resulting Riemannian metric on the state manifold $\mathcal{P}_n$.

Denote by L_ρ (and by R_ρ, respectively) the left and right multiplication operators by ρ. Since the differential of φ in the point ρ is

$$D_\rho \varphi = 2\left(L_\rho^{1/2} + R_\rho^{1/2}\right)^{-1}$$

the pull-back metric on the state manifold $\mathcal{P}_n$ is given by

$$\begin{aligned}
g_\rho^\varphi(A, B) &= g_{\varphi(\rho)}(D_\rho \varphi(A), D_\rho \varphi(B)) \\
&= \langle 2\left(L_\rho^{1/2} + R_\rho^{1/2}\right)^{-1}(A), 2\left(L_\rho^{1/2} + R_\rho^{1/2}\right)^{-1}(B)\rangle.
\end{aligned} \tag{20.3}$$

In particular, the pull back metric takes the form

$$\begin{aligned}
g_\rho^\varphi &(i[\rho, A], i[\rho, A]) \\
&= 4\langle \left(L_\rho^{1/2} + R_\rho^{1/2}\right)^{-1}(i[\rho, A]), \left(L_\rho^{1/2} + R_\rho^{1/2}\right)^{-1}(i[\rho, A])\rangle \\
&= -4\langle \left(L_\rho^{1/2} + R_\rho^{1/2}\right)^{-1}(L_\rho - R_\rho)(A), \left(L_\rho^{1/2} + R_\rho^{1/2}\right)^{-1}(L_\rho - R_\rho)(A)\rangle \\
&= -4\langle \left(L_\rho^{1/2} - R_\rho^{1/2}\right)(A), \left(L_\rho^{1/2} - R_\rho^{1/2}\right)(A)\rangle \\
&= -4\langle [\rho^{1/2}, A], [\rho^{1/2}, A]\rangle = -4\operatorname{Tr}[\rho^{1/2}, A]^2 = 8 I_\rho(A)
\end{aligned}$$

in commutators $i[\rho, A]$ where A is self-adjoint. The quantity

$$I_\rho(A) = -\frac{1}{2}\operatorname{Tr}[\rho^{1/2}, A]^2$$

is the Wigner–Yanase skew information introduced in 1963, and it will be described in detail in Section 20.1.4.

The quantum analogue of the Kullback–Leibler relative entropy is given by the Umegaki relative entropy

$$S(\rho \mid \sigma) = \operatorname{Tr} \rho(\log \rho - \log \sigma).$$

It is a divergence on $\mathcal{P}_n$ in the sense of Section 1.6.1 of Chapter 1. Let us evaluate the associated scalar product. By use of the identity

$$\frac{\partial}{\partial t} \log(L + tK)\Big|_{t=0} = \int_0^\infty (L + s)^{-1} K (L + s)^{-1} ds$$

it is possible to prove the identity

$$-\frac{\partial^2}{\partial t \partial u} S(\rho + tA \mid \rho + uB)\Big|_{t=u=0} = \int_0^\infty \operatorname{Tr} A(\rho + s)^{-1} B(\rho + s)^{-1} \, ds.$$

The right-hand side is the Bogoliubov–Kubo–Mori (BKM) scalar product, and it is very useful in quantum statistical mechanics (Naudts *et al.* 1975, Bratteli and Robinson 1981, Fick and Sauermann 1990). It is however very different from the Wigner–Yanase skew information.

In principle we should not be surprised. It is a common feature of quantum theory that there may exist several quantum analogues to a single classical object. We have demonstrated that there is more than one version of quantum Fisher information corresponding to different characterisations of the classical Fisher information. It is our aim to formulate a coherent, general theory encompassing all possible quantum versions of the classical Fisher information.

20.1.3 Quantum statistics

The aim formulated at the end of the last section is the subject of quantum information geometry.

The geometrical formulation of quantum statistics originates in a study by Čencov of the classical Fisher information. Čencov proved (Čencov 1982) that the Fisher–Rao metric is the only Riemannian metric, defined on the tangent space $T\mathcal{P}_n$, that is decreasing under Markov morphisms. Since Markov morphisms represent coarse graining or randomisation, it means that the Fisher information is the only Riemannian metric possessing the attractive property that distinguishability of probability distributions becomes more difficult when they are observed through a noisy channel.

(Morozova and Čencov 1989) extended the analysis to quantum mechanics by replacing Riemannian metrics defined on the tangent space of the simplex of probability distributions with positive definite sesquilinear (originally bilinear) forms K_ρ defined on the tangent space of a quantum system, where ρ is a positive definite state. Customarily, K_ρ is extended to all operators (matrices) supported by the underlying Hilbert space, cf. (Petz 1996, Hansen 2006) for details. Noisy channels are in this setting represented by stochastic (completely positive and trace preserving) mappings T, and the contraction property by the monotonicity requirement

$$K_{T(\rho)}(T(A), T(A)) \leq K_\rho(A, A)$$

is imposed for every stochastic mapping $T : M_n(\mathbb{C}) \to M_m(\mathbb{C})$. Unlike the classical situation, these requirements no longer uniquely determine the metric. Čencov and Morozova were able to prove that a monotone metric necessarily is given on the form

$$K_\rho(A, B) = \operatorname{Tr} A^* c(L_\rho, R_\rho) B, \tag{20.4}$$

where c is a so-called Morozova–Čencov function and $c(L_\rho, R_\rho)$ is the function taken in the pair of commuting left and right multiplication operators.

(Morozova and Čencov 1989) determined the necessary condition (20.4) with the added information that the function c is symmetric, homogeneous of degree -1 and satisfies $c(x, x) = 1/x$ for $x > 0$. They were in fact unable to prove the existence of even a single Morozova–Čencov function, although they put forward a number of candidates including the functions

$$\frac{2}{x + y}, \qquad \frac{x + y}{2xy}, \qquad \frac{\log x - \log y}{x - y}.$$

Subsequently Petz (Petz 1996) characterised the Morozova–Čencov functions by giving the canonical representation

$$c(x, y) = \frac{1}{y f(xy^{-1})} \qquad x, y > 0, \tag{20.5}$$

where f is a positive operator monotone function defined in the positive half-axis satisfying the functional equation

$$f(t) = t f(t^{-1}) \qquad t > 0. \tag{20.6}$$

By considering (among others) the operator monotone functions

$$\frac{t + 1}{2}, \qquad \frac{2t}{t + 1}, \qquad \frac{t - 1}{\log t} = \int_0^1 t^\alpha \, d\alpha,$$

which all satisfy the functional equation (20.6), Petz proved that the candidates put forward by Čencov and Morozova indeed define monotone metrics. A monotone metric is sometimes called *quantum Fisher information*. The BKM metric corresponds to the function $(t - 1)/\log t$.

A (normalised) Morozova–Čencov function c allows a canonical representation (Hansen 2008, Corollary 2.4) of the form

$$c(x, y) = \int_0^1 c_\lambda(x, y) \, d\mu_c(\lambda) \qquad x, y > 0, \tag{20.7}$$

where μ_c is a probability measure on $[0, 1]$ and, for $\lambda \in [0, 1]$,

$$c_\lambda(x, y) = \frac{1 + \lambda}{2} \left(\frac{1}{x + \lambda y} + \frac{1}{\lambda x + y} \right) \qquad x, y > 0.$$

20.1.4 Measures of quantum information

In (Wigner 1952), Wigner noticed that the obtainable accuracy of the measurement of a physical observable represented by an operator that does not commute with a conserved quantity (observable) is limited by the 'extent' of that non-commutativity. Wigner proved it in the simple case where the physical observable is the x-component of the spin of a spin one-half particle and the z-component of the angular momentum is conserved. Araki and Yanase (Araki and Yanase 1960) demonstrated that this is a general phenomenon and pointed out, following Wigner's example, that under fairly general conditions an approximate measurement may be carried out.

Another difference is that observables that commute with a conserved additive quantity, like the energy, components of the linear or angular momenta, or the electrical charge, can be measured easily and accurately by microscopic apparatuses (the analysis is restricted to one conserved quantity), while other observables can be only approximately measured by a macroscopic apparatus large enough to superpose sufficiently many states with different quantum numbers of the conserved quantity.

In (Wigner and Yanase 1963) Wigner and Yanase proposed to find a measure of our knowledge of a difficult-to-measure observable with respect to a conserved quantity. They discussed a number of postulates that such a measure should satisfy and proposed, tentatively, the so called *skew information* defined by $I_\rho(A) = -\frac{1}{2}\mathrm{Tr}\,[\rho^{\frac{1}{2}}, A]^2$, where ρ is a state (density matrix) and A is an observable (self-adjoint matrix), see the discussion in (Hansen 2008). The postulates Wigner and Yanase discussed were all considered essential for such a measure of information and included the requirement from thermodynamics that knowledge decreases under the mixing of states; or put equivalently, that the proposed measure is a convex function in the state ρ.

The measure should also be additive with respect to the aggregation of isolated subsystems and, for an isolated system, independent of time. These requirements are satisfied by the skew information.

In the process that is the opposite of mixing, the information content should decrease. This requirement comes from thermodynamics where it is satisfied for both classical and quantum mechanical systems. It reflects the loss of information about statistical correlations between two subsystems when they are only considered separately. Wigner and Yanase conjectured that the skew information also possesses this property. They proved it when the state of the aggregated system is pure. We subsequently demonstrated (Hansen 2007) that the conjecture fails for general mixed states.

20.2 Metric adjusted skew information

Wigner and Yanase were aware that other measures of quantum information could satisfy the same postulates, including the measure

$$I_\rho(p, A) = -\frac{1}{2}\mathrm{Tr}\left([\rho^p, A] \cdot [\rho^{1-p}, A]\right) \tag{20.8}$$

with parameter p $(0 < p < 1)$ suggested by Dyson and today known as the Wigner–Yanase–Dyson skew information. Even these measures of quantum information are only examples of a more general class of information measures, the so-called metric adjusted skew informations (Hansen 2008), that all enjoy the same general properties as discussed by Wigner and Yanase for the skew information.

Definition 20.1 (Regular metric) A symmetric monotone metric on the state space of a quantum system is regular, if the corresponding Morozova–Čencov function c admits a strictly positive limit

$$m(c) = \lim_{t \to 0} c(t, 1)^{-1}.$$

We call $m(c)$ the *metric constant*, cf. (Morozova and Čencov 1989, Petz and Sudár 1996).

We also say, more informally, that a Morozova–Čencov function c is regular if $m(c) > 0$. The function $f(t) = c(t, 1)^{-1}$ is positive and operator monotone in the positive half-line and may be extended to the closed positive half-line. Thus the metric constant $m(c) = f(0)$.

Definition 20.2 (metric adjusted skew information) Let c be the Morozova–Čencov function of a regular metric. The metric adjusted skew information $I_\rho^c(A)$ is defined by setting

$$I_\rho^c(A) = \frac{m(c)}{2} K_\rho^c(i[\rho, A], i[\rho, A]) = \frac{m(c)}{2} \operatorname{Tr} i[\rho, A^*] c(L_\rho, R_\rho) i[\rho, A] \qquad (20.9)$$

for every $\rho \in \mathcal{P}_n$ (the manifold of states) and every $A \in M_n(\mathbf{C})$.

The metric adjusted skew information may also be written in the form

$$I_\rho^c(A) = \frac{1}{2} \operatorname{Tr} \rho(A^*A + AA^*) - \frac{m(c)}{2} \operatorname{Tr} A^* d_c(L_\rho, R_\rho) A,$$

where the function d_c given by

$$d_c(x, y) = \frac{x + y}{m(c)} - (x - y)^2 c(x, y) = \int_0^1 xy \cdot c_\lambda(x, y) \frac{(1 + \lambda)^2}{\lambda} \, d\mu_c(\lambda) \qquad (20.10)$$

is operator concave in the first quadrant, and the probability measure μ_c is the representing measure in (20.7) of the Morozova–Čencov function c, cf. (Hansen 2008, Proposition 3.4). It follows, in particular, that the metric adjusted skew information may be extended from the state manifold to the state space (Hansen 2008, Theorem 3.8).

The symmetrised variance of a state ρ with respect to a conserved observable A is defined by

$$\operatorname{Var}_\rho(A) = \frac{1}{2} \operatorname{Tr} \rho(A^*A + AA^*) - |(\operatorname{Tr} \rho A)|^2.$$

It is a concave function in the state variable ρ. We have tacitly extended the definition of the metric adjusted skew information and the symmetrised variance to include the case where A may not be self-adjoint. This does not directly make sense in physical applications, but it is a useful mathematical tool when studying the so-called *dynamical uncertainty principle* (Gibilisco *et al.* 2007, Andai 2008, Audenaert *et al.* 2008).

We collect a number of important properties of the metric adjusted skew information (Hansen 2008, Section 3.1).

Theorem 20.1 *Let c be a regular Morozova–Čencov function.*

(A-1) The metric adjusted skew information is a convex function, $\rho \to I_\rho^c(A)$, on the manifold of states for any $A \in M_n(\mathbf{C})$.

(A-2) For $\rho = \rho_1 \otimes \rho_2$ and $A = A_1 \otimes 1 + 1 \otimes A_2$ we have

$$I_\rho^c(A) = I_{\rho_1}^c(A_1) + I_{\rho_2}^c(A_2).$$

(A-3) If A commutes with an Hamiltonian operator H then

$$I_{\rho_t}^c(A) = I_\rho^c(A) \qquad t \geq 0,$$

where $\rho_t = e^{itH} \rho e^{-itH}$.

(A-4) For any pure state ρ (one-dimensional projection) we have

$$I_\rho^c(A) = \mathrm{Var}_\rho(A)$$

for any $n \times n$ matrix A.

(A-5) For any density matrix ρ and $n \times n$ matrix A we have

$$0 \leq I_\rho^c(A) \leq \mathrm{Var}_\rho(A).$$

The first three items in Theorem 20.1 exhibit that the metric adjusted skew information satisfies the requirements, put forward by Wigner and Yanase, to an effective measure of quantum information. The first item shows that the metric adjusted skew information is decreasing under the mixing of states. The second item shows that it is additive with respect to the aggregation of isolated subsystems, and the third item that, for an isolated system, it is independent of time.

The Wigner–Yanase skew information is obtained as an example of metric adjusted skew information by choosing the Morozova–Čencov function

$$c^{WY}(x,y) = \frac{4}{(\sqrt{x} + \sqrt{y})^2} \qquad x, y > 0$$

in the formula (20.9). The Wigner–Yanase–Dyson skew informations (20.8) with parameter p are more generally obtained by considering the Morozova–Čencov function

$$c^{WYD}(x,y) = \frac{1}{p(1-p)} \cdot \frac{(x^p - y^p)(x^{1-p} - y^{1-p})}{(x-y)^2} \qquad 0 < p < 1$$

with metric constant $m(c^{WYD}) = p(1-p)$, cf. (Hasegawa and Petz 1996, Hansen 2008).

20.2.1 Representations of skew information

The quantum Fisher information coincides with the Fisher information on diagonal matrices, if the representing operator monotone function f satisfies the condition $f(1) = 1$.

Definition 20.3 We denote by $\mathcal{F}_{op}$ the set of functions $f \colon \mathbb{R}_+ \to \mathbb{R}_+$ such that

(A-1) f is operator monotone,
(A-2) $f(t) = tf(t^{-1})$ for all $t > 0$,
(A-3) $f(1) = 1$.

The (normalised) quantum Fisher informations are uniquely represented by the functions in $\mathcal{F}_{op}$. In accordance with the convention for Morozova–Čencov functions we say that a function $f \in \mathcal{F}_{op}$ is regular if $f(0) > 0$, and non-regular if $f(0) = 0$.

The transform of a regular function $f \in \mathcal{F}_{op}$ given by

$$\tilde{f}(t) = \frac{f(0)}{2} d_c(t,1) = \frac{1}{2}\left[(t+1) - (t-1)^2 \frac{f(0)}{f(t)} \right] \qquad t > 0,$$

where c is the Morozova–Čencov function represented by f in (20.5) and d_c is the operator concave function in (20.10), was introduced in (Gibilisco *et al.* 2007, Definition 5.1). It was noted that $\tilde{f}$ is a non-regular function in $\mathcal{F}_{op}$ and that

$$\tilde{f} \le \tilde{g} \quad \Leftrightarrow \quad \frac{f(0)}{f} \ge \frac{g(0)}{g}$$

for regular functions $f, g \in \mathcal{F}_{op}$ (Gibilisco *et al.* 2007, Proposition 5.7). Thus trivially

$$\tilde{f} \le \tilde{g} \quad \Rightarrow \quad I_\rho^{c_f}(A) \le I_\rho^{c_g}(A).$$

In particular, the SLD-information represented by the function $(t+1)/2$ satisfies

$$I_\rho^{\mathrm{SLD}}(A) \ge I_\rho^c(A)$$

for arbitrary state ρ, observable A, and Morozova–Čencov function c.

Subsequently we established (Gibilisco *et al.* 2009) that the correspondence $f \to \tilde{f}$ is a bijection between the regular and the non-regular operator monotone functions in $\mathcal{F}_{op}$. The functions in $\mathcal{F}_{op}$ therefore come in pairs $(f, \tilde{f})$ each consisting of a regular and a non-regular function.

The following result is found in (Audenaert *et al.* 2008), cf. also (Hansen 2008, Hansen 2006).

Theorem 20.2 *A function $f \in \mathcal{F}_{op}$ admits a canonical representation*

$$f(t) = \frac{1+t}{2} \exp \int_0^1 \frac{(\lambda^2 - 1)(1 - t)^2}{(\lambda + t)(1 + \lambda t)(1 + \lambda)^2} h(\lambda) \, d\lambda, \tag{20.11}$$

where the weight function $h : [0,1] \to [0,1]$ is measurable. The equivalence class containing h is uniquely determined by f. Any function on the given form is in $\mathcal{F}_{op}$.

In Theorem 20.3 we exhibit the representing function h in the canonical representation (20.11) of f for a number of important functions in $\mathcal{F}_{op}$.

Theorem 20.3

(A-1) The Wigner–Yanase–Dyson metric induced by the function

$$f_p(t) = p(1-p) \cdot \frac{(t-1)^2}{(t^p - 1)(t^{1-p} - 1)} \qquad 0 < p < 1$$

is represented by

$$h_p(\lambda) = \frac{1}{\pi} \arctan \frac{(\lambda^p + \lambda^{1-p}) \sin p\pi}{1 - \lambda - (\lambda^p - \lambda^{1-p}) \cos p\pi} \qquad 0 < \lambda < 1.$$

Note that $0 \le h \le 1/2$.

(A-2) The Kubo metric induced by the function $f(t) = (t-1)/\log t$ is represented by

$$h(\lambda) = \frac{1}{2} - \frac{1}{\pi} \arctan \left(-\frac{\log \lambda}{\pi} \right)$$

and $0 \le h \le 1/2$.

(A-3) The increasing bridge induced by the functions

$$f_\gamma(t) = t^\gamma \left(\frac{2}{t+1} \right)^{2\gamma - 1} \qquad 0 \le t \le 1, \, 0 \le \gamma \le 1$$

is represented by the functions $h_\gamma(\lambda) = \gamma$. Setting $\gamma = 0$, we obtain that the Bures metric is represented by the zero function.

The following definition introduces an order relation $\preceq$ on the set of functions, $\mathcal{F}_{op}$, representing quantum Fisher information that renders the set into a lattice with a maximal element $(t+1)/2$ and a minimal element $2t/(t+1)$.

Definition 20.4 Let $f, g \in \mathcal{F}_{op}$ and set

$$\varphi(t) = \frac{t+1}{2} \frac{f(t)}{g(t)} \qquad t > 0.$$

We write $f \preceq g$ if the function $\varphi \in \mathcal{F}_{op}$.

Let f and g be functions in $\mathcal{F}_{op}$ with representing functions h_f and h_g according to Theorem 20.2. It is established (Audenaert *et al.* 2008, Theorem 2.4) that $f \preceq g$ if and only if $h_f \ge h_g$ almost everywhere. The order relation subsequently induces an order relation on the set of metric adjusted skew informations that renders also this set into a lattice with the SLD-information (induced by $(t+1)/2$) as maximal element. There is no minimal element.

20.2.2 Optimality of the Wigner–Yanase information

Theorem 20.4 *The functions $h_p(\lambda)$ defined in Theorem 20.3 are decreasing in $p \in (0, 1/2]$ for any λ with $0 < \lambda < 1$.*

The Wigner–Yanase–Dyson skew information is, as a function of the parameter p, increasing in $(0, 1/2]$ and decreasing in $[1/2, 1)$ with respect to the order relation $\preceq$. Maximum is attained in the Wigner–Yanase skew information for $p = 1/2$, cf. (Audenaert *et al.* 2008, Theorem 2.8). More elementary, the function $p \to I_\rho(p, A)$ is increasing in $(0, 1/2]$ and decreasing in $[1/2, 1)$ for fixed ρ and A.

Acknowledgement

The author wants to thank Paolo Gibilisco for many helpful suggestions, in particular with regard to the exposition of the connection between the classical and the quantum formulations of information geometry.

References

Andai, A. (2008).Uncertainty principle with quantum Fisher information, *Journal of Mathematical Physics* **49**, 012106.

Araki, H. and Yanase, M. M. (1960). Measurement of quantum mechanical operators, *Physical Review* **20**, 662–26.

Audenaert, K. and Cai, L. and Hansen, F. (2008). Inequalities for quantum skew information, *Letters in Mathematical Physics* **85**, 135–46.

Bratteli, O. and Robinson, D. W. (1981). *Operator algebras and quantum statistical mechanics II* (New York, Berlin, Heidelberg, Springer-Verlag).

Čencov, N. N. (1982). *Statistical Decision Rules and Optimal Inferences, Transl. Math. Monogr.*, Vol 53. (Providence, RI, American Mathematical Society).

Fick, E. and Sauermann, G. (1990). *The quantum statistics of dynamic processes* (New York, Berlin, Heidelberg, Springer-Verlag).

Gibilisco, P. and Hansen, F. and Isola, T. (2009). On a correspondence between regular and non-regular operator monotone functions, *Linear Algebra and its Applications* **430** (8/9), 2225–32.

Gibilisco, P. and Imparato, D. and Isola, T. (2007). Uncertainty principle and quantum Fisher information II, *Journal of Mathematical Physics* **48**, 072109.

Hansen, F. (2006).Characterizations of symmetric monotone metrics on the state space of quantum systems, *Quantum Information and Computation* **6**, 597–605.

Hansen, F. (2007). The Wigner-Yanase entropy is not subadditive, *Journal of Statistical Physics* **126**, 643–8.

Hansen, F. (2008).Metric adjusted skew information, *Proceedings of the National Academy of Sciences of the United States of America* **105**, 9909–16.

Hasegawa, H. and Petz, D. (1996). On the Rimannian metric of α-entropies of density matrices, *Letters in Mathematical Physics* **38**, 221–5.

Luo, S. (2003a).Wigner-Yanase skew information and uncertainty relations, *Physical Review Letters* **91**, 180403.

Luo, S. (2003b).Wigner-Yanase skew information vs quantum Fisher information, *Proceedings of the American Mathematical Society* **132**, 885–90.

Luo, S. (2005).Quantum versus classical uncertainty, *Theoretical and Mathematical Physics* **143**, 681–88.

Morozova, E. A. and Čencov, N. N. (1989). Markov invariant geometry on state manifolds (Russian). *Itogi Nauki i Techniki* **36**, 69–102. Translated in *Journal of Soviet Mathematics* **56**, 2648–69 (1991).

Naudts, J., Verbeure, A. and Weder, R. (1975). Linear response theory and the KMS condition, *Communications in Mathematical Physics* **44**, 87–99.

Petz, D. (1996). Monotone metrics on matrix spaces, *Linear Algebra and its Applications* **244**, 81–96.

Petz, D. and Sudár, C. (1996). Geometries of quantum states, *Journal of Mathematical Physics* **37**, 2662–73.

Wigner, E. P. (1952). Die Messung quantenmechanischer Operatoren, *Zeitschrift für Physik* **133**, 101–8.

Wigner, E. P. and Yanase, M. M. (1963). Information contents of distributions, *Proceedings of the National Academy of Sciences of the United States of America* **49**, 910–18.

Part IV

Information geometry and algebraic statistics

21

Algebraic varieties vs. differentiable manifolds in statistical models

Giovanni Pistone

Abstract

The mathematical theory of statistical models has a very rich structure that relies on chapters from probability, functional analysis, convex analysis, differential geometry and group algebra. Recently, methods of stochastic analysis and polynomial commutative algebra emerged. Each of these theories contributes a clarification of a relevant statistical object, while the scope of each abstract theory is enlarged by the contribution of mathematical statistics as a highly non-trivial application. In this chapter we will concentrate on the methods based on differential geometry, polynomial commutative algebra and stochastic analysis.

21.1 Background and motivation

Other chapters of the present volume present introductions and new research of some of these topics. Here, we focus mainly on the interplay of geometry and algebra in various contexts, namely general sample spaces, finite sample spaces and Gaussian spaces.

21.1.1 Differential geometry

Some concepts introduced by Ronald Fisher in the theory of inference, such as score, likelihood, information, where shown to be related to the geometry of Riemaniann manifolds with the papers by C. R. Rao (1945) and H. Jeffreys (1946). Both these contributions stem from the remark that the Fisher information matrix is a metric (in the sense of differential geometry) on the set of the densities belonging to a given statistical model. Rao derives his celebrated lower bound for the variance of a unbiased estimator; Jeffreys uses the geometric measure of the Riemaniann metric as an uninformative Bayesian prior.

Exponential models where introduced by B. O. Koopman (1936) to discuss the idea of sufficiency as proposed by Fisher. The relationship between Fisher information and exponential structure were first clarified in seminal works by N. Čencov (1982). The next important step was made by B. Efron (1975) and (1978), where

Algebraic and Geometric Methods in Statistics, ed. Paolo Gibilisco, Eva Riccomagno, Maria Piera Rogantin and Henry P. Wynn. Published by Cambridge University Press. © Cambridge University Press 2010.

the concept of curved exponential model was introduced. In general, exponential models have an intrinsic affine geometry. The geometry of a statistical model which is embedded in a larger exponential model is derived by Efron from the affine structure of the super-model. This type of geometry is different from the Riemaniann geometry induced by the Fisher information because in the latter case the geodesics are not exponential models as they are in the former case. In two discussion papers, Dawid (1975) and (1977) gives the first sketch of a non-parametric geometrical theory of statistical models.

Later, H. Nagaoka and S. Amari (1982) fully developed the idea of many different geometries associated to a parametric statistical model. Each geometry is associated to a specific vector bundle of the statistical manifold. Amari termed this theory Information Geometry (IG) and explained it in detail in (Amari 1985), further developed in (Amari and Nagaoka 2000) (Japanese original edition 1993). A further development consists in the formal use of the notion of Banach manifold in order to avoid use of any specific parametrisation and to rigorously cover the case of infinitely many parameters, e.g. (Pistone and Sempi 1995).

It is our opinion that these studies are interesting from the conceptual point of view and the methodological point of view as well. Examples of this are to be found in Part III of this volume. Moreover, the current literature presents a number of related papers in the areas of graphical models, machine learning and genetic algorithms, to name a few.

21.1.2 Commutative algebra

It has been known for a long time that some statistical models on a finite state space have polynomial invariants, e.g. in contingency table inference. A classical reference is (Fienberg 1980), where generalised odds ratios are discussed. More recently, mainly because of the development of algorithms for the symbolic computation of objects in commutative polynomial algebra, a deeper algebraic perspective emerged, as is well illustrated by Parts I and II of this volume.

In the finite state space case, most statistical models are in fact algebraic varieties and, more precisely, lattice exponential models are toric varieties. Therefore, the existence of polynomial invariants in classical statistical models for contingency tables can be shown to be a general feature of finite state space statistical models of current interest. Also in genetics, the notion of algebraic invariant was used from the very beginning in the Hardy–Weinberg computations; such a term is used explicitly in the literature on phylogenetic inference, see e.g. (Evans and Speed 1993).

We remark that the emergence of new applications of commutative algebra to statistics and computational biology has prompted an intensive activity on the side of the computational algebra research, leading to new improved algorithms. Currently, the contribution to applied statistics is mostly methodological because of the low computational efficiency of the available algorithms, but considerable improvements are expected from the research in progress.

21.1.3 Geometry and algebra

The association of algebraic varieties and differentiable manifolds to study statistical models is not a contingent fact. It is a promising association whose case will be argued in this chapter with the aim to encourage research on the following areas.

(A-1) The algebraic and the geometric descriptions coincide in many cases when the state space is finite. Many theoretical ideas of statistics, such as model invariants or sufficiency, have a computational side that can be approached in ways based more on symbolic computations that on numerical computations.

(A-2) The algebraic geometry presentation has special features, typically related with the notion of algebraic degree, considered as an index of the complexity of the model.

(A-3) The study of tangent bundles and vector bundles that arose in the differential geometry picture has a special structure in the algebraic case.

(A-4) Approximation of continuous state space models with discrete state space models could lead to interesting connections of the algebra and the geometry. This is an area where few rigorous results are known.

21.2 A first example: the Gibbs model

Let Ω be a finite sample space with N points and $E : \Omega \to \mathbb{R}_{\geq 0}$ a function, such that $E(x) = 0$ for some $x \in \Omega$, but not everywhere zero. For each $\beta > 0$ consider the probability density function

$$p(x; \beta) = \frac{e^{-\beta E(x)}}{Z(\beta)}, \quad \text{where} \quad Z(\beta) = \sum_{x \in \Omega} e^{-\beta E(x)}. \tag{21.1}$$

In Statistical Physics, E is called *energy* function, the parameter β *inverse temperature*, the analytic function Z *partition function*, $e^{-\beta E}$ *Boltzmann factor*, and the statistical model $p(\beta)$, $\beta > 0$, is called the *Gibbs model* or *canonical ensemble*. It is a key fact that this set of densities is not weakly closed. Indeed, if $\beta \to \infty$, then $Z(\beta) \to \#\{x : E(x) = 0\}$ and $e^{-\beta E} \to (x : E(x) = 0)$ point-wise. Here, for a set A, $\#(A)$ denotes its count and (A) its indicator function. The weak limit of $p(\beta)$ as $\beta \to \infty$ is the uniform distribution on the states $x \in \Omega$ with zero, i.e. minimal, energy, namely $\Omega_0 = \{E(x) = 0\}$. This limit distribution is not part of the Gibbs model, because it has a smaller support than the full support Ω of all the densities in the Gibbs model (21.1).

An extension of the Gibbs model to negative values of the parameter is given by

$$p(x; \theta) = \frac{e^{-\theta(\max E - E(x))}}{e^{-\theta \max E} Z(-\theta)}, \quad \theta \in \mathbb{R}, \tag{21.2}$$

which is convergent to the uniform distribution on $\{E(x) = \max E\}$ as $\theta \to \infty$.

The exponent of the numerator in (21.1) or (21.2) is defined up to an affine transformation. A canonical presentation of the extended Gibbs model (21.2) is the exponential model

$$p(x; \theta) = e^{\theta u(x) - K(\theta u)} \cdot p(x; 0) \tag{21.3}$$

where $p_0 = p(\cdot; 0)$ is the uniform distribution on Ω, *the random variable u is centred for p_0*, and $K(\theta u) = \mathrm{E}_{p_0}\left[e^{\theta u}\right]$ is the normalising exponent, i.e. the cumulant generating function. The *canonical statistics u* is uniquely defined by

$$\theta u(x) = \log\left(\frac{p(x;\theta)}{p(x;0)}\right) - \mathrm{E}_{p_0}\left[\log\left(\frac{p(x;\theta)}{p(x;0)}\right)\right].$$

We shall derive descriptions of the Gibbs model which are both geometric and algebraic. The geometric picture is useful to further clarify the way in which the limits are obtained. The algebraic description is given by equations that are satisfied by the Gibbs model, by the extended parameter model, and also by the two limits $(\theta \to \pm\infty)$, as we will see below.

Recall that the partition function Z is convex, together with its logarithm, and in the β parametrisation we have

$$\frac{d}{d\beta}\log Z(\beta) = -\mathrm{E}_\beta\left[E\right], \qquad \frac{d^2}{d\beta^2}\log Z(\beta) = \mathrm{Var}_\beta\left(E\right),$$

where expectation and variance are taken w.r.t. $p(.; \beta)$ From

$$-\log p(x; \beta) = \beta E(x) + \log Z(\beta), \tag{21.4}$$

we can write the entropy $S(\beta) = -\mathrm{E}_\beta\left[\log p(x; \beta)\right]$ as

$$S(\beta) = \beta\mathrm{E}_\beta\left[E\right] + \log Z(\beta), \tag{21.5}$$

see (Cover and Thomas 2006). Derivation of Equation (21.4) gives

$$-\frac{d}{d\beta}\log p(x; \beta) = E(x) - \mathrm{E}_\beta\left[E\right]$$

where the right-hand side is a function of the sample point x and the parameter β whose expected value at β is zero, i.e. it is an *estimating function*.

Derivation formulas for $\log Z$ and (21.5) give the following important variational results

$$\frac{d}{d\beta}\mathrm{E}_\beta\left[E\right] = -\mathrm{Var}_\beta\left(E\right), \qquad \frac{d}{d\beta}S(\beta) = -\beta\,\mathrm{Var}_\beta\left(E\right).$$

A number of conclusions concerning the Gibbs model are drawn from the previous equations, e.g., the derivative of the continuous function $\beta \mapsto \mathrm{E}_\beta\left[E\right]$ is negative, therefore the expected value of the energy E decreases monotonically to its minimum value 0 for $\beta \to +\infty$. Furthermore we have that $\lim_{\beta\to\infty}\beta^{-1}S(\beta) = 0$. It is clear that such conclusions are reached using both analytic and geometric arguments.

We now move to the algebra. Let $V = \mathrm{Span}\left(1, E\right)$ and $V^\perp$ the orthogonal space, where $k \in V^\perp$ if, and only if,

$$\sum_{x \in \Omega} k(x) = 0, \qquad \sum_{x \in \Omega} k(x)E(x) = 0. \tag{21.6}$$

From Equations (21.4) and (21.6), it follows that, for each probability density $p = p(.; \beta)$ in the Gibbs model,

$$\sum_{x \in \Omega} k(x)\log p(x) = 0, \qquad k \in V^\perp. \tag{21.7}$$

Conversely, if a strictly positive probability density function p satisfies Equation (21.7) then $\log p = \theta E + C$, for suitable $\theta, C \in \mathbb{R}$, therefore p belongs to the larger model in Equation (21.2). In particular, if $\theta = -\beta$ and $\beta > 0$, then $C = -\log Z(\beta)$ and the Gibbs model is obtained.

For each $k \in V^{\perp}$, we can take its positive part k^+ and its negative part k^-, so that $k = k^+ - k^-$ and $k^+ k^- = 0$ and Equation (21.7) can be rewritten as

$$\prod_{x \in \Omega} p(x)^{k^+(x)} = \prod_{x \in \Omega} p(x)^{k^-(x)}. \tag{21.8}$$

Note that Equation (21.8) does not require the strict positivity of each $p(x)$, $x \in \Omega$. As $\sum_{x \in \Omega} k^+(x) = \sum_{x \in \Omega} k^-(x) = \lambda$, it follows that $r_1 = k^+/\lambda$, and $r_2 = k^-/\lambda$ are probability densities with disjoint support.

When k takes integer values, Equation (21.8) is a *polynomial invariant* for the Gibbs model. It has the form of a binomial with unit coefficients. Again, this equation does not require the strict positivity of the density p and, in fact, the limit densities $p(\pm\infty) = \lim_{\beta \to \pm\infty} p(\beta)$ satisfy it by continuity.

The set of polynomial equations of type (21.8) is not finite, because each equation depends on the choice of a vector k in the orthogonal space. Accurate discussion of this issue requires tools from commutative algebra. If the energy function E takes its values on a lattice, we can choose integer-valued random variables $k_1, \ldots, k_{N-2}$ to be a basis of the orthogonal space $V^{\perp}$. In such a case, we have a finite system of binomial equations

$$\prod_{x \in \Omega} p(x)^{k_j^+(x)} = \prod_{x \in \Omega} p(x)^{k_j^-(x)}, \quad j = 1, \ldots, N-2 \tag{21.9}$$

and every other equation in (21.8) is derived from the system (21.9) in the following sense.

In the polynomial ring $\mathbb{Q}[p(x) \colon x \in \Omega]$, the polynomial invariants of the Gibbs model form a polynomial ideal I, which admits, because of the Hilbert Basis Theorem, a finite generating set. The system of equations (21.9) is one such generating set. The discussion of various canonical forms of such generating sets is one of the issues of Algebraic Statistics. We specialise our discussion with a numerical example.

Example 21.1 Consider $\Omega = \{1, 2, 3, 4, 5\}$ and $E(1) = E(2) = 0$, $E(3) = 1$, $E(4) = E(5) = 2$. The following display shows an integer valued k_j, $j = 1, 2, 3$ of the orthogonal space

	1	E	k_1	k_2	k_3		k_1^+	k_1^-	k_2^+	k_2^-	k_3^+	k_3^-
1	1	0	1	0	1		1	0	0	0	1	0
2	1	0	−1	0	1		0	1	0	0	1	0
3	1	1	0	0	−4		0	0	0	0	0	4
4	1	2	0	1	1		0	0	1	0	1	0
5	1	2	0	−1	1		0	0	0	1	1	0

Equation (21.9) becomes

$$\begin{cases} p(1) = p(2) \\ p(4) = p(5) \\ p(1)p(2)p(4)p(5) = p(3)^4. \end{cases} \tag{21.10}$$

The set of all polynomial invariants of the Gibbs model is a polynomial ideal and Equation (21.10) gives a set of generators of that ideal. The non strictly positive density that is a solution of (21.10) is either $p(1) = p(2) = p(3) = 0$, $p(4) = p(5) = 1/2$, or $p(1) = p(2) = 1/2$, $p(3) = p(4) = p(5) = 0$. These two solutions are the uniform distributions of the sets of values that respectively maximise or minimise the energy function.

Again in the lattice case, a further algebraic representation is possible. In the equation $p(x; \beta) = \mathrm{e}^{-\beta E(x)}/Z(\beta)$ we introduce the new parameters $\zeta_0 = Z(\beta)^{-1}$ and $\zeta_1 = \mathrm{e}^{-\beta}$, so that $p(x; \zeta_0, \zeta_1) = \zeta_0 \zeta_1^{E(x)}$. In such a way, the probabilities are monomials in the parameters ζ_0, ζ_1:

$$\begin{cases} p(1) = p(2) = \zeta_0 \\ p(3) = \zeta_0 \zeta_1 \\ p(4) = p(5) = \zeta_0 \zeta_1^2. \end{cases} \tag{21.11}$$

In algebraic terms, such a model is called a *toric model*. It is interesting to note that in (21.11) the parameter ζ_0 is required to be strictly positive, while the parameter ζ_1 could be zero, giving rise to the uniform distribution on $\{1, 2\} = \{x \colon E(x) = 0\}$. The other limit solutions is not obtained from Equations (21.11).

The algebraic elimination of the indeterminates ζ_0, ζ_1 in (21.11) will produce polynomial invariants. For example, from $(\zeta_0 \zeta_1)^2 = (\zeta_0)(\zeta_0 \zeta_1^2)$, we get $p(3)^2 = p(2)p(5)$.

Next we discuss the uniqueness issue of the monomial parametric representation (21.11), together with the fact that one of the limit solutions is not represented. Let us assume that a generic monomial model $q(x; t) = t^{G(x)}$, where $G(x)$ is integer valued, produces *unnormalised* probability densities that satisfy the binomial system (21.10). Therefore $G(x)$, $x \in \Omega$ is a non-negative integer-valued vector such that $\sum_x G(x)k(x) = 0$ for all k in the orthogonal space $V^\perp = \mathrm{Span}\,(k_1, k_2, k_3)$. The set of all points with non-negative integer-valued coordinates in the kernel of the transpose of $K = [k_1, k_2, k_3]$ is closed under summation and has a unique minimal generating set called a *Hilbert basis* given by the rows G_0, G_1, G_2 of the matrix

$$G = \begin{bmatrix} 1 & 1 & 1 & 1 & 1 \\ 2 & 2 & 1 & 0 & 0 \\ 0 & 0 & 1 & 2 & 2 \end{bmatrix} = \begin{bmatrix} G_0 \\ G_1 \\ G_2 \end{bmatrix}.$$

This computation is implemented in many symbolic software, e.g. CoCoA or 4ti2. See (Schrijver 1986) and (Rapallo 2007) for Hilbert bases and their application to statistical models. A new, canonical, monomial presentation is obtained as

$p(x) = t_0^{G_0(x)} t_1^{G_1(x)} t_2^{G_2(x)}$, i.e.

$$\begin{cases} p(1) = p(2) = t_0 t_1^2 \\ p(3) = t_0 t_1 t_2 \\ p(4) = p(5) = t_0 t_2^2. \end{cases} \qquad (21.12)$$

Given a solution of (21.10), $p(x) \geq 0$, $x \in \Omega$, $\sum_x p(x) = 1$, Equation (21.12) is solvable for $t_0 > 0$ and $t_1, t_2 \geq 0$ by taking $t_0 = 1$, $t_1 = \sqrt{p(1)}$ and $t_2 = \sqrt{p(4)}$. The equations for $p(2)$ and $p(5)$ are satisfied and

$$t_1 t_2 = \sqrt{p(1)}\sqrt{p(4)} = \sqrt[4]{p(1)p(2)p(4)p(5)} = p_3 \quad \text{as } t_0 = 1.$$

Therefore, all solutions of the original binomial equation can be represented by (21.12). Such a description of the closure of the Gibbs model is over-parametrised, but is produced by a canonical procedure, i.e. it is unique, and all limit cases are produced by taking either $t_1 = 0$ or $t_2 = 0$.

Even in this simple example the algebra is intricate, see (Geiger *et al.* 2006), (Rapallo 2007) and Parts I and II of this volume.

21.3 Charts

This section gives an informal presentation of the non-parametric differential geometry of statistical models as it was developed in (Pistone and Sempi 1995, Gibilisco and Pistone 1998, Pistone and Rogantin 1999, Cena 2002, Cena and Pistone 2007). Although this presentation is informal, the core arguments are rigorous; formal statements and proofs are to be found in the aforementioned papers and Chapters 15 and 16 by R. Streater and A. Jenčová in Part III of this volume.

Let $(\Omega, \mathcal{F}, \mu)$ denote a probability space, $\mathcal{M}^1$ the set of its real random variables f such that $\int f \, d\mu = 1$, $\mathcal{M}_\geq$ the cone of non-negative elements of $\mathcal{M}^1$, and $\mathcal{M}_>$ the cone of strictly positive elements. We define the (differential) geometry of these spaces in a way which is meant to be a non-parametric generalisation of the theory presented in (Amari 1985) and (Amari and Nagaoka 2000). We will construct a manifold modelled on an Orlicz space; see the presentation of this theory in (Rao and Ren 2002).

Let Φ be any convex, non-negative, null at zero, real function equivalent to exp at $\pm\infty$, e.g. $\Phi(x) = \cosh(x) - 1$. Let Ψ be a convex, non-negative, null at zero, real function equivalent to the convex conjugate of Φ at $\pm\infty$, e.g. $\Psi(y) = (1 + |y|)\log(1 + |y|) - |y|$. The functions Φ and Ψ are called Young functions. Consider $p \in \mathcal{M}_>$. The relevant Orlicz spaces are the vector spaces of real random variables u such that $\Phi(\alpha u)$ and $\Psi(\alpha u)$ are $p \cdot \mu$-integrable for some $\alpha > 0$. These two spaces, endowed with suitable norms, are denoted by $L^\Phi(p)$ and $L^\Psi(p)$, respectively. We denote by $L_0^\Phi(p)$ and $L_0^\Psi(p)$ the subspaces of $p \cdot \mu$-centred random variables. If the sample space is not finite, then the exponential Orlicz spaces L^Φ are not separable and the closure $M^\Phi(p)$ of the space of bounded random variables is different from $L^\Phi(p)$. There is a natural separating duality between $L^\Phi(p)$ and $L^\Psi(p)$, which is given by the bi-linear form $(u, v) \mapsto \int uvp \, d\mu = \mathrm{E}_p[uv]$. In particular, we exploit the triple of spaces

$$L_0^{\Psi}(p) \hookrightarrow L_0^2(p) \hookrightarrow L_0^{\Phi}(p) \cong L_0^{\Psi}(p)^*, \quad p \in \mathcal{M}_>$$

where '$\hookrightarrow$' denotes continuous and weakly dense inclusion and $*$ denotes the dual space.

Orlicz spaces L^{Φ} and L^{Ψ} appear naturally in statistical models as follows. First, let u be a random variable such that the exponential model $p(\theta) \propto e^{\theta u} \cdot p_0$ is defined on an open interval I containing 0. Therefore, it is possible to prove that $u \in L^{\Phi}(p_0)$, and vice versa. Second, a probability density p in $\mathcal{M}_>$ has finite entropy, $\int p \log p \, d\mu < +\infty$, if, and only if, $p \in L^{\Psi}(\mu)$, where the underlying density is the constant.

At each $f \in \mathcal{M}^1$ we associate the linear fiber $^*T(f) = L_0^{\Psi}(f)$ and at each $p \in \mathcal{M}_>$ we associate the linear fiber $T(f) = L_0^{\Phi}(p)$. Here, linear fiber means a vector space attached to each point of a set. It is a key fact for the construction of the manifold structure, that two Orlicz spaces $L^{\Phi}(p_1)$ and $L^{\Psi}(p_2)$ are equal as vector spaces and homomorphic as Banach spaces if, and only if, the densities p_1 and p_2 are connected by an open one-parameter exponential model. Therefore, $T(p_2)$ is the set of random variables in $L^{\Psi}(p_1)$ which are $p_2 \cdot \mu$-centred.

21.3.1 e-Manifold

For each $p \in \mathcal{M}_>$, consider the chart s_p defined on $\mathcal{M}_>$ by

$$q \mapsto s_p(q) = \log\left(\frac{q}{p}\right) - \mathrm{E}_p\left[\log\left(\frac{q}{p}\right)\right].$$

As $-\mathrm{E}_p\left[\log\left(\frac{q}{p}\right)\right] = D(p\|q)$, the chart s_p maps the density q into the log-likelihood and the KL-divergence (see Section 18.1.4 in this volume).

The random variable $s_p(q)$ is to be seen as the value of the coordinate given to q by the chart at p. The chart is actually well defined for all $q = e^{u - K_p(u)} \cdot p$ such that u belongs to the interior $\mathcal{S}_p$ of the proper domain of $K_p : u \mapsto \log\left(\mathrm{E}_p\left[e^u\right]\right)$ as a convex mapping from $L_0^{\Phi}(p)$ to $\mathbb{R}_{>0} \cup \{+\infty\}$. This domain is called a *maximal exponential model* at p, and it is denoted by $\mathcal{E}(p)$. Each maximal exponential model is closed under mixture and two maximal exponential models are either equal or disjoint (see e.g. Section 19.2 in this volume).

The atlas $(s_p, \mathcal{S}_p)$, $p \in \mathcal{M}_>$ defines a manifold on $\mathcal{M}_>$, called an exponential manifold, e-manifold for short. Its tangent bundle is $T(p)$, $p \in \mathcal{M}_>$. The e-manifold splits into disconnected components consisting of maximal exponential models.

Under this setting, the function K_p is a strictly convex function on the vector space $T(p)$ and the first and second derivatives are given by

$$\mathrm{D}\, K_p(u)\, v = \mathrm{E}_q[v], \tag{21.13}$$

$$\mathrm{D}^2\, K_p(u)\, v \otimes w = \mathrm{Cov}_q(v, w), \tag{21.14}$$

where $s_p(q) = u$.

An analogous theory has been developed by M. Grasselli (2009) for the M^Φ spaces to avoid unpleasant properties of the exponential Orlicz space, e.g. non-separability. The problem of defining information geometry in the framework of Riemaniann manifolds has received much attention, but, to the best of our knowledge, an entirely satisfying solution is still lacking. Classically, the mapping $q \mapsto \sqrt{q}$ has been used to map $\mathcal{M}_>$ into the unit sphere of $L^2(\mu)$, but the Riemaniann structure of the sphere cannot be transported back to $\mathcal{M}_>$, because the set of strictly positive elements of the unit sphere has empty interior. On the other side, this construction works in the case of special distributions. For example, if u has a Gaussian distribution, it follows that $2K_p(u) = \|u\|_{L^2}^2$. In between, the best construction in this direction appears to be that presented in Chapter 18 by K. Fukumizu in Part III of this volume. See also Section 21.4.1.

21.3.2 m-Manifold

For each $p \in \mathcal{M}_>$, consider a second type of chart on $\mathcal{M}^1$

$$l_p : q \to l_p(q) = \frac{q}{p} - 1.$$

The chart is defined for all $q \in \mathcal{M}^1$ such that q/p belongs to $L^\Psi(p)$. Let $\mathcal{L}_p$ be the set of such q's. The atlas $(l_p, \mathcal{L}_p)$, $p \in \mathcal{M}_>$ defines a manifold on $\mathcal{M}^1$, called mixture manifold, m-manifold for short. Its tangent bundle is $^*T(p)$, $p \in \mathcal{M}_>$.

21.3.3 Sub-models and splitting

Given a one-dimensional statistical model $p_\theta \in \mathcal{M}_>$, $\theta \in I$, I open interval, $0 \in I$, the *local representation* in the e-manifold is u_θ with

$$p_\theta = e^{u_\theta - K_p(u_\theta)} \cdot p.$$

The local representation in the m-manifold is

$$l_p(p_\theta) = \frac{p_\theta}{p} - 1.$$

The e-manifold or the m-manifold can be considered as two maximal models in which each specific statistical model is embedded. Each statistical sub-model inherits a geometrical structure from these embeddings. In particular, each model of the form $e^{u - K_p(u)} \cdot p$, where u belongs to some linear subspace V of $L_0^\Phi(p)$ is an *exponential model*. If V happens to be finite dimensional, and $u_1, \ldots, u_n$ is a basis, the exponential model takes the traditional form $\exp\left(\sum_{i=1}^n \theta_i u_i - \psi(\theta_1, \ldots, \theta_n)\right) \cdot p$, where $\psi(\theta_1, \ldots, \theta_n) = K_p\left(\sum_{i=1}^n \theta_i u_i\right)$, see e.g. the 'canonical' presentation of the Gibbs model in (21.3).

Finite-dimensional exponential models have another canonical presentation. Let $U : x \mapsto (u_1, \ldots, u_n)$ and denote by $\tilde{p} \cdot U_* \mu = U_*(p \cdot \mu)$ the U-image of the reference probability. The image of the exponential model is the canonical real exponential model

$$\tilde{p}(y; \theta) = e^{\sum_{i=1}^n \theta_i y_i - \psi(\theta)} \tilde{p}(y)$$

with respect to the real probability measure $U_* \mu$.

All known properties of finite-dimensional exponential models, see (Brown 1986) and (Letac 1992), apply to the infinite-dimensional case, when the proper non-parametric formalism is used.

In the statistical e-manifold, a sub-model should be considered a submanifold. However, this basic idea has to be qualified, because the usual non-parametric definition requires a technical property called *splitting*, which is not always verified unless Ω is finite, see (Lang 1995). For example, the exponential model $\mathcal{E}_V(p)$ has tangent space V at p, but, in general, there is no subspace W such that $T(p) = V \otimes W$. We do not discuss this point here, but we mention that this issue is related to the non-parametric generalisation of the mixed parametrisation of exponential models. See also Section 21.3.5.

21.3.4 Velocity

The velocity at θ of the one-parameter statistical model p_θ, $\theta \in I$, is represented in the s_p chart by $\dot{u}_\theta$, while in the l_p chart the representative is $\dot{p}_\theta/p$. Both representations are related to the derivative of the model as a curve in the probability density simplex. In the first case we have $p_\theta = e^{u_\theta - K_p(u_\theta)} \cdot p$, therefore

$$\dot{p}_\theta = p_\theta \left(\dot{u}_\theta - \mathrm{D}\,K_p(u_\theta)\,\dot{u}_\theta \right) = p_\theta \left(\dot{u}_\theta - \mathrm{E}_\theta[\dot{u}_\theta] \right)$$

so that

$$\frac{\dot{p}_\theta}{p_\theta} = \dot{u}_\theta - \mathrm{E}_\theta[\dot{u}_\theta] \quad \text{and} \quad \dot{u}_\theta = \frac{\dot{p}_\theta}{p_\theta} - \mathrm{E}_p\left[\frac{\dot{p}_\theta}{p_\theta}\right].$$

In the second case $l_p(p_\theta) = p_\theta/p - 1$, so that $\dot{l}_\theta = \dot{p}_\theta/p$.

The two cases are shown to represent the same geometric object by moving to the tangent bundles at p_θ via the two affine connections:

$$T(p) \ni u \mapsto u - \mathrm{E}_{p_\theta}[u] \in T(p_\theta) \quad \text{and} \quad {}^*T(p) \ni v \mapsto \frac{p}{p_\theta} v \in {}^*T(p_\theta).$$

Note that both in the e-manifold and in the m-manifold there is just one chart, that we call *frame*, which is centred at each density. The two representations $\dot{u}_\theta$ and $\dot{l}_\theta$ are equal at $\theta = 0$ and are transported to the same random variable at θ:

$$\frac{\dot{p}_\theta}{p_\theta} = \dot{u}_\theta - \mathrm{E}_\theta[\dot{u}_\theta] = \dot{l}_\theta \frac{p}{p_\theta}.$$

The random variable $\dot{p}_\theta/p_\theta$ is the *Fisher score* at θ of the one-parameter model. The *Fisher information* at θ is the L^2-norm of the velocity vector of the statistical model in the moving frame centred at θ. Moreover, the Fisher information is expressible in terms of the duality between ${}^*T(p)$ and $T(p)$:

$$\mathrm{E}_\theta\left[\left(\frac{\dot{p}_\theta}{p_\theta}\right)^2\right] = \mathrm{E}_\theta\left[\left(\dot{u}_\theta - \mathrm{E}_\theta[\dot{u}_\theta]\right)\left(\dot{l}_\theta \frac{p}{p_\theta}\right)\right] = \mathrm{E}_p\left[\dot{u}_\theta \dot{l}_\theta\right].$$

21.3.5 General Gibbs model as sub-manifold

We discuss in this section a generalisation of the Gibbs model of Section 21.2 and expand the discussion by bringing in new elements of the theory. Let Ω be a finite sample space with N points, p a positive probability density function on Ω and $T_j : \Omega \to \mathbb{R}$, $j = 1,\ldots,m$ non-constant random variables. The probability density function p plays the role of reference measure, as it is the uniform measure in the Gibbs model. Note that in the finite case $L^\Phi = L^2 = L^\Psi$.

For $\theta_j \in \mathbb{R}$, $j = 1,\ldots,m$ we consider the one-parameter family of probability density functions

$$p(x;\theta) = e^{\sum_{j=1}^m \theta_j T_j(x) - \psi(\theta)} \cdot p, \quad e^{\sum_{j=1}^m \theta_j T_j(x)} = \mathrm{E}_p\left[e^{\theta T}\right]. \tag{21.15}$$

Let $V = \mathrm{Span}\,(1, T_j : j = 1,\ldots,m)$ and let $V^\perp$ be its orthogonal space in $L^2(p)$. For each linear basis $k_1,\ldots,k_{N-m-2}$ of $V^\perp$, we consider the system of $N - m - 2$ equations

$$\prod_{x:k_j^+(x)>0} p(x)^{k_j^+(x)} = \prod_{x:k_j^-(x)>0} p(x)^{k_j^-(x)}, \quad j = 1, 2, \ldots, N - m - 2 \tag{21.16}$$

where k_j^+ and k_j^- are the positive and negative part of $(k_j(x))_x$, respectively.

A positive probability density q belongs to the exponential model (21.15) if, and only if, it satisfies the system of equations (21.16). The set of solutions of (21.16) is weakly closed. The set of non-negative solutions is connected by suitable arcs. Indeed, given two non-negative solutions q_1 and q_2 the model $q(\lambda) \propto q_1^{1-\lambda} q_2^\lambda$, $\lambda \in [0, 1]$, is called the *Hellinger arc* from q_1 to q_2. All density functions in the Hellinger arc are solutions of Equation (21.16) and are positive on the set $\Omega_{12} = \{q_1 q_2 > 0\}$ for all $\lambda \in]0, 1[$. This part of the Hellinger arc is a sub-model, possibly reduced to a single point, of an exponential model with reduced support Ω_{12}. See a general discussion in (Csiszár and Matúš 2005).

There is a second important interpretation of the space $V^\perp$ that rejoins arguments used in Statistical Physics. The random variables

$$r_1 = (1 + k_1)p, \ldots, r_{N-m-2} = (1 + k_{N-m-2})p$$

all belong to $\mathcal{M}^1$ and $k_j = l_p(r_j)$, $j = 1,\ldots,N - m - 2$. If $q = e^{u - K_p(u)}p$ is a generic density in $\mathcal{M}_>$, the constraints

$$\mathrm{E}_{r_1}[u] = 0, \ldots, \mathrm{E}_{r_{N-m-2}}[u] = 0$$

are satisfied by $u \in V$, therefore, the constraints

$$\mathrm{E}_{r_j}\left[\log\left(\frac{q}{p}\right)\right] = -K_p(u) = D(p\|q)$$

are satisfied by $u \in V$. This produces a new characterisation of the exponential model (21.15) in terms of the KL-divergence $q \mapsto D(p\|q)$ and the splitting $(V, V^\perp)$.

Moreover, the splitting is related to the so-called *mixed parametrisation* of exponential models. Let q be any density in $\mathcal{M}_>$ and consider the set $\mathcal{Q}$ defined as

$$\{r \in \mathcal{M}_> : \mathrm{E}_r\,[u] = \mathrm{E}_q\,[u]\,, u \in V\} = \left\{ r \in \mathcal{M}_> : \frac{r-q}{p} \in V^\perp \right\}.$$

For each $r = e^{v-K_p(v)} \cdot p \in \mathcal{Q}$, from $r/p - q/p \in V^\perp$ we have $q/p = e^{v-K_p(v)} + w$, where $w \in V^\perp$. Choose r in order to minimise the divergence

$$D(r\|p) = \mathrm{E}_r\left[\log\left(\frac{r}{p}\right)\right] = DK_p(v)v - K_p(v)$$

on $r \in \mathcal{Q} \cap \mathcal{E}(p)$. A standard argument shows that the minimum is unique and it is characterised by the unique $v^* \in V$, such that $r^* = e^{v^*-K_p(v^*)} \cdot p \in \mathcal{Q}$, i.e. $r^* \in \mathcal{E}_V$ and $\mathrm{E}_{r^*}\,[u] = \mathrm{E}_q\,[u]$, $u \in V$. In Example 21.1, we obtain the mixed parametrisation

$$q(x; \zeta_0, \zeta_1, \eta_1, \eta_2, \eta_3) = \zeta_0 \zeta_1^{E(x)} + \eta_1 k_1(x) + \eta_2 k_2(x) + \eta_3 k_3(x).$$

21.3.6 Optimisation

Consider a bounded real function F on Ω, which reaches its maximum on a measurable set $\Omega_{\max} \subset \Omega$. The mapping $\tilde{F} : \mathcal{M}_\geq \ni q \mapsto \mathrm{E}_q\,[F]$ is a regularisation or *relaxation* of the original function F. If F is not constant, i.e. $\Omega \neq \Omega_{\max}$, we have $\tilde{F}(q) = \mathrm{E}_q\,[F] < \max F$, for all $q \in \mathcal{M}_>$. However, if ν is a probability measure such that $\nu(\Omega_{\max}) = 1$ we have $\mathrm{E}_\nu\,[F] = \max F$. This remark suggests to determine $\max F$ by finding a suitable maximising sequence q_n for $\tilde{F}$, see e.g. (Geman and Geman 1984) to name one among many interesting references from various fields. Here we discuss the geometry of this optimisation problem with the tools of information geometry introduced above.

Given any reference probability p, we can represent each positive density q in the maximal exponential model at p as $q = e^{u-K_p(u)} \cdot p$. The expectation of F is an affine function in the m-chart,

$$\mathrm{E}_q\,[F] = \mathrm{E}_p\left[F\left(\frac{q}{p} - 1\right)\right] + \mathrm{E}_p\,[F].$$

In the e-chart the expectation of F is a function of u, $\Phi(u) = \mathrm{E}_q\,[F]$. Equation (21.13) for the derivative of the cumulant function K_p gives

$$\begin{aligned}\Phi(u) = \mathrm{E}_q\,[F] &= \mathrm{E}_q\,[(F - \mathrm{E}_p\,[F])] + \mathrm{E}_p\,[F]\\ &= \mathrm{D}\,K_p\,(u)\,(F - \mathrm{E}_p\,[F]) + \mathrm{E}_p\,[F].\end{aligned}$$

The derivative of this function in the direction v is the Hessian of K applied to $(F - \mathrm{E}_p\,[F]) \otimes v$ and from (21.14) it follows that

$$D^2 K(u)(v, w) = \mathrm{D}\,\Phi(u)\,v = \mathrm{Cov}_q\,(v, F).$$

Therefore, the direction of steepest ascent of the expectation is $F - \mathrm{E}_q\,[F]$.

By the use of both the m- and e-geometry, we have obtained a quite precise description of the setting of this problem.

(A-1) The problem is a convex problem in the m-geometry as the utility function $q \mapsto \mathrm{E}_q\left[F\right]$ is linear and the admissible set $\mathcal{M}^1$ is convex and closed in $L^1(\mu)$. The level sets are affine subspaces in the m-charts.

(A-2) In the e-geometry, given any starting point $q \in \mathcal{M}_>$, the exponential model $e^{\theta F}/\mathrm{E}_q\left[e^{\theta F}\right]$ gives the steepest strict ascent. In fact, on such a statistical model the second derivative of the expected value of F is maximal at each point.

(A-3) If F is continuous and if the exponential model of steepest ascent has a weak limit point whose support belongs to $\Omega_{\max}$, then $\lim_{\theta \to \infty} \int F e^{\theta F}/\mathrm{E}_p\left[e^{\theta F}\right]\,d\mu = \max F$.

21.3.7 Exercise: location model of the Cauchy distribution

The following example shows the mechanism of the e-chart and the m-charts. The position model for the Cauchy distribution is

$$f(x;\theta) = f(x - \theta), \quad f(x) = \frac{1}{\pi(1 + x^2)}, \quad \theta \in \mathbb{R}.$$

If f is the reference density, $p = f$, such a model is a curve $(p_\theta)_{\theta \in \mathbb{R}}$ in the manifold $\mathcal{M}_>$, and

$$p(x;\theta) = \frac{1 + x^2}{1 + (x - \theta)^2}\, p(x), \quad \theta \in \mathbb{R}.$$

Therefore, the m-coordinate is

$$\frac{p(x;\theta)}{p(x)}\,1 = \frac{1 + x^2}{1 + (x - \theta)^2} - 1 = \frac{2\theta x - \theta^2}{1 + (x - \theta)^2}.$$

For all $\theta \in \mathbb{R}$, $p(x;\theta)$ is uniformly bounded and, therefore, $p(x;\theta) = e^{v_\theta} \cdot p$, with $v_\theta \in M^{\Phi}(p)$ and specifically

$$v(x;\theta) = \log\left(\frac{1 + x^2}{1 + (x - \theta)^2}\right).$$

The expression of the model as a sub-model of the maximal exponential model is

$$p_\theta = e^{u_\theta - K_p(u_\theta)} \cdot p,$$

where the e-coordinate is

$$u(x;\theta) = \log\left(\frac{1 + x^2}{1 + (x - \theta)^2}\right) - \int \log\left(\frac{1 + x^2}{1 + (x - \theta)^2}\right) \frac{1}{\pi(1 + x^2)}\,dx$$

and

$$K_p(u_\theta) = -\int \log\left(\frac{1 + x^2}{1 + (x - \theta)^2}\right) \frac{1}{\pi(1 + x^2)}\,dx.$$

The point-wise derivative with respect to θ of $v(x;\theta)$ is $\frac{d}{d\theta}v(x;\theta) = \frac{2(x-\theta)}{1+(x-\theta)^2}$, which is bounded by 1 in absolute value. Therefore, $\theta \mapsto v$ is differentiable as a mapping with values in $L^{\Phi_1}(p)$ and

$$\frac{d}{d\theta}K_p(u_\theta) = -\int \frac{2(x-\theta)}{1+(x-\theta)^2}\frac{1}{\pi(1+x^2)}\,dx.$$

The partial fraction expansion of the integrand is

$$\frac{2(x-\theta)}{(1+x^2)(1+(x-\theta)^2)} = \frac{1}{\theta(\theta^2+4)}\left(\frac{2\theta x - 2(\theta^2-2)}{1+x^2} + \frac{-2\theta x - 2(\theta^2+2)}{1+(x-\theta)^2}\right)$$

and its primitive function is

$$\frac{1}{\theta(\theta^2+4)}\left(\theta\log\left(\frac{1+x^2}{1+(x-\theta)^2}\right) - 2(\theta^2-2)\arctan(x) - 4(\theta^2+1)\arctan(x-\theta)\right).$$

Therefore,

$$\frac{d}{d\theta}K_p(u_\theta) = \frac{6\theta}{\theta^2+4} \quad \text{and} \quad K_p(u_\theta) = 3\log\left(\theta^2+4\right).$$

This model is in fact a solution of a differential equation on the exponential manifold. This introduces Section 21.4 below.

21.4 Differential equations on the statistical manifold

A *vector field* U of the m-bundle $^*T(p) = L_0^\Psi(p)$, $p \in \mathcal{M}_>$, is a mapping defined on some connected open domain $D \subset \mathcal{M}_>$, with values in $L^\Psi(p)$, and is a *section* of the m-bundle, that is $U(p) \in {}^*T(p)$, for all $p \in D \subset \mathcal{M}_>$. This geometric language is a specification of the statistical notion of *estimating function*, i.e. a family U_p of random variables such that for all densities p in a statistical model D, it holds $E_p[U_p] = 0$. Given a sample point $\bar{x}$, p is estimated by a $\hat{p} \in D$ such that $U_{\hat{p}}(\bar{x}) = 0$, see the discussion in (Amari and Nagaoka 2000).

In analogy with our discussion in Section 21.3.4 of the velocity of a one-parameter statistical model, we say that a one-parameter statistical model in $\mathcal{M}_>$, $p(\theta)$, $\theta \in I$, I open real interval such that $0 \in I$, solves the differential equation associated to the vector field U if for all $\theta \in I$ the following equality holds true in $^*T(p(\theta))$

$$\frac{\dot{p}(\theta)}{p(\theta)} = U(p(\theta)), \quad \text{for all } \theta \in I. \tag{21.17}$$

Equation (21.17) is written with respect to the moving frame at p_θ. In terms of estimating functions, the solution of (21.17) is a one-dimensional statistical model whose score statistics $\frac{d}{d\theta}\log p(\theta)$ is given by the estimating function $U(p(\theta))$. With respect to a fixed frame at p, we should write

$$\dot{u}_\theta = F(p(\theta)) - E_p[F(p(\theta))] \quad \text{e-connection, assuming } \dot{u}_\theta \in T(p_\theta)$$

$$\dot{i}_\theta = \frac{p}{p(\theta)}F(p(\theta)) \quad \text{m-connection.} \tag{21.18}$$

These two equations represent the same differential equation as (21.17). In the fixed frame there are two different representations of the same equation. In the moving frame the two representations coincide.

Existence and uniqueness for differential equations of the type (21.17) are to be discussed in the framework of differential equations on a differentiable manifold. The basic general method consists in the reduction to one of the two chart representations (21.18), which are evolution equations in Banach spaces.

Example 21.2 (Exponential models) Consider the exponential model introduced in Section 21.3.6, $p_\theta = e^{\theta F}/\mathrm{E}_p\left[e^{\theta F}\right]$, $\theta \in \mathbb{R}$. In this case the velocity in the moving frame is

$$\frac{\dot{p}_\theta}{p_\theta} = F - \mathrm{E}_{p_\theta}\left[F\right].$$

In this case the vector field is $p \mapsto F - \mathrm{E}_p\left[F\right]$. In general, exponential models are solutions of the differential equation for a constant vector field; that is, a vector field whose unique dependence on p is the centring operation. In the fixed frame at $p(0)$, the equation is $\dot{u}_\theta = F - \mathrm{E}_{p(0)}\left[F\right]$, whose solution is $u_\theta = \theta\left(F - \mathrm{E}_{p(0)}\left[F\right]\right) + u_0$. All one-dimensional exponential models are solutions of such equations.

Example 21.3 (Location model) Consider a simple non-exponential example with $\Omega = \mathbb{R}$ and D the class of positive densities p with logarithmic derivative $p'/p \in L_0^\Psi(p)$. For such densities, the mapping $U : p \mapsto -p'/p$ is a vector field. We can therefore consider the differential equation (21.17). Let us find the solution. If $f \in D$, the location model $p_\theta(x) = f(x - \theta)$ is such that the score is

$$\frac{\dot{p}_\theta(x)}{p_\theta(x)} = -\frac{f'(x - \theta)}{f(x - \theta)} = F(f(\cdot - \theta))(x)$$

and the translation model is a solution of the differential equation. The classical Pearson classes of distributions, such as the Cauchy distribution, are special cases of this construction. For details on the Pearson class see (Johnson *et al.* 1995). In the fixed frame the equation is

$$\dot{u}_\theta(x) = -\partial_x u_\theta(x) - \int u_\theta(x)\partial_x u_\theta(x)\ dx.$$

Example 21.4 (Group model) More generally, any semi group τ_t on the space of positive densities, with infinitesimal generator A, i.e. $(d/dt)\tau_t p = A\tau_t p$, on some domain D will produce the same situation. The model $p_\theta = \tau_\theta f$, $f \in D$ has score

$$\frac{\dot{p}_\theta}{p_\theta} = \frac{A\tau_\theta f}{\tau_\theta f} = U(p_\theta)$$

where the vector field is defined by $U(q) = A(q)/q$, $q \in D$.

Example 21.5 (Heat equation) The heat equation

$$\frac{\partial}{\partial t}p(t, x) - \frac{\partial^2}{\partial x^2}p(t, x) = 0$$

is an interesting example of a differential equation in $\mathcal{M}_>$. In fact, we can consider the vector field

$$U_p(x) = \frac{\frac{\partial^2}{\partial x^2} p(x)}{p(x)}.$$

Upon division of both sides of the heat equation by $p(t, x)$, we obtain an equation of the form (21.17), whose solution is the solution of the heat equation, i.e. the model obtained by the action of the heat kernel on the initial density. Moreover, the heat equation has a variational form. For each $v \in D$

$$\mathrm{E}_p\left[F(p)v\right] = \int p''(x)v(x)\ dx = -\int p'(x)v(x)\ dx = -\mathrm{E}_p\left[\frac{p'}{p}v\right]$$

from which we derive the weak form of the differential equation as

$$\mathrm{E}_{p_\theta}\left[\frac{\dot{p}_\theta}{p_\theta}v\right] + \mathrm{E}_{p_\theta}\left[F_0(p_\theta)v\right] = 0 \quad v \in D$$

where F_0 is the vector field associated to the translation model. The geometry associated to heat equations and generalisations are discussed in detail in (Otto 2001).

Example 21.6 (Optimisation on an exponential model) As a last example we reconsider the framework of Section 21.3.6. Practical computational implementations of these schemes look for maximising sequences in $\mathcal{M}_>$ that belong to a restricted subclass of densities, usually an exponential model. For a discussion of a class of genetic algorithms along these lines, see (Malagò *et al.* 2008). Let V be a linear subspace of $T(p_0)$ and let us denote by V_p the linear space of random variables in V, re-centred at p. Assume that the orthogonal projection F_p of F onto V_p is well defined for all p in the exponential model $\mathcal{E}_V$. Then $U(p) = F_p$ is a vector field defined on a domain including $\mathcal{E}_V$ and we can consider the differential equation $\dot{p}_\theta/p_\theta = U(p_\theta)$. By construction, the solution is a sub-model of the given exponential model, and the velocity vector is parallel to the direction of steepest ascent of the expectation of F. Critical points of the equation are characterised by $\mathrm{E}_p[F] = 0$, $p \in \mathcal{E}_V$.

21.4.1 Deformed exponentials

The theory of non-parametric Information Geometry and its algebraic counterpart, as they were described in the previous sections, are not really restricted to exponential models. Various generalisations, based on transformations with functions other than the couple exp and log, have been proposed, see e.g. (Naudts 2002, Naudts 2004). We observe that: (1) this is an interesting area of application outside mainframe statistics; (2) the construction could lead to an interesting generalisation of e-manifold and m-manifold to new types of model Banach space; (3) these types of models are used in an area where the algebraic features of statistical models, in the sense we are discussing, have not been considered yet.

As an example, we discuss the features of one of the proposals, see (Kaniadakis 2001, Kaniadakis 2005, Pistone 2009). The real function

$$\exp_{\{\kappa\}}(x) = \left(\kappa x + \sqrt{1 + \kappa^2 x^2}\right)^{\frac{1}{\kappa}}, \quad -1 < \kappa < 1, \quad x \in \mathbb{R},$$

maps $\mathbb{R}$ unto $\mathbb{R}_>$ and is strictly increasing and strictly convex. Its inverse

$$\ln_{\{\kappa\}}(y) = \frac{y^\kappa - y^{-\kappa}}{2\kappa}, \quad y > 0,$$

is strictly increasing and strictly concave. The *deformed exponential and logarithm functions* $\exp_{\{\kappa\}}$ and $\ln_{\{\kappa\}}$ reduce to the ordinary exp, ln functions in the limit $\kappa \to 0$. Moreover,

$$\exp_{\{\kappa\}}(x)\exp_{\{\kappa\}}(-x) = 1, \quad \ln_{\{\kappa\}}(y) + \ln_{\{\kappa\}}\left(y^{-1}\right) = 0.$$

It is possible to define group operations $(\mathbb{R}, \oplus)$ and $(\mathbb{R}_>)$, such that

$$\exp_{\{\kappa\}}(x_1 \oplus x_2) = \exp_{\{\kappa\}}(x_1)\exp_{\{\kappa\}}(x_2),$$
$$\exp_{\{\kappa\}}(x_1 + x_2) = \exp_{\{\kappa\}}(x_1) \otimes \exp_{\{\kappa\}}(x_2).$$

We refer to the literature for the discussion of the convex duality features of $\exp_{\{\kappa\}}$ and $\ln_{\{\kappa\}}$.

Given positive density functions q and p such that $(q/p)^\kappa, (p/q)^\kappa \in L^1(q)$, the *Kaniadakis divergence* or κ-divergence is

$$D_\kappa(q\|p) = \mathrm{E}_q\left[\ln_{\{\kappa\}}\left(\frac{q}{p}\right)\right] = \frac{1}{2\kappa}\mathrm{E}_q\left[\left(\frac{q}{p}\right)^\kappa - \left(\frac{p}{q}\right)^\kappa\right].$$

The properties of $-\ln_{\{\kappa\}}$ imply

$$D_\kappa(q\|p) = \mathrm{E}_q\left[-\ln_{\{\kappa\}}\left(\frac{p}{q}\right)\right] \geq -\ln_{\{\kappa\}}\left(\mathrm{E}_q\left[\frac{p}{q}\right]\right) = \ln_{\{\kappa\}}(1) = 0.$$

Let us define a statistical manifold modelled on a Lebesgue space by a slight variation of the tricks used in the standard exponential case. We discuss the case $1/k = 2$. The symmetrisation of $\exp_{\{1/2\}}$ gives

$$\frac{\exp_{\{1/2\}}(x) + \exp_{\{1/2\}}(-x)}{2} = \frac{\left(\frac{x}{2} + \sqrt{1 + \frac{x^2}{4}}\right)^2 + \left(-\frac{x}{2} + \sqrt{1 + \frac{x^2}{4}}\right)^2}{2} = 1 + \frac{x^2}{2}$$

so that the relevant Young function is $x \mapsto x^2/2$ and the associated Banach spaces are the L^2-spaces.

Given $u \in L_0^a(p)$, the real function $K \mapsto \mathrm{E}_p\left[\exp_{\{\kappa\}}(u - K)\right]$ is strictly monotone from $+\infty$ to 0, then there exists a unique $K_{1/2,p}(u)$ such that

$$q = \exp_{\{\kappa\}}(u - K_{\kappa,p}(u))\,p, \quad \kappa = 1/2,$$

is a density in $\mathcal{M}_>$. Vice versa, let $p \in \mathcal{M}_>$ be the reference density and consider the set $\mathcal{Q} = \left\{ q : (q/p)^\kappa, (p/q)^\kappa \in L^1(p) \right\}$, $\kappa = 1/2$. On $\mathcal{Q}$ the $\frac{1}{2}$-divergence $D_{1/2}(p\|q)$ is defined and for each $q \in \mathcal{Q}$ define

$$u = \ln_{\{\kappa\}}\left(\frac{q}{p}\right) - \mathrm{E}_p\left[\ln_{\{\kappa\}}\left(\frac{q}{p}\right)\right] = \ln_{\{\kappa\}}\left(\frac{q}{p}\right) + D_\kappa(p\|q),$$

$\kappa = 1/2$. Therefore, $q = \exp_{\{\kappa\}}(u + D_\kappa(p\|q))\,p$. We do not discuss further this construction, especially the issue of the existence of the atlas of charts, one for each reference density p, cf. the general parametric construction in (Ohara and Wada 2008),

We can define $\frac{1}{2}$-exponential models as

$$q = \exp_{\{1/2\}}\left(u - K_{1/2,p}(u)\right) \cdot p, \quad u \in V, \tag{21.19}$$

where V is a linear subspace of $L_0^2(p)$. If $V^\perp$ is the orthogonal of V as a linear subspace of $L_0^2(p)$, the implicit representation of the exponential model (21.19) is $\mathrm{E}_p\left[\ln_{\{1/2\}}\left(\frac{q}{p}\right)k\right] = 0, \quad k \in V^\perp$.

We conclude by remarking that we could derive, as we did in the case of a finite state space, non-deformed exponential, lattice-valued canonical variables, the relevant binomial-type equations based on the deformed product operation $\otimes$. If $k = 1/2$, $a \otimes b = \exp_{\{1/2\}}\left(\ln_{\{1/2\}} a + \ln_{\{1/2\}} b\right)$. This function is not algebraic in the usual sense, but it is algebraic with respect to the commutative group $(\mathbb{R}_{>0}, \otimes)$.

21.5 Abstract Wiener space

The maximal exponential model of a Gaussian reference measure has special features related to the fact that all moments exist. We discuss here the general case of an infinite-dimensional Gaussian space, in particular some classes of polynomial models. Polynomial models are interesting *per se* from an algebraic viewpoint. Moreover, they have been suggested as approximation tools in Statistical Physics and Mathematical Finance.

Let $(\Omega, \mathcal{F}, \mathbb{P})$ be a probability space. A Gaussian subspace $\mathcal{H}$ of $L^2(\Omega, \mathcal{F}, \mathbb{P}) = L^2$ is a closed subspace of Gaussian random variables, such that $\sigma(\mathcal{H}) = \mathcal{F}$. Assume that H is a separable Hilbert space and $\delta : H \to \mathcal{H}$ a mapping such that $\langle \delta(h_1), \delta(h_2)\rangle_{\mathcal{H}} = \langle h_1, h_2\rangle_H$. This setting is called an abstract Wiener space. We refer mainly to the recent textbook by D. Nualart (2006). Notice that for each $X, Y \in \mathcal{H}$ the sum is normally distributed, $X + Y \sim \mathrm{N}(0, \|X + Y\|_{L^2}^2)$, and that the mapping δ is a linear and surjective isometry of H unto $\mathcal{H}$ called *divergence* or *abstract Wiener integral*.

Example 21.7 (Discrete white noise) Let $X_1, X_2, \ldots$ be a Gaussian white noise (GWN) on the canonical space $(\mathbb{R}^{\mathbb{N}}, \mathcal{F}, \nu^{\otimes \mathbb{N}})$, $\nu(dx) = (2\pi)^{-1/2} \exp\left(-x^2/2\right) dx$. The Hilbert space of square-summable sequences $H = \ell^2$ is the domain of a divergence because the mapping $\delta : a \mapsto \sum_{i=1}^\infty a(i) X_i$, $a \in H$, is a linear isometry between H and the closure $\mathcal{H}$ of $\mathrm{Span}\,(X_i : i = 1, 2, \ldots)$.

Example 21.8 (Continuous white noise) Let μ be the Wiener probability measure on the space of continuous trajectories $(C[0,1], \mathcal{B})$, W_t, $t \in [0,1]$, namely the canonical process. A divergence is defined on $H = L^2[0,1]$ by the Wiener integral $h\colon \int_0^1 h(s)dW_s$. This follows by the identity $\left\langle \int_0^1 h_1(s)dW_s, \int_0^1 h_2(s)dW_s \right\rangle_{\mathcal{H}} = \langle h_1, h_2 \rangle_H$.

21.5.1 Polynomial random variables

Consider the $\mathbb{R}$-algebra of polynomials $\mathbb{R}[\delta(h)\colon h \in H] = \mathbb{R}[\delta]$. Note that there is an infinite number of indeterminates, but in each polynomial only a finite number appear, i.e. if $F \in \mathbb{R}[\delta]$, there exist a finite sequence $h_1, \ldots, h_n \in H$ and a real polynomial $f \in \mathbb{R}[x_1, \ldots, x_n]$ such that $F = f(\delta(h_1), \ldots, \delta(h_n))$. As the Gaussian distribution has moments of all orders, it follows that $F \in L^2$.

The $\mathbb{R}$-algebra $\mathbb{R}[\delta]$ is a dense subspace of L^2. Indeed, if e_i, $i = 1, 2, \ldots$, is an orthonormal basis of H, then an orthonormal basis of L^2 is given by the family $H_{k_1}(\delta(e_{i_1})) \cdots H_{k_n}(\delta(e_{i_n}))$, $n, k_1, \ldots, k_n \geq 0$, $1 \leq i_1 < \cdots < i_n$, where $H_k(x)$ is the real Hermite polynomial of order k. If the space H is infinite dimensional, a second orthonormal basis is formed by the square-free monomials $\delta(e_{i_1}) \cdots \delta(e_{i_n})$, for all integers $n \geq 1$ and $1 \leq i_1 < \cdots < i_n$. For example, let us consider $\delta(e_1)^2$. As H is infinite dimensional, for all integers $n \geq 1$ we can write $e_1 = \sum_{i=1}^n h_i$ such that $\langle h_i, h_j \rangle_H = 0$ for $i \neq j$ and $1/n$ if $i = j$. We have

$$\delta(e_1)^2 = \left(\sum_{i=1}^n \delta(h_i) \right)^2 = \sum_{i \neq j} \delta(h_i)\delta(h_j) + \sum_i \delta(h_j)^2$$

$$= \left(\sum_{i \neq j} \delta(h_i)\delta(h_j) + 1 \right) + \left(\sum_i \delta(h_j)^2 - 1 \right)$$

and

$$\mathrm{E}\left(\left(\sum_i \delta(h_j)^2 - \frac{1}{n} \right)^2 \right) = \frac{2}{n} \to 0 \quad \text{as } n \to \infty.$$

The square-free monomial $\delta(e_{i_1}) \cdots \delta(e_{i_n})$ is also called the Wiener–Ito (symmetric, multiple) integral of $e_{i_1} \otimes \cdots \otimes e_{i_n}$. The mapping $I_n : e_{i_1} \otimes \cdots \otimes e_{i_n} \mapsto \delta(e_{i_1}) \cdots \delta(e_{i_n})$ extends to a one-to-one mapping from $H^{\otimes n}$ to a subspace of L^2 denoted $\mathcal{H}_n$. The space $\mathcal{H}_n$ is the space of n-order interactions and $\mathcal{H} = \bigoplus \mathcal{H}_n$ is an orthogonal decomposition.

Conditioning is especially simple in the case of square-free monomials. Let $Y = \delta(h_{i_1}) \cdots \delta(h_{i_n})$, $h_1, \ldots, h_n \in H$ and H_0 a closed subspace of H. Let $\bar{h}_j$ be the projection of h_j on H_0, $j = 1, \ldots, n$. Therefore, $\mathrm{E}(Y|\delta(h)\colon h \in H_0)$ is a linear combination of square-free monomials in the $\delta(\bar{h}_i)$'s.

For $h_1, \ldots, h_n \in H$, $F_1, \ldots, F_n \in \mathbb{R}[\delta]$, $n = 1, 2, \ldots$, take objects of the form $\sum_{i=1}^n F_i h_i$ and form the tensor product denoted by $\mathbb{R}[\delta] \otimes_{\mathbb{R}} H$. In the discrete white noise case, $\sum_{i=1}^n F_i h_i = (\sum_{i=1}^n F_i h_i(k))_{k=1}^\infty$ is a discrete stochastic process of second order. In the continuous white noise case, $\sum_{i=1}^n F_i h_i$, $h_i \in C[0,1]$, $i = 1, \ldots, n$, is a continuous second-order stochastic process.

The polynomial representation $F = f(\delta(h_i)\colon i = 1, \ldots, n)$ is not unique. In particular, we can represent $h_1, \ldots, h_n$ as a linear transformation of an orthonormal sequence $e_1, \ldots, e_m \in H$, so that $F = f \circ A(\delta(e_j) : j = 1, \ldots, m)$, where A is an $n \times m$ real matrix and $f \circ A \in \mathbb{R}[y_1, \ldots, y_m]$. Let $I(\delta)$ be the ideal generated by $\delta(\alpha h_i + \beta h_2) - \alpha \delta(h_1) - \beta \delta(h_2)$, $h_1, h_2 \in H$, $\alpha, \beta \in \mathbb{R}$. One can show that $F, G \in \mathbb{R}[\delta]$ are equal as random variables if, and only if, $F - G \in I(\delta)$. Therefore, the class of polynomial type random variables is the quotient $\mathbb{R}$-algebra $\mathrm{Poly}(\delta) = \mathbb{R}[\delta]/I(\delta)$.

In Stochastic Analysis, a derivative operator ∇ is defined as a closed operator whose domain is a Sobolev-type space denoted by $\mathbb{D}_1^2$. Here, we restrict our attention to the derivative of a polynomial random variable $F \in \mathrm{Poly}(\delta)$. Define $\nabla \colon \mathrm{Poly}(\delta) \to \mathrm{Poly}(\delta) \otimes_{\mathbb{R}} H$ for $F = f(\delta(h_i) : i = 1, \ldots, n)$ by

$$\nabla F = \sum_{i=1}^{n} \frac{\partial}{\partial x_i} f(\delta(h_i) : i = 1, \ldots, n) h_i.$$

One can check that the equality $F = G$ implies $\nabla F = \nabla G$, or, in algebraic terms, if $F - G \in I(\delta)$, then $\nabla F - \nabla G \in I(\delta) \otimes_{\mathbb{R}} H$. The ∇ of a polynomial random variable is a stochastic process. The operator ∇ is a derivation of the $\mathbb{R}$-algebra $\mathrm{Poly}(\delta)$ because it is linear and

$$\nabla(FG) = G\nabla F + F\nabla G.$$

Moreover, ∇ can be considered a gradient, because for $F = f(\delta(e_i) : i = 1, \ldots, n)$ and $h \in H$, we have

$$\frac{d}{dt} f\left(\delta(e_i) + t \langle e_i, h \rangle_H\right)\Big|_{t=0} = \langle \nabla F, h \rangle_H.$$

Example 21.9 (∇^* of a constant) Let F be a monomial with respect to an orthonormal sequence $e_1, \ldots, e_n \in H$, e.g. $F = \delta(e_1)^{\alpha_1} \cdots \delta(e_n)^{\alpha_n}$ with $\alpha_i \in \mathbb{Z}_{\geq 0}$, $i = 1, \ldots, n$. The set of such random variables is a linear basis of $\mathrm{Poly}(\delta)$. For $h \in H$, let $\bar{h} = \sum_{i=1}^{n} \langle e_i, h \rangle_H$. Recall that $Z \sim \mathrm{N}(0,1)$, implies $\mathrm{E}\left(Z^{\alpha+1}\right) = \alpha \mathrm{E}\left(Z^{\alpha-1}\right)$. Therefore,

$$\langle F, \delta(h) \rangle_{L^2} = \mathrm{E}\left(\delta(e_1)^{\alpha_1} \cdots \delta(e_n)^{\alpha_n} \delta(\bar{h})\right)$$

$$= \sum_{i=1}^{n} \mathrm{E}\left(\delta(e_1)^{\alpha_1} \cdots (\delta(e_i)^{\alpha_i + i}) \cdots \delta(e_n)^{\alpha_n}\right) \langle h, e_i \rangle_H$$

$$= \sum_{i=1}^{n} \mathrm{E}\left(\delta(e_1)^{\alpha_1}\right) \cdots \mathrm{E}\left(\delta(e_i)^{\alpha_i + i}\right) \cdots \mathrm{E}\left(\delta(e_n)^{\alpha_n}\right) \langle h, e_i \rangle_H$$

$$= \sum_{i=1}^{n} \mathrm{E}\left(\delta(e_1)^{\alpha_1}\right) \cdots \mathrm{E}\left(\delta(e_i)^{\alpha_i + 1}\right) \cdots \mathrm{E}\left(\delta(e_n)^{\alpha_n}\right) \langle h, e_i \rangle_H$$

$$= \sum_{i=1}^{n} \mathrm{E}\left(\delta(e_1)^{\alpha_1}\right) \cdots \mathrm{E}\left(\alpha\delta(e_i)^{\alpha_i - 1}\right) \cdots \mathrm{E}\left(\delta(e_n)^{\alpha_n}\right) \langle h, e_i \rangle_H$$

$$= \sum_{i=1}^{n} \mathrm{E}\left(\delta(e_1)^{\alpha_1} \cdots (\alpha\delta(e_i)^{\alpha_i - 1}) \cdots \delta(e_n)^{\alpha_n}\right) \langle h, e_i \rangle_H$$

$$= \left\langle \nabla F, \bar{h} \right\rangle_{L^2 \otimes H} = \left\langle \nabla F, h \right\rangle_{L^2 \otimes H}.$$

Indeed, the value at h of the adjoint of ∇ is $\nabla^*(h) = \delta(h)$.

It is possible to prove that the adjoint of ∇ is defined on $\mathrm{Poly}(\delta) \otimes_{\mathbb{R}} H$ and for $F = \delta(e_1)^{\beta_1} \cdots \delta(e_n)^{\beta_n}$ and $G = Fh$, we have

$$\nabla^* G = -\langle \nabla F, h \rangle_H + \delta(h)F.$$

As ∇^* extends δ, it is denoted by $\delta = \nabla^*$ and it is called the *divergence*.

21.5.2 Models of polynomial type

In the context of an abstract Wiener space $(\Omega, \mathcal{F}, \mathbb{P}, H, \delta)$, we want to discuss the densities in $\mathcal{E}(1)$, i.e. the densities of the form $F = \exp(U - K(U))$, $\mathrm{E}(U) = 0$. It has been suggested in various contexts the approximation of general exponential models with polynomial exponential models. We consider two cases: a polynomial form for U or a polynomial form for F. In the first case the main issue is the exponential integrability of U. In the second case the main issue is the positivity of the polynomial random variable.

Recall a few elementary results for Gaussian random variables. If $Z \sim \mathrm{N}(0,1)$, the moment generating function of Z^2 is $\mathrm{E}\left(e^{\theta Z^2}\right) = (1-\theta)^{-1/2}$, $\theta < 1$. If Z_1, Z_2 are independent $\mathrm{N}(0,1)$, the moment generating function of the product $Z_1 Z_2$ is $\mathrm{E}\left(e^{\theta Z_1 Z_2}\right) = (1-\theta^2/2)^{-1/2}$, $|\theta| < \sqrt{2}$.

Example 21.10 (Exponential integrability) Assume that the random variable u has a density p_U with respect to the Gaussian measure $\nu(dx) = (2\pi)^{-1/2}e^{-x^2/2}\,dx$, so that

$$\mathrm{E}\left(e^{tU}\right) = \int_{-\infty}^{+\infty} e^{tx} p_U(x)\nu(dx) = e^{\frac{t^2}{2}} \int_{-\infty}^{+\infty} p_U(x)(2\pi)^{-1/2} e^{-\frac{(x-t)^2}{2}}\,dx$$

$$= e^{\frac{t^2}{2}} p_U \star \nu(t)$$

is finite for $t \in \mathbb{R}$ if the convolution $p_U \star \nu$ is well defined. Various sufficient conditions, based on the evaluation of ∇U, are available for that.

The product of three independent Gaussian variables is not exponentially integrable. This implies that a polynomial random variable is exponentially integrable if, and only if, it is of degree no larger than two. The closure of the set of such polynomials in L^2 is $\mathbb{R} \oplus \mathcal{H}_1 \oplus \mathcal{H}_2$. A general discussion of the exponential integrability of a random variable $U \in \mathbb{D}_1^2$ based on a bound on ∇U is in (Üstünel 1995, Chapter VIII). The same reference contains a result useful for the proof of the exponential convergence in the same class of random variables.

Example 21.11 (Quadratic exponential models) The exponential model whose canonical statistics are $\delta(e_1), \delta(e_2), \delta(e_1)\delta(e_2)$ has the form

$$F_{\theta_1, \theta_2, \theta_{12}} = \exp\left(\theta_1 \delta(e_1) + \theta_2 \delta(e_2) + \theta_{12}\delta(e_1)\delta(e_2) - \psi(\theta_1, \theta_2, \theta_{12})\right).$$

The cumulant function is

$$\psi(\theta_1, \theta_2, \theta_{12}) = \frac{1}{2} \frac{\theta_1^2 + \theta_2^2 + 2\theta_1\theta_2\theta_{12}}{1-\theta_{12}^2} - \frac{1}{2}\log\left(1-\theta_{12}^2\right),$$

with $\theta_1, \theta_2 \in \mathbb{R}$, $\theta_{12}^2 < 1$. Note that the expectation parameters are

$$\eta_1 = \frac{\theta_1 + \theta_2 \theta_{12}}{1 - \theta_{12}^2}$$

$$\eta_2 = \frac{\theta_2 + \theta_1 \theta_{12}}{1 - \theta_{12}^2}$$

$$\eta_{12} = \frac{\theta_1 \theta_2 (1 + \theta_{12}^2) + (\theta_1^2 + \theta_2^2)\theta_{12} + \theta_{12}(1 - \theta_{12}^2)}{\left(1 - \theta_{12}^2\right)^2}.$$

The expectation parameters are rational functions of the canonical parameters. Let V be the linear space of polynomial random variables of degree 2 at most. The orthogonal space $V^\perp$ is the closure in $^*T(1)$ of $\oplus_{n>2}\mathcal{H}_n$. A positive density q belongs to the exponential model $\mathcal{E}_V(1)$ if, and only if, $\mathrm{E}\left(\log\left(q\right)h\right) = 0$ for all multivariate Hermite polynomials of degree larger than two.

Let $F \in \mathrm{Poly}(\delta)$ be such that $F > 0$ and $\mathrm{E}(F) = 1$, $m = \min F > 0$. Polynomial perturbation of the Gaussian density is a classical subject, see (Johnson and Kotz 1970, Chapter 12, Section 4.2) for the univariate case. Here, the polynomial F is taken to be an element of $\mathcal{M}_>$ and can be written as $F = \exp(V)$ for some random variable V. If $\alpha > 0$, $\mathrm{E}(\exp(\alpha V)) = \exp(F^\alpha) < +\infty$ because $F \in L^\alpha$ for $\alpha \geq 1$ and $\mathrm{E}(F^\alpha) \leq 1$ for $\alpha < 1$. For the negative case $-\alpha$ $(\alpha > 0)$, we have $\mathrm{E}(\exp(-\alpha V)) < e^{-\alpha m}$. It follows that $V = \log F \in L^\Phi(\mu)$ and, moreover, V has Laplace transform defined everywhere, and it is the limit, in the L^Φ-convergence, of bounded random variables. The exponential model in standard form is $F = \exp(U - K(U))$, $U = \log F - \mathrm{E}(\log F)$.

Example 21.12 (Polynomial density with two parameters) If $\theta_1^2 + \theta_2^2 < 1$, then

$$F_{\theta_1,\theta_2} = 1 - (\theta_1^2 + \theta_2^2) + (\theta_1 \delta(e_1) + \theta_2 \delta(e_2))^2$$

satisfies the assumptions with $ma = 1 - (\theta_1^2 + \theta_2^2)$. The centering is

$$\mathrm{E}(\log(F_{\theta_1,\theta_2})) = -\log\left(1 - (\theta_1^2 + \theta_2^2)\right) - \sum_{k=1}^{\infty} \frac{2k!!}{k}\left(\frac{\theta_1^2 + \theta_2^2}{1 - (\theta_1^2 + \theta_2^2)}\right)^k.$$

We refer to (Üstünel 1995, Chapter VI) for a discussion of the convergence of polynomial densities to a limit density based on the convergence of distributions on an abstract Wiener space.

Gaussian quadrature formulas, see (Abramovitz and Stegun 1965, 25.4) suggest a connection between computations on polynomial random variables of the abstract Wiener space and the polynomial description of design, see the papers in Part II. Let us illustrate this point in the simplest case.

If f is a one-dimensional polynomial in $\mathbb{R}[x]$ with degree less than or equal to $2n-1$, the remainder formula for the ideal generated by the n-th Hermite polynomial $H_n = \delta^n(1)$ gives the decomposition

$$f(x) = f_{n-1}(x) + g_{n-1}(x)H_n(x)$$

where f_{n-1}, g_{n-1} are polynomials of degree not greater than $n-1$. For a standand Gaussian random variable Z with distribution ν we have

$$\mathrm{E}\left(f(Z)\right) = \mathrm{E}\left(f_{n-1}(Z)\right) + \mathrm{E}\left(g_{n-1}(Z)H_n(Z)\right) = \mathrm{E}\left(f_{n-1}(Z)\right)$$

because $\mathrm{E}\left(g_{n-1}(Z)H_n(Z)\right) = \langle g_{n-1}, \delta^n 1\rangle_\nu = \langle d^n g_{n-1}, 1\rangle_\nu = 0$. The polynomial f_{n-1} is equal to f on the zero-set of H_n, therefore

$$\mathrm{E}\left(f(Z)\right) = \sum_{x\,:\,H_n(x)=0} w_n(x)f(x), \quad w_n(x) \propto H_{n-1}^{-2}(x).$$

This induces a correspondence of some functionals of the abstract Wiener space with a discrete model with uniform distribution and suggests an interesting concept of approximation of general functionals via the Fourier–Hermite series. Vice versa, concepts from design of experiments can be lifted to the abstract Wiener space.

21.6 Discussion and acknowledgements

In this chapter we have presented, mainly informally, a number of thoughts raised by considering contemporary algebraic and geometric methods in statistics. Part of the presented material is a summary of or a comment on current research as it appears in this volume. Another part of the material is even more adventurous and it points in directions of research that this author considers promising. Here is a summary list.

The algebraic and geometric pictures are of interest in fields usually considered far from statistics, such as Statistical Physics and Machine Learning. Generalised exponential models on finite state space are of special interest in the connection between algebraic statistics and information geometry. We believe that a special focus on approximation methods to deal with computationally intractable models would be promising.

The differential geometric picture has been recently studied in connection to nonlinear evolution equations, both deterministic and stochastic. Again, approximation methods inspired by the intrinsic geometry of the problem are of special interest.

Some models used in Stochastics, such as abstract Wiener spaces, have special tools to deal with the existence of densities and their algebraic computation.

Much effort has been delivered by the editors of this volume to produce an up-to-date and usable collection. Over the years, we have been jointly dedicated to the effort to explore the beauties of algebraic and geometrical methods in statistics. On behalf of the readers of this volume, I wish to warmly thank all of them.

References

4ti2 Team (2006). *4ti2 – A software package for algebraic, geometric and combinatorial problems on linear spaces* (available at `www.4ti2.de`).

Abramovitz, M. and Stegun, I. A. (eds.) (1965). *Handbook of Mathematical Functions* (New York, Dover).

Amari, S. (1982). Differential geometry of curved exponential families. Curvature and information loss, *Annals of Statistics* **10**(2), 357–87.

Amari, S. (1985). *Differential-geometrical Methods in Statistics* (New York, Springer-Verlag).

Amari, S. and Nagaoka, H. (2000). *Method of Information Geometry* (Providence, RI, American Mathmatical Society).

Brown, L. D. (1986). *Fundamentals of Statistical Exponential Families with Applications in Statistical Decision Theory* (Hayward, CA, Institute of Mathematical Statistics).

Cena, A. (2002). Geometric structures on the non-parametric statistical manifold, PhD thesis, Dottorato in Matematica, Università di Milano.

Cena, A. and Pistone, G. (2007). Exponential statistical manifold. *Annals of the Institute of Statistical Mathematics* **59**, 27–56.

Čencov, N. N. (1982). *Statistical Decision Rules and Optimal Inference.* (Providence, RI, American Mathematical Society).

CoCoATeam (2007). *CoCoA, a system for doing Computations in Commutative Algebra*, 4.7 edn (available at `http://cocoa.dima.unige.it`).

Cover, T. M. and Thomas, J. A. (2006). *Elements of Information Theory* 2nd edn (Hoboken, NJ, John Wiley & Sons).

Csiszár, I. and Matúš, F. (2005). Closures of exponential families, *Annals of Probability* **33**(2), 582–600.

Dawid, A. P. (1975). Discussion of a paper by Bradley Efron, *Annals of Statistics* **3**(6), 1231–4.

Dawid, A. P. (1977). Further comments on a paper by Bradley Efron, *Annals of Statistics* **5**(6), 1249.

Efron, B. (1975). Defining the curvature of a statistical problem (with applications to second-order efficiency) (with discussion), *Annals of Statistics* **3**(6), 1189–242.

Efron, B. (1978). The geometry of exponential families, *Annals of Statistics* **6**(2), 362–76.

Evans, S. N. and Speed, T. P. (1993). Invariants of some probability models used in phylogenetic inference, *Annals of Statistics* **21**(1), 355–77.

Fienberg, S. E. (1980). *The Analysis of Cross-classified Categorical Data* 2nd edn (Cambridge, MA, MIT Press).

Geiger, D., Meek, C. and Sturmfels, B. (2006). On the toric algebra of graphical models, *Annals of Statistics* **34**, 1463–92.

Geman, S. and Geman, D. (1984). Stochastic relaxation, Gibbs distributions, and the bayesian restoration of images, *IEEE Transactions on Pattern Analysis and Machine Intelligence* **6**(6), 721–41.

Gibilisco, P. and Pistone, G. (1998). Connections on non-parametric statistical manifolds by Orlicz space geometry, *Infinite Dimensional Analysis, Quantum Probability and Related Topics* **1**(2), 325–47.

Grasselli, M. R. (2009). Dual connections in nonparametric classical information geometry. *Annals of the Institute for Statistical Mathematics* (to appear) (available at arXiv:math-ph/0104031v1).

Jeffreys, H. (1946). An invariant form of the prior probability in estimation problems, *Proceedings of the Royal Society of London Series A* **186**, 453–61.

Johnson, N. L. and Kotz, S. (1970). *Distributions in Statistics. Continuous Univariate Distributions. 1.* (Boston, MA, Houghton Mifflin Co.).

Johnson, N. L., Kotz, S. and Balakrishnan, N. (1995). *Continuous univariate distributions vol. 2* 2nd edn (New York, John Wiley & Sons).

Kaniadakis, G. (2001). Non-linear kinetics underlying generalized statistics, *Physica A* **296**(3–4), 405–25.

Kaniadakis, G. (2005). Statistical mechanics in the context of special relativity II, *Physical Review E* **72**(3), 036108.

Koopman, B. O. (1936). On distributions admitting a sufficient statistic, *Transactions of the American Mathematical Society* **39**(3), 399–409.

Lang, S. (1995). *Differential and Riemannian manifolds* 3rd edn (New York, Springer-Verlag).

Letac, G. (1992). *Lectures on Natural Exponential Families and Their Variance Functions* (Instituto de Matemática Pura e Aplicada (IMPA), Rio de Janeiro).

Malagò, L., Matteucci, M. and Dal Seno, B. (2008). An information geometry perspective on estimation of distribution algorithms: boundary analysis. In *Proc. GECCO '08* (New York, ACM), 2081–8.

Naudts, J. (2002). Deformed exponentials and logarithms in generalized thermostatistics, *Physica A* **316**(1-4), 323–34.

Naudts, J. (2004). Estimators, escort probabilities, and ϕ-exponential families in statistical physics, *Journal of Inequalities in Pure and Applied Mathematics* **5**(4), Article 102.

Nualart, D. (2006). *The Malliavin Calculus and Related Topics* 2nd edn (Berlin, Springer-Verlag).

Ohara, A. and Wada, T. (2008). Information geometry of q-Gaussian densities and Behaviours of solutions to related diffusion equations (available at arXiv: 0810.0624v1).

Otto, F. (2001). The geometry of dissipative evolution equations: the porous medium equation, *Communications in Partial Differential Equations* **26**(1-2), 101–74.

Pistone, G. (2009). κ-exponential models from the geometrical viewpoint, *The European Physical Journal B*, **70**(1), 29–37.

Pistone, G. and Rogantin, M. P. (1999). The exponential statistical manifold: mean parameters, orthogonality and space transformations, *Bernoulli* **5**(4), 721–60.

Pistone, G. and Sempi, C. (1995). An infinite-dimensional geometric structure on the space of all the probability measures equivalent to a given one, *Annals of Statistics* **23**(5), 1543–61.

Rao, C. R. (1945). Information and accuracy attainable in the estimation of statistical parameters, *Bullettin of Calcutta Mathematical Society* **37**, 81–9.

Rao, M. M. and Ren, Z. D. (2002). *Applications of Orlicz Spaces* (New York, Marcel Dekker).

Rapallo, F. (2007). Toric statistical models: Parametric and binomial representations, *Annals of the Institute of Statistical Mathematics* **59**(4), 727–40.

Schrijver, A. (1986). *Theory of Linear and Integer Programming* (Chichester, John Wiley & Sons).

Üstünel, A. S. (1995). *An Introduction to Analysis on Wiener Space* (Berlin, Springer-Verlag).